# Dietary Antioxidants and Cosmetics

# Dietary Antioxidants and Cosmetics

Editors

**Irene Dini**
**Sonia Laneri**

MDPI

Basel • Beijing • Wuhan • Barcelona • Belgrade • Novi Sad • Cluj • Manchester

*Editors*
Irene Dini
University of Naples Federico II
Napoli
Italy

Sonia Laneri
University of Naples Federico II
Napoli
Italy

*Editorial Office*
MDPI
St. Alban-Anlage 66
4052 Basel, Switzerland

This is a reprint of articles from the Special Issue published online in the open access journal *Antioxidants* (ISSN 2076-3921) (available at: https://www.mdpi.com/journal/antioxidants/special_issues/Dietary_Cosmetic).

For citation purposes, cite each article independently as indicated on the article page online and as indicated below:

Lastname, A.A.; Lastname, B.B. Article Title. *Journal Name* **Year**, *Volume Number*, Page Range.

**ISBN 978-3-7258-0055-1 (Hbk)**
**ISBN 978-3-7258-0056-8 (PDF)**
**doi.org/10.3390/books978-3-7258-0056-8**

# Contents

antioxidants

*Editorial*

# Dietary and Cosmetic Antioxidants

**Irene Dini**

Department of Pharmacy, University of Naples Federico II, Via Domenico Montesano 49, 80131 Napoli, Italy; irdini@unina.it

Spices, herbs, fruits, whole grains, vegetables, and sea organisms contain antioxidant molecules that can scavenge free radicals and reduce their development, quenching the reactive oxygen and nitrogen species [1]. Even agri-food waste can contain bioactive compounds with antioxidant, anti-inflammatory, and antimicrobial activities [2], which could be exploited to formulate natural and sustainable cosmetics; in this industry, the skin's well-being and environmental problems are two inextricably linked realities [3].

This Special Issue includes research articles and reviews addressing bioactive compounds in food and organic waste with potential hydrating, protective, repairing, regenerating, and whitening effects on the skin and delivery systems capable of improving their performance in cosmetics.

Dini reviewed the potential for nanotechnologies, in supplements and cosmetics formulations, to enhance the performance of nutricosmeceutic products.

Gu et al. studied the potential for alginate–chitosan-coated nanoliposomes to improve the bioavailability of bamboo leaf flavonoids. Moccia et al. investigated transfersomes' efficacy as carriers of the ellagic acid obtained from a chestnut-wood mud industrial byproduct. Dini and Mancusi reviewed the potential role of biopeptides as antimicrobial, antioxidant, antiaging, and anti-inflammatory molecules. Kirindage et al. examined the antioxidative and anti-inflammatory actions of fucosterol extracted from the brown algae *Sargassum horneri*. Mansinhos et al. evaluated the whitening and sun-protective potential of Thymus lotocephalus extracts. Gigliobianco et al. investigated pomegranate and its waste products' phenol components profile, cell viability, and antioxidant and antibacterial activities. Yang et al. determined the possibility of developing cosmetic materials with skin-whitening and anti-inflammatory functions using fermented maca root extracts with Lactobacillus strains. Aiello et al. studied the effect of carnosine on UVA-induced changes. Yang et al. investigated the ability of an enzyme (carbohydrase celluclast) to improve the cosmeceutical potential of *Ishige okamurae*. Cardeira et al. evaluated the potential anti-inflammatory and antimicrobial activity of protein-rich extracts from sardine waste and codfish frames. Choi et al. assessed the skin-whitening action of the nomilin extracted from discarded yuzu byproducts. Finally, Mahendra et al. reviewed the anti-inflammatory, antioxidant, antibacterial, wound-healing, and skin-whitening potential of *Swietenia macrophylla*.

The articles in this Special Issue confirm the strong potential of foods (vegetable and animal) and organic waste as sources of bioactive compounds with a lightening, moisturizing, and protective action on the skin and as preservatives to increase the shelf life of cosmetics. They examine nanotechnologies to improve the bioavailability of bioactive compounds, ensure their release at the site of action, cover their unpleasant taste, and extend their expiry date. Nevertheless, the authors agree that there is a need for further toxicology studies to guarantee consumer safety and for experimental research on the large-scale recovery of bioactives from organic matrices before new cosmetics can be developed.

**Citation:** Dini, I. Dietary and Cosmetic Antioxidants. *Antioxidants* **2024**, *13*, 228. https://doi.org/10.3390/antiox13020228

Received: 24 January 2024
Revised: 8 February 2024
Accepted: 11 February 2024
Published: 13 February 2024



**Funding:** This research received no external funding.

**Conflicts of Interest:** The author declares no conflicts of interest.

**List of Contributions:**

1. Dini I. Contribution of Nanoscience Research in Antioxidants Delivery Used in Nutricosmetic Sector. *Antioxidants* **2022**, *11*, 563. https://doi.org/10.3390/antiox11030563.
2. Gu, Y.; Zhao, Z.; Xue, F.; Zhang, Y. Alginate-Chitosan Coated Nanoliposomes as Effective Delivery Systems for Bamboo Leaf Flavonoids: Characterization, In Vitro Release, Skin Permeation and Anti-Senescence Activity. *Antioxidants* **2022**, *11*, 1024. https://doi.org/10.3390/antiox11051024.
3. Dini, I.; Mancusi, A. Food Peptides for the Nutricosmetic Industry. *Antioxidants* **2023**, *12*, 788. https://doi.org/10.3390/antiox12040788.
4. Kirindage, K.G.I.S.; Jayasinghe, A.M.K.; Han, E.-J.; Jee, Y.; Kim, H.-J.; Do, S.G.; Fernando, I.P.S.; Ahn, G. Fucosterol Isolated from Dietary Brown Alga *Sargassum horneri* Protects TNF-α/IFN-γ-Stimulated Human Dermal Fibroblasts Via Regulating Nrf2/HO-1 and NF-κB/MAPK Pathways. *Antioxidants* **2022**, *11*, 1429. https://doi.org/10.3390/antiox11081429.
5. Moccia, F.; Liberti, D.; Giovando, S.; Caddeo, C.; Monti, D.M.; Panzella, L.; Napolitano, A. Chestnut Wood Mud as a Source of Ellagic Acid for Dermo-Cosmetic Applications. *Antioxidants* **2022**, *11*, 1681. https://doi.org/10.3390/antiox11091681.
6. Mansinhos, I.; Gonçalves, S.; Rodríguez-Solana, R.; Duarte, H.; Ordóñez-Díaz, J.L.; Moreno-Rojas, J.M.; Romano, A. Response of *Thymus lotocephalus* In Vitro Cultures to Drought Stress and Role of Green Extracts in Cosmetics. *Antioxidants* **2022**, *11*, 1475. https://doi.org/10.3390/antiox11081475.
7. Gigliobianco, M.R.; Cortese, M.; Nannini, S.; Di Nicolantonio, L.; Peregrina, D.V.; Lupidi, G.; Vitali, L.A.; Bocchietto, E.; Di Martino, P.; Censi, R. Chemical, Antioxidant, and Antimicrobial Properties of the Peel and Male Flower By-Products of Four Varieties of *Punica granatum* L. Cultivated in the Marche Region for Their Use in Cosmetic Products. *Antioxidants* **2022**, *11*, 768. https://doi.org/10.3390/antiox11040768.
8. Yang, J.; Cho, H.; Gil, M.; Kim, K.E. Anti-Inflammation and Anti-Melanogenic Effects of Maca Root Extracts Fermented Using *Lactobacillus* Strains. *Antioxidants* **2023**, *12*, 798. https://doi.org/10.3390/antiox12040798.
9. Aiello, G.; Rescigno, F.; Meloni, M.; Zoanni, B.; Aldini, G.; Carini, M.; D'Amato, A. The Effect of Carnosine on UVA-Induced Changes in Intracellular Signaling of Human Skin Fibroblast Spheroids. *Antioxidants* **2023**, *12*, 300. https://doi.org/10.3390/antiox12020300.
10. Yang, F.; Hyun, J.; Nagahawatta, D.P.; Kim, Y.M.; Heo, M.-S.; Jeon, Y.-J. Cosmeceutical Effects of *Ishige okamurae* Celluclast Extract. *Antioxidants* **2022**, *11*, 2442. https://doi.org/10.3390/antiox11122442.
11. Cardeira, M.; Bernardo, A.; Leonardo, I.C.; Gaspar, F.B.; Marques, M.; Melgosa, R.; Paiva, A.; Simões, P.; Fernández, N.; Serra, A.T. Cosmeceutical Potential of Extracts Derived from Fishery Industry Residues: Sardine Wastes and Codfish Frames. *Antioxidants* **2022**, *11*, 1925. https://doi.org/10.3390/antiox11101925.
12. Choi, M.-H.; Yang, S.-H.; Kim, N.D.; Shin, H.-J. Nomilin from Yuzu Seed Has In Vitro Antioxidant Activity and Downregulates Melanogenesis in B16F10 Melanoma Cells through the PKA/CREB Signaling Pathway. *Antioxidants* **2022**, *11*, 1636. https://doi.org/10.3390/antiox11091636.
13. Mahendra, C.K.; Goh, K.W.; Ming, L.C.; Zengin, G.; Low, L.E.; Ser, H.-L.; Goh, B.H. The Prospects of *Swietenia macrophylla* King in Skin Care. *Antioxidants* **2022**, *11*, 913. https://doi.org/10.3390/antiox11050913.

## References

1. Dini, I. An overview of functional beverages. In *Functional and Medicinal Beverages*; Grumezescu, A.M., Holban, A.M., Eds.; Academic Press: Cambridge, MA, USA, 2019; Volume 11, pp. 1–40.
2. Singh, N.; Yadav, S.S. A review on health benefits of phenolics derived from dietary spices. *Curr. Res. Food Sci.* **2022**, *5*, 1508–1523. [CrossRef] [PubMed]
3. Fonseca, S.; Amaral, M.N.; Reis, C.P.; Custódio, L. Marine Natural Products as Innovative Cosmetic Ingredients. *Mar. Drugs* **2023**, *21*, 170. [CrossRef] [PubMed]

*antioxidants*

MDPI

*Review*

# Contribution of Nanoscience Research in Antioxidants Delivery Used in Nutricosmetic Sector

**Irene Dini**

Pharmacy Department, "Federico II" University, Via D. Montesano, 49, 80131 Naples, Italy; irdini@unina.it

**Abstract:** Nanoscience applications in the food and cosmetic industry offer many potential benefits for consumers and society. Nanotechnologies permit the manipulation of matter at the nanoscale level, resulting in new properties and characteristics useful in food and cosmetic production, processing, packaging, and storage. Nanotechnology protects sensitive bioactive compounds, improves their bioavailability and water solubility, guarantees their release at a site of action, avoids contact with other constituents, and masks unpleasant taste. Biopolymeric nanoparticles, nanofibers, nanoemulsions, nanocapsules, and colloids are delivery systems used to produce food supplements and cosmetics. There are no barriers to nanoscience applications in food supplements and cosmetic industries, although the toxicity of nano-sized delivery systems is not clear. The physicochemical and toxicological characterization of nanoscale delivery systems used by the nutricosmeceutic industry is reviewed in this work.

**Keywords:** nanotechnology; nutraceutic; nutricosmetic; nanoceutic; phytochemical delivery; nanoemulsion; polymeric nanoparticles; edible nanocoating; nanocosmeceuticals

**Citation:** Dini, I. Contribution of Nanoscience Research in Antioxidants Delivery Used in Nutricosmetic Sector. *Antioxidants* **2022**, *11*, 563. https://doi.org/10.3390/antiox11030563

Academic Editor: Stanley Omaye

Received: 22 February 2022
Accepted: 15 March 2022
Published: 16 March 2022



## 1. Introduction

Nutricosmetics is a new sector of cosmetics that aims to optimize the use of cosmetic products and food supplements such as micronutrients (minerals, vitamins), macronutrients (peptides, essential fatty acids), and botanicals (herbal and fruit extracts) to nourish the skin and reduce skin aging through an integrated "In and Out" approach [1–5]. By 2030, the world's population over 60 years will grow to 1.4 billion [6], so it is reasonable to assume that the number of people who will buy cosmetics in hopes of maintaining a youthful appearance will grow in the coming years. In this scenario, nutricosmetic products should find a large market as they are natural products, improve health, and are considered free of side effects. The skin is the first line of defense between our body and the world [7]. It maintains the balance of liquids by binding water, preventing its loss, and promoting perspiration. The skin is subjected to multiple stressors that lead to premature skin aging. Free radicals produced by air pollution, cold, and UV rays induce inflammatory processes and accelerate skin aging by altering our body's DNA, lipids, and proteins [8]. Sportswear can produce dryness and irritable skin, increasing friction. Frequent showers and detergents modify the hydrolipidic film and the ability to regulate liquids and the skin's elasticity. The nutricosmetic approach repairs the skin barrier [9], improves skin hydration, fights inflammation, and protects the skin from damage caused by the sun's rays, combining food supplements that intervene from the inside with cosmetic products for topical use, which interfere from the outside [10]. Nanotechnological systems are enjoying great success in the food and cosmetic fields. In recent years, new delivery systems for bioactive compounds have been investigated. The purpose of this work is to review the physicochemical and toxicological impact of nanoscale delivery systems used by the nutricosmeceutic industry.

## 2. Nanocosmetics and Nanonutraceuticals Delivery Systems

Nanochemicals are formulations containing nanotechnology as delivery systems to improve the bioactive components' performance [11–13]. REACH (Registration, Evaluation,

Authorisation, and Restriction of Chemicals) regulates the exposure and hazards of nanochemicals [14].

Nanocosmetics is the cosmetic field where nanomaterials/nanoparticles are used to develop cosmetic products. The international guidelines (EC Regulation 1223/2009) that guarantee the protection and safety of cosmetic products defined nanomaterials as only the "material insoluble or bio-persistent (e.g., metal oxides, metals, etc.) and intentionally manufactured with one or more external dimensions, or an internal structure, on the scale from 1 to 100 nm", excluding materials that are soluble, degradable, and/or non-persistent in biological systems (e.g., liposomes, plant-derived vesicles, emulsions, etc.) [15].

Nanonutraceuticals are nanotechnology delivery systems used to improve the performance of bioactive components in foods, including food supplements. An edible delivery system must be realized with GRAS (Generally Recognized as Safe) ingredients by using processing operations that conform to good manufacturing practices. It must have a high loading capacity, encapsulation efficiency, and retention efficiency. It must have the capacity to protect chemically labile encapsulated compounds from chemical degradation (e.g., oxidative degradation) [16]. It must be compatible with the food or beverage matrix that it will be incorporated into, without causing any adverse effects on product appearance, texture, mouthfeel, flavor, or shelf-life [17]. It must be resistant to environmental stresses during production, storage, transport, and utilization (e.g., thermal processing, light exposure, mechanical agitation, chilling, freezing, or dehydration) [17]. It must be designed to control the release and/or absorption of the bioactive lipophilic component of a particular site within the gastrointestinal tract, such as the mouth, stomach, small intestine, or large intestine [17].

In the nutricosmetic field, nanotechnology is used to prepare sunscreens, barrier creams, antiacne, moisturizers, antiaging, antioxidants, hair, nails, lip, and skin cosmetics. The industry has created many nanoscale delivery systems that transport each bioactive based on its nature (lipophilic and hydrophilic) and chemical–physical properties. Nanostructures may have one-dimension, two-dimensions (e.g., nanotubes), or three dimensions (nanoparticles) at the nanometer level [18]. Nanostructures that protect and deliver lipophilic compounds include simple oil in water (O/W) emulsions, water-in-oil-in-water (W/O/W) double emulsions, capsules, liposomes, and colloidosomes. The systems able to protect and deliver hydrophilic bioactive components are gelled networks (hydrogels), gel particles/fluid gels (gelled nanoparticulates), water in oil emulsions (W/O), and protein–polysaccharide structures (self-assembled structures).

### 2.1. Nanodispersions

#### 2.1.1. Nanoemulsions

Nanoemulsions (also known as miniemulsions or submicron emulsions) are metastable colloidal dispersions with average droplet radii ranging from 10 to 100 nm. They can be made to be highly viscous or gel-like. They are optically transparent, since their small dimensions disperse light waves imperceptibly [19]. Two types of nanoemulsions exist: O/W (oil/water) able to encapsulate, protect, and deliver hydrophobic functional components such as lipophilic vitamins (e.g., β-carotene), omega-3, and nutraceutical (e.g., plant sterols, carotenoids, etc.) essential oils; and W/O (water/oil) used to encapsulate water-soluble active agents such as polyphenols, emulsifiers (phospholipids, Tweens, proteins), texture modifiers (pectin, sodium alginate, carrageenan), and preservatives (parabens) [20]. One or more emulsifiers is used to decrease the energy required to form smaller droplets, reduce the interfacial tension, and prevent or slow down the aggregation of particles of the dispersed phase by increasing repulsion forces between them [21]. Examples of food-grade emulsifiers are proteins and polysaccharides [22]. The silicone oil-in-water nanoemulsions are used to enhance the silicone oil deposition on the hair surface to preserve the moisture and lubrication of hair [23]. The droplet size depends on the approach, the operating conditions (e.g., energy intensity, time, and temperature), and the system's composition (e.g., interfacial tension and viscosity). Nanoemulsions may be formed using either high or low energy (phase inversion and spontaneous phase separation) techniques. The high-energy approaches employ homogenizers, microfluidizers, and sonicators [24]. The low energy methods (physicochemical approaches) are PIT (phase inversion temperature), PIC (phase inversion composition), and spontaneous emulsification.

In the cosmetic field, nanoemulsions are used to increase the delivery of active compounds in the skin, avoid cream in fluid products, and maintain the transparency and gloss after spreading [25]. Examples are O/W nanoemulsions containing *Opuntia ficus indica* (L.) extract used as moisturizing agents [26], nanoemulsions containing hydroalcoholic extracts of *Vellozia squamata* leaves employed as antiaging agents [27], and pomegranate seed oil nanoemulsions prepared to enhance antioxidant activities in the skin and to protect against photodamage [28].

Resveratrol is encapsulated in nanoemulsions in food supplements to protect it from UV exposure [29].

(1) The high-energy methods used to produce nanoemulsions

The high-energy methods are high-pressure homogenization, microfluidification, and sonication.

High-pressure homogenization (HPH) is a purely mechanical process induced by forcing a fluidic product through the homogenizing nozzle at high pressure (150–200 MPa, or 350–400 MPa for ultra-high-pressure homogenization (UHPH)) [30]. HPH is used to make O/W liquid nanoemulsion in which the oil phase is less than 20%. It can decrease the polydispersity of oil droplets and the droplet size [31]. It is inappropriate to formulate creamy or high viscosity nanoemulsions whose droplet diameters are below 200 nm [32]. Tiny oil droplets of an extract of jackfruit, obtained by HPH at 800 bar, were used to formulate a cream with low viscosity and high stability [33].

The microfluidizer works on the principle of pressurized stream. It uses a pump to force a coarse emulsion pre-mix through a narrow orifice at high pressures to facilitate droplet disruption. The channel is designed to split the coarse emulsion into two streams made to impinge on each other at high velocity in an interaction chamber [34]. Smaller emulsions can be made by enhancing the pressure up to ~700 MPa [35]. It produces smaller and narrower particle sizes of nanoemulsions than the HPH [36].

Sonication is a process in which sound waves (high-intensity ultrasonic wave frequency >20 kHz) are used to agitate particles in solution. Such disruptions can be used to form emulsions containing very fine droplets, mix solutions, and speed the dissolution of a solid into a liquid. A nanoemulsion containing avocado oil droplets obtained by the ultrasonication technique was used by Silva et al. (2013) to formulate a sun protector [37].

(2) Low-energy methods used to produce nanoemulsions

Low energy methods are based on the spontaneous dissolution of hydrophobic substances (oil, lipophilic surfactant, and water-miscible solvent) in an organic solvent, which is further emulsified with an aqueous solution (hydrophilic surfactant and water) [38].

In the PIT method, the nanoemulsions are spontaneously made by changing the temperature profile of the components. Low interfacial tensions (102–105 $mNm^{-1}$) are used to promote the emulsification process [39].

In the PIC method, a phase inversion occurs when the continuous phase is mixed over the component that will make the dispersed phase, and the chemical energy from the reaction of the components forms a fine dispersion [40].

The spontaneous emulsification is obtained, putting a lipid phase containing the surfactant in the aqueous phase under nonstop magnetic stirring and removing the aqueous phase under reduced pressure [41].

Low-energy methods were used to make a micellar formulation based on an extract of *Vellozia squamata* with antioxidant properties [27].

#### 2.1.2. Nano-Double Emulsions

The double emulsions are compartmentalized liquid dispersions in which the droplets of the dispersed phase contain smaller droplets of similar composition (but not necessarily identical) as the continuous phase. The double emulsion structures can be water-in-oil-in-water (W/O/W) or oil-in-water-in-oil (O/W/O). Concerning the W/O/W type, the three distinct phases consist of internal water droplets dispersed in an oil phase, then enclosed in a continuous water phase. An emulsifier with low interfacial tension stabilizes the interface between two immiscible liquids. The W/O/W emulsions can carry both polar and nonpolar bioactive compounds. They are used as a delivery system of flavors, phytochemicals, probiotics, hydrophilic (i.e., water-soluble vitamins, minerals), and hydrophobic (i.e., polyunsaturated fatty acids) nutrients [42,43]. Finally, the multiple emulsions O/W/O and W/O/W are used to separate two aqueous phase components that might adversely react with each other if they were present in the same aqueous phase and to protect and to release an aqueous phase component locked in the inner phase to an exact site, such as the mouth, stomach, or small intestine. Recently, an anti-pollution cosmetic containing D-biotin based on a W/O/W multiple emulsion was formulated for skin protection by Ali et al. [44].

(1) Methods used to produce W/O/W nanoemulsions

- Bulk emulsification methods

  W/O/W nanoemulsions are produced using a two-step emulsifying process. The first step creates the smallest possible internal droplets (water-in-oil emulsion) by high shear conditions. The second step is carried out under lower energetic conditions to prevent rupture of the primary emulsion. It consists in dispersing and emulsifying the mixture in water, using a combination of surfactants and shear to create a stable, aqueous emulsion. Two types of surfactants are needed: a low hydrophilic–lipophilic balance (HLB) for the interface between the interior water droplet and the encompassing oil droplet; and a high HLB for the interface between the oil droplet and continuous water phase [45]. The multiple emulsions are thermodynamic unstable. Biopolymers are used to minimize the leakage of the encapsulates from the internal aqueous phase, flocculation of the droplets, or phase separation during processing and storage of the Pickering particles. Biopolymers such as gelatin, caseinate, whey protein, bean protein, gum acacia, xanthan gum, and gelled starch stabilize the droplets in the internal phase of W/O/W food emulsions. Instead, polysaccharides such as carrageenan, locust bean gum, xanthan gum, pectin, gum arabic, whey protein isolate, sodium caseinate, egg white powder, and microcrystalline cellulose are utilized to stabilize the droplets of secondary emulsions [46]. Other stabilizers used are texture modifier sugars (e.g., sucrose, HFCS), polyols (e.g., glycerol, sorbitol), polysaccharides (e.g., xanthan, pectin, carrageenan, alginate), and proteins (e.g., whey protein isolate, gelatin); weighting agents such as dense lipophilic materials (e.g., brominated vegetable oil, sucrose acetate isobutyrate, ester gums); and ripening retarder such as lipophilic materials with very low water-solubility (e.g., ester gums) [47,48]. The W/O/W double Pickering emulsions containing silica particles and carboxymethyl cellulose as stabilizers were used in polymer particles with closed-cell porous and uniform size [49].

  A phase-inversion method was optimized to create a multicore W/O/W nanoemulsion for dermal delivery of acyclovir [49].

- Microfluidic techniques

  The microfluidic device's technology is proposed to produce a double emulsion as an alternate and versatile route. Capillary microfluidic devices consist of coaxial assemblies of glass capillaries on glass slides. One of the advantages of these devices is that their irritability can be easily and precisely controlled by a surface reaction with an appropriate surface modifier. In contrast to bulk emulsification methods, the emulsion in a microfluidic device is made by precisely fabricating one drop at a time. This process results in a highly monodispersed emulsion. One of the most attractive features of microfluidic techniques is that they enable the fabrication of double, triple, and even higher-order emulsions. The size and number of the encapsulated droplets can be manipulated (droplet sizes from 2 μm to 250 μm in diameter). Hydrophilic and hydrophobic surfaces, as well as surfactants, stabilize the two-fluid interfaces, extending manufactured products' stability. The flowing conditions can control the dispersity and size of droplets. The productivity is limited [50]. Lee et al. prepared a W/O/W double Pickering emulsion by a capillary microfluidic device in which the ellipsoidal double emulsion droplet formed a peanut-like colloidosome, enhancing the number of inner droplets [51].

#### 2.1.3. Association Colloids

The association colloids such as surfactant micelles, vesicles, bilayers, reverse micelles, and liquid crystals are stable micro heterogeneous systems (dimension range of 5 to 100 nm) in which the particles of the dispersed colloidal phase are typically transparent solutions. They are formed by molecules or ions dissolved in the dispersion medium containing small polar, nonpolar, and/or amphiphilic functional particles.

The micelles are relatively simple spherical or rod-like structures that consist of a phospholipid double layer. The encapsulation in O/W micelles allows for the inclusion of lipophilic compounds into aqueous systems. The encapsulation of compounds in W/O nanoemulsions or reversed micelles introduces hydrophilic compounds into a lipophilic environment. Their formation is spontaneous (thermodynamically favorable), and it is determined by the hydrophobic effect obtained by reducing the contact area between the nonpolar groups of the surfactant and water. The temperature, ionic strength, pH, and concentration and molecular characteristics of the surfactant and cosurfactant affect the type of colloid formed. Colloidal delivery systems can be classified into self-assembled molecules, solid-in-liquid, and liquid-in-liquid dispersions. The bioactive compounds can be incorporated as a solution (soluble form) or dispersion (insoluble form). The chemical reactivity of the soluble component increases its absorption, bio-accessibility, and taste characteristics [52]. Recent advances showed the utility of using colloids (lipid-origin carriers and nanoemulsion) as photosensitive nanocarriers in cosmetic formulations [53].

### 2.2. *Nanoparticles*

Nanoparticles are colloidal-sized particles composed of polymers with diameters ranging from 10 to 1000 nm. They are grouped into solid lipid nanoparticles, polymeric nanoparticles, ceramic nanoparticles, hydrogel nanoparticles, copolymerized peptide nanoparticles, nanocrystals and nanosuspensions, nanotubes and nanowires, functionalized nanocarriers, nanospheres, and nanocapsules [54].

The nanoparticle delivery systems improve skin penetration, release, and surface functionalization of active compounds. They may contain metallic ions. Metallic nanoparticles, which contain iron oxide, gold, silver, gadolinium, and nickel, are used to formulate cosmetics, sunscreens, and personal care products [54]. For example, sunscreens containing ZnO or $TiO_2$ nanoparticles are more transparent than micron-sized formulations [55]. It was shown that the safranal-nanoparticles (solid–lipid nanoparticles) improve sun screening activity in the range of 103–230 nm [56]. Nail paints having nanoparticles enhance mar resistance, toughness, and impact resistance of the mammalian nails [57].

In nutraceutical preparations (including food supplements), nanoparticles are used to improve the bioaccessibility of resveratrol [58] and folic acid [59] and the antioxidant properties of bioactive compounds [60].

#### Methods Used to Produce Nanoparticles

(1) Desolvation

Desolvation is a thermodynamically driven, self-assembly process for polymeric materials. The addition of desolvating agents (i.e., salts, ethanol, acetone) separates and coacervates the polymeric molecules in the aqueous phase. The self-assembly of the polymer molecules occurs with electrostatic interactions, since the overall free energy in the system is minimized during desolvation. The polymeric molecules form particles of different shapes and sizes depending on the preparation conditions. A balance between attractive and repulsive forces is necessary for fabricating particles of an appropriate size. The suppression and expression of hydrophobic interactions provide a way to control the size of polymeric particles during desolvation [61].

Duclairoir et al. made nanoparticles based on gliadins with vitamin E to improve the interaction between vitamin E and epidermal keratin [62].

(2) Electrospray drying technique

A syringe pump slowly injects a conducting liquid through a needle on which an electrical potential is applied. The electric stress accelerates the liquid away from the needle. The nanoparticles are produced when the solvent is evaporated [63].

Zaeim et al. used the electrospray technique to encapsulate probiotics and showed that the acacia gum concentration affects the viscosity, and the concentration of Tween-80 affects the form of the particles [64].

### 2.3. *Nanocapsules*

Nanoencapsulation is a technology used to package flavoring agents (e.g., sweeteners, seasonings, spices, essential oils), food acids and bases (e.g., citric acid, sodium bicarbonate), lipids (e.g., vegetable oils, milk fat), food additives (e.g., preservatives, pigments), minerals (e.g., calcium and iron salts), vitamins (e.g., carotene), colors, omega-3 oils, phytochemicals, and probiotic bacteria. The industry produces single-core and multiple-core nanocapsules. Single-core nanocapsules have high core loading (e.g., 90% of the total capsule weight). They are obtained by complex coacervation, fluidized bed drying, droplet coextrusion, and molecular inclusion. The storage stability is obtained using high-pressure homogenization and/or a surfactant.

In cosmetics, nanocapsules are employed in antiaging and moisturizing creams [65]. They can be applied to the skin or incorporated in semisolid formulations. After topical application on the skin, nanocapsules create a thin film with water evanescence responsible for the long-term delivery and increased storage capacity of the bioactive compound [65]. The most used biodegradable polymers in skincare formulations are aliphatic polyesters (i.e., poly(ε-caprolactone) (PCL) and poly(lactic acid) (PLA)), the chitin a polysaccharide found in the exoskeleton of crustaceans and insects, the gelatin which consists of a mixture of high molecular weight proteins, and the hyaluronic acid [66]. The multiple core nanocapsules are principally produced by spray drying, the core material is dispersed throughout the wall material, and the central area is occupied by the void resulting from the expansion of particles during the later drying stages.

Examples of nano-capsular vehicles for nutraceuticals are casein micelles used to deliver calcium phosphate and protein, stabilize hydrophobic substances, and enrich non-fat or low-fat food products; and poly(lactide-co-glycolide) micelles employed to carrier essential oils [67].

#### Nanoencapsulation Techniques

The nanoencapsulation techniques can be used to encapsulate hydrophilic and lipophilic bioactive compounds. Top-down and bottom-up approaches are employed to make nanoparticles.

The top-down approach is a mechanical process that uses shear or particle collisions as the energy source to break down larger entities into smaller aggregates [68].

The bottom-up (or self-assembly method) uses chemical or physical forces operating at the nanoscale to assemble lipids and proteins into larger structures. The pH, temperature, concentration, and ionic strength affect the process [68]. The emulsification, coacervation, and supercritical fluid techniques are used to encapsulate the hydrophilic and lipophilic compounds. The inclusion complexation, emulsification–solvent evaporation, and nanoprecipitation techniques are mainly used for lipophilic compounds [68].

Both top-down and bottom-up approaches have been utilized to deliver carotenoids in food and cosmetic formulations [69].

(1) Emulsification Technique

Emulsion technology is applied to encapsulate bioactive compounds in aqueous solutions. Nanoemulsions are made using high-energy emulsification methods. They possess high kinetic stability due to tiny emulsion droplet sizes. Nanoemulsions can either be used directly in the liquid state or be dried to powder form using drying techniques such as spray drying and freeze-drying after emulsification [70].

(2) Emulsification–solvent evaporation

In the solvent evaporation process, the polymer is dissolved in a suitable water-immiscible solvent, and the supplement or the nutrient is dispersed or dissolved in this polymeric solution. The organic solvent diffuses into the aqueous phase and evaporates at the water/air interface. The solution or dispersion is emulsified in a continuous aqueous phase to form the droplets. The nanospheres can be obtained after filtration and drying [70].

(3) Supercritical fluid technique

Supercritical fluids are used for the encapsulation of thermally sensitive compounds. This technique employs low critical temperature and low organic solvent levels. The bioactive compound and the polymer are solubilized in a supercritical fluid. The solution is expanded through a nozzle, the supercritical fluid is evaporated in the spraying process, and solute particles are precipitated. The most common processing techniques involving supercritical fluids are supercritical antisolvent (SAS) and rapid expansion of critical solution (RESS). The process of SAS uses a liquid solvent to dissolve the solute (e.g., methanol) utterly miscible with the supercritical fluid and a supercritical fluid. The $CO_2$ is the most widely used supercritical fluid because of its mild critical conditions (Tc = 31.1 °C, Pc = 73.8 bars), nontoxicity, nonflammability, and low price. The extract of the liquid solvent by supercritical fluid leads to the instantaneous precipitation of the solute (the solute is insoluble in the supercritical fluid), resulting in the formation of nanoparticles. The RESS differs from the SAS process because its solute is dissolved in a supercritical fluid (such as supercritical methanol), and the solution rapidly expands through a small nozzle into a region at lower pressure. In this procedure, the precipitate is solvent-free [71]. Dai et al. used nanoparticles containing curcumin in zein–lecithin composite obtained by antisolvent coprecipitation to improve the stability of curcumin against UV, thermal treatment, and high ionic strength [72].

(4) Complex coacervation

Complex coacervation is a spontaneous phase separation process involving two liquid phases (acetone or ethanol) in colloidal systems, which results from the interaction of two oppositely charged polyelectrolytes upon mixing in an aqueous protein solution. The formation of coacervates depends on the substrates, pH, temperature, molecular weight, ionic strength, and polyelectrolyte concentration and are limited by the addition of a crosslinking agent, such as glutaraldehyde. There are simple and complex coacervation methods. In simple coacervation, a solute phase is moved to the coacervation phase by changing the temperature, ionic strength, molecular weight, pH, and electrostatic interaction. Complex coacervation is obtained by mixing two ions of opposite charge into two immiscible liquid phases. The electrostatic interactions control the structure during the synthesis. The coacervation methods use a cationic (e.g., calcium chloride) or polyanionic (e.g., sodium tripolyphosphate) counter-ion, and a biodegradable hydrophilic polymer (e.g., chitosan, sodium alginate, and gelatin) [73] with a positive or negative charge in which the bioactive component is included. A polymer is added that allows polyelectrolyte complexation and nanocapsule formation [74].

Polysaccharide–protein complexes are used as carriers for encapsulating active compounds in supplement and cosmetic formulations. Coacervation is used to encapsulate polyphenols and essential oils [75].

(5) Inclusion complexation

Inclusion complexation encapsulates a supra-molecular association of a ligand (encapsulated ingredient) into a cavity-bearing substrate (shell material) through hydrogen bonding, van der Waals forces, or an entropy-driven hydrophobic effect. Inclusion complexation is used for lipophilic molecules (i.e., essential oils and vitamins) [76]. Only a few particular molecular compounds, such as β-cyclodextrin and β-lactogloglobulin, are suitable for encapsulation through this method [77].

Scalia et al. used this technique to produce a sunscreen entrapping 2-ethyl-hexyl-p-dimethylaminobenzoate-DMAB into the hydroxypropyl-β-cyclodextrin cavity to improve its photostability [78].

(6) Nanoprecipitation

The nanoprecipitation method is also called solvent displacement. This method is used to encapsulate lipophilic compounds. It yields stable aqueous suspensions of nanoparticles of about 300–320 nm based on the spontaneous emulsification of water [79].

Jummes et al. formulated poly-ε-caprolactone nanoparticles loaded with *Cymbopogon martinii* essential oil to use as antioxidants in the cosmetic field [80].

### 2.4. Nanohydrogels

Hydrogels are hydrated polymer gels formed by three-dimensional macromolecular networks that swell but do not dissolve in water. They are obtained from natural polysaccharides such as dextran, pullulan, or cholesterol-containing polysaccharide. Nanohydrogels have tiny dimensions, usually 20–30 nm. The reduced size of nanohydrogels enables the dispersion of water-insoluble additives (e.g., flavors, colors, and preservatives) [81]. Nanohydrogels can be produced either by physical or chemical gelation. The physical hydrogels (also called "reversible" or "pseudo" gels, physical nanohydrogels) exhibit high water sensitivity (degrade and even disintegrate entirely in water) and thermo-reversibility (melt to a polymer solution when exposed to heat) [82]. The chemical nanohydrogels (also called "irreversible" or "permanent" gels) are networks of polymer chains covalently linked at strategic connection sites. The chemical nanohydrogels neither disintegrate nor dissolve in an aqueous solution and have viscoelastic properties. They act as foaming and emulsifying agents.

Protein–polysaccharide nanohydrogels are used to encapsulate hydrophobic nutraceuticals [82]. A protein-based nanohydrogel was proposed by Bourbon et al. to encapsulate lipophilic compounds such as curcumin and hydrophilic compounds such as caffeine [83].

### 2.5. Solid Lipid Nanocarriers (SLNs) and Lipid Nanocarriers (NLCs)

SLNs are prepared from solid lipids (e.g., glyceryl behenate, stearic triglyceride, acetyl palmitate, and glycerol tripalmitate), while NLCs are composed of solid and a tiny amount of liquid lipids. Generally, biomolecules are incorporated between the fatty acid chains, lipid layers, or amorphous clusters. Both the SLNs and NLCs are stabilized by surfactants and can improve the bioavailability of highly lipophilic molecules. They are valuable in preparing the topical and transdermal application of poorly water-soluble and/or unstable compounds [84]. SLNs have a skin hydration effect and enhance the penetration through the stratum corneum since the surfactant(s) act as a permeation enhancer. NLCs have higher loading capacity and physicochemical stability during storage than SLNs [85,86]. Neem oil–solid lipid nanoparticles are formulated to be antibacterial on acne microbes [87]. Tretinoin- and behenate solid lipid nanoparticle-based gels have been made to reduce skin irritation and improve occlusive effects in cosmetic preparations for treating acne [88]. Solid lipid nanoparticles and nanostructured lipid carriers containing ascorbyl palmitate are utilized to improve moisturizing effects [89].

Lipid nano-delivery systems containing genistein, resveratrol, curcumin, quercetin, and epigallocatechin-3-gallate modulate chronic inflammation, oxidative stress, and aging-associated disorders [90]. Nano-lipoidal carriers of isotretinoin are used in cosmetics to combat aging [91].

Many techniques are used to make SLNs and NLCs. In some instances, different methods are employed together to prepare the nanoparticles. The high-pressure homogenization (HPH) and microemulsion techniques are the most used [84].

### 2.6. Nanoliposomes

Liposomes are spherical colloidal structures (0.05–5 μm) created with one or more lipid bilayers, (phospholipids, such as lecithin) by supplying energy (by sonication, homogenization, heating, etc.), which contain an aqueous solution in the core. Stabilizing agents, such as sterols (e.g., cholesterol), are used to increase liposomes' stability. A significant advantage of nanoliposomes is that they can simultaneously incorporate and release two materials with different solubilities (bifunctional liposomes). They can be utilized in the entrapment, delivery, and release of water-soluble and lipid-soluble functional components such as peptides, enzymes, vitamins, and flavors. Nanoliposomes can be made by supercritical fluid ($CO_2$), high-pressure homogenization, and micro fluidization techniques [92].

In cosmetic products, nanoliposomes are used to formulate antiaging creams, sun lotions, moisturizers, lipsticks, treatment of hair loss, and facial beauty masks [12]. Tan et al. showed that lutein, β-carotene, lycopene, and canthaxanthin encapsulation into liposomes enhanced their antioxidant activity [93]. The lipsticks containing rice bran oil liposomes showed better stability and antioxidant properties than conventional ones [94].

In nutraceutical preparations, nanoliposomes improve curcumin stability at the alkaline condition and temperature variation [95] and the antioxidant activity of phenolic compounds extracted from olive leaves [96].

### 2.7. Nanoclay

"Nanoclay" is obtained by clay minerals or layered silicates with metal oxides and organic matter traces. The clay minerals are hydrous aluminum phyllosilicates with variable magnesium, iron, alkaline earth, alkali metals, and other cations. Synthetic and natural clays are commercially available. Natural clays are made with "$SiO_2$" and "$AlO_6$" units [97]. For example, kaolite is made with "$SiO_2$" and "$AlO_6$" units in ratios of 1:1, montmorillonite and vermiculite in ratios of 2:1, and chlorite in a ratio of 2:2 [97]. Clay minerals are used in wound healing formulations [98]. Mg-rich smectite clay mineral helps in collagen synthesis and angiogenesis on skin wounds [99]. Palygorksite and sepiolite showed anti-inflammatory properties [100]. Clay minerals are used to formulate thermal muds (nanoclay/spring water hydrogel) [101] employed in skincare products for degreasing, cleansing, exfoliating, invigorating, hydrating, and firming activities [102].

## 3. The Use of Antioxidants in Nutricosmetic Products

Nutricosmetics is an umbrella term for food supplements with aesthetic benefits beyond their primary nutritional value. They are considered nonpharmaceutical and nonmedicinal products, although they are sold in capsules, tablets, syrups, gels, solutions, and extracts. Nutricosmetic supplements can contain nutrients and secondary plant metabolites (also known as phytochemicals or botanicals) [103,104]. Phytochemicals are non-nutritive plant chemicals with protective or disease preventive properties such as antioxidant activity, antimicrobial effects, hormone metabolism modulation, immune system stimulation, and anti-aggregate action. They are considered non-essential nutrients since the human body does not require them for sustaining life. The great changeability of phytochemical compounds determines a significant variation in their physicochemical properties (e.g., solubility in water or oil medium) [105]. The health effects depend on absorption, distribution, metabolism, and excretion. The absorption depends on the dose, the matrix in which they are ingested, and the presence of compounds able to bind or solubilize phytochemicals, reducing their bioactivity or product stability. Some phytochemicals are present in plant foods, such as glycosides or other conjugates, and must be hydrolyzed to be absorbed. Their metabolism may be affected by environmental exposures, stability, activity, gut microbials, and variations in levels of endogenous compounds that modulate biotransformation pathways [106,107]. In particular, among phytochemicals, antioxidant compounds such as vitamins (i.e., E, A, and C), tocopherols, carotenoids, methylxanthines (theophylline, caffeine, and theobromine), and phenols have been shown to improve our aesthetic well-being, making anti-inflammatory, antioxidant, photoprotective, antiaging, antiviral, and antibacterial effects [10,108,109]. The combination of topical application cosmetics and oral intake products enhances the results [110]. Both synthetic and natural molecules are employed in nutricosmetic products [111]. Butylated hydroxytoluene (BHT), butylated hydroxyl anisole (BHA), and propyl gallate are examples of synthetic antioxidants. Some synthetic antioxidants are obtained from natural ones. Polyphenols, mineral antioxidants (i.e., selenium, iron, copper, manganese, and zinc), vitamins, and phyto-antioxidants are natural compounds used in cosmetic products. Synthetic and natural antioxidants can be used together to produce synergistic stabilization effects [112]. Antioxidants can be grouped into non-enzymatic and enzymatic compounds [113]. Generally, their levels depend on the types of skin cells. For example, melanocytes do not contain enzymatic antioxidants [114]. The biopharmaceutical classification of antioxidants is based on their permeability and solubility. Four classes of antioxidants are estimated: high solubility–high permeability (i.e., vitamin C, are located in cellular fluids); low solubility–low permeability; low solubility–high permeability; and high solubility–low permeability (i.e., vitamin E, are present in cell membranes) [115]. The administration of antioxidant compounds involves overcoming different obstacles depending on whether they are administered for oral or topical use. The biological activity of the antioxidants administered orally is negatively influenced by the low solubility in the gastrointestinal fluids and aqueous media, instability at physiological pH, and degradation due to enzymes and light. Efficient delivery systems which can enhance their bioavailability are micelles, nanoemulsions, nanoparticles, nanocochleates, nanocapsules, nanocrystals, etc.

In cases where they are used in preparations for topical use, the main problems are instability, low permeability, and water-solubility. The instability is due to environmental stress (i.e., air, light, moisture, heat, oxygen, metal ions, and alkalinity) and determines the shelf life of the products [112]. The low permeability and water-solubility negatively affect their ability to enter into more profound layers of the skin and arrive at the target tissue [116]. For example, the use of resveratrol is limited in cosmetic formulations due to instability [117]. The microencapsulation techniques [118] and some biodegradable polymer-based delivery systems such as liposomes, solid lipid nanoparticles, nanostructured lipid carriers, and emulsions are employed to improve the antioxidants' bioactivity in cosmetic products [65].

### 3.1. Vitamin C

Vitamin C, also known as L-ascorbic acid, is a water-soluble vitamin used in skincare and skin lightening products such as antioxidants, skin repair, and connective tissue repair [119,120]. It, especially when exposed to air, oxidizes to dehydroascorbic acid and hydrolyzes to form 2,3-L-diketogulonate (at alkaline pH) [121] (Figure 1).

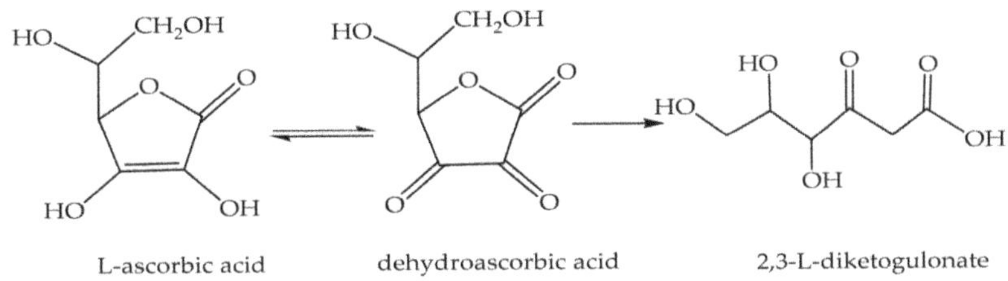


**Figure 1.** Ascorbic acid oxidation reaction.

Therefore, in cosmetic and food fields, some preferred ester derivatives (i.e., retinyl ascorbate, ascorbyl palmitate) encapsulate into microemulsions, polymeric nanoparticles, bilayer vesicles, and solid lipid nanoparticles [122]. Some applications use phosphatidylcholine/lecithin liposomes [123] to enhance the penetration of vitamin C.

### 3.2. Vitamin A

Vitamin A is a fat-soluble vitamin belonging to the group of substances structurally related to the retinol (Figure 2). Vitamin A and retinoids improve wound healing; prevent acne, skin aging, and psoriasis; affect keratinization, keratinocyte proliferation, and epidermal differentiation; and reduce oxidation and inflammation [124]. The retinoids are encapsulated into caprolactone-based nanocapsules and liposomes to protect the bioactive compounds from photodegradation [125,126]. Retinol is encapsulated into silicon particles to prepare antiaging and antiacne formulations [127]. Nanospheres are optimized for the encapsulation of adapalene (third-generation retinoid) to deliver the bioactive to hair follicles [128].

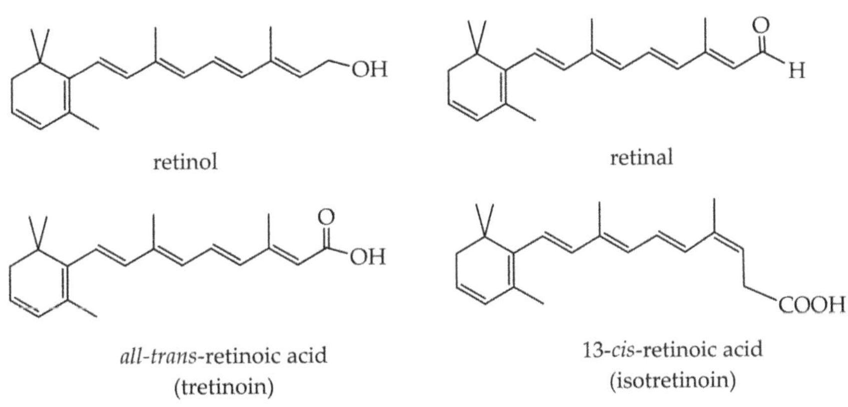


**Figure 2.** Structures of the most widespread retinoids.

### 3.3. Phenolic Compounds

Phenolics can be divided into flavonoids (including flavonols, flavones, catechins, flavanones, anthocyanidins, and isoflavones), phenolic acids, stilbenes, coumarins, and tannins (Figure 3).

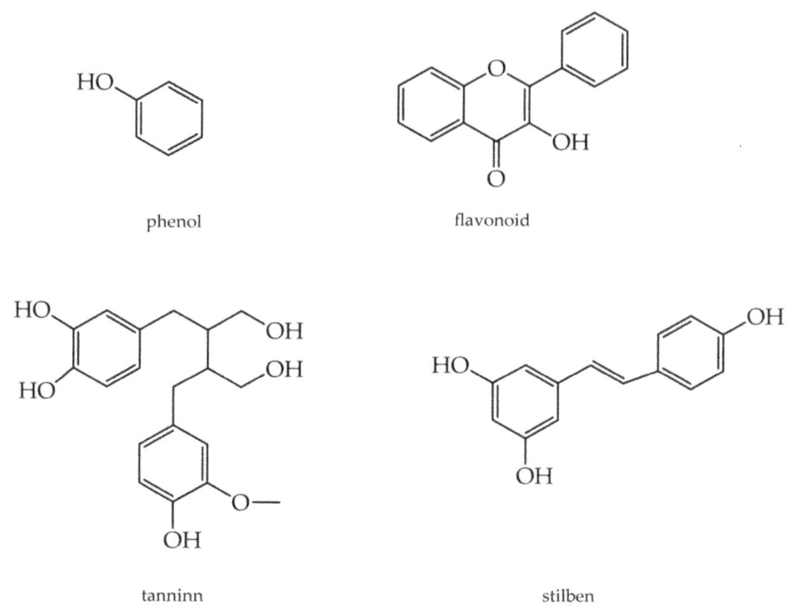


**Figure 3.** Structures of the most widespread phenols.

Phenolic compounds control skin inflammation, wound healing, and barrier homeostasis [129,130]. Phenolics can scavenge metal ions, block pathogenic free radicals, and induce the expression of protective genes against oxidative stress [10]. For example, the isoflavone genistein is used as an anti-wrinkle to protect and hydrate the skin and decrease the oxygen free radical, the expression of MMP-1, and inducible nitrogen oxide synthase [131,132].

Unfortunately, low levels of aqueous solubility, poor gastrointestinal stability, low absorption due to passive diffusion and active efflux in the gastrointestinal tract, and lack of target specificity in the human body limit their application. Examples of nanosystems used to enhance the delivery and bioavailability of polyphenols are nanocapsules, solid lipid nanoparticles, niosomes, and microemulsions. MethoxyPEG-palmitate nanocapsules and chitosan particles are used to load curcumin, the chitosan-tripolyphosphate nanoparticles to encapsulate epigallocatechin-3-gallate [133], and polymeric nanocapsule suspensions are employed in topical formulations directly applied on the skin or as an ingredient in semisolid formulations. The nanocapsules have a pH (slightly acid) similar to the skin pH; they can form a thin film, which causes long-term delivery [134]. The solid lipid nanoparticles are employed to decrease resveratrol photodegradation, improve its cellular uptake, and internalization in keratinocytes [135]. Niosomes improve the antioxidant activity and the skin permeation of ascorbic acid, ellagic acid, curcumin, and resveratrol [136]. Microemulsions, lipid nanoparticles, and liposomes are exploited to enhance the anti-skin-aging properties of the soy's isoflavones [85]. Some polyphenols used in the cosmetic field have been encapsulated in nanoparticles made by ionic gelation and microspheres produced by spray-drying [137].

### 3.4. Coenzyme Q10

Coenzyme Q10 (CoQ10) is a lipophilic antioxidant synthesized in humans employed in cosmetics formulation as an anti-photo-aging bioactive. It improves fibroblast proliferation and protects lipid membranes and DNA against oxidative damage, acting as a radical scavenger [138]. Liposomes, lipid nanoparticles, and solid nanoparticles are used to improve penetration in the deeper layers of the skin [139]. Farbound et al. incorporated CoQ10 into SLNs and then into a semisolid emulsion in a cosmetic formulation to improve skin elasticity and hydration [140]. El-Leithy et al. formulated a nanoemulsion containing CoQ10 as an anti-wrinkle cosmetic product [141]. Pegoraro et al. formulated a topical cosmetic product to decrease the effects of UVB radiation, encapsulating CoQ10 and vitamin E in caprolactone nanocapsules and incorporating them into hydrogels [142].

### *3.5. Terpenoids (Also Called Isoprenoids)*

The terpenoid subclasses are tocotrienols and tocopherols (Figure 4). They are derived from five-carbon isoprene units, which differ in carbon skeletons and functional groups. Most are multicyclic structures.

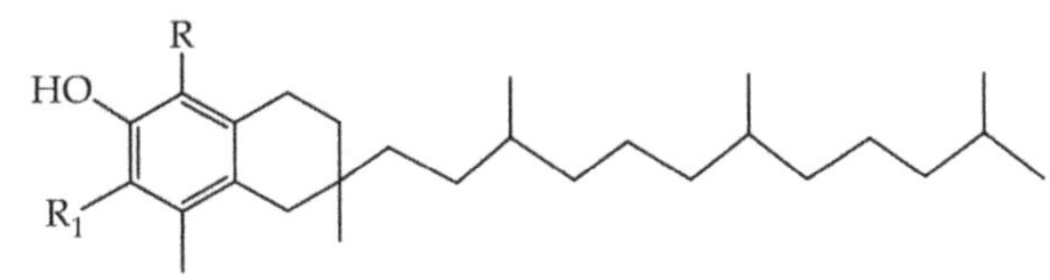


tocopherol

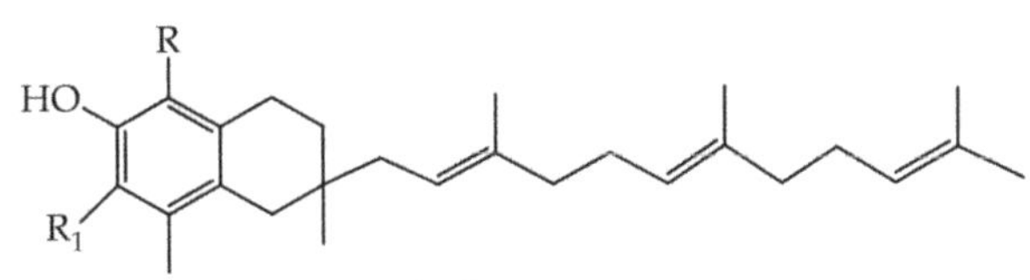


tocotrienol

| R | R1 | |
|---|---|---|
| $CH_3$ | $CH_3$ | α |
| $CH_3$ | H | β |
| H | $CH_3$ | γ |
| H | H | δ |

**Figure 4.** Tocopherol and tocotrienol.

The terpenes react with free radicals by partitioning themselves into fatty membranes. The tocopherols are used in cosmetic formulations as skin-conditioning agents or antioxidants [143] to control wound healing, skin inflammation, and barrier homeostasis [129,130]. Tocopherol is a skin irritant and light-sensitive liquid. Therefore, it is placed into nanocarriers to produce a cosmetically appealing formulation. The best performances are obtained using nanostructured lipid carriers and retinol-encapsulated chitosan nanoparticles (i.e., zein-chitosan and succinic-chitosan nanoparticles). The antioxidant activity of the encapsulated retinol is significantly greater than pure retinol [144–147]. The tocotrienols are oral care agents, light stabilizers, and skin-conditioning agents; a-tocopherol acetate is the most widespread vitamin E used in commercial sunscreen and skincare products. It is used as a preservative. Both α-tocopheryl acetate and a-tocopherol are used to make supplements, since they are generally recognized as safe (GRAS) food ingredients. Their daily intake varies from 0.15 to 2 mg/kg for DL-α-tocopherol and D-α-tocopherol [143]. Fullerene (carbon-based vehicle) nanocapsules with ascorbic acid and vitamin E have been employed to improve skin protective activity against premature aging [148].

### *3.6. Carotenoids (Alpha-Carotene, Beta-Carotene, Lycopene, Phytoene, Phytofluene)*

The carotenoids (i.e., lutein, zeaxanthin, α-carotene, and lycopene) (Figure 5) are natural colorants that can reduce oxygen singlets, stabilize other antioxidants, and block free radicals.

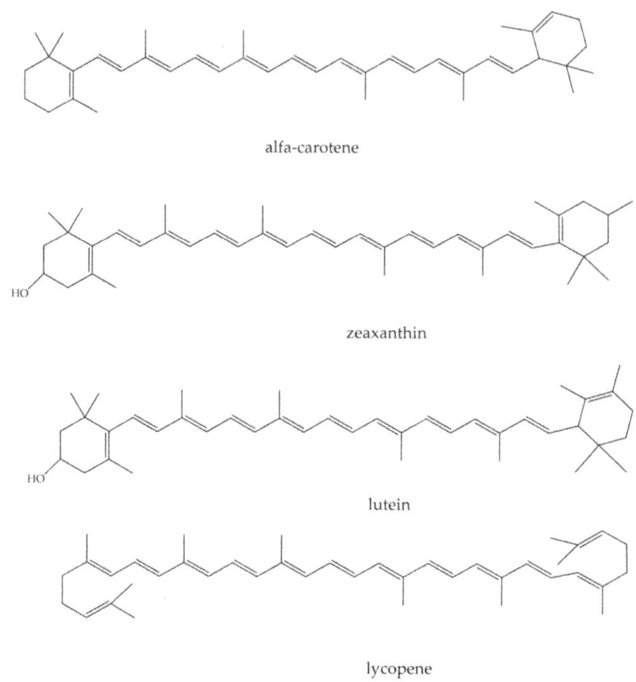


**Figure 5.** Structures of the most widespread carotenoids.

The carotenoids protect against cancers (i.e., uterine, prostate, breast, colorectal, and lung), cardiovascular illnesses, and eye (i.e., age-related macular degeneration and cataracts) and skin disorders [149–153]. The retinoic acid and derivatives modulate, at the skin level, the expression of genes implicated in cellular differentiation and proliferation. The carotenoids act as photoprotective agents against UV radiation. For example, the oral use of the astaxanthin increases skin condition and decreases skin hyper-pigmentation, improving some antioxidant enzymes (i.e., superoxide dismutase, catalases enzyme) activity, and suppressing tyrosinase action [153,154]. The β-carotene protects from sunburn diseases (inhibiting metalloproteinase activation and increasing 5-α-hydroperoxide synthesis) and prevents wrinkle formation and skin flaccidity [155,156]. Carotenoid intake improves skin yellowness and pigmentation [157]. Zeaxanthin enhances the antioxidative defense system and the skin's hydration and elasticity [158–160]. Phytoene and phytofluene have skin whitening effects [152]. Astaxanthin (a xanthophyll carotenoid) was loaded in poly(D,L-lactic-co-glycolic acid) (PLGA) nanoparticles by using the emulsion solvent evaporation technique to improve its antioxidant and anti-wrinkle effects [161].

### *3.7. Organosulphur Compounds*

The organosulphur compounds are organic substances classified according to the functional groups containing sulfur atoms in thiols, sulfides, sulfoxides, sulfones, thiosulfinates, sulfimides, sulfoximides, sulfonediimines, glucosinolates, thioketones, thioaldehydes, S-oxides and S,S-dioxides of thiocarbonyl compounds, thiocarboxylic acids, thioamides, sulfonic, sulfinic, sulfenic acids, and related compounds, sulfuranes, and persulfuranes (Figure 6) [162].

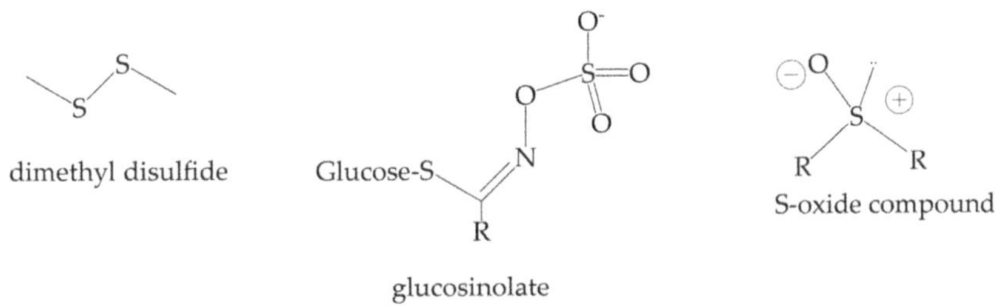


**Figure 6.** Structures of the most widespread organosulphur compounds.

They have anti-atherosclerosis, antifungal, antimicrobial, immunostimulating, and antithrombotic activities [162]. In cosmetic formulations, they are used as antioxidants and skin-conditioning agents. Some organo-sulfur compounds (i.e., 1-propylmercaptan, 2,5-dimethylthiophene, dimethyl disulfide, diallyl disulfide, and propyl disulfide) decrease tyrosinase activity and melanin development, reducing the reactive oxygen species and improving the glutathion/glutathione disulfide ratio in B16 cells [163]. The administration of broccoli extract (rich in glucosinolates) decreases skin lesions and improves keratin production. Sulforaphane decreases the risk of skin lesions caused by UV radiation [164]. Allicin cures premature aging, inhibiting a leukocyte elastase [165]. The pegylated liposomes and nanoparticles (e.g., diallytrisulfide-polybutyl cyanoacrylate-nanoparticles) have been employed to deliver the organosulphur compounds [166]. An anti-dandruff shampoo was formulated using SLNs of garlic as an antifungal agent [136].

### *3.8. Methylxanthines (Theophylline, Caffeine, and Theobromine)*

The methylxanthines are alkaloids (purine base) used in cosmetic formulations as scavengers of reactive oxygen species generated by UV exposure, antiaging, and anti-cellulitis agents (Figure 7). Caffeine is an antiaging, photoprotective, antioxidant, anticellulite, and antiacne compound [167]. Unfortunately, it can precipitate and form clumps when used in topical applications. [168]. Solid lipid nanoparticles have been designed to solve these problems [169].

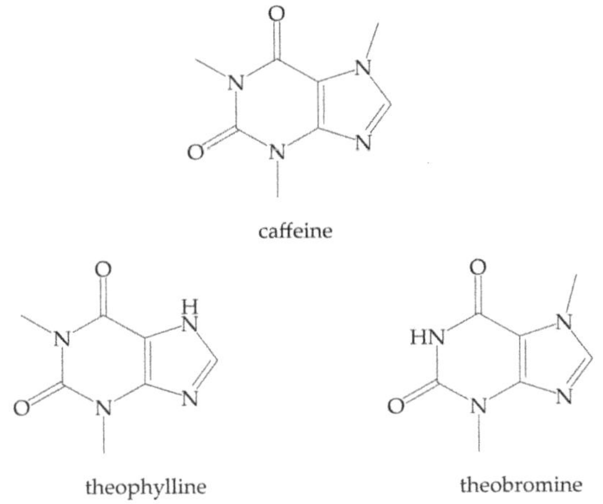


**Figure 7.** Structures of the most widespread methylxanthines.

## 4. Nanodelivery System Toxicity

The nanodelivery systems' different particle sizes, surface groups, zeta potentials, and aggregation states can determine different bioavailability and toxic reactions than conventional ones [170]. The smaller size of nanomaterials allows them easier access into cells, tissues, and organs, decreasing the influence of intestinal clearance mechanisms and protracting their stay in the gastrointestinal tract [171]. Therefore, the tolerable upper intake levels (UL) and recommended daily allowance (RDA) of nutrients need reevaluation [172]. Furthermore, the nanoparticles interact with various immune system components, breaking up immunostimulation or immunosuppression, promoting inflammation and autoimmune disorders, or increasing the host's susceptibility to infections. They can interact with the innate and adaptive immune systems [173]. The innate immune system is our first line of defense against invading organisms and can immediately respond to any stress. It consists of cells (e.g., physical epithelial barriers, phagocytic leukocytes, dendritic cells, natural killer cells) and proteins (e.g., circulating plasma proteins) that are always present and ready to mobilize and fight microbes at the site of infection. The innate immune system is nonspecific and has no memory. The adaptive immune system acts as a second line of defense and can respond efficiently to re-exposure to the same pathogen. The components of the adaptive immune system are generally silent; however, when activated, these components "adapt" to the presence of infectious agents by activating, proliferating, and creating potent mechanisms for neutralizing or eliminating the microbes. There are two adaptive immune responses: humoral immunity, mediated by antibodies produced by B lymphocytes, and cell-mediated immunity, mediated by T lymphocytes. The nanoparticles can elicit an immune response by directly immunostimulating antigen-presenting cells or delivering antigen to specific cellular compartments. Their compatibility with the immune system depends on size, surface charge, hydrophobicity/hydrophilicity, and steric effects of the particle coating. Predicting nanoparticles' innate responses (in vitro or in vivo) is still challenging [10]. Some nanomaterials induce a (pro)inflammatory response and are taken up by phagocytic cells, whereas others seem to reduce these activities, reducing the ability of these immune cells to fight (e.g., bacteria).

Moreover, nanomaterials can affect the adaptive immune response, disrupting the Th1/Th2 balance, influencing cytokine production in peripheral blood mononuclear cells, and overproducing TNF-$\alpha$ (tumor necrosis factor) and INF$\gamma$ interferon, decreasing levels of IL-10 (interleukin) and IL-2. It is possible to show an overt immune response when nanoparticles are designed with poly(ethylene glycol) (PEG) or other types of polymers to provide a hydrophilic environment [174,175]. The silver, titanium dioxide, zinc, and zinc oxide nanoparticles are highly toxic, since their high surface area increases the contact with biomolecules and triggers adverse responses. Cationic charge determines a high affinity towards the negatively charged plasma membrane, which determines retention of one $Cl^-$ ion and one water molecule per proton and consequent lysosomal swelling and rupture [176]. The toxicity of nanoceuticals was studied mainly in animal experiments. More human and clinical trials should be carried out to know the potential positive and negative effects of nanoceutics and/or nanocosmetics on human health [177].

## 5. Conclusions

The use of nanotechnologies represents an excellent opportunity to improve the performance of nutricosmeceutic products. Nanodelivery systems improve bioactive compounds' bioavailability and water solubility, guarantee their release at the site of action, mask their unpleasant taste, and prolong their expiration date. Unfortunately, only a few studies on animals and very limited on humans have investigated their potential toxic effects. More human and clinical trials should be carried out to protect consumer health.

**Funding:** This research received no external funding.

**Conflicts of Interest:** The author declares no conflict of interest.

## References

1. Dini, I.; Laneri, S. Nutricosmetics: A brief overview. *Phytoth. Res.* **2019**, *33*, 3054–3063. [CrossRef] [PubMed]
2. Laneri, S.; Di Lorenzo, R.; Sacchi, A.; Dini, I. Dosage of Bioactive Molecules in the Nutricosmeceutical Helix aspersa Muller Mucus and Formulation of New Cosmetic Cream with Moisturizing Effect. *Nat. Prod. Commun.* **2019**, *14*, 1934578X19868606. [CrossRef]

3. Laneri, S.; Di Lorenzo, R.M.; Bernardi, A.; Sacchi, A.; Dini, I. Aloe barbadensis: A plant of nutricosmetic interest. *Nat. Prod. Commun.* **2020**, *15*, 1934578X20932744. [CrossRef]
4. Laneri, S.; Dini, I.; Tito, A.; Di Lorenzo, R.; Bimonte, M.; Tortora, A.; Zappelli, C.; Angelillo, M.; Bernardi, A.; Sacchi, A.; et al. Plant cell culture extract of Cirsium eriophorum with skin pore refiner activity by modulating sebum production and inflammatory response. *Phytoth. Res.* **2021**, *35*, 530–540. [CrossRef]
5. Dini, I.; Falanga, D.; Di Lorenzo, R.; Tito, A.; Carotenuto, G.; Zappelli, C.; Grumetto, L.; Sacchi, A.; Laneri, S.; Apone, F. An Extract from Ficus carica Cell Cultures Works as an Anti-Stress Ingredient for the Skin. *Antioxidants* **2021**, *10*, 515. [CrossRef]
6. WHO. Ageing. Available online: https://www.who.int/news-room/fact-sheets/detail/ageing-and-health (accessed on 4 October 2021).
7. Dini, I.; Laneri, S. The New Challenge of Green Cosmetics: Natural Food Ingredients for Cosmetic Formulations. *Molecules* **2021**, *26*, 3921. [CrossRef]
8. Dini, I. Spices and herbs as therapeutic foods. In *Food Quality: Balancing Health and Disease*; Holban, A.M., Grumezescu, A.M., Eds.; Academic Press: London, UK; Elservier: London, UK, 2018; pp. 433–469.
9. Wertz, P.W. Lipids and barrier function of the skin. *Acta Derm. Venereol.* **2000**, *208*, 7–11. [CrossRef]
10. Dini, I.; Laneri, S. Spices, Condiments, extra virgin olive oil and aromas as not only flavorings, but precious allies for our wellbeing. *Antioxidants* **2021**, *10*, 868. [CrossRef]
11. Garcia, L.B.; Pires, G.A.; Oliveira, D.A.J.; Silva, L.A.O.; Gomes, A.F.; Amaral, J.G.; Pereira, G.R.; Ruela, A.L.M. Industrial Crops & Products Incorporation of glycolic extract of cocoa beans (*Theobroma cacao* L.) into microemulsions and emulgels for skincare. *Ind. Crop. Prod.* **2021**, *161*, 113181.
12. Kaul, S.; Gulati, N.; Verma, D.; Mukherjee, S.; Nagaich, U. Role of Nanotechnology in cosmeceuticals: A review of recent advances. *J. Pharm.* **2018**, *2018*, 3420204. [CrossRef]
13. Arshad, R.; Gulshad, L.; Haq, I.U.; Farooq, M.A.; Al-Farga, A.; Siddique, R.; Manzoor, M.F.; Karrar, E. Nanotechnology: A novel tool to enhance the bioavailability of micronutrients. *Food Sci. Nutr.* **2021**, *9*, 3354–3361. [CrossRef]
14. CEC Regulation (EC.). No. 1907/2006 of the European Parliament and of the Council of 18 December 2006 Concerning the Registration, Evaluation, Authorisation and Restriction of Chemicals (REACH). 2006. Available online: https://www.legislation.gov.uk/eur/2006/1907/contents (accessed on 9 April 2021).
15. Regulation (EC.). No. 1223/2009 of the European Parliament and of the Council of 30 November 2009 on Cosmetic Products. Available online: https://ec.europa.eu/health/sites/health/files/endocrine_disruptors/docs/cosmetic_1223_2009_regulation_en.pdf (accessed on 30 April 2019).
16. Hoet, P.H.M. Opinion of the Scientific Committee on Consumer Safety (SCCS)–Revision of the opinion on the safety of the use of Silica, Hydrated Silica, and Silica Surface Modified with Alkyl Silylates (nano form) in cosmetic products. *Regul. Toxicol. Pharmacol.* **2016**, *74*, 79–80.
17. Paolino, D.; Mancuso, A.; Cristiano, M.C.; Froiio, F.; Lammari, N.; Celia, C.; Fresta, M. Nanonutraceuticals: The New Frontier of Supplementary Food. *Nanomaterials* **2021**, *11*, 792. [CrossRef]
18. Barhoum, A.; García-Betancourt, M.L.; Jeevanandam, J.; Hussien, E.A.; Mekkawy, S.A.; Mostafa, M.; Omran, M.M.; Abdalla, M.S.; Bechelany, M. Review on Natural, Incidental, Bioinspired, and Engineered Nanomaterials: History, Definitions, Classifications, Synthesis, Properties, Market, Toxicities, Risks, and Regulations. *Nanomaterials* **2022**, *12*, 177. [CrossRef]
19. McClements, D.J. Edible nanoemulsions: Fabrication, properties, and functional performance. *Soft Matter* **2011**, *7*, 2297–2316. [CrossRef]
20. Marzuki, N.H.C.; Wahab, R.A.; Hamid, M.A. An overview of nanoemulsion: Concepts of development and cosmeceutical applications. *Biotechnol. Biotechnol. Equip.* **2019**, *33*, 779–797. [CrossRef]
21. Gasa-Falcon, A.; Odriozola-Serrano, I.; Oms-Oliu, G.; Martín-Belloso, O. Nanostructured Lipid-Based Delivery Systems as a Strategy to Increase Functionality of Bioactive Compounds. *Foods* **2020**, *9*, 325. [CrossRef]
22. Aswathanarayan, J.B.; Vittal, R.R. Nanoemulsions and Their Potential Applications in Food Industry. *Front. Sustain. Food Syst.* **2019**, *3*, 95. [CrossRef]
23. Da Silva, A.C.F.; Costa, A.M.; Ascenso, A.; Ribeiro, H.M.; Marto, J.; Gonçalves, L.M.; Carvalheiro, M.; Simões, S. Nanoemulsions for cosmetic products. In *Nanocosmetics*; Nanda, A., Nanda, S., Nguyen, T.A., Rajendran, S., Slimani, Y., Eds.; Micro and Nano Technologies; Elsevier: Amsterdam, The Netherlands, 2020; pp. 59–77.
24. Kumar, M.; Bishnoi, R.S.; Shukla, A.K.; Jain, C.P. Techniques for Formulation of Nanoemulsion Drug Delivery System: A Review. *Prev. Nutr. Food Sci.* **2019**, *24*, 225–234. [CrossRef]
25. Bolzinger, M.A.; Briançon, S.; Pelletier, J.; Chevalier, Y. Penetration of drugs through skin, a complex rate-controlling membrane. *Curr. Opin. Colloid Interface Sci.* **2012**, *17*, 156–165. [CrossRef]
26. De Azevedo Ribeiro, R.C.; Barreto, S.M.A.G.; Ostrosky, E.A.; Da Rocha-Filho, P.A.; Veríssimo, L.M.; Ferrari, M. Production and characterization of cosmetic nanoemulsions containing *Opuntia ficus-indica* (L.) Mill extract as moisturizing agent. *Molecules* **2015**, *20*, 2492–2509. [CrossRef]
27. Quintão, F.J.O.; Tavares, R.S.N.; Vieira-Filho, S.A.; Souza, G.H.B.; Santos, O.D.H. Hydroalcoholic extracts of *Vellozia squamata*: Study of its nanoemulsions for pharmaceutical or cosmetic applications. *Braz. J. Pharmacogn.* **2013**, *23*, 101–107. [CrossRef]

28. Baccarin, T.; Mitjans, M.; Ramos, D.; Lemos-Senna, E.; Vinardell, M.P. Photoprotection by Punica granatum seed oil nanoemulsion entrapping polyphenol-rich ethyl acetate fraction against UVB-induced D.N.A. damage in human keratinocyte (HaCaT) cell line. *J. Photochem. Photobiol. B* **2015**, *153*, 127–136. [CrossRef]
29. Mc clements, D.J. Nanoscale nutrient delivery systems for food applications: Improving bioactive dispersibility, stability, and bioavailability. *J. Food Sci.* **2015**, *80*, N1602–N1611. [CrossRef]
30. Comuzzo, P.; Calligaris, S. Potential Applications of High Pressure Homogenization in Winemaking: A Review. *Beverages* **2019**, *5*, 56. [CrossRef]
31. Sakulku, U.; Nuchuchua, O.; Uawongyart, N.; Puttipipatkhachorn, S.; Soottitantawat, A.; Ruktanonchai, U. Characterization and mosquito repellent activity of citronella oil nanoemulsion. *Int. J. Pharm.* **2009**, *372*, 105–111. [CrossRef]
32. Sharma, N.; Mishra, S.; Sharma, S.; Deshpande, R.D.; Sharma, R.K. Preparation and optimization of nanoemulsions for targeting drug delivery. *Int. J. Drug Dev. Res.* **2013**, *5*, 37–48.
33. Ruiz-Montañez, G.; Ragazzo-Sanchez, J.A.; Picart-Palmade, L.; Calderón-Santoyo, M.; Chevalier-Lucia, D. Optimization of nanoemulsions processed by high-pressure homogenization to protect a bioactive extract of jackfruit (*Artocarpus heterophyllus Lam*). *Innov. Food Sci. Emerg. Technol.* **2017**, *40*, 35–41. [CrossRef]
34. Villalobos-Castillejos, F.; Granillo-Guerrero, V.G.; Leyva-Daniel, D.E.; Alamilla-Beltrán, L.; Gutiérrez-López, G.F.; Monroy-Villagrana, A.; Jafari, S.M. Fabrication of Nanoemulsions by Microfluidization. In *Nanoemulsions*; Jafari, S.M., Ed.; McCl: Toronto, ON, Canada, 2018; pp. 207–232.
35. Thakur, A.; Walia, M.K.; Kumar, S. Nanoemulsion in enhancement of bioavailability of poorly soluble drugs: A review. *Pharmacophore* **2013**, *4*, 15–25.
36. Lee, L.; Norton, I.T. Comparing droplet breakup for a high-pressure valve homogenizer and a Microfluidizer for the potential production of food-grade nanoemulsions. *J. Food Eng.* **2013**, *114*, 158–163. [CrossRef]
37. Silva, F.F.; Ricci-Júnior, E.; Mansur, C.R. Nanoemulsions containing octyl methoxycinnamate and solid particles of $TiO_2$: Preparation, characterization and in vitro evaluation of the solar protection factor. *Drug Dev. Ind. Pharm.* **2013**, *39*, 1378–1388. [CrossRef] [PubMed]
38. Date, A.A.; Desai, N.; Dixit, R.; Nagarsenker, M. Self-nanoemulsifying drug delivery systems: Formulation insights, applications and advances. *Nanomedicine* **2010**, *5*, 1595–1616. [CrossRef] [PubMed]
39. Izquierdo, P.; Esquena, J.; Tadros, T.F.; Dederen, C.; Garcia, M.J.; Azemar, N.; Solans, C. Formation and stability of nano-emulsions prepared using the phase inversion temperature method. *Langmuir* **2002**, *18*, 26–30. [CrossRef]
40. Solè, I.; Pey, C.M.; Maestro, A.; González, C.; Porras, M.; Solans, C.; Gutiérrez, J.M. Nano-emulsions prepared by the phase inversion composition method: Preparation variables and scale up. *J. Colloid Interface Sci.* **2010**, *344*, 417–423. [CrossRef]
41. Izquierdo, P.; Feng, J.; Esquena, J.; Tadros, T.F.; Dederen, J.C.; Garcia, M.J.; Azemar, N.; Solans, C. The influence of surfactant mixing ratio on nanoemulsion formation by the pit method. *J. Colloid Interface Sci.* **2005**, *285*, 388–394. [CrossRef]
42. Dini, I. Chapter 14—Use of Essential Oils in Food Packaging. In *Essential Oils in Food Preservation, Flavor and Safety*; Academic Press: Cambridge, MA, USA, 2016; pp. 139–147.
43. Lamba, H.; Sathish, K.; Sabikhi, L. Double Emulsions: Emerging Delivery System for Plant Bioactives. *Food Bioprocess Technol.* **2015**, *8*, 709–728. [CrossRef]
44. Ali, A.; Iqbal, S.; Ilyas, A.; Khan, H.; Asad, M.H.H.B.; Fatima, N.; Akhtar, N. Anti-pollution cosmetic-based one-step formation of w/o/w multiple emulsion containing D-biotin for skin protection: Fabrication and in vitro and in vivo evaluation. *Drug Deliv. Transl. Res.* **2019**, *9*, 1117–1132. [CrossRef]
45. Tadros, T.F. Future developments in cosmetic formulations. *Int. J. Cosmet. Sci.* **1992**, *14*, 93–111. [CrossRef]
46. Dickinson, E. Double emulsions stabilized by food biopolymers. *Food Biophys.* **2010**, *6*, 1–11. [CrossRef]
47. Klojdová, I.; Štětina, J.; Horáčková, Š. W/O/W multiple emulsions as the functional component of dairy products. *Chem. Eng. Technol.* **2019**, *42*, 715–727. [CrossRef]
48. Zhi, Z.; Liu, R.; Wang, W.; Dewettinck, K.; Van Bockstaele, F. Recent progress in oil-in-water-in-oil (O/W/O) double emulsions. *Crit. Rev. Food Sci. Nutr.* **2022**, 1–12. [CrossRef]
49. Nair, M.; Lusignan, C.P.; Boris, D.C. Preparation of uniform micrometer-sized polymer particles with closed-cell porous architecture made by limited coalescence of a double emulsion. *Colloids Surf. Asp.* **2014**, *443*, 583–595. [CrossRef]
50. Zhao, H.; Yang, Y.; Chen, Y.; Li, J.; Wang, L.; Li, C. A review of multiple Pickering emulsions: Solid stabilization, preparation, particle effect, and application. *Chem. Eng. Sci.* **2022**, *248*, 117085. [CrossRef]
51. Lee, D.; Weitz, D.A. Nonspherical colloidosomes with multiple compartments from double emulsions. *Small* **2009**, *5*, 1932–1935. [CrossRef]
52. Ali, A.; Ahmad, U.; Akhtar, J.; Khan, M.M. Engineered nano scale formulation strategies to augment efficiency of nutraceuticals. *J. Funct. Foods* **2019**, *62*, 103554. [CrossRef]
53. Pucek, A.; Tokarek, B.; Waglewska, E.; Bazylińska, U. Recent Advances in the Structural Design of Photosensitive Agent Formulations Using "Soft" Colloidal Nanocarriers. *Pharmaceutics* **2020**, *12*, 587. [CrossRef]
54. Hameed, A.; Fatima, G.R.; Malik, K.; Muqadas, A.; Fazal-ur-Rehman, M. Scope of nanotechnology in cosmetics: Dermatology and skin care products. *J. Med. Chem. Sci.* **2019**, *2*, 9–16.
55. Smijs, T.G.; Pavel, S. Titanium dioxide and zinc oxide nanoparticles in sunscreens: Focus on their safety and effectiveness. *Nanotechnol. Sci. Appl.* **2011**, *4*, 95. [CrossRef]

56. Khameneh, B.; Halimi, V.; Jaafari, M.R.; Golmohammadzadeh, S. Safranal-loaded solid lipid nanoparticles: Evaluation of sunscreen and moisturizing potential for topical applications. *Iran. J. Basic Med. Sci.* **2015**, *18*, 58–63.
57. Amato, S.W.; Farer, A.; Hoyte, W.M.; Pavlovsky, M.; Smith, R.; Valdiviezo, G. Coatings for Mammalian Nails that Include Nanosized Particles. US Patent 12/162,311, 5 August 2010.
58. Davidov-Pardo, G.; Pérez-Ciordia, S.; Marín-Arroyo, M.R.; Mcclements, D.J. Improving resveratrol bioaccessibility using biopolymer nanoparticles and complexes: Impact of protein-carbohydrate maillard conjugation. *J. Agric. Food Chem.* **2015**, *63*, 3915–3923. [CrossRef]
59. Penalva, R.; Esparza, I.; Agüeros, M.; Gonzalez-Navarro, C.J.; Gonzalez-Ferrero, C.; Irache, J.M. Casein nanoparticles as carriers for the oral delivery of folic acid. *Food Hydrocoll.* **2015**, *4*, 399–406. [CrossRef]
60. Liang, J.; Yan, H.; Wang, X.; Zhou, Y.; Gao, X.; Puligundla, P.; Wan, X. Encapsulation of epigallocatechin gallate in zein/chitosan nanoparticles for controlled applications in food systems. *Food Chem.* **2017**, *231*, 19–24. [CrossRef]
61. Sailaja, A.K.; Amareshwar, P. Preparation of alginate nanoparticles by desovaltion technique using acetone as desolvating agent. *Asian J. Pharm. Clin. Res.* **2012**, *5*, 132–134.
62. Duclairoir, C.; Orecchioni, A.M.; Depraetere, P.; Nakache, E. Alpha-tocopherol encapsulation and in vitro release from wheat gliadin nanoparticles. *J. Microencapsul.* **2002**, *9*, 53–60. [CrossRef]
63. Sundar, S.; Kundu, J.; Kundu, S.C. Biopolymeric nanoparticles. *Sci. Technol. Adv. Mater.* **2010**, *11*, 014104. [CrossRef]
64. Zaeim, D.; Sarabi-Jamab, M.; Ghorani, B.; Kadkhodaee, R.; Tromp, R.H. Electrospray-assisted drying of live probiotics in acacia gum microparticles matrix. *Carbohydr. Polym.* **2018**, *183*, 183–191. [CrossRef]
65. Casanova, F.; Santos, L. Encapsulation of cosmetic active ingredients for topical application—A review. *J. Microencapsul.* **2016**, *33*, 1–17. [CrossRef]
66. Calamak, S. Micro/Nanoencapsulation of Active Food Compounds: Encapsulation, Characterization and Biological Fate of Encapsulated Systems. In *Sustainable Agriculture Reviews*; Maurya, V.K., Gothandam, K.M., Ranjan, S., Dasgupta, N., Lichtfouse, E., Eds.; Springer: Cham, Germany, 2021; Volume 55, pp. 93–114.
67. Arpagaus, C. Production of food bioactive-loaded nanoparticles by nano spray drying. In *Nanoencapsulation of Food Ingredients by Specialized Equipment: Volume 3 in the Nanoencapsulation in the Food Industry Series*; Academic Press: Cambridge, MA, USA, 2019; Volume 3, pp. 151–211.
68. Pateiro, M.; Gómez, B.; Munekata, P.E.S.; Barba, F.J.; Putnik, P.; Kovačević, D.B.; Lorenzo, J.M. Nanoencapsulation of Promising Bioactive Compounds to Improve Their Absorption, Stability, Functionality and the Appearance of the Final Food Products. *Molecules* **2021**, *26*, 1547. [CrossRef]
69. Rehman, A.; Tong, Q.Y.; Jafari, S.M.; Assadpour, E.; Shehzad, Q.; Aadil, R.M.; Iqbal, M.W.; Rashed, M.M.A.; Mushtaq, B.S.; Ashraf, W. Carotenoid-loaded nanocarriers: A comprehensive review. *Adv. Colloid. Interface Sci.* **2020**, *275*, 102048. [CrossRef] [PubMed]
70. Panigrahi, S.S.; Syed, I.; Sivabalan, S.; Sarkar, P. Nanoencapsulation strategies for lipid-soluble vitamins. *Chem. Pap.* **2018**, *73*, 1–16. [CrossRef]
71. Liu, G.; Li, J.; Deng, S. Applications of Supercritical Anti-Solvent Process in Preparation of Solid Multicomponent Systems. *Pharmaceutics* **2021**, *13*, 475. [CrossRef] [PubMed]
72. Dai, L.; Sun, C.; Li, R.; Mao, L.; Liu, F.; Gao, Y. Structural characterization, formation mechanism and stability of curcumin in zein-lecithin composite nanoparticles fabricated by antisolvent co-precipitation. *Food Chem.* **2017**, *237*, 1163–1171. [CrossRef] [PubMed]
73. Devi, N.; Sarmah, M.; Khatun, B.; Maji, T.K. Encapsulation of active ingredients in polysaccharide–protein complex coacervates. *Adv. Colloid Interface Sci.* **2017**, *239*, 136–145. [CrossRef] [PubMed]
74. Schmitt, C.; Turgeon, S.L. Protein/polysaccharide complexes and coacervates in food systems. *Adv. Colloid Interface Sci.* **2011**, *167*, 63–70. [CrossRef]
75. Fang, Z.; Bhandari, B. Encapsulation of polyphenols—A review. *Trends Food Sci. Technol.* **2010**, *21*, 510–523. [CrossRef]
76. Reis, C.P.; Neufeld, R.J.; Ribeiro, A.J.; Veiga, F. Nanoencapsulation I. Methods for preparation of drugloaded polymeric nanoparticles. *Nanomed. Nanotechnol. Biol. Med.* **2006**, *2*, 8–21. [CrossRef]
77. Uppal, S.; Kaur, K.; Mehta, S.K. Inclusion Complexation and Coacervation to Fabricate Nanoencapsulated Foods. In *Nano-Food Engineering*; Hebbar, U., Ranjan, S., Mishra, R.K., Eds.; Springer: Cham, Germany, 2020; pp. 97–123.
78. Scalia, S.; Villani, S.; Casolari, A. Inclusion Complexation of the Sunscreen Agent 2-Ethylhexyl-p-dimethylaminobenzoate with Hydroxypropyl-β-cyclodextrin: Effect on Photostability. *J. Pharm. Pharmacol.* **1999**, *51*, 1367–1374. [CrossRef]
79. Lammari, N.; Louaer, O.; Meniai, A.H.; Elaissari, A. Encapsulation of Essential Oils via Nanoprecipitation Process: Overview, Progress, Challenges and Prospects. *Pharmaceutics* **2020**, *12*, 431. [CrossRef]
80. Jummes, B.; Sganzerla, W.G.; da Rosa, C.G.; Noronha, C.M.; Nunes, M.R.; Bertoldi, F.C.; Barreto, P.L.M. Antioxidant and antimicrobial poly-ε-caprolactone nanoparticles loaded with *Cymbopogon martinii* Roxb. essential oil. *Biocatal. Agric. Biotechnol.* **2020**, *23*, 101499. [CrossRef]
81. Mitura, S.; Sionkowska, A.; Jaiswal, A. Biopolymers for hydrogels in cosmetics: Review. *J. Mater. Sci. Mater. Med.* **2020**, *31*, 50. [CrossRef]
82. Varaprasad, K.; Raghavendra, G.M.; Jayaramudu, T.; Yallapu, M.M.; Sadiku, R. A mini review on hydrogels classification and recent developments in miscellaneous applications. *Mater. Sci. Eng. C* **2017**, *79*, 958–971. [CrossRef]

83. Bourbon, A.I.; Cerqueira, M.A.; Vicente, A.A. encapsulation and controlled release of bioactive compounds in lactoferrin-glycomacropeptide nanohydrogels: Curcumin and caffeine as model compounds. *J. Food Eng.* **2016**, *180*, 110–119. [CrossRef]
84. Zambrano-Zaragoza, M.L.; González-Reza, R.; Mendoza-Muñoz, N.; Miranda-Linares, V.; Bernal-Couoh, T.F.; Mendoza-Elvira, S.; Quintanar-Guerrero, D. Nanosystems in Edible Coatings: A Novel Strategy for Food Preservation. *Int. J. Mol. Sci.* **2018**, *19*, 705. [CrossRef]
85. Kim, M.-H.; Jeon, Y.-E.; Kang, S.; Lee, J.-Y.; Lee, K.W.; Kim, K.-T.; Kim, D.-D. Lipid Nanoparticles for Enhancing the Physicochemical Stability and Topical Skin Delivery of Orobol. *Pharmaceutics* **2020**, *12*, 845. [CrossRef]
86. Pavlou, P.; Siamidi, A.; Varvaresou, A.; Vlachou, M. Skin Care Formulations and Lipid Carriers as Skin Moisturizing Agents. *Cosmetics* **2021**, *8*, 89. [CrossRef]
87. Vijayan, V.; Shaikh, A.; Saktivhel, S.; Reddy, K.R. Formulation and characterization of solid lipid nanoparticles loaded Neem oil for topical treatment of acne. *J. Acute Dis.* **2013**, *2*, 282–286. [CrossRef]
88. Ahmad, J. Lipid Nanoparticles Based Cosmetics with Potential Application in Alleviating Skin Disorders. *Cosmetics* **2021**, *8*, 84. [CrossRef]
89. Üner, M.; Wissing, S.A.; Yener, G.; Müller, R.H. Skin moisturizing effect and skin penetration of ascorbyl palmitate entrapped in Solid Lipid Nanoparticles (SLN) and Nanostructured Lipid Carriers (NLC) incorporated into hydrogel. *Pharmazie* **2005**, *60*, 751–755.
90. Lushchak, O.; Karpenko, R.; Zayahckivska, A.; Koliada, A.; Vaiserman, A. Lipid-Based Nano-delivery of Phytobioactive Compounds in Anti-aging Medicine. In *Systemic Delivery Technologies in Anti-Aging Medicine: Methods and Applications*; Lai, W.-F., Ed.; Springer: Cham, Switzerland, 2020; pp. 221–245.
91. Raza, K.; Singh, B.; Singla, N.; Negi, P.; Singal, P.; Katare, O.R. Nano-lipoidal carriers of isotretinoin with antiaging potential: Formulation, characterization and biochemical evaluation. *J. Drug Target* **2013**, *21*, 435–442. [CrossRef]
92. Dos Santos, P.P.; Andrade, L.D.A.; Flôres, S.H.; Rios, A.D.O. Nanoencapsulation of carotenoids: A focus on different delivery systems and evaluation parameters. *J. Food Sci. Technol.* **2018**, *55*, 3851–3860. [CrossRef]
93. Tan, C.; Xue, J.; Lou, X.; Abbas, S.; Guan, Y.; Feng, B.; Zhang, X.; Xia, S. Liposomes as delivery systems for carotenoids: Comparative studies of loading ability, storage stability and in vitro release. *Food Funct.* **2014**, *5*, 1232. [CrossRef]
94. Amnuaikit, T.; Pinsuwan, S.; Ingkatawornwong, S.; Worachotekamjorn, K. Development of lipsticks containing rice bran oil liposome. *Planta Med.* **2008**, *74*, PD14. [CrossRef]
95. Chen, X.; Zou, L.Q.; Niu, J.; Liu, W.; Peng, S.F.; Liu, C.M. The stability, sustained release and cellular antioxidant activity of curcumin nanoliposomes. *Molecules* **2015**, *20*, 14293–14311. [CrossRef]
96. Tavakoli, H.; Hosseini, O.; Jafari, S.M.; Katouzian, I. Evaluation of physicochemical and antioxidant properties of yogurt enriched by olive leaf phenolics within nanoliposomes. *J. Agric. Food Chem.* **2018**, *66*, 9231–9240. [CrossRef]
97. Kotal, M.; Bhowmik, A.K. Polymer nanocomposites from modified clays: Recent advances and challenges. *Prog. Polym. Sci.* **2015**, *51*, 127–187. [CrossRef]
98. Murray, H.H. Overview—clay mineral applications. *Appl. Clay Sci.* **1991**, *5*, 379–395. [CrossRef]
99. Kommireddy, D.S.; Ichinose, I.; Lvov, Y.M.; Mills, D.K. Nanoparticle Multilayers: Surface Modification for Cell Attachment and Growth. *J. Biomed. Nanotechnol.* **2006**, *1*, 286–290. [CrossRef]
100. Sasaki, Y.; Sathi, G.A.; Yamamoto, O. Wound healing effect of bioactive ion released from Mg-smectite. Mater. *Sci. Eng. C. Mater. Biol. Appl.* **2017**, *77*, 52–57. [CrossRef]
101. Cervini-Silva, J.; Nieto- Camacho, A.; Ramírez-Apan, M.T.; Gómez-Vidales, V.; Palacios, E.; Montoya, A.; Ronquillo de Jesús, E. Anti-inflammatory, anti-bacterial, and cytotoxic activity of fibrous clays. *Colloids Surf. B Biointerfaces* **2015**, *129*, 1–6. [CrossRef]
102. Sánchez-Espejo, R.; Aguzzi, C.; Salcedo, I.; Cerezo, P.; Viseras, C. Clays in complementary and alternative medicine. *Mater. Technol.* **2014**, *29*, B78–B81. [CrossRef]
103. García-Villén, F.; Sánchez-Espejo, R.; Borrego-Sánchez, A.; Cerezo, P.; Perioli, L.; Viseras, C. Safety of Nanoclay/Spring Water Hydrogels: Assessment and Mobility of Hazardous Elements. *Pharmaceutics* **2020**, *12*, 764. [CrossRef] [PubMed]
104. Dini, I.; Izzo, L.; Graziani, G.; Ritieni, A. The Nutraceutical Properties of "Pizza Napoletana Marinara TSG" a Traditional Food Rich in Bioaccessible Antioxidants. *Antioxidants* **2021**, *10*, 495. [CrossRef] [PubMed]
105. Cavallo, P.; Dini, I.; Sepe, I.; Galasso, G.; Fedele, F.L.; Sicari, A.; Bolletti Censi, S.; Gaspari, A.; Ritieni, A.; Lorito, M.; et al. An Innovative Olive Pâté with Nutraceutical Properties. *Antioxidants* **2020**, *9*, 581. [CrossRef] [PubMed]
106. Fytianos, G.; Rahdar, A.; Kyzas, G.Z. Nanomaterials in Cosmetics: Recent Updates. *Nanomaterials* **2020**, *10*, 979. [CrossRef]
107. Dey, P. Gut microbiota in phytopharmacology: A comprehensive overview of concepts, reciprocal interactions, biotransformations and mode of actions. *Pharmacol. Res.* **2019**, *147*, 104367. [CrossRef]
108. van Duynhoven, J.P.M.; van Velzen, E.J.J.; Westerhuis, J.A.; Foltz, M.; Jacobs, D.M.; Smilde, A.K. Nutrikinetics: Concept, technologies, applications, perspectives. *Trends Food Sci. Technol.* **2012**, *26*, 4–13. [CrossRef]
109. Dini, I. Bio Discarded from Waste to Resource. *Foods* **2021**, *10*, 2652. [CrossRef]
110. Dini, I. The Potential of Dietary Antioxidants. *Antioxidants* **2021**, *10*, 1752. [CrossRef]
111. Faria-Silva, C.; Ascenso, A.; Costa, A.M.; Marto, J.; Calvarheiro, M.; Ribeiro, H.M.; Simoes, S. Feeding the skin: A new trend in food and cosmetics convergence. *Trends Food Sci. Technol.* **2020**, *95*, 21–32. [CrossRef]
112. Van Tran, V.; Moon, J.Y.; Lee, Y.C. Liposomes for delivery of antioxidants in cosmeceuticals: Challenges and development strategies. *J. Control. Release* **2019**, *300*, 114–140. [CrossRef]

113. Nimse, S.B.; Pal, D. Free radicals, natural antioxidants, and their reaction mechanisms. *RSC Adv.* **2015**, *5*, 27986–28006. [CrossRef]
114. Yohn, J.J.; Norris, D.A.; Yrastorza, D.G.; Buno, I.J.; Leff, J.A.; Hake, S.S.; Repine, J.E. Disparate antioxidant enzyme activities in cultured human cutaneous fibroblasts, keratinocytes, and melanocytes. *J. Investig. Dermatol.* **1991**, *97*, 405–409. [CrossRef]
115. Fda, C.; Purdie, F.P. *Waiver of In Vivo Bioavailability and Bioequivalence Studies for Immediate-Release Solid Oral Dosage Forms Based on a Biopharmaceutics Classification System Guidance for Industry*; US Food and Drug Administraiton: Silver Spring, MD, USA, 2017.
116. Oresajo, C.; Pillai, S.; Manco, M.; Yatskayer, M.; Mcdaniel, D. Antioxidants and the skin: Understanding formulation and efficacy. *Dermatol. Ther.* **2012**, *25*, 252–259. [CrossRef]
117. Boo, Y.C. Human Skin Lightening Efficacy of Resveratrol and Its Analogs: From in Vitro Studies to Cosmetic Applications. *Antioxidants* **2019**, *8*, 332. [CrossRef]
118. Martins, I.M.; Barreiro, M.F.; Coelho, M.; Rodrigues, A.E. Microencapsulation of essential oils with biodegradable polymeric carriers for cosmetic applications. *Chem. Eng. J.* **2014**, *245*, 191–200. [CrossRef]
119. Enescu, C.D.; Bedford, L.M.; Potts, G.; Fahs, F. A review of topical vitamin C derivatives and their efficacy. *J. Cosmet Dermatol.* **2021**, 1–11. [CrossRef]
120. Baek, J.; Ramasamy, M.; Willis, N.C.; Kim, D.S.; Anderson, W.A.; Tam, K.C. Encapsulation and controlled release of vitamin C in modified cellulose nanocrystal/chitosan nanocapsules. *Curr. Res. Food Sci.* **2021**, *4*, 215–223. [CrossRef]
121. Simpson, G.L.W.; Ortwerth, B.J. The non-oxidative degradation of ascorbic acid at physiological conditions. *Biochim. Biophys. Acta* **2000**, *1501*, 12–24. [CrossRef]
122. Mahamongkol, H.; Bellantone, R.A.; Stagni, G.; Plakogiannis, F.M. Permeation study of five formulations of alpha-tocopherol acetate through human cadaver skin. *J. Cosmet. Sci.* **2005**, *56*, 91–103.
123. Caritá, A.C.; Fonseca-Santos, B.; Shultz, J.D.; Michniak-Kohn, B.; Chorilli, M.; Leonardi, G.R. Vitamin C: One compound, several uses. Advances for delivery, efficiency and stability. *Nanomed. Nanotechnol.* **2020**, *24*, 102117. [CrossRef]
124. Zasada, M.; Budzisz, E. Retinoids: Active molecules influencing skin structure formation in cosmetic and dermatological treatments. *Postepy Dermatol. Alergol.* **2019**, *36*, 392–397. [CrossRef]
125. Ourique, A.F.; Pohlmann, A.R.; Guterres, S.S.; Beck, R.C.R. Tretinoin-loaded nanocapsules: Preparation, physicochemical characterization, and photostability study. *Int. J. Pharm.* **2008**, *352*, 1–4. [CrossRef]
126. Ioele, G.; Cione, E.; Risoli, A.; Genchi, G.; Ragno, G. Accelerated photostability study of tretinoin and isotretinoin in liposome formulations. *Int. J. Pharm.* **2005**, *293*, 251–260. [CrossRef]
127. Shields, C.W.; White, J.P.; Osta, E.G.; Patel, J.; Rajkumar, S.; Kirby, N.; Therrien, J.; Zauscher, S. Encapsulation and controlled release of retinol from silicone particles for topical delivery. *J. Control. Release* **2018**, *278*, 37–48. [CrossRef]
128. Harde, H.; Agrawal, A.K.; Katariya, M.; Kale, D.; Jain, S. Development of a topical adapalene-solid lipid nanoparticle loaded gel with enhanced efficacy and improved skin tolerability. *RSC Adv.* **2015**, *5*, 43917–43929. [CrossRef]
129. Parish, W.E.; Read, J.; Paterson, S.E. Changes in basal cell mitosis and transepidermal water loss in skin cultures treated with vitamins C and E. *Exp. Dermatol.* **2005**, *14*, 684–691. [CrossRef]
130. De Freitas Cuba, L.; Braga Filho, A.; Cherubini, K.; Salum, F.G.; Figueiredo, M.A. Topical application of Aloe vera and vitamin E on induced ulcers on the tongue of rats subjected to radiation: Clinical and histological evaluation. *Support. Care Cancer* **2016**, *24*, 2557–2564. [CrossRef]
131. Irrera, N.; Pizzino, G.; D'Anna, R.; Vaccaro, M.; Arcoraci, V.; Squadrito, F.; Altavilla, D.; Bitto, A. Dietary management of skin health: The role of genistein. *Nutrients* **2017**, *9*, 622. [CrossRef]
132. Liu, T.; Li, N.; Yan, Y.Q.; Liu, Y.; Xiong, K.; Liu, Y.; Xia, Q.M.; Zhang, H.; Liu, Z.D. Recent advances in the antiaging effects of phytoestrogens on collagen, water content, and oxidative stress. *Phytother. Res.* **2020**, *34*, 435–447. [CrossRef]
133. Maya, S.; Sabitha, M.; Nair, S.V.; Jayakumar, R. Phytomedicine-loaded polymeric nanomedicines: Potential cancer therapeutics. Multifaceted Development and Application of Biopolymers for Biology. *Biomed. Nanotechnol.* **2012**, *254*, 203–239.
134. Guterres, S.S.; Alves, M.P.; Pohlmann, A.R. Polymeric nanoparticles, nanospheres and nanocapsules for cutaneous applications. *Drug Target Insights* **2007**, *2*, 147–157. [CrossRef] [PubMed]
135. Carlotti, M.E.; Sapino, S.; Ugazio, E.; Gallarate, M.; Morel, S. Resveratrol in solid lipid nanoparticles resveratrol in solid lipid nanoparticles. *J. Dispers. Sci. Technol.* **2012**, *33*, 465–471. [CrossRef]
136. Sharma, N.; Singh, S.; Kanojia, N.; Grewal, A.S.; Arora, S. Nanotechnology: A Modern Contraption in Cosmetics and Dermatology. *Appl. Clin. Res. Clin. Trials Regul. Aff.* **2018**, *5*, 147–158. [CrossRef]
137. Harris, R.; Lecumberri, E.; Mengíbar, M.; Heras, A. Chitosan nanoparticles and microspheres for the encapsulation of natural antioxidants extracted from Ilex paraguariensis. *Carbohydr. Polym.* **2011**, *84*, 803–806. [CrossRef]
138. Muta-Takada, K.; Terada, T.; Yamanishi, H.; Ashida, Y.; Inomata, S.; Nishiyama, T.; Amano, S. Coenzyme Q10 protects against oxidative stress-induced cell death and enhances the synthesis of basement membrane components in dermal and epidermal cells. *Biofactors* **2009**, *35*, 435–441. [CrossRef]
139. Puglia, C.; Bonina, F. Lipid nanoparticles as novel delivery systems for cosmetics and dermal pharmaceuticals. *Expert Opin. Drug Deliv.* **2012**, *9*, 429–441. [CrossRef]
140. Farboud, E.S.; Nasrollahi, S.A.; Tabbakhi, Z. Novel formulation and evaluation of a Q10-loaded solid lipid nanoparticle cream: In vitro and in vivo studies. *Int. J. Nanomed.* **2011**, *6*, 611–617. [CrossRef]
141. El-Leithy, E.S.; Makky, A.M.; Khattab, A.M.; Hussein, D.G. Optimization of nutraceutical coenzyme Q10 nanoemulsion with improved skin permeability and anti-wrinkle efficiency. *Drug Dev. Ind. Pharm.* **2018**, *44*, 316–328. [CrossRef]

142. Pegoraro, N.S.; Barbieri, A.V.; Camponogara, C.; Mattiazzi, J.; Brum, E.S.; Marchiori, M.C.L.; Oliveira, S.M.; Cruz, L. Nanoencapsulation of coenzyme Q10 and vitamin E acetate protects against UVB radiation-induced skin injury in mice. *Colloids Surf. B Biointerfaces* **2017**, *150*, 32–40. [CrossRef]
143. Fiume, M.M.; Bergfeld, W.F.; Belsito, D.V.; Ronald, A.H.; Klaassen, C.D.; Liebler, D.C.; Marks, G.J.J.; Shank, R.C.; Slaga, T.J.; Snyder, P.W.; et al. Safety Assessment of Tocopherols and Tocotrienols as Used in Cosmetics. *Int. J. Toxicol.* **2018**, *37*, 61S–94S. [CrossRef]
144. . Ben-shabat, S.; Kazdan, Y.; Beit-yannai, E.; Sintov, A.C. Use of alpha-tocopherol esters for topical vitamin E treatment: Evaluation of their skin permeation. *J. Pharm. Pharmacol.* **2013**, *65*, 652–658. [CrossRef]
145. Kim, D.; Jeong, Y.; Choi, C.; Roh, S.; Kang, S.; Jang, M.; Nah, J. Retinol-encapsulated low molecular water-soluble chitosan nanoparticles. *Int. J. Pharm.* **2006**, *319*, 130–138. [CrossRef]
146. Park, C.; Park, D.; Kim, B. Effects of a chitosan coating on properties of retinol- encapsulated Zein nanoparticles. *Food Sci. Biotechnol.* **2015**, *24*, 1725–1733. [CrossRef]
147. Huang, S.; Sun, S.; Chiu, C. Retinol-encapsulated water-soluble succinated chitosan nanoparticles for antioxidant applications. *Aust. J. Biol. Sci.* **2013**, *24*, 315–329. [CrossRef]
148. Ito, S.; Itoga, K.; Yamato, M.; Akamatsu, H.; Okano, T. The co-application effects of fullerene and ascorbic acid on UV-B irradiated mouse skin. *Toxicology* **2010**, *267*, 27–38. [CrossRef]
149. Britton, G.; Liaaen-Jensen, S.; Pfander, H. *Carotenoids. Volume 5: Nutrition and Health*; Birkhäuser: Basel, Switzerland, 2009.
150. Krinsky, N.I.; Johnson, E.J. Carotenoid actions and their relation to health and disease. *Mol. Aspects Med.* **2005**, *26*, 459–516. [CrossRef]
151. Krinsky, N.; Mayne, S.T.; Sies, H. *Carotenoids in Health and Disease*; Marcel Dekker: New York, NY, USA, 2004.
152. Rodriguez-Concepcion, M.; Avalos, J.; Bonet, M.L.; Boronat, A.; Gomez-Gomez, L.; Hornero-Mendez, D.; Limon, M.C.; Meléndez-Martínez, A.J.; Olmedilla-Alonso, B.; Palou, A.; et al. A global perspective on carotenoids: Metabolism, biotechnology, and benefits for nutrition and health. *Prog. Lipid Res.* **2018**, *70*, 62–93. [CrossRef]
153. Thomas, N.V.; Kim, S.K. Beneficial effects of marine algal compounds in cosmeceuticals. *Mar. Drugs* **2013**, *11*, 146–164. [CrossRef]
154. Mathews-Roth, M. Treatment of erythropoietic protoporphyria with beta–carotene. *Photo–Dermatol.* **1984**, *1*, 318–321. [CrossRef]
155. Bin–Jumah, M.; Alwakeel, S.S.; Moga, M.; Buvnariu, L.; Bigiu, N.; Zia-Ul-Haq, M. Application of Carotenoids in Cosmetics. In *Carotenoids: Structure and Function in the Human Body*; Zia-Ul-Haq, M., Dewanjee, S., Riaz, M., Eds.; Springer Nature: Basingstoke, UK, 2021; pp. 747–756.
156. Heinrich, U.; Tronnier, H.; Stahl, W.; Béjot, M.; Maurette, J.M. Antioxidant supplements improve parameters related to skin structure in humans. *Skin Pharmacol. Physiol.* **2006**, *19*, 224–231. [CrossRef]
157. Meléndez-Martínez, A.J.; Stinco, C.M.; Mapelli-Brahm, P. Skin Carotenoids in Public Health and Nutricosmetics: The Emerging Roles and Applications of the UV Radiation-Absorbing Colourless Carotenoids Phytoene and Phytofluene. *Nutrients* **2019**, *11*, 1093. [CrossRef]
158. Darvin, M.; Patzelt, A.; Gehse, S.; Schanzer, S.; Benderoth, C.; Sterry, W.; Lademann, J. Cutaneous concentration of lycopene correlates significantly with the roughness of the skin. *Eur. J. Pharm. Biopharm.* **2008**, *69*, 943–947. [CrossRef]
159. Singh, P.; Rani, B.; Chauhan, A.; Maheshwari, R. Lycopene's antioxidant activity in cosmetics meadow. *Int. Res. J. Pharm.* **2012**, *3*, 46–47.
160. Havas, F.; Krispin, S.; Meléndez-Martínez, A.J.; von Oppen-Bezalel, L. Preliminary data on the safety of phytoene and phytofluene-rich products for human use including topical application. *J. Toxicol.* **2018**, *2018*, 5475784. [CrossRef]
161. Hu, F.; Liu, W.; Yan, L.; Kong, F. Optimization and characterization of poly (lactic-coglycolic acid) nanoparticles loaded with astaxanthin and evaluation of anti-photodamage effect in vitro. *R. Soc. Open. Sci.* **2019**, *6*, 191184. [CrossRef]
162. Cremlyn, R.J.; Cremlyn, R.J.W. *An Introduction to Organosulfur Chemistry*; John Wiley & Sons: Hoboken, NJ, USA, 1996; pp. 2–264.
163. Chu, H.L.; Wang, B.S.; Duh, P.D. Effects of selected organo-sulfur compounds on melanin formation. *J. Agric. Food Chem.* **2009**, *57*, 7072–7077. [CrossRef] [PubMed]
164. Hahm, E.-R.; Singh, S.V. Diallyl trisulfide inhibits estrogen receptor-α activity in human breast cancer cells. *Breast. Cancer Res. Treat.* **2014**, *144*, 47–57. [CrossRef] [PubMed]
165. Pangastuti, A.; Indriwati, S.E.; Amin, M. Investigation of the antiaging properties of allicin from Allium sativum L bulb extracts by a reverse docking approach. *Tropical. J. Pharm. Res.* **2018**, *17*, 635–639. [CrossRef]
166. Kuran, D.; Pogorzelska, A.; Wiktorska, K. Breast Cancer Prevention-Is there a Future for Sulforaphane and Its Analogs? *Nutrients* **2020**, *12*, 1559. [CrossRef] [PubMed]
167. Rosado, C.; Tokunaga, V.K.; Sauce, R.; De Oliveira, C.A.; Sarruf, F.D.; Parise-Filho, R.; Maurício, E.; De Almeida, T.S.; Velasco, M.V.R.; Baby, A.R. Another Reason for Using Caffeine in Dermocosmetics: Sunscreen Adjuvant. *Front. Physiol.* **2019**, *10*, 519. [CrossRef]
168. Dos Santos, É.M.; de Macedo, L.M.; Tundisi, L.L.; Ataide, J.A.; Camargo, G.A.; Alves, R.C.; Oliveira, M.B.; Mazzola, P.G. Coffee by-products in topical formulations: A review. *Trends Food Sci. Technol.* **2021**, *111*, 280–291. [CrossRef]
169. Puglia, C.; Offerta, A.; Tirendi, G.G.; Tarico, M.S.; Curreri, S.; Bonina, F.; Perrotta, R.E. Design of solid lipid nanoparticles for caffeine topical administration. *Drug Deliv.* **2016**, *23*, 36–40. [CrossRef]
170. Jain, K.; Mehra, N.K.; Jain, N.K. Nanotechnology in drug delivery: Safety and toxicity issues. *Curr. Pharm. Des.* **2015**, *21*, 4252–4261. [CrossRef]

171. Chen, L.Y.; Remondetto, G.E.; Subirade, M. 2. Food protein-based materials as nutraceutical delivery systems. *Trends Food Sci. Technol.* **2006**, *17*, 272–283. [CrossRef]
172. Livney, Y.D. Nanostructured delivery systems in food: Latest developments and potential future directions. *Curr. Opin. Food Sci.* **2015**, *3*, 125–135. [CrossRef]
173. Dobrovolskaia, M.A.; Aggarwal, P.; Hall, J.B.; McNeil, S.E. Preclinical studies to understand nanoparticle interaction with the immune system and its potential effects on nanoparticle biodistribution. *Mol. Pharm.* **2008**, *5*, 487–495. [CrossRef]
174. De Oliveira Viana, I.M.; Roussel, S.; Defrêne, J.; Lima, E.M.; Barabé, F.; Bertrand, N. Innate and adaptive immune responses toward nanomedicines. *Acta Pharm. Sin. B* **2021**, *11*, 852–870. [CrossRef]
175. Ernst, L.M.; Casals, E.; Italiani, P.; Boraschi, D.; Puntes, V. The Interactions between Nanoparticles and the Innate Immune System from a Nanotechnologist Perspective. *Nanomaterials* **2021**, *11*, 2991. [CrossRef]
176. Khan, I.; Saeed, K.; Khan, I. Nanoparticles: Properties, applications and toxicities. *Arab. J. Chem.* **2017**, *12*, 908–931. [CrossRef]
177. Rachmawati, H.; Larasati, A.; Adi, A.C.; Shegokar, R. Role of nanocarriers and their surface modification in targeting delivery of bioactive compounds. In *Nanopharmaceuticals*; Ranjita Shegokar, R., Ed.; Elsevier: Amsterdam, The Netherlands, 2020; pp. 17–43.

*antioxidants*

MDPI

*Article*

# Alginate-Chitosan Coated Nanoliposomes as Effective Delivery Systems for Bamboo Leaf Flavonoids: Characterization, In Vitro Release, Skin Permeation and Anti-Senescence Activity

**Yanpei Gu, Zhenlei Zhao †, Fan Xue and Ying Zhang ***

Zhejiang Key Laboratory for Agro-Food Processing, Zhejiang Engineering Center for Food Technology and Equipment, College of Biosystems Engineering and Food Science, Zhejiang University, Hangzhou 310058, China; yanpeigu@zju.edu.cn (Y.G.); zhenlei-zhao@foxmail.com (Z.Z.); xf21713038@zju.edu.cn (F.X.)

* Correspondence: yzhang@zju.edu.cn

† Current address: Zhejiang Provincial Key Lab of Geriatrics & Geriatrics, Institute of Zhejiang Province, Department of Geriatrics, Zhejiang Hospital, Zhejiang University School of Medicine, Hangzhou 310058, China.

**Abstract:** The use of bamboo leaf flavonoids (BLF) as functional food and cosmetic ingredients is limited by low bioavailability and difficulty in being absorbed by the intestine or skin. The aim of this study was to prepare BLF-loaded alginate-chitosan coated nanoliposomes (AL-CH-BLF-Lip) to overcome these challenges. The nanocarriers were characterized by dynamic light scattering, high performance liquid chromatography, Fourier transform infrared spectroscopy and differential scanning calorimetry. The biological activity was analyzed by in vitro antioxidant activity, transdermal absorption, cytotoxicity and AAPH induced HaCaT cell senescence model. The results showed that the size of nanocarriers ranged from 152.13 to 228.90 nm and had a low polydispersity index (0.25–0.36). Chitosan (CH) and alginate (AL) were successfully coated on BLF-loaded nanoliposomes (BLF-Lip), the encapsulation efficiency of BLF-Lip, BLF-loaded chitosan coated nanoliposomes (CH-BLF-Lip) and AL-CH-BLF-Lip were 71.31%, 78.77% and 82.74%, respectively. In addition, BLF-Lip, CH-BLF-Lip and AL-CH-BLF-Lip showed better in vitro release and free radical scavenging ability compared with naked BLF. In particular, the skin permeability of BLF-Lip, CH-BLF-Lip, and AL-CH-BLF-Lip increased 2.1, 2.4 and 2.9 times after 24 h, respectively. Furthermore, the use of nanoliposomes could significantly improve the anti-senescence activity of BLF ($p < 0.01$). Conclusively, alginate-chitosan coated nanoliposomes are promising delivery systems for BLF that can be used in functional foods and cosmetics.

**Keywords:** bamboo leaf flavonoids; nanoliposome; biopolymer conjugation; in vitro release; skin permeability; anti-senescence activity

**Citation:** Gu, Y.; Zhao, Z.; Xue, F.; Zhang, Y. Alginate-Chitosan Coated Nanoliposomes as Effective Delivery Systems for Bamboo Leaf Flavonoids: Characterization, In Vitro Release, Skin Permeation and Anti-Senescence Activity. *Antioxidants* **2022**, *11*, 1024. https://doi.org/10.3390/antiox11051024

Academic Editors: Irene Dini and Sonia Laneri

Received: 20 April 2022
Accepted: 12 May 2022
Published: 23 May 2022



## 1. Introduction

Flavonoids are a class of secondary metabolites widely existing in nature and that have a variety health and medicinal values. The main active components of bamboo leaf flavonoids (BLF) isolated from bamboo leaves are four carbon glycoside flavonoids, namely orientin, isoorientin, vitexin and isovitexin (Supplementary Figure S1) [1]. Studies have shown that BLF have a variety of biological effects such as antioxidant, lipid-lowering, anti-radiation, anti-bacterial, anti-inflammatory and anti-aging, and have a broad application prospect in functional foods and cosmetics [2–7]. However, similar to other flavonoids, BLF have a darker color, are difficult to be absorbed, and are sensitive to factors such as light, oxygen and high temperature, which limit their applications. The use of nano carriers can effectively improve the stability and bioavailability of the embedded compounds [8,9]. Among them, nanoliposomes have been widely used due to the advantages of sustained release, cell affinity, histocompatibility, low toxicity and targeting. Usually composed of natural phospholipids and cholesterol, nanoliposomes have one or more closed vesicles with lipid bilayer structure that can encapsulate lipophilic and hydrophilic compounds, in which hydrophobic compounds are inserted into lipid bilayer membranes and hydrophilic compounds are encapsulated in the internal aqueous phase [10,11]. However, nanoliposomes have low physical stability, and are easily affected by light, acid, alkali and other factors during long-term storage, resulting in aggregation and fusion of vesicles, and ultimately lead to the leakage of embedded compounds. The oxidation and hydrolysis of phospholipids is one of the main reasons for the chemical instability of nanoliposomes [12,13].

Layer by layer self-assembly refers to the formation of self-assembled multilayers by alternating deposition in polyelectrolyte solutions with opposite charges [14,15]. Studies have shown that the biopolymer coated nanoliposomes prepared by layer-by-layer self-assembly can reduce oxidative damage and hydrolysis reaction. In addition, an electrostatic bridge can be formed between phospholipid and biopolymer to minimize the permeability of the phospholipid bilayer, thereby improving the stability of nanoliposomes [10,16]. At present, a large number of studies on biopolymer coated nanoliposomes have been reported.

Chitosan (CH) is a cationic polysaccharide obtained by deacetylation of chitin. Under acidic condition, positively charged CH binds to negatively charged liposomes through electrostatic interaction. At the same time, the residual acyl groups of CH can be inserted into the phospholipid membrane bilayer of liposomes [17,18]. Shishir et al. [18] found that CH coated nanoliposomes had higher stability, better sustained-release and improved cellular uptake. Alginate (AL) is a non-toxic inert polyanionic polysaccharide that is widely used as dietary fiber, thickener, emulsifier and stabilizer in food industry. Liu et al. [16,19] found that AL-CH-coated nanoliposomes had higher physical and digestive stability compared with uncoated nanoliposomes. Cui et al. [20] proved that AL and CH could improve the stability and transmembrane transport efficiency of chito-oligosaccharides. In addition, CH-AL nanogel carriers were able to increase the skin penetration and sustained release of pirfenidone [21].

In this study, we used the thin layer dispersion method combined with sonication to prepare BLF-loaded nanoliposomes (BLF-Lip), and further prepared biopolymer coated BLF-Lip by electrostatic layer by layer self-assembly. The carrier systems were characterized by dynamic light scattering (DLS), high performance liquid chromatography (HPLC), Fourier transform infrared (FTIR) spectroscopy and differential scanning calorimetry (DSC). We studied the effects of encapsulation of BLF on skin permeability and in vitro release. In addition, changes in antioxidant activity and anti-senescence activity were also evaluated when encapsulating BLF. The above research aimed to improve the bioavailability, transdermal absorption and anti-senescence activity of BLF, and expand their application value in functional foods and cosmetics.

## 2. Materials and Methods

### 2.1. Materials

Bamboo-leaf flavonoids (the mass fraction of four carbon glycoside flavonoids was 80.66%, Supplementary Figure S2) were prepared in laboratory. L-α-Phosphatidylcholine, cholesterin, CH, AL, 2–2′azino-bis-(3-ethylbenzothiazoline-6-sulfonic acid) diammonium salt (ABTS), 2,2-Diphenyl-1-picrylhydrazyl (DPPH), 2,2′-Azobis (2-methylpropionamidine) dihydrochloride (AAPH), 3-(4,5-dimthyl-2-thiazolyl)-2,5-diphenyl-2-H-tetrazolium bromide (MTT), 4′,6-diamidino-2-phenylindole (DAPI) were obtained from Sigma Aldrich (St. Louis, MO, USA). Reagents required for cell culture were purchased from HyClone (Logan, UT, USA) and Gibco (Grand Island, NY, USA), respectively. K9M-H3 and Alexa Fluor 488-conjugated Goat Anti-Rabbit IgG (H + L) antibodies were purchased from ABclonal Technology (Wuhan, Hubei, China); p21, p16 and Goat Anti-Rabbit secondary antibodies were obtained from Cell Signaling Technology (Beverly, MA, USA). HPLC-grade methanol was obtained from TEDIA (Cincinnati, OH, USA). The 10 kDa dialysis bag was provided by Spectrum (San Jose, CA, USA). Other chemicals and reagents were purchased from Sinopharm (Shanghai, China).

### 2.2. Preparation of BLF-Lip

The thin layer dispersion method combined with sonication was used to prepare BLF-Lip [18,19]. First, phosphatidylcholine, cholesterol, Tween-80 and vitamin E were dissolved in absolute ethanol at a mass ratio of 6:1:1:8:0.12 and stirred evenly with a magnetic stirrer. Then, the ethanol was removed under vacuum at 55 °C by a rotary evaporator. After the dried lipid film was formed, we slowly added phosphate buffer solution (PBS; pH 7.4, 0.05 M) dissolved with 1.5 mg/mL BLF. Then after being fully washed for 1 h under low-speed rotation condition, the primary BLF loaded liposomes with lipid concentration of 8 mg/mL were obtained. The primary BLF loaded liposomes were sonicated using SK8210HP sonicator (Kudo, Shanghai, China) for 8 min (75% of ultrasonic time set) at a power of 120 W to obtain BLF-Lip. The whole preparation process was protected from light, and the obtained BLF-Lip were stored at 4 °C under dark condition.

### 2.3. Preparation of Biopolymer Coated Nanoliposomes

A total of 0.1 g CH and AL were dissolved in 100 mL of 1% glacial acetic acid solution and deionized water, respectively, stirred overnight and then had the pH adjusted to 5.5. After centrifugation and filtration, 0.1% CH and AL solution were obtained. CH and AL were conjugated to the surface of BLF-Lip in two steps according to the layer by layer self-assembly approach [16]. In short, the above BLF-Lip were placed in a burette and added dropwise to 0.1% CH solution (1:1, $v/v$), incubated for 1 h under gentle stirring to fully deposit CH on the surface of BLF-Lip, then adjusted pH to 5.5. The same procedure was used to add the positively charged BLF-loaded chitosan coated nanoliposomes (CH-BLF-Lip) to the negatively charged AL solution, then adjusted pH to 5.5 to obtain BLF-loaded alginate-chitosan coated nanoliposomes (AL-CH-BLF-Lip). The solution was stored at 4 °C for further analysis.

### 2.4. Particle Size, Polydispersity Index (PDI) and Zeta Potential Measurement

AL-CH-BLF-Lip, CH-BLF-Lip and BLF-Lip were diluted 2.5, 5 and 10 times with ultrapure water, respectively. A Nano-ZS90 dynamic light scattering instrument (Malvern, United Kingdom) was used to measure the particle size, PDI and zeta potential at 25 °C, the scattering angle was 90°, the refractive index was 1.330, and the viscosity was 0.933. Each sample was determined at least three times in parallel.

### 2.5. Standard Curve and Encapsulation Efficiency (EE) Determination

The BLF were accurately weighed in a 10 mL flask, dissolved in methanol and prepared solutions of 400, 200, 100, 50, 25, 10, 5 μg/mL, respectively. HPLC with diode array detector (Dionex ultimate 3000, ThermoFisher Scientific, Waltham, MA, USA) was used to analyze the amount of BLF. Samples were injected into Luna $C_{18}$ column (250 mm × 4.60 mm × 5 μm) after filtration with 0.22 μm membrane for analysis according to the following chromatographic conditions: scanning wavelength range of 200–400 nm, quantitative wavelength of 330 nm, column temperature of 40 °C, flow rate of 0.5 mL/min, injection volume of 10 μL, and mobile phase consisted of acetonitrile −0.1% trifluoroacetic acid (20:80, $v/v$). The standard curve was drawn by peak area normalization method.

The EE was determined by ultrafiltration centrifugation [22]. A 50% methanol aqueous solution containing 1% acetic acid was added to the nanoliposomes (10:1, $v/v$), sonicated for 10 min, and centrifuged at 10,000 rpm for 10 min to obtain total BLF. An equal amount of nanoliposomes were added to the inner tube of ultrafiltration tubes (MWCO 10 kDa, Millipore, Billerica, MA, USA), centrifuged at 4000 rpm for 40 min, and 1 mL PBS was added to repeat the above operation once. The free BLF was obtained in the outer tube of the ultrafiltration tube.

$$\text{EE (\%)} = (\text{Total amount of BLF} - \text{free amount of BLF})/\text{Total amount of BLF} \times 100 \quad (1)$$

### 2.6. FTIR

In order to confirm that the biopolymers were coated on the surface of BLF-Lip, FTIR analysis was performed using an FTIR Avatar 370 spectroscopy (Nicolet, Madison, WI, USA). BLF, CH, AL and lyophilized BLF-loaded nanoliposomes were scanned at 4000 to 400 $cm^{-1}$ with a resolution of 4 $cm^{-1}$ using the KBr tablet method. OMNIC software version 8 was used to analyze data.

### 2.7. DSC

A total of 2–5 mg lyophilized BLF-loaded nanoliposomes were placed in an alumina crucible and put in the sample chamber of DSC (Mettler Toledo, Zurich, Switzerland). The blank alumina crucible was used as a reference, the heating rate was 10 °C/min, the nitrogen flow rate was 30 mL/min, the scanning temperature range was 20–250 °C, and the thermodynamic curve was recorded.

### 2.8. In Vitro Release

The dialysis method was used to measure the in vitro release of BLF and BLF-loaded nanoliposomes, mainly referring to the method of Lopes et al., with some modifications [23]. One mL of naked BLF, BLF-Lip, CH-BLF-Lip and AL-CH-BLF-Lip were placed in a 10 kDa dialysis bag, then suspended in 50 mL PBS (pH6.0), and stirred in a magnetic stirring apparatus at a speed of 100 rpm under 37 °C. One mL of the medium outside dialysis bag was sampled after 0.5, 1, 2, 4, 6, 8, 12 and 24 h, and replaced with the same volume of PBS at the same time. The cumulative release (CR) was calculated by the following formula.

$$CR(\%) = \frac{C_n \times V + \sum_{i=1}^{n-1} C_i \times V_i}{C_0} \times 100. \quad (2)$$

$C_0$, $C_i$ and $C_n$ were the concentrations of BLF (μg/mL) measured at the 0, $i$ and $n$th sampling, respectively. $V$ and $V_i$ were the receiving liquid volume and sampling volume, respectively.

### 2.9. In Vitro Skin Permeation Study

The Franz diffusion cell method was used to determine the in vitro skin permeation [24]. C57BL/6J mice were anesthetized and sacrificed by neck amputation. The hair on the back skin of mice was removed with depilatory cream, and the connective tissue and subcutaneous fat were peeled off. The skin tissue was fixed on the diffusion cell, and 2 mL naked BLF and BLF-loaded nanoliposomes were added to the supply chamber. The effective diffusion area was 4.9 $cm^2$, the volume of the receiving chamber was 7 mL, and the receiving solution was PBS (pH 7.4). During the whole experiment, the diffusion cell was stirred in a magnetic stirring apparatus at a speed of 100 rpm under 37 °C. A total of 0.4 mL receiving solution was sampled after 0.5, 1, 2, 4, 6, 8, 12 and 24 h, and the same volume of the receiving solution was supplemented at the same time. The cumulative permeation (CP) was calculated by the following formula.

$$CP = \frac{C_n \times V + \sum_{i=1}^{n-1} C_i \times V_i}{A}. \quad (3)$$

$C_i$ and $C_n$ were the concentrations of BLF (μg/mL) measured at the $i$ and $n$th sampling, respectively. $A$ was the effective diffusion area. $V$ and $V_i$ were the receiving liquid volume and sampling volume, respectively.

### 2.10. ABTS Radical Cation ($ABTS^+$) Assay

The free radical scavenging activity of naked BLF and BLF-loaded nanoliposomes was studied by $ABTS^+$ assay [18]. A total of 7.4 mM ABTS and 2.6 mM potassium persulfate solution were mixed in a volume ratio of 1:1 and incubated at room temperature in dark for 12 h to obtain $ABTS^+$ stock solution, diluted with absolute ethanol to an absorbance at 734 nm of 0.70 ± 0.02 to obtain $ABTS^+$ working solution. Then, 200 μL naked BLF and BLF-loaded nanoliposomes were mixed with 0.8 mL $ABTS^+$ working solution, and incubated for 6 min at room temperature, the control group was replaced with the same volume of ethanol. The absorbance was measured at 734 nm using a bio-Tek microplate reader (Winooski, VT, USA).

$$\text{ABTS scavenging capacity (\%)} = (A_{control} - A_{sample})/A_{control} \times 100. \quad (4)$$

*2.11. DPPH Assay*

DPPH assay followed the previously described procedure with some modifications [25]. One mL of 0.2 mmol/L DPPH prepared with anhydrous ethanol was shaken with 0.2 mL naked BLF and BLF-loaded nanoliposomes, and then incubated for 30 min at room temperature in dark. The absorbance was measured at 517 nm using a bio-Tek microplate reader (Winooski, VT, USA). The control group was replaced with the same volume of anhydrous ethanol, and the positive control was Trolox.

$$\text{DPPH scavenging capacity (\%)} = (A_{control} - A_{sample})/A_{control} \times 100. \quad (5)$$

*2.12. Cell Culture*

HaCaT cells were obtained from the Cell Bank of the Chinese Academy of Sciences. The resuscitated HaCaT cells were cultured in DMEM high glucose medium containing 15% fetal bovine serum and 1% antibiotics (penicillin, streptomycin) and placed in a humidified incubator at 37 °C and 5% $CO_2$. Approximately, 80% confluent HaCaT cells were digested with a 0.25% Trypsin/0.02% EDTA solution and passaged, and the logarithmic growth phase cells were taken for experiments.

*2.13. Cytotoxicity Study*

HaCaT cells were seeded into the 96-well cell culture plates at a density of $5 \times 10^3$ cells/well. Naked BLF and BLF-loaded nanoliposomes of different doses were added after the cells were adherent overnight. At the same time, a blank group containing only the medium and a negative control group containing only cells were set. After being incubated for 24 h, we absorbed the supernatant and added the fresh medium containing 10 μL MTT solution to each well. After incubation in the incubator for 4 h, the culture medium was absorbed and shaken with DMSO solution for 10 min. The absorbance of each well was measured at a wavelength of 490 nm using a bio-Tek microplate reader (Winooski, VT, USA).

*2.14. Anti-Senescence Activity*

2.14.1. Assessment of Cell Proliferation

In order to find the protective effect of naked BLF and BLF-loaded nanoliposomes on AAPH-induced senescence, HaCaT cells were cultured overnight and pretreated with samples for 4 h, then co-incubated with AAPH for 48 h. At the same time, a blank group containing only the medium and a negative control group containing only cells without AAPH and samples were set. The absorbance of each well was measured at a wavelength of 490 nm using a bio-Tek microplate reader (Winooski, VT, USA).

2.14.2. Confocal Fluorescence Microscopy

To detect senescence-associated heterochromatin foci (SAHF) formation in cultured cells by immunofluorescence analysis, HaCaT cells were fixed and permeabilized at room temperature with 4% paraformaldehyde and 0.5% Triton X-100 solution, respectively. After washing with PBS for three times, cells were blocked by PBS containing 3% goat serum and 1% bovine serum albumin for 1 h, then followed by adding the prepared primary antibody specific for K9M-H3 and incubating at 4 °C overnight. Cells were then incubated with a secondary antibody specific for Alexa Fluor 488-conjugated Goat Anti-Rabbit IgG (H + L), and nuclei were marked by 1 μg/mL DAPI solution at the same time. Cells with condensed K9M-H3 that co-localized with DAPI in the nuclei were considered as SAHF positive, and the percentage of SAHF positive cells relative to the total number of cells was calculated.

2.14.3. Western Blotting

RIPA Lysis buffer supplemented with protease inhibitors (Beyotime, Shanghai, China) was used to extract the total protein of each group of HaCaT cells. The protein concentration was quantified using the BCA protein assay kit (Beyotime, Shanghai, China). After separation by polyacrylamide gel electrophoresis, the protein was transferred to PVDF membrane. Blots were incubated with K9M-H3, p21 and p16 antibodies (1:1000), washed, and incubated with a horseradish peroxidase (HRP)-labeled secondary antibody. Visualizing the bands by using ECL commercial kit (Beyotime, Shanghai, China) for chemiluminescence. Taking GAPDH as the reference protein, Image J software was used to analyze the gray value to quantify the protein expression.

### *2.15. Statistical Analysis*

All experiments were repeated at least three times, and the data were expressed as mean ± standard deviation (SD). Statistical analyses were performed by SPSS 19.0 (IBM, Chicago, USA), significant differences between the groups were determined by using a one-way analysis of variance (ANOVA), taking $p < 0.05$ as significant difference, and $p < 0.01$ as extremely significant difference.

## 3. Results and Discussion

### *3.1. Particle Size, PDI, Zeta Potential and EE*

The characteristics of BLF-Lip, CH-BLF-Lip and AL-CH-BLF-Lip were summarized in Table 1. The particle size, zeta potential and PDI of BLF-Lip were 152.13 ± 5.20 nm, −3.81 ± 0.79 mV and 0.25 ± 0.06, respectively. The particle size of CH-BLF-Lip and AL-CH-BLF-Lip was increased to 194.63 ± 4.25 nm and 228.90 ± 4.89 nm, which preliminarily proved that biopolymers were coated on BLF-Lip. This phenomenon could be explained by the electrostatic interaction between the phospholipid head group $PO_4^{-3}$ and the specific functional groups $NH_3^+$ and $COO^-$ of CH and AL. The charged nanoliposomes can reduce aggregation and fusion to improve stability. Zeta potential is an important indicator to measure the amount of charge [26]. The zeta potential of CH-BLF-Lip was 8.43 ± 1.56 mV, which changed to −27.77 ± 0.45 mV after coating AL. Alternating positive and negative charges further confirmed that CH and AL were successfully coated on the surface of BLF-Lip [27,28]. PDI defines the homogeneity of size dispersion in nanoparticles. Compared with BLF-Lip, the PDI value of CH-BLF-Lip and AL-CH-BLF-Lip had been increased to a certain extent, but was still less than 0.36, which could be considered as moderately homogeneous dispersion [28]. The EE detected by HPLC also confirmed that BLF was encapsulated in nanoliposomes. The EE of BLF-Lip was determined to be 71.31 ± 1.67%, increased to 78.77 ± 1.59% and 82.74 ± 0.75% after coating CH and AL, respectively. The possible reason for the increase of EE is that the free BLF are also encapsulated on the surface of nanoliposomes during the modification of polysaccharides. In addition, the dense bilayer formed by CH and AL on the surface of Lip prevents the fast release of BLF.

**Table 1.** Characterization of BLF-loaded nanoliposomes. BLF, bamboo leaf flavonoids; BLF-Lip, BLF-loaded nanoliposomes; CH-BLF-Lip, BLF-loaded chitosan coated nanoliposomes; AL-CH-BLF-Lip, BLF-loaded alginate-chitosan coated nanoliposomes.

| Liposomes | Particle Size (nm) | Polydispersity Index (PDI) | Zeta Potential (mV) | Encapsulation Efficiency (%) |
|---|---|---|---|---|
| BLF-Lip | 152.13 ± 5.20 | 0.25 ± 0.06 | −3.81 ± 0.79 | 71.31 ± 1.67 |
| CH-BLF-Lip | 194.63 ± 4.25 | 0.31 ± 0.04 | 8.43 ± 1.56 | 78.77 ± 1.59 |
| AL-CH-BLF-Lip | 228.90 ± 4.89 | 0.36 ± 0.03 | −27.77 ± 0.45 | 82.74 ± 0.75 |

### 3.2. FTIR Analysis

The FTIR spectra of BLF, BLF-Lip, CH, AL, CH-BLF-Lip and AL-CH-BLF-Lip were shown in Figure 1. The characteristic peaks of BLF were found at the wavenumbers of 3386 $cm^{-1}$ (O-H stretching vibration), 1717 $cm^{-1}$ (C=O stretching vibration), 1612 and 1517 $cm^{-1}$ (the aromatic ring C=C skeleton stretching vibration) and 1116 $cm^{-1}$ (C-O stretching vibration). The characteristic peaks of nanoliposomes mainly including 3389 $cm^{-1}$ (O-H stretching of hydroxy bond of cholesterol in liposome and water vapor association during pressing), 2926 $cm^{-1}$ and 2854 $cm^{-1}$ (asymmetric and symmetric stretching vibration of C-H bonds in $CH_2$), 1736 $cm^{-1}$ (C=O stretching vibration in phospholipid), 1467 $cm^{-1}$ (asymmetric stretching vibration of $CH_3$), 1237 $cm^{-1}$ and 1089 $cm^{-1}$ ($PO_2^-$ symmetric and asymmetric stretching vibration) and 971 $cm^{-1}$ (asymmetric stretching vibration of C-C-$N^+$). After BLF were encapsulated in nanoliposomes, the FTIR spectrum of BLF-Lip was similar to nanoliposomes and showed almost no characteristic peaks of BLF. The result indicated that BLF might be well encapsulated in nanoliposomes.

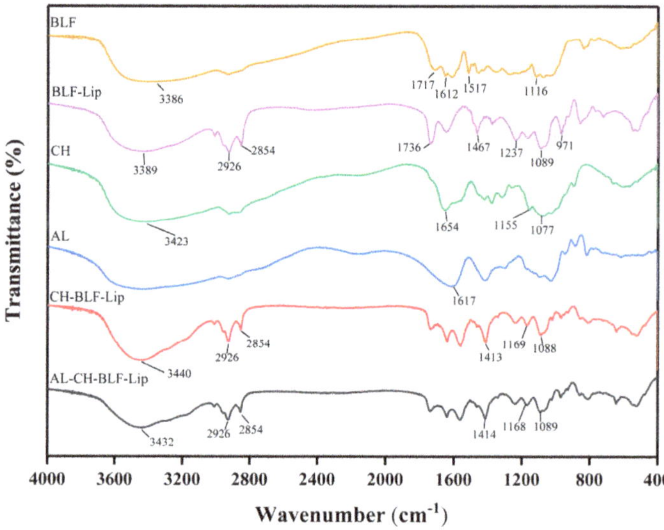


**Figure 1.** Fourier transform infrared (FTIR) spectra of bamboo leaf flavonoids (BLF), BLF−loaded nanoliposomes (BLF−Lip), chitosan (CH), alginate (AL) and BLF−loaded chitosan coated nanoliposomes (CH−BLF−Lip) and BLF−loaded alginate-chitosan coated nanoliposomes (AL−CH−BLF−Lip).

The wavenumber at 3423 $cm^{-1}$ (O-H stretching vibration), 1654 $cm^{-1}$ (C=O stretching vibration), 1155 and 1077 $cm^{-1}$ (C-O stretching vibration) were the mainly characteristic peaks of CH. The peaks of 1736 $cm^{-1}$, 1467 $cm^{-1}$ and 1237 $cm^{-1}$ belonging to liposomes were shifted in CH-BLF-Lip and AL-CH-BLF-Lip, indicating the formation of new hydrogen bonds or the strengthening of hydrogen bonds. In addition, the 1654 $cm^{-1}$ of CH and the 1617 $cm^{-1}$ (carboxyl asymmetric stretching vibration) of AL disappeared. In summary, the alteration of characteristic peaks confirmed the existence of strong electrostatic interaction between AL, CH and BLF-Lip, thus forming AL-CH-BLF-Lip.

### 3.3. DSC Analysis

DSC was used to understand the effects of polysaccharides on the thermal behavior of BLF-Lip. Enthalpy energy (ΔH) and phase transition temperature ($T_c$) were measured (Figure 2). BLF-Lip, CH-BLF-Lip, AL-CH-BLF-Lip showed fully wide endothermic peaks (69.76, 77.1 and 84.4 °C, respectively) during phase transition, indicating poor crystallinity of all samples. Compared with BLF-Lip, the $T_c$ and ΔH of CH-BLF-Lip and AL-CH-BLF-

Lip increased, which suggested that the layer-by-layer modification of polysaccharides improved the thermal stability of BLF-Lip. The reasonable explanation is that CH and AL affect the membrane fluidity of nanoliposomes, which is consistent with the results of other flavonoids [29]. Finally, based on the above results, a schematic illustration for the layer-by-layer deposition of CH and AL onto the surface of BLF-NL is shown in Figure 3.

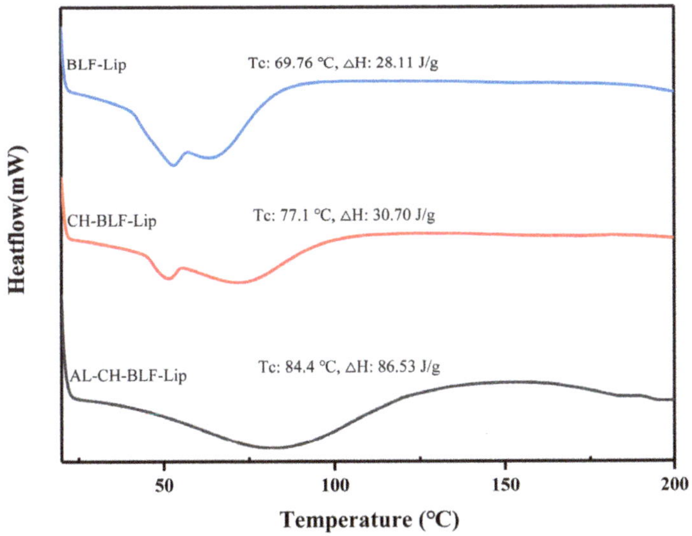


**Figure 2.** DSC analysis. BLF-Lip, Tc, Phase transition temperature; ΔH, Enthalpy variation. BLF-Lip, BLF-loaded nanoliposomes; CH-BLF-Lip, BLF-loaded chitosan coated nanoliposomes; AL-CH-BLF-Lip, BLF-loaded alginate-chitosan coated nanoliposomes.

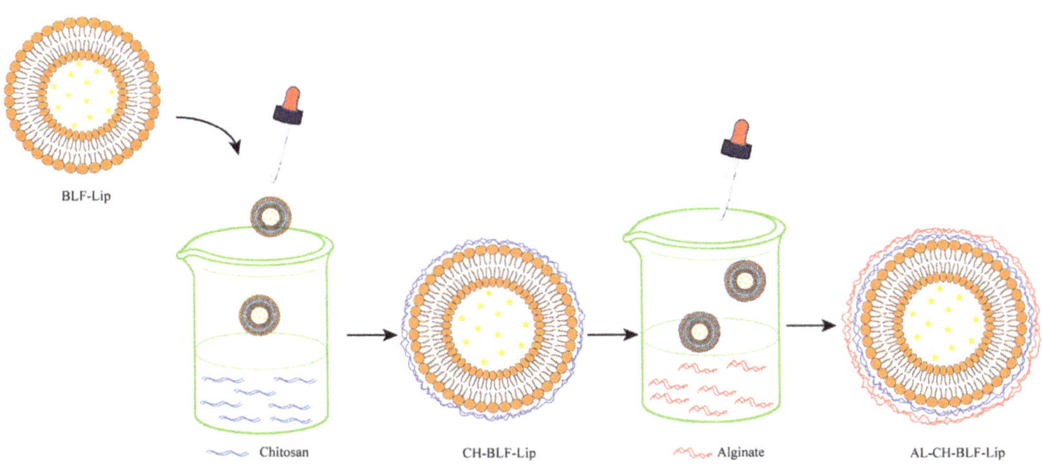


**Figure 3.** Illustration of the electrostatic deposition of chitosan and alginate layer onto the surface of nanoliposomes. BLF-Lip, BLF-loaded nanoliposomes; CH-BLF-Lip, BLF-loaded chitosan coated nanoliposomes; AL-CH-BLF-Lip, BLF-loaded alginate-chitosan coated nanoliposomes.

### 3.4. Antioxidant Activity

BLF are recognized as dietary antioxidants with a variety of biological activities. In this study, the antioxidant activity of naked BLF and BLF-loaded nanoliposomes was analyzed by ABTS free radical cation decolorization and DPPH free radical scavenging assay. Data in Figure 4A showed that the ABTS radical scavenging rate of naked BLF was 40.46 ± 4.61%. BLF-Lip, CH-BLF-Lip and AL-CH-BLF-Lip showed higher antioxidant activity compared with naked BLF (60.30 ± 2.23%, 59.53 ± 2.6% and 54.65 ± 3.91%, respectively). It should be noted that the differences between the three BLF-loaded nanoliposomes were not significant. The DPPH free radical scavenging rate of BLF-Lip, CH-BLF-Lip and AL-CH-BLF-Lip on DPPH free radicals was 27.95 ± 2.1%, 33.59 ± 3.18%, and 36.77 ± 4.11%, respectively, which were significantly higher than naked BLF (18.32 ± 2.83%). Compared with BLF-Lip, the DPPH free radical scavenging ability was significantly improved after coating CH and AL (Figure 4B). The possible explanation is the formation or enhancement of hydrogen bonds during the encapsulation process and the synergistic effect of phospholipid, CH and AL [30]. In conclusion, BLF-loaded nanoliposomes can significantly improve the antioxidant capacity of BLF, especially after coating polysaccharides.

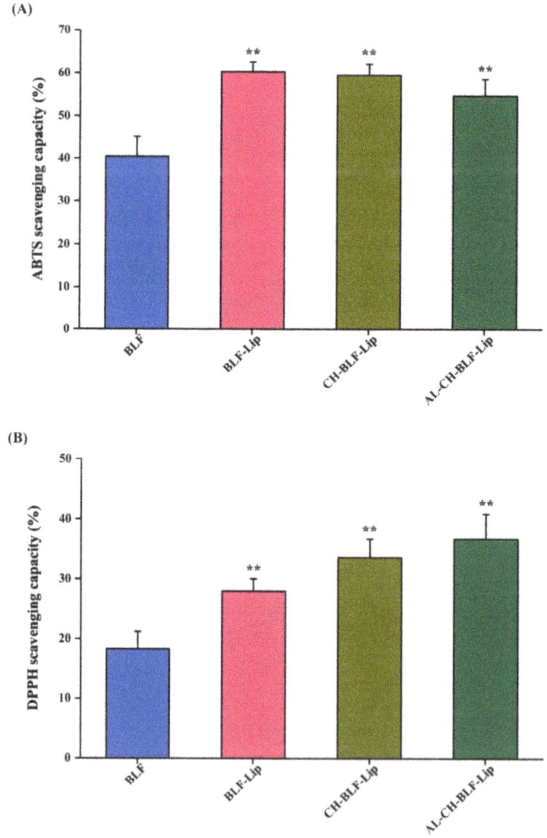


**Figure 4.** (**A**) ABTS radical scavenging capacity. (**B**) DPPH radical scavenging capacity. Data are presented as mean ± SD of three independent replications, ** $p < 0.01$ vs. BLF group. BLF, bamboo leaf flavonoids; BLF-Lip, BLF-loaded nanoliposomes; CH-BLF-Lip, BLF-loaded chitosan coated nanoliposomes; AL-CH-BLF-Lip, BLF-loaded alginate-chitosan coated nanoliposomes.

### 3.5. In Vitro Release Study

Sustained release of nanoliposomes means that when nanoliposomes enter the body, the encapsulated compounds cannot be released quickly due to hydrogen bonding or the protective effect of liposome membrane, so they can last for a long time in the body, which is an important factor affecting their potential applications. In this study, an in vitro release model was used to partially reflect the prolonged residence time. The CR of naked BLF and BLF releasing from BLF-loaded nanoliposomes in dialysis bag was measured within 24 h (Figure 5). The CR ratio of naked BLF exceeded 60% after 2 h, while BLF-Lip, CH-BLF-Lip and AL-CH-BLF-Lip were only 21.92, 12.38 and 11.89%, respectively. After incubation for 24 h, naked BLF was almost completely released, and the CR ratio of BLF-Lip, CH-BLF-Lip and AL-CH-BLF-Lip were 55.33, 45.39 and 40.32%, respectively. Therefore, the polysaccharides coating can further prolong the release of BLF and achieve the effect of controlled release.

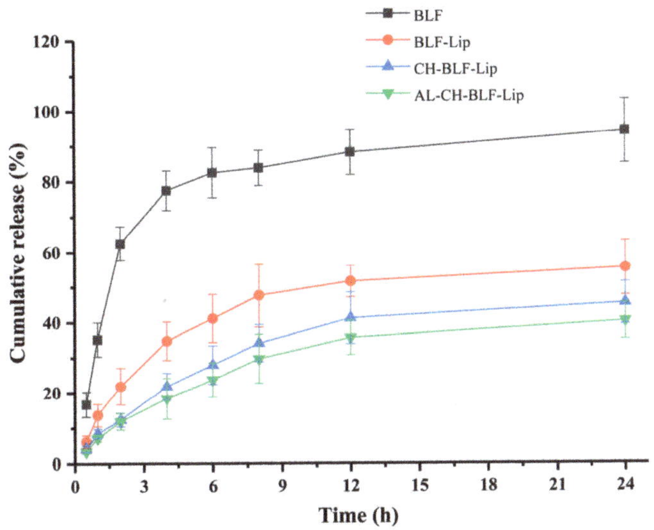


**Figure 5.** In vitro release behavior of naked BLF and BLF releasing from different BLF-loaded nanoliposomes. Data were presented as mean ± SD of three independent replications. BLF, bamboo leaf flavonoids; BLF-Lip, BLF-loaded nanoliposomes; CH-BLF-Lip, BLF-loaded chitosan coated nanoliposomes; AL-CH-BLF-Lip, BLF-loaded alginate-chitosan coated nanoliposomes.

### 3.6. In Vitro Skin Permeation Study

An in vitro skin penetration assay was performed to evaluate the transdermal efficiency of naked BLF and different BLF-loaded nanoliposomes. Mouse skin has similar barrier characteristics to human skin, and can be used as a substitute in in vitro transdermal studies [31]. The transdermal efficiency was calculated by the cumulative transdermal volume per $cm^2$ (Figure 6). BLF and BLF-loaded nanoliposomes showed similar trends. After 6 h, the cumulative transdermal volume of naked BLF tended to be stable, while BLF-Lip, CH-BLF-Lip and AL-CH-BLF-Lip still increased significantly. The cumulative transdermal volume of BLF-Lip, CH-BLF-Lip and AL-CH-BLF-Lip after 24 h were 18.75 ± 1.98, 21.57 ± 1.88 and 25.91 ± 1.73 $\mu g/cm^2$, respectively, which were significantly higher than naked BLF (8.93 ± 1.17 $\mu g/cm^2$).

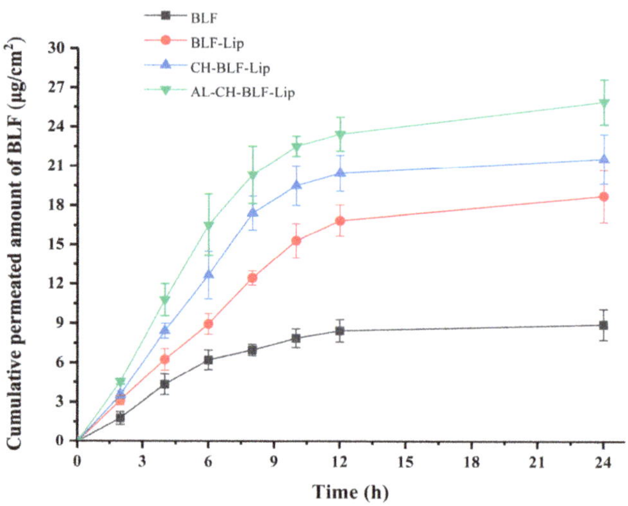


**Figure 6.** In vitro skin penetration of naked BLF and BLF releasing from different BLF-loaded nanoliposomes. Data were presented as mean ± SD of three independent replications. BLF, bamboo leaf flavonoids; BLF-Lip, BLF-loaded nanoliposomes; CH-BLF-Lip, BLF-loaded chitosan coated nanoliposomes; AL-CH-BLF-Lip, BLF-loaded alginate-chitosan coated nanoliposomes.

The stratum corneum is considered to be the main barrier to the transdermal delivery of compounds [32]. Liposomes can significantly improve the transdermal permeability of compounds, which may be due to the ability to mimic the cell membrane constituents. Generally, the smaller the particle size of the nanoparticles, the better the permeability [33,34]. However, in this study, CH-BLF-Lip and AL-CH-BLF-Lip with larger particle size showed better permeability, suggesting that there may be other factors affecting their transdermal absorption ability. One possible explanation is that CH and AL have good bioadhesivity, which can increase the contact time between the skin and nanoliposomes and promote the skin penetration ability of BLF [12]. The surface charge is also an important factor. Studies have shown that the greater the absolute value of zeta potential, the better the permeability [31]. In addition, negatively charged nanoliposomes promote skin penetration more effectively than positively charged nanoliposomes [12]. In summary, the result indicates that BLF-Lip and polysaccharide coated BLF-Lip have better skin permeability than naked BLF.

### 3.7. The Cytotoxicity of Naked BLF and Different BLF-Loaded Nanoliposomes

We further studied the cytotoxicity of naked BLF and BLF-loaded nanoliposomes in HaCaT cells at different BLF concentrations (10, 20, 40, 80, and 160 μg/mL). A cell viability above 90% indicates that the compound is non-toxic at the specified concentration [30]. The results showed that at the concentration of 10–80 μg/mL, naked BLF and BLF-loaded nanoliposomes had no cytotoxicity on HaCaT cells. When up to 160 μg/mL, the cell viability was reduced to 81.41%. However, encapsulation of BLF in nanoliposomes could reduce cytotoxicity (Figure 7A).

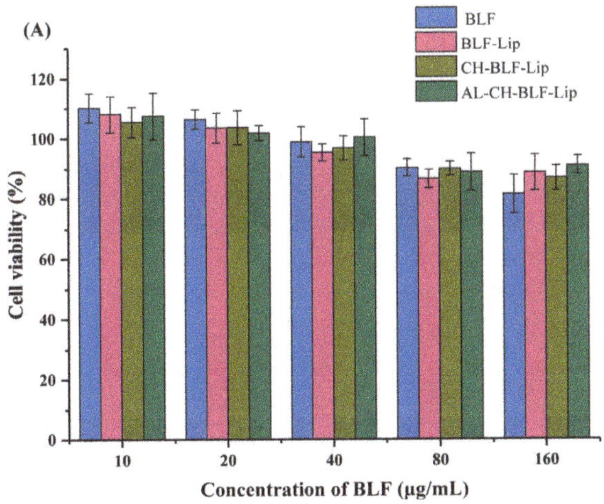


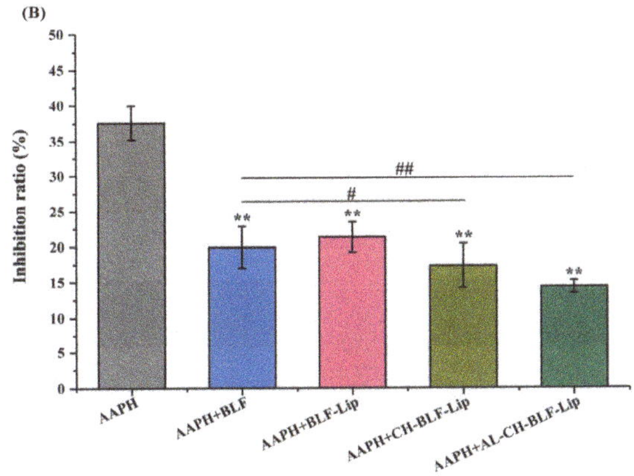


**Figure 7.** (**A**) The cytotoxicity at different concentrations (10–160 μg/mL BLF) of naked BLF and different BLF-loaded nanoliposomes in HaCaT cells. (**B**) The restoration effect of naked BLF and different BLF-loaded nanoliposomes (10 μg/mL BLF) on the inhibition of proliferation induced by AAPH in HaCaT cells. Data were presented as mean ± SD of three independent replications. ** $p < 0.01$ vs. AAPH group, # $p < 0.05$, ## $p < 0.01$ vs. AAPH + BLF group. BLF, bamboo leaf flavonoids; BLF-Lip, BLF-loaded nanoliposomes; CH-BLF-Lip, BLF-loaded chitosan coated nanoliposomes; AL-CH-BLF-Lip, BLF-loaded alginate-chitosan coated nanoliposomes; AAPH, 2,2′-Azobis (2-methylpropionamidine) dihydrochloride.

### *3.8. The Anti-Senescence Effect of BLF and BLF-Loaded Nanoliposomes*

Next, we analyzed the inhibitory effect of naked BLF and BLF-loaded nanoliposomes on cellular senescence. Oxidative stress is a key factor in skin aging [35]. We used oxidant AAPH to treat HaCaT cells to establish a cellular senescence model [36].

#### 3.8.1. Assessment of Cell Proliferation

Impaired cell proliferation is an important characteristic of cellular senescence [37,38]. The results showed that 10 μg/mL BLF and BLF-loaded nanoliposomes significantly reduced the inhibitory effect of AAPH on the proliferation of HaCaT cells. The inhibition rate of BLF-Lip, CH-BLF-Lip and Al-CH-BLF-Lip on cell proliferation decreased from 37.58% to 21.33%, 17.29% and 14.31%, respectively. It is noteworthy that the effects of

CH-BLF-Lip and AL-CH-BLF-Lip were significantly different from naked BLF, indicating that polysaccharides coating could better restore cell proliferation (Figure 7B).

3.8.2. SAHF

Cellular senescence is often accompanied by the appearance of punctate heterochromatin structure, known as SAHF, which can be visualized by staining with DAPI and antibody against K9M-H3 [38]. As shown in Figure 8, compared with the control group, 1 mM AAPH could significantly increase the proportion of K9M-H3 aggregated cells. After co-treatment with naked BLF and BLF-loaded nanoliposomes, the proportion of K9M-H3 aggregated cells decreased significantly ($p < 0.01$). Compared with naked BLF, the effects of BLF-Lip, CH-BLF-Lip and AL-CH-BLF-Lip were significantly different ($p < 0.01$).

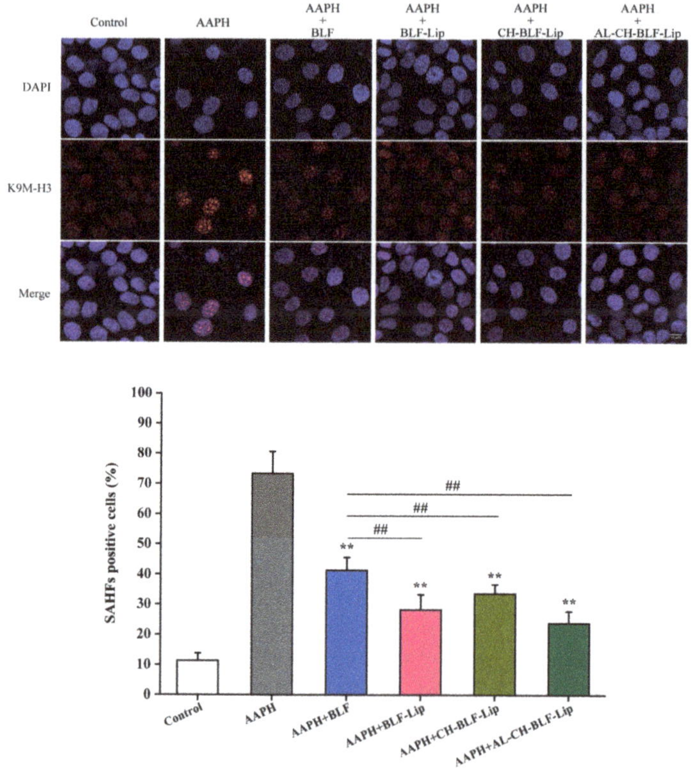


**Figure 8.** Representative images of SAHF formation. Scale bar = 10 μm. The percentage of SAHF positive cells was shown below. Data were presented as mean ± SD of three independent replications. ** $p < 0.01$ vs. AAPH group, ## $p < 0.01$ vs. AAPH + BLF group. BLF, bamboo leaf flavonoids; BLF-Lip, BLF-loaded nanoliposomes; CH-BLF-Lip, BLF-loaded chitosan coated nanoliposomes; AL-CH-BLF-Lip, BLF-loaded alginate-chitosan coated nanoliposomes; AAPH, 2,2′-Azobis (2-methylpropionamidine) dihydrochloride.

3.8.3. Expression of K9M-H3, p21 and p16 Proteins

Next, we detected the expression of K9M-H3 protein, and the results were shown in Figure 9. We found that after treating HaCaT cells with 1 mM AAPH, K9M-H3 protein not only aggregated, but also overexpressed. Naked BLF and BLF-loaded nanoliposomes could significantly reduce the expression level of K9M-H3 protein ($p < 0.01$). Compared with naked BLF, the effects of BLF-Lip and CH-BLF-Lip were significantly different ($p < 0.05$). Cell cycle regulators such as p16 and p21 are often used as biomarkers to detect senescent cells [38]. Naked BLF and BLF-loaded nanoliposomes could significantly reduce the overexpression of p16 and p21 proteins induced by AAPH in HaCaT cells. Compared with naked BLF, the effects of BLF-Lip, CH-BLF-Lip and AL-CH-BLF-Lip were significantly different ($p < 0.01$). In summary, BLF-loaded nanoliposomes have a stronger ability to inhibit cellular senescence than naked BLF.

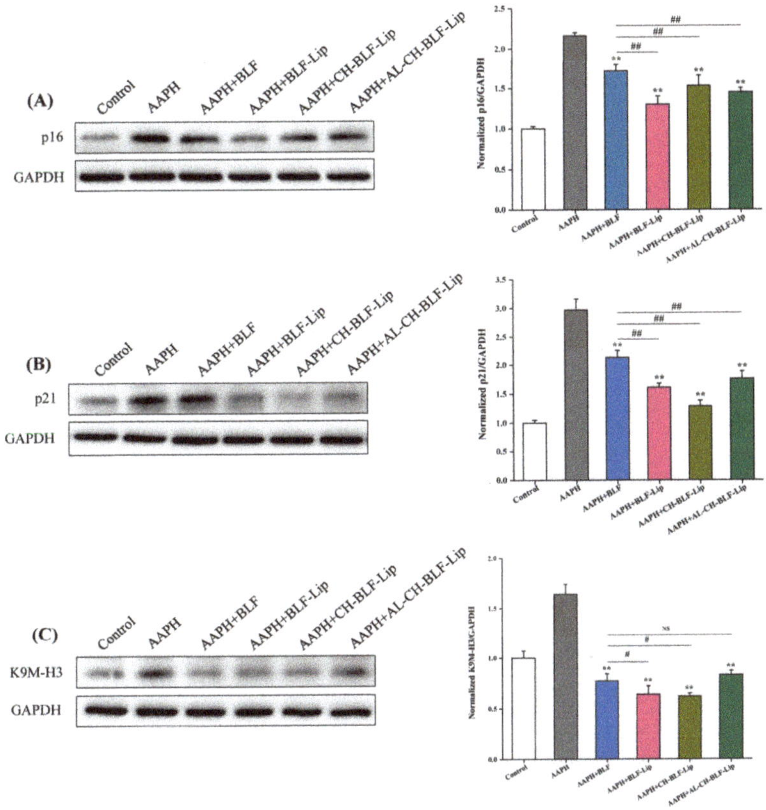


**Figure 9.** Protein expression of p16 (**A**), p21(**B**), and K9M-H3 (**C**) determined by western blotting. Quantitation was shown on the right. Data were presented as mean ± SD of three independent replications. ** $p < 0.01$ vs. AAPH treated group. # $p < 0.05$, ## $p < 0.01$ vs. AAPH + BLF treated group. NS, no significant difference.

## 4. Conclusions

In summary, based on positively charged CH and negatively charged AL deposition on the surface of BLF-Lip, this study has successfully prepared AL-CH-BLF-Lip. The increase of particle size, the positive and negative alternation of zeta potential, the increase of EE, the change of FTIR spectra and the increase of $T_c$ and $\Delta H$ can prove this result. In addition, after encapsulating BLF in nanoliposomes, the antioxidant activity, in vitro release, skin permeation and anti-senescence ability have been significantly improved, especially CH-BLF-Lip and AL-CH-BLF-Lip. In short, alginate-chitosan coated nanoliposomes can be considered as effective delivery systems for improving the controlled release and skin permeability of BLF. In the future, BLF-loaded nanoliposomes can be used as potential therapeutics for skin aging, and the effectiveness needs to be verified by further clinical trials.

**Supplementary Materials:** The following supporting information can be downloaded at: https://www.mdpi.com/article/10.3390/antiox11051024/s1, Figure S1: The chemical structure of orientin, isoorientin, vitexin and isovitexin; Figure S2: Liquid chromatogram of bamboo leaf flavonoids prepared in our laboratory.

**Author Contributions:** Y.Z. conceived the project and acted as the supervisor. Y.G. designed and performed the most of experiments and wrote the manuscript. Z.Z., F.X. analyzed the data and participated in part of the experiments. Z.Z. revised the manuscript. All authors have read and agreed to the published version of the manuscript.

**Funding:** This research was funded by Key R & D projects of Guangdong province (2019B020212001) and Science and technology innovation activity plan for college students in Zhejiang province (new seedling talent plan) (2020R401231).

**Institutional Review Board Statement:** Not applicable.

**Informed Consent Statement:** Not applicable.

**Data Availability Statement:** Data is contained within the article and supplementary material.

**Conflicts of Interest:** The authors declare no conflict of interest.

## References

1. Yang, J.-P.; He, H.; Lu, Y.-H. Four flavonoid compounds from Phyllostachys edulis leaf extract retard the digestion of starch and its working mechanisms. *J. Agric. Food Chem.* **2014**, *62*, 7760–7770. [CrossRef] [PubMed]
2. Gong, J.; Xia, D.; Huang, J.; Ge, Q.; Mao, J.; Liu, S.; Zhang, Y. Functional components of bamboo shavings and bamboo leaf extracts and their antioxidant activities in vitro. *J. Med. Food* **2015**, *18*, 453–459. [CrossRef] [PubMed]
3. Nirmala, C.; Bisht, M.S.; Bajwa, H.K.; Santosh, O. Bamboo: A rich source of natural antioxidants and its applications in the food and pharmaceutical industry. *Trends Food Sci. Technol.* **2018**, *77*, 91–99. [CrossRef]
4. Yang, C.; Yifan, L.; Dan, L.; Qian, Y.; Ming-Yan, J. Bamboo Leaf Flavones and Tea Polyphenols Show a Lipid-lowering Effect in a Rat Model of Hyperlipidemia. *Drug Res.* **2015**, *65*, 668–671. [CrossRef] [PubMed]
5. Wedler, J.; Daubitz, T.; Schlotterbeck, G.; Butterweck, V. In vitro anti-inflammatory and wound-healing potential of a Phyllostachys edulis leaf extract–identification of isoorientin as an active compound. *Planta Med.* **2014**, *80*, 1678–1684. [CrossRef] [PubMed]
6. Yu, Y.; Li, Z.; Cao, G.; Huang, S.; Yang, H. Bamboo Leaf Flavonoids Extracts Alleviate Oxidative Stress in HepG2 Cells via Naturally Modulating Reactive Oxygen Species Production and Nrf2-Mediated Antioxidant Defense Responses. *J. Food Sci.* **2019**, *84*, 1609–1620. [CrossRef]
7. Shu, G.; Kong, F.; Xu, D.; Yin, L.; He, C.; Lin, J.; Fu, H.; Wang, K.; Tian, Y.; Zhao, X. Bamboo leaf flavone changed the community of cecum microbiota and improved the immune function in broilers. *Sci. Rep.* **2020**, *10*, 12324. [CrossRef]
8. Jones, D.; Caballero, S.; Davidov-Pardo, G. Bioavailability of nanotechnology-based bioactives and nutraceuticals. *Adv. Food Nutr. Res.* **2019**, *88*, 235–273. [CrossRef]
9. Gigliobianco, M.R.; Casadidio, C.; Censi, R.; Di Martino, P. Nanocrystals of Poorly Soluble Drugs: Drug Bioavailability and Physicochemical Stability. *Pharmaceutics* **2018**, *10*, 134. [CrossRef]
10. Islam Shishir, M.R.; Karim, N.; Gowd, V.; Zheng, X.; Chen, W. Liposomal delivery of natural product: A promising approach in health research. *Trends Food Sci. Technol.* **2019**, *85*, 177–200. [CrossRef]
11. Liu, W.; Liu, J.; Salt, L.J.; Ridout, M.J.; Han, J.; Wilde, P.J. Structural stability of liposome-stabilized oil-in-water pickering emulsions and their fate during in vitro digestion. *Food Funct.* **2019**, *10*, 7262–7274. [CrossRef] [PubMed]
12. Van Tran, V.; Moon, J.-Y.; Lee, Y.-C. Liposomes for delivery of antioxidants in cosmeceuticals: Challenges and development strategies. *J. Control. Release* **2019**, *300*, 114–140. [CrossRef] [PubMed]
13. Deodhar, S.; Dash, A.K. Long circulating liposomes: Challenges and opportunities. *Ther. Deliv.* **2018**, *9*, 857–872. [CrossRef] [PubMed]
14. Huang, Z.; Li, M.; Li, N.; Tang, X.; Ouyang, Z. Antibacterial Properties Enhancement of Layer-by-Layer Self-Assembled Nanofiltration Membranes. *J. Nanosci. Nanotechnol.* **2018**, *18*, 4524–4533. [CrossRef] [PubMed]
15. Li, S.; Sun, J.; Yan, J.; Zhang, S.; Shi, C.; McClements, D.J.; Liu, X.; Liu, F. Development of antibacterial nanoemulsions incorporating thyme oil: Layer-by-layer self-assembly of whey protein isolate and chitosan hydrochloride. *Food Chem.* **2021**, *339*, 128016. [CrossRef]
16. Liu, W.; Liu, W.; Ye, A.; Peng, S.; Wei, F.; Liu, C.; Han, J. Environmental stress stability of microencapsules based on liposomes decorated with chitosan and sodium alginate. *Food Chem.* **2016**, *196*, 396–404. [CrossRef]
17. Cuomo, F.; Cofelice, M.; Venditti, F.; Ceglie, A.; Miguel, M.; Lindman, B.; Lopez, F. In-vitro digestion of curcumin loaded chitosan-coated liposomes. *Colloids Surf. B Biointerfaces* **2018**, *168*, 29–34. [CrossRef]
18. Shishir, M.R.I.; Karim, N.; Gowd, V.; Xie, J.; Zheng, X.; Chen, W. Pectin-chitosan conjugated nanoliposome as a promising delivery system for neohesperidin: Characterization, release behavior, cellular uptake, and antioxidant property. *Food Hydrocoll.* **2019**, *95*, 432–444. [CrossRef]

19. Liu, W.; Liu, J.; Liu, W.; Li, T.; Liu, C. Improved physical and in vitro digestion stability of a polyelectrolyte delivery system based on layer-by-layer self-assembly alginate-chitosan-coated nanoliposomes. *J. Agric. Food Chem.* **2013**, *61*, 4133–4144. [CrossRef]
20. Cui, T.; Jia, A.; Yao, M.; Zhang, M.; Sun, C.; Shi, Y.; Liu, X.; Sun, J.; Liu, C. Characterization and Caco-2 Cell Transport Assay of Chito-Oligosaccharides Nano-Liposomes Based on Layer-by-Layer Coated. *Molecules* **2021**, *26*, 4144. [CrossRef]
21. Abnoos, M.; Mohseni, M.; Mousavi, S.A.J.; Ashtari, K.; Ilka, R.; Mehravi, B. Chitosan-alginate nano-carrier for transdermal delivery of pirfenidone in idiopathic pulmonary fibrosis. *Int. J. Biol. Macromol.* **2018**, *118*, 1319–1325. [CrossRef] [PubMed]
22. Zhou, F.; Xu, T.; Zhao, Y.; Song, H.; Zhang, L.; Wu, X.; Lu, B. Chitosan-coated liposomes as delivery systems for improving the stability and oral bioavailability of acteoside. *Food Hydrocoll.* **2018**, *83*, 17–24. [CrossRef]
23. Lopes, N.A.; Pinilla, C.M.B.; Brandelli, A. Pectin and polygalacturonic acid-coated liposomes as novel delivery system for nisin: Preparation, characterization and release behavior. *Food Hydrocoll.* **2017**, *70*, 1–7. [CrossRef]
24. Abd, E.; Gomes, J.; Sales, C.C.; Yousef, S.; Forouz, F.; Telaprolu, K.C.; Roberts, M.S.; Grice, J.E.; Lopes, P.S.; Leite-Silva, V.R.; et al. Deformable liposomes as enhancer of caffeine penetration through human skin in a Franz diffusion cell test. *Int. J. Cosmet. Sci.* **2021**, *43*, 1–10. [CrossRef]
25. Bao, T.; Xu, Y.; Gowd, V.; Zhao, J.; Xie, J.; Liang, W.; Chen, W. Systematic study on phytochemicals and antioxidant activity of some new and common mulberry cultivars in China. *J. Funct. Foods* **2016**, *25*, 537–547. [CrossRef]
26. Heurtault, B.; Saulnier, P.; Pech, B.; Proust, J.-E.; Benoit, J.-P. Physico-chemical stability of colloidal lipid particles. *Biomaterials* **2003**, *24*, 4283–4300. [CrossRef]
27. Mo, L.; Song, J.G.; Lee, H.; Zhao, M.; Kim, H.Y.; Lee, Y.J.; Ko, H.W.; Han, H.-K. PEGylated hyaluronic acid-coated liposome for enhanced in vivo efficacy of sorafenib via active tumor cell targeting and prolonged systemic exposure. *Nanomedicine* **2018**, *14*, 557–567. [CrossRef]
28. Jeon, S.; Yoo, C.Y.; Park, S.N. Improved stability and skin permeability of sodium hyaluronate-chitosan multilayered liposomes by Layer-by-Layer electrostatic deposition for quercetin delivery. *Colloids Surf. B Biointerfaces* **2015**, *129*, 7–14. [CrossRef]
29. Pagnussatt, F.A.; de Lima, V.R.; Dora, C.L.; Costa, J.A.V.; Putaux, J.-L.; Badiale-Furlong, E. Assessment of the encapsulation effect of phenolic compounds from Spirulina sp. LEB-18 on their antifusarium activities. *Food Chem.* **2016**, *211*, 616–623. [CrossRef]
30. Chen, Y.; Xia, G.; Zhao, Z.; Xue, F.; Gu, Y.; Chen, C.; Zhang, Y. 7,8-Dihydroxyflavone nano-liposomes decorated by crosslinked and glycosylated lactoferrin: Storage stability, antioxidant activity, in vitro release, gastrointestinal digestion and transport in Caco-2 cell monolayers. *J. Funct. Foods* **2020**, *65*, 103742. [CrossRef]
31. Zhou, W.; Liu, W.; Zou, L.; Liu, W.; Liu, C.; Liang, R.; Chen, J. Storage stability and skin permeation of vitamin C liposomes improved by pectin coating. *Colloids Surf. B. Biointerfaces* **2014**, *117*, 330–337. [CrossRef] [PubMed]
32. Sun, M.-C.; Xu, X.-L.; Lou, X.-F.; Du, Y.-Z. Recent Progress and Future Directions: The Nano-Drug Delivery System for the Treatment of Vitiligo. *Int. J. Nanomed.* **2020**, *15*, 3267–3279. [CrossRef] [PubMed]
33. Verma, D.D.; Verma, S.; Blume, G.; Fahr, A. Particle size of liposomes influences dermal delivery of substances into skin. *Int. J. Pharm.* **2003**, *258*, 141–151. [CrossRef]
34. Vogt, A.; Wischke, C.; Neffe, A.T.; Ma, N.; Alexiev, U.; Lendlein, A. Nanocarriers for drug delivery into and through the skin—Do existing technologies match clinical challenges? *J. Control. Release* **2016**, *242*, 3–15. [CrossRef]
35. Bardaweel, S.K.; Gul, M.; Alzweiri, M.; Ishaqat, A.; ALSalamat, H.A.; Bashatwah, R.M. Reactive Oxygen Species: The Dual Role in Physiological and Pathological Conditions of the Human Body. *Eurasian J. Med.* **2018**, *50*, 193–201. [CrossRef]
36. Gu, Y.; Han, J.; Xue, F.; Xiao, H.; Chen, L.; Zhao, Z.; Zhang, Y. 4,4′-Dimethoxychalcone protects the skin from AAPH-induced senescence and UVB-induced photoaging by activating autophagy. *Food Funct.* **2022**, *13*, 4114–4129. [CrossRef]
37. Campisi, J.; Kapahi, P.; Lithgow, G.J.; Melov, S.; Newman, J.C.; Verdin, E. From discoveries in ageing research to therapeutics for healthy ageing. *Nature* **2019**, *571*, 183–192. [CrossRef]
38. Gu, Y.; Han, J.; Jiang, C.; Zhang, Y. Biomarkers, oxidative stress and autophagy in skin aging. *Ageing Res. Rev.* **2020**, *59*, 101036. [CrossRef]

*antioxidants*

MDPI

*Review*

# Food Peptides for the Nutricosmetic Industry

**Irene Dini [1,*] and Andrea Mancusi [2]**

1 Department of Pharmacy, University of Naples Federico II, Via Domenico Montesano 49, 80131 Napoli, Italy
2 Department of Food Microbiology, Istituto Zooprofilattico Sperimentale del Mezzogiorno, Via Salute 2, 80055 Portici, Italy
* Correspondence: irdini@unina.it

**Abstract:** In recent years, numerous reports have described bioactive peptides (biopeptides)/hydrolysates produced from various food sources. Biopeptides are considered interesting for industrial application since they show numerous functional properties (e.g., anti-aging, antioxidant, anti-inflammatory, and antimicrobial properties) and technological properties (e.g., solubility, emulsifying, and foaming). Moreover, they have fewer side effects than synthetic drugs. Nevertheless, some challenges must be overcome before their administration via the oral route. The gastric, pancreatic, and small intestinal enzymes and acidic stomach conditions can affect their bioavailability and the levels that can reach the site of action. Some delivery systems have been studied to avoid these problems (e.g., microemulsions, liposomes, solid lipid particles). This paper summarizes the results of studies conducted on biopeptides isolated from plants, marine organisms, animals, and biowaste by-products, discusses their potential application in the nutricosmetic industry, and considers potential delivery systems that could maintain their bioactivity. Our results show that food peptides are environmentally sustainable products that can be used as antioxidant, antimicrobial, anti-aging, and anti-inflammatory agents in nutricosmetic formulations. Biopeptide production from biowaste requires expertise in analytical procedures and good manufacturing practice. It is hoped that new analytical procedures can be developed to simplify large-scale production and that the authorities adopt and regulate use of appropriate testing standards to guarantee the population's safety.

**Keywords:** food antioxidant peptides; food analytical methods; large-scale biopeptide production; supplements; delivery systems; nutricosmetic; cosmeceutical; circular economy; waste recycling; anti-aging; skincare market

**Citation:** Dini, I.; Mancusi, A. Food Peptides for the Nutricosmetic Industry. *Antioxidants* **2023**, *12*, 788. https://doi.org/10.3390/antiox12040788

Academic Editor: Stanley Omaye

Received: 27 February 2023
Revised: 18 March 2023
Accepted: 21 March 2023
Published: 23 March 2023



## 1. Introduction

The cosmetic industry considers food peptides as innovative bioactive compounds for cosmetics market growth. According to the Food and Drug Administration (FDA; responsible for health products' regulation in the USA), peptides are defined as amino acid polymers with a specific sequence and less than 40 amino acids in total [1]. According to their intended action mechanism, cosmetic peptides can be categorized into: signal peptides (which stimulate matrix protein production, cell growth, and other cell metabolic functions); carrier peptides (which help transport of active or trace elements inside the cell); neurotransmitter-inhibiting peptides (which inhibit acetylcholine release that may lead to expression wrinkles); and enzyme-inhibiting peptides (which decrease the activity of enzymes related to skin aging) [2]. Peptides have gained worldwide attention for their sustainability, with no toxic side effects [3]. The global bioactive peptide market was USD 4960.4 million in 2022 with an expected compound annual growth rate (CAGR) of 9.4% in 2022–2030 [4]. The growing number of biopeptides listed in the "European glossary of common ingredient names for use in the labeling of cosmetic products" (there were 2698 entries with the word peptide [5] in the 2022 revision compared to 848 entries in the 2019 revision [6]) demonstrates the market interest in these bioactive compounds. As a result, much research has been performed to optimize biopeptide production from natural sources (e.g., food products and protein-rich by-products of the food industries) and to examine their bioactivity in vitro (cell culture and biochemical assays) and in vivo (animal and human tests). Traditional medicine and modern scientific research consider bioactive peptides useful for formulating food supplements and cosmetic products. Bioactivity, interaction with skin cells by multiple mechanisms, high potency at a low dosage, and size compatible with penetration into the upper skin layers seem to confirm this hypothesis [7]. However, some questions remain unclear. Large-scale production, dietary interactions, and human absorption are the most significant problems to solve. This review summarizes the latest knowledge on purification and identification methods used to obtain natural peptides and the approaches used to improve their bioavailability, hoping to provide a basis for their application in the nutricosmetic market as well as a starting point for further studies.

For this purpose, systematic bibliometric analyses were performed using bibliometric records published between 1993 and 2023 in Scopus and Web of Science. These two central citation databases rank journals entries based on productivity, influence, and prestige.

## 2. Production Methods for Natural Biopeptides

Natural peptides can be obtained by enzymatic hydrolysis, fermentation, and chemical-physical processes (alkaline or acidic treatments and use of microwaves, ultrasonics, hydrostatic pressure, and pulsed electric fields). Electrophoresis, membrane separation, or chromatography techniques (gel permeation chromatography, ion-exchange chromatography, reversed-phase high-performance liquid chromatography, etc.) can be used for their isolation and spectroscopic technologies (i.e., MS or NMR) as characterization techniques (Figure 1).

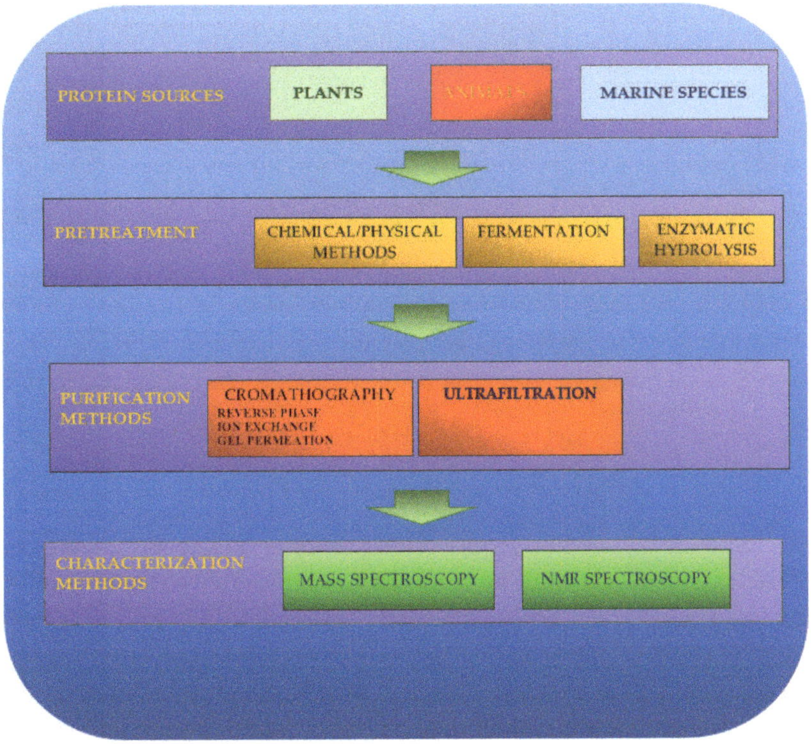


**Figure 1.** Production of natural peptides.

### 2.1. Pretreatment

#### 2.1.1. Chemical Processes

A basic or acidic environment and high temperatures hydrolyze the protein structure [8,9].

Alkaline hydrolysis is a nonspecific protein hydrolysis that cleavages amide bonds. The disadvantages of this method are the lack of specificity in the cleavage of peptide chains [10] and the loss of some amino acids (i.e., lysine, cysteine, serine, arginine, isoleucine, and threonine) [8,10,11].

Acidic hydrolysis can transform asparagine into aspartic acid and glutamine into glutamic acid and can damage tryptophan, cysteine, and methionine. The high level of salts generated by the neutralization process can also affect the antioxidant peptides' bioactivities. Therefore, desalination is necessary before bioassay (Figure 2) [9–11].

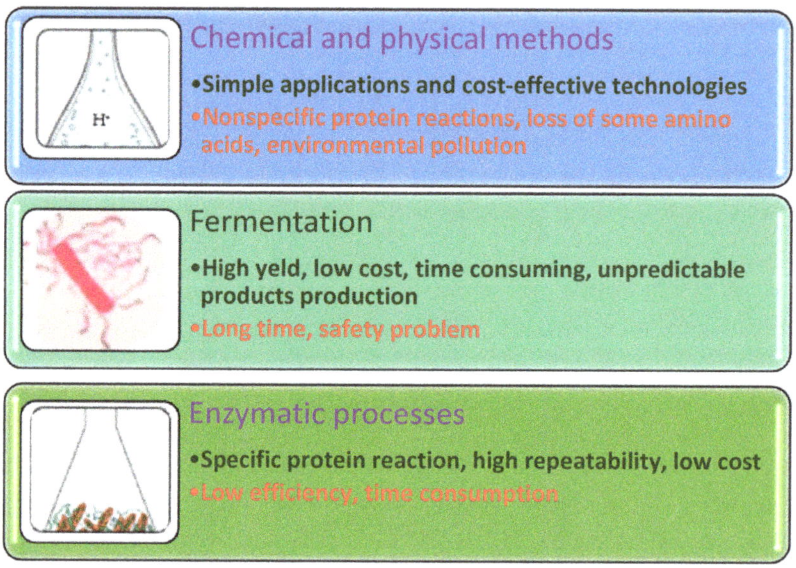


**Figure 2.** Natural source pretreatments to prepare peptides. Advantages are shown in green. Disadvantages are given in red.

#### 2.1.2. Physical Methods

The ultrasonic and microwave techniques are simple applications and cost-effective technologies used at the industrial level [8,12]. Ultrasonication systems can break molecular bonds and produce cavitational bubbles that collapse. The high temperatures, high pressure, and shock waves can damage the cell membrane and make the small molecules available [8,9]. Ultrasound increases the release of antioxidant [13] and ACE-inhibitory peptides by increasing surface hydrophobicity [14].

The microwave-assisted processing uses electromagnetic radiation (ranging from 300 MHz to 300 GHz) to extract biopeptides from protein resources. The energy is transferred through molecular interactions in the material by ionic conduction mechanisms and dipolar rotation. The extraction of peptides is due to the collision of charged ions through inter- and intra-molecular friction, resulting in thermal energy production that breaks the membranes and protein cell walls. Microwaving improves enzymatic proteolysis and hydrolysate properties (e.g., antioxidant properties), decreases hydrolysis time, alters protein conformation, and enhances enzyme accessibility [15,16].

Hydrostatic pressure uses isostatic pressures (ranging from 100 and 1000 MPa), with or without heat treatment. This technology's advantage is minimal damage to the biopeptides due to the low temperature applied [17]. Guan et al. [18] used hydrostatic pressure (200 MPa for 4 h) to obtain biopeptides from soy protein hydrolysates with higher antioxidant activities and ACE inhibitory properties.

Pulsed electric field (PEF) technology employs short pulses of electric fields with intensity ranging from 10 and 80 kV/cm and lasting for micro-or milliseconds. PEF improves the production of biopeptides by denaturation, unfolding, or gelation. Electric field strength, pulse shape, number of pulses, and treatment time affect the process quality [14]. The use of physical methods to obtain antioxidant peptides is limited since they can lead to unpredictable reactions and force-sensitive residues (Figure 2).

#### 2.1.3. Fermentation

Microbial fermentation is an eco-friendly, cost-effective, but time-consuming method of obtaining biopeptides using bacteria or yeast; their proteases hydrolyze natural proteins into protein hydrolysates. Fermentation can make unpredictable products [19]. The microbial strain, type of protein, fermentation time, and temperature conditions can affect the yield and quality of hydrolysis [10,20]. Generally, fermentation positively impacts organoleptic and physicochemical final product quality [21] (Figure 2).

#### 2.1.4. Enzymatic Processes

Enzymatic processes use single, double, or multiple proteases sourced from plants (such as papain), from microorganisms (such as Flavourzyme, Aalcalase, and Protamex), or from animals (such as pepsin and trypsin) to cleave proteins and release bioactive peptides [22]. The enzyme/substrate ratio, temperature, pH, and hydrolysis time can affect hydrolysis [8]. For example, pepsin, trypsin/pancreatin, and α-chymotrypsin (gastrointestinal digestion enzymes) can be used to produce inhibitory peptides (SSTY-hydrolysate-derived DPP-IV) [23], and soy flour hydrolysate can be employed to produce antioxidant peptides [24]. The main advantages are control, definition, and short reaction time [19] (Figure 2). The enzyme choice must be based on the intended hydrolysate products (i.e., trypsin enzyme for hydrolyzing casein). Generally, the peptides' activities and molecular weights decrease when the hydrolysis degree is enhanced. However, the product activity may decline when a certain amount of enzymatic hydrolysis occurs since peptides are transformed into amino acids or their active groups are destroyed. Therefore, it is necessary to strictly control the degree of enzymatic hydrolysis to ensure that the peptides formed have high activity [25,26].

### *2.2. Purification Technologies*

The separation of peptides in protein hydrolysate can be achieved by using ultrafiltration, chromatography, supercritical fluid extraction, and electrophoresis (Figure 3).

#### 2.2.1. Ultrafiltration

Membrane separation is a technique that saves energy, does not require extra chemical agents, does not pollute, and can be combined with other techniques [12]. Permeable membranes with different molecular weights are employed to purify peptides. In recent years, microfiltration, nanofiltration, ultrafiltration, and affinity membrane filtration technologies have been developed [27].

Microfiltration is a precision filtration technique using membranes with pores between 0.1 and 1 μm. Organic (e.g., cellulose acetate, polycarbonate, polyamide, and polypropylene) and inorganic (e.g., metal and ceramic) microfiltration membranes are available commercially. Microfiltration membranes separate the proteins and peptides in the brine [28]. Ultrafiltration is a membrane filtration process employed to purify and concentrate the components of a solution using membranes with pores between 1 nm and 0.5 μm. Ultrafiltration membranes can separate molecular weight ranges between 10,000–300,000 Da.

Nanofiltration membranes contain nanoscale pores that can separate molecular weights between those intercepted in reverse osmosis and ultrafiltration techniques with a diameter of ~1 nm. The separation mechanism depends on pore dimensions, charge, three-dimensional obstruction, and electrostatic repulsions. Active peptides with similar molecular weights, but different isoelectric points, can be separated using nanofiltration. The nanofiltration membranes can trap only ions and not electrically neutral molecules [29].

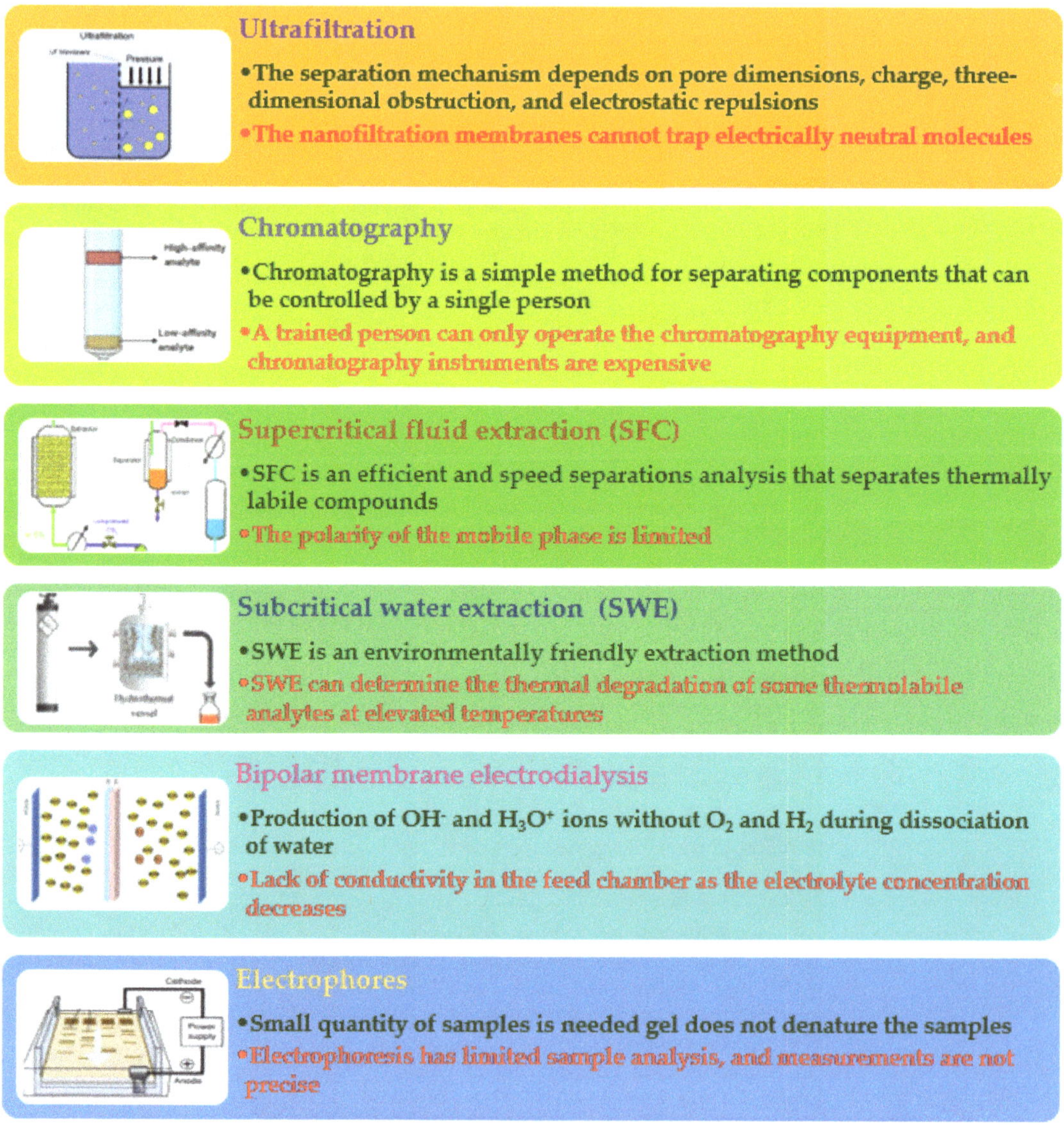


**Figure 3.** Peptide purification methods. Advantages are reported in green. Disadvantages are given in red.

2.2.2. Chromatography

Chromatographic techniques are powerful tools for separating complex peptide mixtures. They are often associated with bioassay to reduce the number of peptides that coexist in the active fraction based on their chromatographic characteristics and biological potentiality [22,29]. For example, Shih et al. [30] employed cation-exchange liquid chromatography and an ACE inhibitory assay to purify biopeptides from *Cassia obtusifolia*. The chromatographic techniques most commonly used to separate biopeptides are high-speed countercurrent chromatography (CC), magnetic solid-phase extraction, and high-performance liquid chromatography (HPLC).

The high-speed countercurrent chromatography is a carrier-free liquid-liquid partitioning chromatography that needs a large amount of equipment and time. It gives high recovery rates and does not suffer from interference caused by the carrier; however, it has poor separation efficiency.

The magnetic solid-phase extraction separates and purifies compounds from a complex matrix dissolved in a liquid. A magnet adsorbs the analyte, a magnetic separator allows their recovery, and a solvent elutes the analytes. Yu et al. [31] used magnetic solid-phase extraction to purify $\gamma$-glutamyl peptides from garlic.

The HPLC is a fast separation technique that permits good separations and high recovery rates based on hydrophobic properties (reverse-phase chromatography, RP), hydrophilic properties (hydrophilic interacting liquid chromatography, HILIC), size (size exclusion chromatography, SEC), and charge (ion exchange chromatography, IEX). The HPLC can be combined with spectroscopic instruments (e.g., mass) to purify and identify peptides.

(RP)-HPLC separates peptides based on the amino acids' hydrophobic properties. It employs a polar mobile phase and a nonpolar stationary phase. The (RP)-HPLC gives good separation, high resolution, and high recoveries. Hara et al. [32] showed that adding trifluoroethanol (10%–16%) to the mobile phase significantly increases the separation of peptides. A disadvantage of (RP)-HPLC is that it cannot be used for hydrophilic peptides [33]. The (RP)-HPLC was applied to separate wheat germ protein hydrolysates [34], cow milk products [35], turmeric and ginger [36].

Hydrophilic interacting liquid chromatography (HILIC) or normal-phase (NP) chromatography employs a hydrophobic organic mobile phase (e.g., silica stationary phases with siloxanes, silanols, and with (or without) a small number of metals, polysulfoethyl A (derivatized silica), PolyWAX (weak anion exchanger), Polycat A (weak cation exchanger), (ZIC)-HILIC (zwitterionic)), and a hydrophilic stationary phase (e.g., 70% acetonitrile, methanol or isopropanol) to separate biopeptides [37]. The HILIC technique was employed to purify biopeptides in homogenized milk [38].

Ion-exchange chromatography (IEX) separates biopeptides depending on their charge. It uses a resin that contains ionic groups (anions or cations) as a fixed phase and a polar solvent as a mobile phase. The disadvantage of this technique is the high salt concentration in the final product. A desalting step is necessary to minimize interference during peptide identification or bioassay [12].

Ion exchange chromatography is simple, resistant to alkalis and acids, and gives a high resolution. IEX was applied to separate bioactive peptides from *Boletus* mushrooms [39], sea cucumber [40], and watermelon [41].

Gel chromatography, also called size exclusion chromatography, is a fast and straightforward separation technique that uses a gel with a network of pores as a fixed phase. Solutes are eluted in order from large to small molecular size. The mobile phase viscosity must be such as to cross the column [8]. The disadvantages of this technique are low resolution, a limited peak capacity, and large eluent volumes [12]. Gel chromatography was applied to separate bioactive peptides from animal muscle proteins [42].

#### 2.2.3. Supercritical Fluid Extraction

Supercritical fluid extraction uses fresh fluid through a sample (pressure and temperature controlled) that is recovered from the extract by depressurization. This technique allows a faster extraction than traditional methods since the supercritical fluids can penetrate a porous solid more easily. Supercritical fluid can be recycled or reused, diminishing waste. Supercritical carbon dioxide ($CO_2$) is used to obtain bioactive peptides from various sources (e.g., *Chenopodium quinoa*, fruit waste, *Ganoderma lucidum*) [43–45]. Under low-temperature conditions, the proteins cannot denature (they cannot make polypeptides and free amino acids), the amino acids cannot oxidize, and the amino acids and carbohydrates cannot generate the Maillard browning products [46]. Unfortunately, the apparatus has a high cost, solvent compression needs recirculation measures to decrease energy costs, and the modifiers used to alter the polarity of the $CO_2$ require a subsequent separation process. Supercritical fluid extraction can be coupled with enzymatic hydrolysis to produce biopeptides in less time and at low cost [47].

#### 2.2.4. Subcritical Water Extraction (Pressurized Hot Water or Hot-Compressed Water)

Subcritical water employs water in a liquid state under high pressure and temperature between 100 and 374 °C. It can be used for polar and nonpolar compounds. Under these conditions, the dielectric constant reduces, the ionic product constant (Kw) increases, and the hydrogen bonds deteriorate, making subcritical water similar to methanol and ethanol (less-polar solvents). The high pressure and high temperature used separately can denature the proteins [48,49]. Subcritical water conditions (the combined effects of high pressure and temperature) break down proteins into peptides and free amino acids [50] with an irreversible first-order reaction. Hydronium and hydroxyl ion reactions confer bicatalytic characteristics on water (weak acid or base) [50]. High temperature and pressure disrupt weak interactions (e.g., hydrogen bonds) and the loss of quaternary, tertiary, and secondary structures [49,51,52]. The union of $H^+$ (derived from the hydronium ion) to the N-terminal generates atom excitation and breaks the peptide bond. Finally, the $OH^-$ links with the new carbon cation of the C-terminal [53]. The degradation of Maillard's amino acids and reaction products improves the medium's pH at temperatures above 200 °C [54–56]. Subcritical water was used as a hydrolysis medium to obtain protein from vegetal meal (e.g., soybean, rice bran, deoiled Oryza sativa bran) [57,58], animal sources (e.g., ice-cream wastewater, African snail *Achatina fulica*) [59] and other sources (e.g., laver *Pyropia yezoensis*) [60], obtaining hydrolysates with a strong free-radical scavenging capacity and antioxidant activity. The disadvantages of subcritical water are the lack of selectivity and cluster formation during the overheating process [61].

#### 2.2.5. Bipolar Membrane Electrodialysis

Bipolar membrane electrodialysis (EDBM) technology does not use chemicals to separate peptides. However, it employs monovalent anion exchange and cation-selective permeation membranes to separate ions and bipolar membranes that allow the production of $H^+$ and $OH^-$ ions from water under the current application. Mikhaylin et al. [62] used EDBM to separate casein from milk.

#### 2.2.6. Electrophores

Sodium dodecyl sulfate-polyacrylamide gel electrophoresis (SDS-PAGE) is an analytical technique for separating biopeptides based on their molecular weight. In this technique, the migration rate through the gel matrix is affected by the peptides' size (smaller peptides migrate faster than larger peptides due to less resistance from the gel matrix), charge, and chain length. Siow et al. used SDS-PAGE to obtain peptides from *Parkia speciosa* seeds [63].

Capillary electrophoresis is a liquid-phase separation technology in which a capillary tube separates biopeptides using a high-voltage electric field as a driving force. Capillary electrophoresis can separate analytes in various separation modes (e.g., capillary zone electrophoresis, capillary isoelectric, and capillary gel electrophoresis).

Capillary zone electrophoresis separates the biopeptides based on the different charges. The speed affects the separation process [64]. The capillary isoelectric focusing electrophoresis separates analytes based on their different isoelectric points. It is generally used to separate peptide isomers [65–67].

Capillary gel electrophoresis separates analytes based on molecular sieving (molecular shapes and weights). It is used to separate biopeptides with many hydrophobic side chains.

### 2.3. *Identification of Peptide Sequences*

#### 2.3.1. Mass Spectroscopy

Mass spectrometry (MS) is a versatile analysis method for biopeptide characterization by molecular weight. MS/MS is an alternative strategy to obtain mass spectra of peptide fragment ions from a particular precursor ion. The MS can identify and quantify the peptides, ionizing them in various modes.

Electrospray ionization (ESI) employs electrical energy to transfer ions from an initial solution into a gas stage. This ionization into gas needs the dispersion of charged droplets, solvent evaporation, and expulsion of ions from the highly charged droplets [68]. ESI-MS detects femtomole quantities of multiple compounds, including non-volatile and thermolabile analytes such as peptides. It can analyze the peptides' intact mass and amino acid sequence by MS/MS technology. The peptide bond (-CO-$NH^-$) is the most common source of fragmentation. It produces y-ions (C-terminal fragment ions) and b-ions (N-terminal fragment ions). The amino acid sequence is deduced from the peptide's fragmentation along its backbone and around the peptide bond that generates peptide fragment types. In the case of isoleucine and leucine (that have identical molecular weight), w-ions allow the characterization (the R-group is different for isoleucine ($CH_2CH_3$) and leucine ($CH_3CHCH_3$). Worsztynowicz et al. [69] used MS/MS technology to identify biopeptides from whey proteins, Karami et al. [34] from wheat germ, and Zanoni et al. from hempseed [70]. MALDI-TOF spectra analysis of peptides usually identifies only the molecular ion [M+H]+ (it is an advantage since it maximizes the signal), but gives no evidence of structure. Daughter ions formed during the application of the extraction pulse increase the background noise. This approach is poorly sensitive and selective [71]. Only highly-abundant proteins generate patterns of peptides for their unambiguous identification. The signals of co-migrating peptides are suppressed. Ayala-Niño et al. [72] studied biopeptides from amaranth seed proteins using MALDI TOF. MALDI-TOF/TOF instrumentation can be applied to obtain more specific and reliable results [73]. Cakir et al. [74] examined proteins in black cumin seeds by MALDI-TOF/TOF-MS analysis.

#### 2.3.2. NMR

The molecule momentum, in the presence of a magnetic field, can align in the same or opposite direction to the field. Two states separated by an energy gap (resonance frequency) are formed. The nucleus' chemical environment in a molecule and the magnetic field strength affect this difference in resonance frequency [75]. The structures of 131 peptides from 17 fungi genera were unambiguously characterized using 1D and 2D NMR [76].

## 3. The Human Skin

The human skin is a complex organ with an exceptional structure. It is made up of diverse cell types and compartments with distinctive functions. The epidermis (outermost layer) contains four sublayers (strata corneum, granulosum, spinosum, and basalis) and four major cell types (melanocytes, keratinocytes, Langerhans and Merkel cells). The epidermal–dermal junction (border between the epidermis and dermis) constitutes the basement membrane (an aggregation of proteins and structures). Under the basement membrane is the underlying dermis, which contains dendritic cells, mast cells, macrophages, fibroblasts, elastic fibers, collagen, hair follicles, blood vessels, nerves, lymph vessels, and sweat glands.

The dermis allows nutrients to reach the skin and has a structural support function [77]. Aging affects all skin layers, altering their structure and function [77]. The organism's aging is inevitably a progressive process. The consequences of aging are changes at the tissue and cellular levels. During physiological skin aging, the number of keratinocytes at the epidermis level gradually decreases. The epithelium layers atrophy. The reproductive layer cell division activity, Langerhans cell and melanocyte numbers decrease. The dermis' connective tissue also atrophies (its cellular and extracellular matrix components diminish). The fibroblasts synthesize collagen, but the fibers are less elastic and efficient, and protein fibers, already existing, are subject to degeneration. Only the corneocyte number in the dead stratum corneum shows no changes. The accumulation of damage reduces cells' ability to renew [78]. The decrease in lipid and CD44 glycoprotein (regulator of keratinocyte proliferation) levels, the loss of hyaluronic acid homeostasis, and the reduced cell proliferation in the basal layer contribute to this decline [79,80].

Moreover, the contact surface area between the epidermis and dermis becomes thinner, resulting in a weakened epidermis nutrition supply and a further decline in basal cell proliferation ability [81,82]. The epidermal–dermal junction and dermis also become thinner, causing wrinkle formation since there are fewer cells, with less oxygen and less nutrition. The dermal extracellular matrix (ECM) accumulates type I and type III collagens [83], and there is a decrease in the synthesis of type I/III [84], altering the elastic fiber organization [85]. The low fibroblast levels increase wrinkling and reduce elasticity [86]. Skin aging is associated with extrinsic (external) factors, e.g., UVA, UVB, temperature, environmental pollution, nutritional factors, cigarette smoke, lack of sleep, and stress [87] and with intrinsic (endogenous) factors, e.g., genetic factors, chronological time, hormones, decreased age-related antioxidant capacity, and increase in reactive oxygen species [88,89]. The principal consequences are blemished, dry, pale skin with rugged texture, visible pores, redness, small actinic keratomes, gradual loss of elasticity, and fine wrinkles [90]. The intrinsic clinical skin aging signs caused by intrinsic factors are xerosis (dry skin), fine lines, and laxity [91]. The aging signs caused by extrinsic factors are irregular pigmentation, coarse wrinkles, and lentigines (or age spots). The photo-exposed areas (e.g., the face, hands, and neck) show a more visible occurrence of these changes. The duration and intensity of exposure to environmental factors and the skin type affect the occurrence of extrinsic skin aging signs [91]. Skin aging impacts human aesthetics and increases susceptibility to infections and chronic wounds (e.g., venous, pressure, or diabetic foot ulcers, dermatitis, and melanoma) [92,93].

## 4. Biopeptides' Potential in Cosmeceutical Applications

The increased demand for natural cosmetics has led to the formulation of a new generation of cosmetics based on active compounds obtained from natural sources such as biopeptides. Biopeptides can enhance skin health (acting against aging-related enzymes) and decrease the harmful effect of agents that produce skin injuries (acting as antioxidant, antimicrobial, and anti-inflammatory agents). Multifunctional biopeptides, which can simultaneously start, modulate, or impede multiple physiological pathways, are preferred to single-activity peptides [94]. The problems associated with using biopeptides in cosmetics concern the yields of the techniques with which biopeptides are produced (in terms of production quantity and concentration of biopeptides capable of expressing desirable bioactivity) and the biopeptides' structural stability and bioactivity during product manufacturing and storage. Biopeptide activity is affected by pH, interactions with other components, temperature, water activity, and formulation processes (e.g., concentration, delivery of the active compounds, and packaging) [95,96].

For example, when used in gels, creams, or lotions, the parameters to consider are sensitivity to temperature and pH to guarantee the peptide bioactivity at the action site. Moreover, it is essential to realize that only the bioactive peptides with low molecular weight penetrate the skin. Biopeptides with high molecular weight, hydrophobic character, and poor aqueous solubility at high concentrations require carriers to permit their release when needed [97]. When biopeptides are administrated orally, their bioavailability (integrity during digestion, intestinal absorption, and transport) must be controlled [97] since they are exposed to gastric, pancreatic, small intestinal enzymes, and acidic stomach conditions, meaning that only minimal biopeptide levels (nano-molar or pico-molar concentrations) reach the action site [98]. Finally, some protein-derived peptides have a bitter taste. Therefore, they must be subjected to processes to debitter them and/or mask the bitter taste to enhance the sensory properties of the final product [99]. Transport systems studied to overcome these problems include liposomes, biopolymer microgel emulsions, and solid–lipid nanoparticles [100]. Some natural lipid-based systems (e.g., chitosan fabricated nanocarriers, soy lecithin-derived nanoliposome, and microgels from alginates and methacrylate) were suggested as potential inclusion complexes for biopeptides [101]. Another limitation is the risk of allergens since most peptide preparations are produced as unpurified mixtures of several components. Plant hydrolysates may contain allergens and potentially toxic contaminating compounds (environmental pollutants) [102]. Therefore, peptide preparations from plant tissue cultures (grown in the laboratory, under axenic and controlled conditions) are preferred today [103].

### *4.1. Biopeptides with Anti-Aging Properties*

Some natural peptides (e.g., snake venom, yeast, skin frog, toads, spirulina, and fish) have anti-aging properties [104]. They can inhibit key physiological enzymes such as elastase, tyrosinase, collagenase, and hyaluronidase, which are involved in the degradation of the skin protein matrix and are overproduced when intrinsic or chronological aging occurs [105]. Some of these biopeptides are under patent protection, such as the pentapeptide-3 (GPRPA) (from snake venom), which decreases skin roughness and wrinkles [106], and the hexapeptide11 (FVAPFP) (from yeast) that improves skin firmness [107].

#### 4.1.1. Biopeptides That Decrease Collagenase Activity

Collagen is the most widely distributed protein in mammals. It confers support and strength to human skin and can restore flexibility and elasticity [108]. Collagen has a role in the structural integrity and strength of connective tissues (e.g., tendons, teeth, and skin) [109]. There are different forms of collagen: type I (found in the skin, bone tissues, and tendons and widely used in cosmetic formulations), type II (found in cartilage), and type III (found in vasculature and skin) [110]. The collagen-derived peptides and collagen hydrolysates positively improve skin conditions [111]. Under heat treatment, collagen is converted into water-soluble gelatin, from which can be obtained collagen peptides (by enzymolysis). Collagen peptides are antioxidant compounds that prevent dermal collagen decomposition, negatively affect collagenase and gelatinase activity, decrease skin moisture loss and reduce wrinkling; they increase skin hydration and elasticity, and address collagen degradation and elastic fiber abnormalities due to UV radiation [112]. They can improve the hyaluronic acid content in skin tissue by enhancing the expression of hyaluronic acid synthase mRNA and filaggrin and decreasing the expression of hyaluronidase mRNA. Collagen and collagen peptides can be obtained from animal tissue, poultry, livestock, fish (bones, scales, and skin), and vegetables (spirulina) [113]. The source of collagen peptides affects their anti-skin-ageing effect (Table 1) [112]. In vivo studies showed that women given oral supplementation of collagen hydrolysate showed improvements in skin hydration, elasticity, wrinkling [114], dermal thickness, firmness [115], and texture [116] and a lessening of skin pores [117].

**Table 1.** The anti-skin-ageing effect of collagen peptides made from a natural source.

| Peptides | Source | Skin-Aging Effect | Biblio |
|---|---|---|---|
| Type I collagen-derived collagen peptide Chicken collagen | Pig collagen | Enhancement of skin collagen content by changing the ratio of type I and type III collagen. No effect on skin moisturizing. | [118] |
| High tripeptide-containing collagen hydrolysate (HTC-col) has high tripeptides comprising the Gly-X-Y sequence. | Porcine skin | Anti-photoaging action. Skin dryness improvement. | [119] |
| Chicken-derived collagen peptide | Chicken collagen | Anti-inflammatory. Antioxidant. Collagen I synthesis. Improve cell proliferation on human skin fibroblasts | [120] |
| YGDEY (Tyr-Gly-Asp-Glu-Tyr) from. | Tilapia collagen hydrolysate | Prevention of ultraviolet (UVB)-induced damage to cells Inhibition of UVB-mediated photoaging of the skin. Improvement of the glutathione and superoxide dismutase expression. Enhancement of type I procollagen. Reduction of the ROS in keratinocytes. Prevention of DNA oxidative damage. Inhibition of the collagenase and gelatinase expression. | [121] |
| Ala-Tyr dipeptide | Carp skin hydrolysate | Antioxidant activity | [122] |
| Hydrolyzed collagen | *Prionace glauca* | Stimulation of the collagen type I mRNA by fibroblasts. mRNA production improvement. | [123] |
| Hydrolyzed collagen with neutrase | Alaska pollock | Antioxidant activity | [124] |
| Hydrolyzed collagen with pepsin under acidic conditions | *Rana chensinensis* | Antioxidant activity | [125] |
| Hydrolyzed collagen with pepsin, subtilisin A, and both enzymes | *Arthrospira maxima* (spirulina) | Peptides obtained from PHS showed the highest collagenase inhibition activity | [126] |
| Peptides | *Tetraselmis suecica Dunaliella tertiolecta*, and *Nannochloropsis* | Decrease in hyaluronidase enzyme | [127] |

#### 4.1.2. Biopeptides That Decrease Hyaluronidase Activity

Hyaluronic acid (HA) is an anionic, non-sulfated linear glycosaminoglycan [128]. It is a component of the dermis extracellular matrix in many human body tissues (e.g., synovial fluid, gum, eyes, heart valves, and skeletal tissues). It can maintain skin moisture (since it can bind water) [129], improve skin rejuvenation and viscosity, and decreases extracellular fluid permeability [130]. The concentration of HA in the skin naturally declines with age. The hyaluronidase enzyme degrades it producing a loss of skin strength, flexibility, and moisture. Peptides from three microalgae (*Tetraselmis suecica, Dunaliella tertiolecta,* and *Nannochloropsis* sp.) can decrease the hyaluronidase enzyme [127]. The cosmetics industry proposes products containing hyaluronic acid for topical application, for which inhibition of hyaluronic acid degradation is crucial to avoid the inflammatory process related to the exogenous application of hyaluronic acid [130].

#### 4.1.3. Biopeptides That Decrease Tyrosinase Action

Tyrosinase is a metalloenzyme (whose active site includes two copper ions) responsible for melanin (the pigment that controls skin color) production [131]. Tyrosinase overproduction causes skin hyperpigmentation, leading to a darker skin appearance (dark brown spots and irregular grey patches) [132]. Tyrosinase activity is inhibited by compounds that block the active site or chelate copper ions [131] (Table 2). High amounts of serine can bind copper [133] and affect the C-terminal tyrosine residue [134] and amino acids with hydroxyl function [133].

**Table 2.** Summary of recent studies on biopeptides that decrease tyrosinase activity.

| Peptides | Source | Activity | Biblio |
|---|---|---|---|
| Skin collagen peptides (3–10 kDa fraction) | *Todarodes pacificus* | Copper-chelation | [135] |
| Albumin peptide obtained using papain | Rice bran | Tyrosinase inhibition, copper-chelation | [133] |
| HGGEGGRPY, LQPSHY, and HPTSEVY | Rice | Tyrosinase inhibition | [134] |
| Peptides | Faba bean (*Vicia faba*) | Tyrosinase inhibition | [136] |
| Water and ethanol extracts from soy milk fermented with lactic acid bacteria strains, | Soy milk | Tyrosinase inhibition | [137] |

#### 4.1.4. Biopeptides That Decrease Elastase Action

Elastin is an extracellular matrix of highly polymerized protein that gives elasticity to connective tissues. Arteries, skin, and lungs contain elastin [138]. It maintains skin elasticity and firmness. Elastin contains two amino-acid sequences, one responsible for crosslinking and the other for hydrophobicity. The extensive crosslinking in elastin determines insolubility and durability [139]. Fibroblasts and vascular smooth muscle cells synthesize elastin until puberty and stop when the body matures. Overproduction of the enzyme elastase decreases the elastin fibers' production [140]. Elastin-derived peptides may prevent and regulate skin photoaging (decreasing elastase activities and fibroblast apoptosis and improving the hydroxyproline content), water content, and fibroblast proliferation [141]. Aging increases elastin degradation and elastin-derived peptide (EDP) levels, enhancing the affinity and deposition of calcium [142]. Two elastin hydrolysate-derived peptides (TGVLTVM and NHIINGW) from the skipjack have shown protective effects against skin damage due to UVA irradiation through the attenuation of oxidative stress and mitochondrial damage [143]. One elastase inhibitory peptide Phe-Phe-Val-Pro-Phe (FFVPF), with significant stability in the gastric environment, was obtained from walnut meal protein hydrolysates [144]. Norzagaray-Valenzuela et al. found peptides in microalgae (*Dunaliella tertiolecta, Tetraselmis suecica,* and *Nannochloropsis* sp) with elastase inhibitory effects [127].

### *4.2. Biopeptides with Antioxidant Properties Derived from Foods*

In the cells, oxidative stress is generated by the imbalance between the endogenous antioxidant defense system ability and free radicals, which can produce oxidants. The reactive oxygen species (ROS, e.g., hydroxyl radical ($^{\bullet}OH$), superoxide anion radical ($O_2^{\bullet -}$), lipid radical ($ROO^{\bullet}$)) and reactive nitrogen species (e.g., nitrogen oxide ($NO^{\bullet}$)) are responsible for degenerative changes in the aging process, heart disease, arteriosclerosis, stroke, cancer, and diabetes [145]. Antioxidant molecules (synthetic and natural) decrease the risk of chronic diseases (e.g., cardiovascular pathologies, diabetes, arthritis, Alzheimers and cancer) and skin aging related to oxidative stress and control the oxidation of food nutrients [146]. Natural antioxidants are heterogeneous secondary metabolites such as phenols, vitamins (e.g., E and C), carotenoids, glutathione, biopeptides, and some enzymes such as glutathione peroxidase, superoxide dismutase, and catalase [147]. Synthetic antioxidants (i.e., propyl gallate, t-butyl hydroquinone, and butyl hydroxyanisole) have the disadvantage of high costs and potential toxicity risk [7]. Antioxidants can donate electrons, catalyze oxide-reductive reactions (e.g., antioxidant enzyme), and prevent the interaction of transition metals (e.g., copper and iron) with hydrogen peroxide and superoxide binding proteins [147]. The research on safe and high-efficiency antioxidants from natural products (especially foods) has attracted widespread attention. Vitamins, carotenoids, bioflavonoids, and peptides have attractive antioxidant potential [148–150].

Peptidic antioxidants (PAs) can chelate metal ions, scavenge radicals, and quench singlet oxygen. They can be ingested safely and sometimes act as antibacterial, antihypertensive, and hypocholesterolemic molecules [151]. Peptide sequences with antioxidant activity are found in food proteins and biowaste proteins, where they occur as inactive sequences. Gastrointestinal digestion, enzymatic hydrolysis, and microbial fermentation can release biopeptides from precursor proteins [152]. Biowaste peptides are attractive for industrial applications and, at the same time, promote environmental protection. The liberated peptides must be purified before determining the sequences [153,154].

Peptides with molecular weights less than 3 kDa, containing 2–20 amino acids, among which are hydrophobic amino acids (e.g., tryptophan, phenylalanine, valine, histidine, glycine, isoleucine, lysine, and proline), and having extra aromatic rings, hydrophobic properties, and donor electrons have potential antioxidant activity [155,156].

The aromatic ring guarantees that the loss of electrons will not transform the peptide into free radicals. The extra electrons can deactivate the free radicals [157]. The hydrophobic properties permit the accessible entrance of the antioxidant peptides into target organs through hydrophobic interactions with membrane lipid bilayers [158]. Nevertheless, according to Chen et al. and Tironi et al., the antioxidant capacity of peptides declines after hydrolysis [159,160].

Chen et al. [160] found that the concentration of antioxidant amino acids and the peptide sequence might affect the antioxidant potential [161]. The amino acid sequence order of antioxidant activity is Pro-Tyr-Ser-Phe-Lys > Gly-Phe-Gly-Pro-Glu-Leu > Val-Gly-Gly-Arg-Pro, when DPPH, ABTS, and OH radical assays are used to measure it [162] and the order is Trp-Pro-Pro > Gln-Pro if the hydroxyl radical scavengers are evaluated [163]. The peptides' synergistic influence improves their antioxidant potential and avoidance of gastrointestinal proteolysis [164–168]. The peptides' antioxidant properties involve free radical scavenging, metal ion chelation, singlet oxygen quenching, and lipid peroxidation inhibition (enzymatic and non-enzymatic) [169]. PAs were found in plants (corn, rapeseed, cacao seed, rice, rye, wheat, soybean, pea, and hemp seed) [170–172], milk [173–175], marine organisms (algae, mackerel, horse mackerel bonito, yellowfin, monkfish, oysters, tuna, salmon, mussel, catfish, sardine, eel, squid, and tilapia) [176–178], eggs (ovalbumin, yolk, and white lysozyme) [179] and animals (porcine myofibrils, skin, buffalo horn, and dry-cured ham) [180].

Methods Used to Evaluate the Antioxidant Potential

The most popular protocols used to test antioxidant activities employ spectrophotometric tests. They can evaluate the hydrogen atom transfer (HAT) mechanism, single electron transfer (SET) mechanism, transient metal chelation, and in the cellular system, the Nrf2/Keap1pathway.

The ORAC (Oxygen Radical Absorbance Capacity), TRAP (Total Radical Trapping Antioxidant Parameter), CBA (Crocin Bleaching Assay), and LPA (Lipid Peroxidation Assay) evaluate HAT-based reactions.

The TEAC (Trolox Equivalent Antioxidant Capacity, also known as ABTS), DPPH (2,2-diphenyl-1-picrylhydrazyl radical scavenging activity), and CRC (Copper II Reduction Capacity assay) measure the SET-based reactions [177,181].

The EECC (EDTA Equivalent Iron Chelation Capacity) and CECC (Carnosine Equivalent Iron Chelation Capacity) test the ion chelating capacity [182].

The ABTS assay's pH strongly affects the antioxidant potential of tryptophan, tyrosine, and their derivate peptides [183].

### 4.3. Peptides with Antimicrobial Activity

The skin is constantly exposed to microbial agents. Skin aging decreases the cutaneous production of antimicrobial peptides [184]. The bioactive peptides with antimicrobial activity against *Staphylococcus aureus, Propionibacterium acnes, Pseudomonas aeruginosa, Enterococcus faecium, Acinetobacter baumannii, Klebsiella pneumoniae, Propionibacterium acnes,* and *Enterobacter species*, are promising functional ingredients in food supplements and cosmeceuticals (Table 3) [185–188]. The antimicrobial activity of biopeptides is ascribed to the formation of transmembrane channels (by polymerization or self-aggregation), which lead to cytoplasm leakage and/or cell death, and/or inhibition of cell division, protein-folding, cell wall and protein biosynthesis, nucleic acid synthesis, and lipopolysaccharide formation [189].

**Table 3.** Summary of recent studies on antimicrobial biopeptides.

| Peptides | Source | Activity | Biblio |
|---|---|---|---|
| TITLDVEPSDTIDGVK ILVLQSNQIR<br>ISGLIYEETR MALSSLPR<br>ISAILPSR<br>LPDAALNR IGNGGELPR<br>QVHPDTGISK EAESSLTGGNGCAK | *Saccharina longicruris* | *Staphylococcus aureus* | [191] |
| MDN<br>ELAAAC<br>LRDDF<br>GNAPGAVA<br>ALRMSG<br>RDRFL | *Alfalfa RuBisCo* | *Listeria innocua* | [192] |
| QAIIHNEKVQAHGKKVL | *Crocodylus siamensis* | *Escherichia coli, Staphylococcus aureus, Klebsiella pneumoniae* and *Pseudomonas aeruginosa.* | [193,194] |
| Cationic peptides | Rice bran | *Propionibacterium acnes* JCM 6473 | [190] |
| Peptides generated by *Aspergillus oryzae*,<br>*Aspergillus flavipes* proteases | Bovine milk | *Listeria monocytogenes*<br>*Staphylococcus aureus*<br>*Salmonella enterica* Enteritidis<br>*Escherichia coli*<br>*Pseudomonas aeruginosa* | [191] |

It seems that positive charges (ranging from +2 to +9), small size (15–50 amino acids residues), and an amphipathic structure (ca. 50% hydrophobic residues) facilitate the biopeptide's interaction with the negatively charged membrane of some microorganisms [190].

### 4.4. *Peptides with Anti-Inflammatory Activity*

Inflammation is how the body restores itself after injury, replaces damaged tissue and combats pathogens [195]. Inflammation may be acute or chronic. It can remain for a few minutes to weeks or years. Inducing inflammatory process factors are lipopolysaccharide, dextran sodium sulfate, and other toxicants. Inflammatory processes induce the production of cytokines (IL1α, IL1β, IL2, IL6, IL8, IL12, TNFα, and IFNγ) by T lymphocytes cells and macrophages [196]. Moreover, inflammation involves immune systems cells such as mitogen-activated protein kinases (MAPK; intracellular serine/threonine protein kinases) [197], nuclear factor kappa B (NF-κB; which binds nucleotropic DNA and regulates the expression of inflammatory factors) [198], and phosphatidylinositol 3-kinase/protein kinase B (PI3K/Akt) [199].

The inflammation processes can contribute to aging diseases [200] and impact the pathophysiology of cancer, rheumatoid arthritis, atherosclerosis, asthma, ulcerative colitis, and type-2 diabetes [201]. Biopeptides' anti-inflammatory effect is affected by low molecular weights (less than 1 kDa ca. 500 Da, composed of 2–6 amino acids) (Table 4) [202] and depends on the amino acid composition (number, quality, and positions). Low molecular weight peptides can reach their target place intact (they have few cleavage sites for endopeptidase enzymes) and have specific transport modes. Peptide transport modes include PepT1 transporters, cytokinesis, cellular bypass, and passive diffusion; dipeptides and tripeptides can also be in the PepT1 category [195]. Regardless of amino acid composition, highly hydrophobic biopeptides containing leucine, tryptophan, and phenylalanine have anti-inflammatory potential. Leucine and isoleucine can act on PI3K (Akt kinases in the PI3K/Akt signaling pathway) and ERK kinases (in MAPK pathways) [203]. They can mitigate the damage caused by inflammatory factors by decreasing the kinases' phosphorylation and changing macrophages from M1 to M2 [204]. Moreover, highly hydrophobic biopeptides can avoid lipopolysaccharide (LPS)-stimulated inflammatory responses by forming peptide-lipopolysaccharide complexes and scavenging LPS through cell membrane charge exchange [205]. Positively charged amino acids (e.g., lysine, histidine, and arginine) positively influence the biopeptides' anti-inflammatory potential, affecting the inflammatory response linked to the activation of cascade pathways and improving their absorption in the intestine [206]. Lysine can regulate the kinase ERK's phosphorylation and the nuclear transcription factor NF-κB's translocation in the MAPK signaling pathways [207]. Arginine can reduce p38 and ERK kinases phosphorylation (in the MAPK pathways) and decreases the expression of TLR4 receptors and transcription factor p65's nuclear translocation (by constraining IκB kinase phosphorylation in the NF-κB pathway) [208]. The presence of glycine and glutamine also influences the anti-inflammatory activity of biopeptides. Glycine has a high affinity for calcium, interferes with $Ca^{2+}$ signaling [209], modulates the NF-κB signaling pathway [210], downregulates inflammatory factors (TNF-α, IL-1β, IL-8, and IL-6), and modulates the MAPK pathways (JNK, ERK, and p38) [211]. Finally, the anti-inflammatory properties of biopeptides are related to the amino acid positions. Hydrophobic amino acids situated at the peptide chain's N-terminus and charged amino acid C-terminal ends have a positive anti-inflammatory impact [195].

**Table 4.** Summary of recent studies on anti-inflammatory biopeptides.

| Peptides | Source | Activity | Biblio |
|---|---|---|---|
| LDAVNR (686 Da) and MMLDF (655 Da) [35] | Spirulina | IL-8 produced by endothelial cells EA.hy926 | [212] |
| FLWGKSY | Spent hen muscle | IL-6 | [200] |
| VLER, WVGK, VVRP, VLLF, VALVR, LFGK, FGPK | Millet bran | TNF-α, IL-1β, PGE2 | [213] |
| DQWL | Whey | IL-1β, COX-2, and TNF-α, and the secretion of IL-1β and TNF-α proteins in LPS-induced RAW 264.7 | [214] |
| YFVP, SGRDP, MVWGP, TGSYTEGWS | Sunflower | IL-1β | [215]. |
| KLRSRNLLHPT, TNGRHSAKKH | Bee pollen | COX-2, IL-6, iNOS, TNF-α | [216] |

## 5. Peptide Delivery Systems

Chemical, physical, and biological variability can degrade biopeptides, decrease storage life, and limit their application in different formulations. Chemical instability is due to oxidation reactions, deamination, etc. Physical instability is mainly produced by aggregation, denaturation, and surface adsorption. Biological instability is due to cell enzymes, which may cause degradation or inactivation of the active molecule and loss of biological activity [217]. Using nanocarriers can enhance the biopeptide's stability and limit side effects. Nanocarriers (e.g., liposomes, niosomes, novasomes, transferosomes, ethosomes, cubosomes, ultrasomes, photosomes, polymerosomes, nanofibres, metal nanoparticles, dendrimers, nanocrystals, carbon nanotubes, fullerene, cyclodextrin nanosponges, solid lipid nanoparticles), hydrogels, and nanoemulsions are carrier systems used for biopeptide delivery.

Liposomes are sphere-shaped vesicles with a hydrophilic core enclosed by at least one phospholipid bilayer. They can enter the skin by merging with the lipids of the stratum corneum or via the sebaceous glands [218]. The liposomes can be made with food-grade materials (biodegradable and non-toxic) and can encapsulate nonpolar, polar, and amphiphilic amino acids [219,220]. Mechanical methods (e.g., sonication, film formation, microfluidization, and extrusion), solvent replacement methods (reverse phase evaporation, injecting ethanol, and proliposome techniques), or detergent removal methods can be used to produce them [221]. The liposomes are used in lipsticks, antiperspirants, creams, deodorants, moisturizers, and hair care formulations. They are employed to improve the solubility of vitamins (e.g., A, E, and K), antioxidants (e.g., lycopene, coenzyme Q10, carotenoids, etc.), and other active biomolecules in water, facilitate the skin's hydration and restore the skin's epidermal layers by incorporating lipid compounds (e.g., cholesterols, and ceramides) [222]. They can deliver biopeptides in moisturizing, anti-aging creams, body sprays, deodorants, lotions, sunscreens, fragrances, shampoos, conditioning agents, etc. High production cost and osmotic stability limit their use in cosmetic products [223].

Niosomes contain one to seven bi-lipid layers, a non-ionic surfactant (spans, tweens, alkyl amides, brijs, polyoxyethylene alkyl ethers, and sorbitan esters), and an amorphous central core [224,225]. They are obtained by mixing free fatty acids, cholesterol, and a non-phospholipid surfactant.

Novasomes can deliver hydrophilic and hydrophobic molecules, have a lower production cost than liposomes [220], improve the biopeptides residence time on the dermal layers and skin penetration, decrease the horny layer barrier's resistance and the biopeptides' systemic absorption [226]. Novasomes have high molecule entrapment efficiency and a much lower production cost than liposomes. They have a little higher deposition volume on the skin than niosomes [227]. Moreover, they are stable at pH changes between 2 and 13 and temperatures between 0 °C and 100 °C.

Ethosomes are vesicles containing phospholipids with a high concentration of ethanol (20–50%) which improve the bioactive peptides' permeation across the skin, mediating the disruption of the skin's lipid layers. Ethasomes with niacinamide are used to decrease aging, pigmentation, skin blotches, and acne [228,229].

Transferosomes are deformable vesicles containing phospholipids and an edge activator (e.g., sodium chlorate, tween 80, and span 80). They can be used as curcumin, capsaicin, and resveratrol vehicles in transdermal skin layers [230], in antiwrinkle [231], and anti-aging cosmetics [232].

Cubosomes are self-assembling honeycomb-shaped liquid crystalline lipid nanoparticles (3D structures obtained from a bi-continuous cubic liquid phase with two aqueous channels divided by a surfactant bilayer) which can contain lipophilic, hydrophilic, and amphiphilic molecules [233,234]. They are used to absorb pollutants and as stabilizers for oil-in-water emulsions [230].

Ultrasomes are liposomes that contain a UV-endonuclease enzyme that repairs UV-damaged DNA and decreases the expression of pro-inflammatory cytokines [235].

Photosomes are liposomal formulations of photolyase. They are incorporated in sunscreen products [236,237].

Polymersomes are artificial vesicular systems containing block copolymers encapsulating lipophilic and/or lipophobic molecules. They have higher stability than liposomes because of their thick and rigid bilayer structure [238,239]. They enhance skin elasticity and increase the skin cells' activation energy [240].

Biopolymer microgels are small particles comprising a cross-linked polymer molecule network [241]. They can contain natural, synthetic, or bio-polymers (e.g., chitosan, hyaluronic acid, collagen, gelatin, and polyvinyl alcohol), polyacrylamide, xanthan gum, polyethylene glycol, pectin, starch, cellulose, alginate). They can be obtained by coacervation, antisolvent precipitation, and emulsion. Unfortunately, porous microgels can diffuse small peptides. Biopolymer hydrogels are used to produce "beauty masks" [242,243].

Solid lipid particles (SLN) are a colloidal delivery system formed by crystallized lipid particles in an aqueous medium [244]. SLNs are used in cosmetic creams, lotions, and sunscreens [243].

Nanostructured lipid carriers (NLCs) are a mixture of solid and liquid lipids with a less ordered structure that load more active molecules than SLN into their pockets. NLCs are suitable carriers for volatile essential oils [245].

Nanofibers are one-dimensional nanomaterials (e.g., collagen, silk, PVP, and PVA) having a high surface area to volume ratio, high bioactive loading capacity, small diameters, and excellent absorbing capacity. They can be used for production of cleansers, face masks, and skin healing products [246].

Inorganic nanocosmetics are nanoparticles containing metals (e.g., gold, silver, aluminum, platinum, titanium) or metalloids (e.g., silica and selenium). Among metal-based nanoparticles, gold and silver are the most used. Gold has high stability and penetrability, is inert, and is non-cytotoxic. Gold nanoparticles have antioxidant and anti-aging effects, enhance skin elasticity, skin firmness, and blood circulation, and have antibacterial, antifungal, and antiseptic properties [247].

Silver has antimicrobial properties against many microbial species and is an anti-inflammatory agent. Silver nanoparticles (AgNPs) are used in lotions, skin cleansers, creams, shampoos, deodorants, and toothpaste [248].

$ZnO_2$ and $TiO_2$ nanoparticles are used mainly in sunscreen for UV-A and UV-B filters [249,250].

Inorganic metalloid silica and selenium are the most used in the cosmetic field. Silica has a feel-good texture and excellent penetrability and can enclose hydrophilic and hydrophobic molecules. Silicone-based vesicles are used to deliver vitamins A, C, and E and oils such as jojoba and lanolin, in emollients and creams [251,252].

Silica nanoparticles are employed in lipsticks to homogenize lipstick pigments, in anti-aging/anti-wrinkle creams, and in hair and nail cosmetic products. They can improve cosmetic products' texture, effectiveness, and shelf-life and act as an anti-caking agent. Moreover, they have high photostability and protect against UV radiation [253].

Dendrimers are macromolecular organic nanocarriers with a network of symmetric branches (the number of branches required determines the production process) arising from a central core, with functional groups attached at their terminal ends [223]. Polyvalence, solubility, monodispersity, low cytotoxicity, self-assembling, chemical stability, and electrostatic interactions are key factors responsible for their high selectivity and precision in the biopeptides' delivery [254]. Biodegradable polymers (e.g., polysaccharides, poly α-esters, poly alkyl cyanoacrylates, and poly amidoamine dendrimers) are used in cosmetic formulations to benefit hair (e.g., hair-styling gels and shampoos), skin (e.g., anti-acne cream) and nails (e.g., nail polishes), and as sunscreens. Dendrimers were developed to improve resveratrol and vitamins A and B6 (PAMAM dendrimer) solubility and skin infiltration [221] and give a glossy appearance to the skin and hair (carbosiloxane dendrimer able to resist oil and water) [255].

Nanocrystals are clusters of thousands of active agents linked together in a fixed pattern to form a group (sizes ranging from 10 to 400 nm) having a very high surface area to volume ratio and high solubility and bioavailability. They facilitate biopeptide absorption into the skin by creating a high biological adhesion and concentration gradient on the skin surface for long periods. They are usually utilized to administer poorly soluble active compounds [256]. Undissolved nanocrystals can aggregate in hair follicles to produce an active molecules reservoir in addition to intracellular and intercellular pathways [257–259].

Fullerenes (or buckyballs) are spherical structures with many carbon atoms [260]. They can deliver biopeptides in cosmetics (e.g., anti-wrinkle, anti-acne, lightening toner, pore reduction, and moisturizing creams) and sunscreen [261,262].

Cyclodextrin nanosponges are natural oligosaccharides (containing 6–8 glucopyranose molecules) with a truncated cone-shaped structure [263]. Cyclodextrin's lipophilic cavity can encapsulate aromatic molecules, aliphatic hydrocarbons, and vitamins [264]. They are used in perfumes, tanning products, deodorants, laundry detergents, odor removers, underarm odor shields, etc. [265].

Microemulsions (diameter 10 to 100 nm) are classified as water-in-oil (W/O) and oil-in-water (O/W) based on the predominant system's components. The W/O microemulsions are thermodynamically stable, have noninvasive administration, high solubilization capacity, and are easily formulated but require high concentrations of surfactants to stabilize them [244] and can be only employed in oral formulations that contain mainly oil (e.g., oil-filled soft capsules). Water-dispersible forms can be formulated by homogenizing the W/O microemulsion with water and a hydrophilic emulsifier to form a W/O/W type system. Mortazavi et al. used W/O microemulsion to encapsulate PKEK, a tetrapeptide that can decrease the pigmentation process [266].

The O/W microemulsion can encapsulate hydrophobic biopeptides mixed with a hydrophobic surfactant and a co-surfactant [267,268].

Water-in-oil-in-water (W/O/W) systems are used to encapsulate the water-soluble peptides. They are multicompartment liquid dispersions where the dispersed phase is an emulsion [269]. The double emulsion can mask flavor and odor and regulate bioactive ingredients released during digestion. The type of oil used significantly affects the formation and structure of multiple emulsions and the skin barrier function [270]. Their use is limited by instability [271]. The $W_1/O/W_2$ double emulsion system is a helpful delivery matrix for hydrophilic biopeptides, as shown by Ying et al. [272], who prepared applications of $W_1/O/W_2$ double emulsions containing soy peptides by a two-step emulsification process and Giroux et al. [273] who encapsulated β-lactoglobulin hydrolysate using a $W_1/O/W_2$ emulsion system, obtaining a peptides' release inversely correlated to the oil's viscosity and peptides' hydrophobicity.

## 6. Conclusions

Foods and food waste are promising sources of biopeptides for the nutricosmetic industry. Using food waste for the production of biopeptides may contribute to sustainable development and represent economic advantages. Conversion of highly abundant, inexpensive and renewable biomass to obtain biopeptides for nutricosmetic formulation is dependent on purification processes and "tailor-made" manipulation of the precursor structures. This requires expertise in analytical procedures and good manufacturing practice to ensure the population's safety. Therefore, before thinking of recovering biopeptides on a large scale from food waste, technologies are required that can produce peptides industrially as well as regulated analytical tests to ensure consumer safety.

Nanotechnology is becoming a crucial tool for developing new cosmetic and personal care products including biopeptides in their industrial formulation. Inspection of the extensive literature shows that nanomaterials that can deliver biopeptides differ in physical and chemical properties, biocompatibility, stability, site-specificity, and biopeptide-loading capability. Additional research on biopeptide applications in final products is needed to understand their potential risks and consumer acceptance.

**Author Contributions:** Methodology, investigation, resources, writing—original draft preparation, I.D.; methodology, investigation, resources, writing—original draft preparation, A.M. All authors have read and agreed to the published version of the manuscript.

**Funding:** This research received no external funding.

**Institutional Review Board Statement:** Not applicable.

**Informed Consent Statement:** Not applicable.

**Data Availability Statement:** Not applicable.

**Conflicts of Interest:** The authors declare no conflict of interest.

## References

1. Cosmetic Ingredient Review. Safety Assessment of Soy Proteins and Peptides as Used in Cosmetics. Available online: https://www.cir-safety.org/sites/default/files/soypep092015final.pdf (accessed on 30 April 2020).
2. Ferreira, M.S.; Magalhães, M.C.; Sousa-Lobo, J.M.; Almeida, I.F. Trending Anti-Aging Peptides. *Cosmetics* **2020**, *7*, 91. [CrossRef]

3. Tkaczewska, J. Peptides and protein hydrolysates as food preservatives and bioactive components of edible films and coatings—A review. *Trends Food Sci. Technol.* **2020**, *106*, 298–311. [CrossRef]
4. Bioactive Peptides Market Analysis. Available online: https://www.coherentmarketinsights.com/market-insight/bioactive-peptide-market-3018 (accessed on 1 September 2022).
5. European Commission. Commission Decision (EU) 2022/677 of 29 April 2022 establishing a glossary of common ingredient names for use in the labelling of cosmetic products. *Off. J. Eur. Union* **2022**, *127*, 1–448.
6. European Commission. Commission Decision (EU) 2019/701 of 5 April 2019 establishing a glossary of common ingredient names for use in the labelling of cosmetic products. *Off. J. Eur. Union* **2019**, *121*, 1–370.
7. Wen, C.; Zhang, J.; Zhang, H.; Duan, Y.; Ma, H. Plant protein-derived antioxidant peptides: Isolation, identification, mechanism of action and application in food systems: A review. *Trends Food Sci. Technol.* **2020**, *105*, 308–322. [CrossRef]
8. Zaky, A.A.; El-Aty, A.A.; Ma, A.; Jia, Y. An overview on antioxidant peptides from rice bran proteins: Extraction, identification, and applications. *Crit. Rev. Food Sci. Nutr.* **2020**, *62*, 1350–1362. [CrossRef]
9. Pal, G.K.; Suresh, P. Sustainable valorisation of seafood by-products: Recovery of collagen and development of collagen-based novel functional food ingredients. *Innov. Food Sci. Emerg. Technol.* **2016**, *37*, 201–215. [CrossRef]
10. Tadesse, S.A.; Emire, S.A. Production and processing of antioxidant bioactive peptides: A driving force for the functional food market. *Heliyon* **2020**, *6*, e04765. [CrossRef]
11. Villamil, O.; Váquiro, H.; Solanilla, J.F. Fish viscera protein hydrolysates: Production, potential applications and functional and bioactive properties. *Food Chem.* **2017**, *224*, 160–171. [CrossRef]
12. Cermeño, M.; Kleekayai, T.; Amigo-Benavent, M.; Harnedy-Rothwell, P.; FitzGerald, R.J. Current knowledge on the extraction, purification, identification, and validation of bioactive peptides from seaweed. *Electrophoresis* **2020**, *41*, 1694–1717. [CrossRef] [PubMed]
13. Alizadeh, O.; Aliakbarlu, J. Effects of ultrasound and ohmic heating pretreatments on hydrolysis, antioxidant and antibacterial activities of whey protein concentrate and its fractions. *Food Sci. Technol.* **2020**, *131*, 109913. [CrossRef]
14. Ulug, S.K.; Jahandideh, F.; Wu, J. Novel technologies for the production of bioactive peptides. *Trends Food Sci. Technol.* **2021**, *108*, 27–39. [CrossRef]
15. Garcia-Vaquero, M.; Ummat, V.; Tiwari, B.; Rajauria, G. Exploring Ultrasound, Microwave and Ultrasound–Microwave Assisted Extraction Technologies to Increase the Extraction of Bioactive Compounds and Antioxidants from Brown Macroalgae. *Mar. Drugs* **2020**, *18*, 172. [CrossRef] [PubMed]
16. Uluko, H.; Zhang, S.; Liu, L.; Tsakama, M.; Lu, J.; Lv, J. Effects of thermal, microwave, and ultrasound pretreatments on antioxidative capacity of enzymatic milk protein concentrate hydrolysates. *J. Funct. Foods* **2015**, *18*, 1138–1146. [CrossRef]
17. Naderi, N.; House, J.D.; Pouliot, Y.; Doyen, A. Effects of high hydrostatic pressure processing on hen egg compounds and egg products. *Compr. Rev. Food Sci. Food Saf.* **2017**, *16*, 707–720. [CrossRef]
18. Guan, H.; Diao, X.; Jiang, F.; Han, J.; Kong, B. The enzymatic hydrolysis of soy protein isolate by Corolase PP under high hydrostatic pressure and its effect on bioactivity and characteristics of hydrolysates. *Food Chem.* **2018**, *245*, 89–96. [CrossRef]
19. Saleh, A.S.; Wang, P.; Wang, N.; Yang, S.; Xiao, Z. Technologies for enhancement of bioactive components and potential health benefits of cereal and cereal-based foods: Research advances and application challenges. *Crit. Rev. Food Sci. Nutr.* **2019**, *59*, 207–227. [CrossRef]
20. Daliri, E.B.-M.; Oh, D.H.; Lee, B.H. Bioactive peptides. *Foods* **2017**, *6*, 32. [CrossRef]
21. Najafian, L.; Babji, A.S. Purification and identification of antioxidant peptides from fermented fish sauce (Budu). *J. Aquat. Food Prod. Technol.* **2019**, *28*, 14–24. [CrossRef]
22. Nong, N.T.P.; Hsu, J.-L. Characteristics of Food Protein-Derived Antidiabetic Bioactive Peptides: A Literature Update. *Int. J. Mol. Sci.* **2021**, *22*, 9508. [CrossRef]
23. Nong, N.T.P.; Chen, Y.-K.; Shih, W.-L.; Hsu, J.-L. Characterization of Novel Dipeptidyl Peptidase-IV Inhibitory Peptides from Soft-Shelled Turtle Yolk Hydrolysate Using Orthogonal Bioassay-Guided Fractionations Coupled with In Vitro and In Silico Study. *Pharmaceuticals* **2020**, *13*, 308. [CrossRef] [PubMed]
24. Cavaliere, C.; Montone, A.M.I.; Aita, S.E.; Capparelli, R.; Cerrato, A.; Cuomo, P.; Laganà, A.; Montone, C.M.; Piovesana, S.; Capriotti, A.L. Production and characterization of medium-sized and short antioxidant peptides from soy flour-simulated gastrointestinal hydrolysate. *Antioxidants* **2021**, *10*, 734. [CrossRef] [PubMed]
25. Idowu, A.T.; Benjakul, S. Bitterness of fish protein hydrolysate and its debittering prospects. *J. Food Biochem.* **2019**, *43*, e12978. [CrossRef]
26. Venkatesan, J.; Anil, S.; Kim, S.K.; Shim, M.S. Marine Fish Proteins and Peptides for Cosmeceuticals: A Review. *Mar. Drugs* **2017**, *15*, 143. [CrossRef] [PubMed]
27. Zhang, N.; Zhang, C.; Chen, Y.; Zheng, B. Purification and Characterization of Antioxidant Peptides of *Pseudosciaena crocea* Protein Hydrolysates. *Molecules* **2017**, *22*, 57. [CrossRef]
28. Szymczak, M.; Felisiak, K.; Szymczak, B. Characteristics of herring marinated in reused brines after microfiltration. *J. Food Sci. Technol.* **2018**, *55*, 4395–4440. [CrossRef]
29. Saidi, S.; Saoudi, M.; Amar, R. Valorisation of tuna processing waste biomass: Isolation, purification and characterisation of four novel antioxidant peptides from tuna by-product hydrolysate. *Environ. Sci. Pollut. Res. Int.* **2018**, *25*, 17383–17392. [CrossRef]

30. Shih, Y.-H.; Chen, F.-A.; Wang, L.-F.; Hsu, J.-L. Discovery and study of novel antihypertensive peptides derived from *Cassia obtusifolia* seeds. *J. Agric. Food Chem.* **2019**, *67*, 7810–7820. [CrossRef]
31. Yu, X.; Lim, C.Y.X.; Dong, B.; Hadinoto, K. Development of magnetic solid phase extraction platform for the purification of bioactive γ-glutamyl peptides from garlic (*Allium sativum*). *LWT* **2020**, *127*, 109410. [CrossRef]
32. Hara, T.; Huang, Y.; Ito, A.; Kawakami, T.; Hojo, H.; Murata, M. Trifluoroethanol-containing RP-HPLC mobile phases for the separation of transmembrane peptides human glycophorin-A, integrin alpha-1, and p24: Analysis and prevention of potential side reactions due to formic acid. *J. Pept. Sci.* **2015**, *21*, 61–70. [CrossRef]
33. Maux, S.L.; Nongonierma, A.B.; Murray, B.; Kelly, P.M.; FitzGerald, R.J. Identification of short peptide sequences in the nanofiltration permeate of a bioactive whey protein hydrolysate. *Food Res. Int.* **2015**, *77*, 534–539. [CrossRef]
34. Karami, Z.; Peighambardoust, S.H.; Hesari, J.; Akbari-Adergani, B.; Andreu, D. Antioxidant, anticancer and ACE-inhibitory activities of bioactive peptides from wheat germ protein hydrolysates. *Food Biosci.* **2019**, *32*, 100450. [CrossRef]
35. Alu'datt, M.H.; Al-u'datt, D.a.G.F.; Alhamad, M.N.; Tranchant, C.C.; Rababah, T.; Gammoh, S.; Althnaibat, R.M.; Daradkeh, M.G.; Kubow, S. Characterization and biological properties of peptides isolated from dried fermented cow milk products by RP-HPLC: Amino acid composition, antioxidant, antihypertensive, and antidiabetic properties. *J. Food Sci.* **2021**, *86*, 3046–3060. [CrossRef]
36. Sompinit, K.; Lersiripong, S.; Reamtong, O.; Pattarayingsakul, W.; Patikarnmonthon, N.; Panbangred, W. In vitro study on novel bioactive peptides with antioxidant and antihypertensive properties from edible rhizomes. *LWT* **2020**, *134*, 110227. [CrossRef]
37. Boersema, P.J.; Mohammed, S.; Heck, A.J.R. Hydrophilic interaction liquid chromatography (HILIC) in proteomics. *Anal. Bioanal. Chem.* **2008**, *391*, 151–159. [CrossRef]
38. Giacometti, J.; Buretić-Tomljanović, A. Peptidomics as a tool for characterizing bioactive milk peptides. *Food Chem.* **2017**, *230*, 91–98. [CrossRef]
39. Kaprasob, R.; Khongdetch, J.; Laohakunjit, N.; Selamassakul, O.; Kaisangsri, N. Isolation and characterization, antioxidant, and antihypertensive activity of novel bioactive peptides derived from hydrolysis of King Boletus mushroom. *LWT* **2022**, *160*, 113287. [CrossRef]
40. Safari, R.; Yaghoubzadeh, Z. Antioxidant activity of bioactive peptides extracted from sea cucumber (*Holothuria leucospilata*). *Int. J. Pept. Res. Ther.* **2020**, *26*, 2393–2398. [CrossRef]
41. Mirzapour-Kouhdasht, A.; Garcia-Vaquero, M.; Eun, J.-B.; Simal-Gandara, J. Influence of Enzymatic Hydrolysis and Molecular Weight Fractionation on the Antioxidant and Lipase/α-Amylase Inhibitory Activities In Vitro of Watermelon Seed Protein Hydrolysates. *Molecules* **2022**, *27*, 7897. [CrossRef] [PubMed]
42. Islam, M.S.; Wang, H.; Admassu, H.; Sulieman, A.A.; Wei, F.A. Health benefits of bioactive peptides produced from muscle proteins: Antioxidant, anti-cancer, and anti-diabetic activities. *Process Biochem.* **2022**, *116*, 116–125. [CrossRef]
43. Olivera-Montenegro, L.; Bugarin, A.; Marzano, A.; Best, I.; Zabot, G.L.; Romero, H. Production of Protein Hydrolysate from Quinoa (*Chenopodium quinoa* Willd.): Economic and Experimental Evaluation of Two Pretreatments Using Supercritical Fluids' Extraction and Conventional Solvent Extraction. *Foods* **2022**, *11*, 1015. [CrossRef] [PubMed]
44. Ray, A.; Dubey, K.K.; Marathe, S.J.; Singhal, R. Supercritical fluid extraction of bioactives from fruit waste and its therapeutic potential. *Food Biosci.* **2023**, *52*, 102418. [CrossRef]
45. Sun, J.; He, H.; Xie, B.J. Novel Antioxidant Peptides from Fermented Mushroom *Ganoderma lucidum*. *J. Agric. Food Chem.* **2004**, *52*, 6646–6652. [CrossRef]
46. Lamoolphak, W.; Goto, M.; Sasaki, M.; Suphantharika, M.; Muangnapoh, C.; Prommuag, C.; Shotipruk, A. Hydrothermal decomposition of yeast cells for production of proteins and amino acids. *J. Hazard. Mater.* **2006**, *137*, 1643–1648. [CrossRef]
47. Rivas-Vela, C.I.; Amaya-Llano, S.L.; Castaño-Tostado, E.; Castillo-Herrera, G.A. Protein Hydrolysis by Subcritical Water: A New Perspective on Obtaining Bioactive Peptides. *Molecules* **2021**, *26*, 6655. [CrossRef] [PubMed]
48. Abdelmoez, W.; Yoshida, H. Production of amino and organic acids from protein using sub-critical water technology. *Int. J. Chem. React. Eng.* **2013**, *11*, 369–384. [CrossRef]
49. Rodiles-Lopez, J.O.; Arroyo-Maya, I.J.; Jaramillo-Flores, M.E.; Gutierrez-Lopez, G.F.; Hernandez-Arana, A.; Barbosa-Canovas, G.V.; Niranjan, K.; Hernandez-Sanchez, H. Effects of high hydrostatic pressure on the structure of bovine α-lactalbumin. *J. Dairy Sci.* **2010**, *93*, 1420–1428. [CrossRef]
50. Rogalinski, T.; Liu, K.; Albrecht, T.; Brunner, G. Hydrolysis kinetics of biopolymers in subcritical water. *J. Supercrit. Fluids* **2008**, *46*, 335–341. [CrossRef]
51. Di Domenico Ziero, H.; Buller, L.S.; Mudhoo, A.; Ampese, L.C.; Mussatto, S.I.; Carneiro, T.F. An overview of subcritical and supercritical water treatment of different biomasses for protein and amino acids production and recovery. *J. Environ. Chem. Eng.* **2020**, *8*, 104406. [CrossRef]
52. Tabilo-Munizaga, G.; Gordon, T.A.; Villalobos-Carvajal, R.; Moreno-Osorio, L.; Salazar, F.N.; Pérez-Won, M.; Acuña, S. Effects of high hydrostatic pressure (HHP) on the protein structure and thermal stability of Sauvignon blanc wine. *Food Chem.* **2014**, *155*, 214–220. [CrossRef]
53. Brunner, G. Near critical and supercritical water. Part I. Hydrolytic and hydrothermal processes. *J. Supercrit. Fluids* **2009**, *47*, 373–381. [CrossRef]
54. Koh, B.B.; Lee, E.J.; Ramachandraiah, K.; Hong, G.P. Characterization of bovine serum albumin hydrolysates prepared by subcritical water processing. *Food Chem.* **2019**, *278*, 203–207. [CrossRef] [PubMed]

55. Melgosa, R.; Trigueros, E.; Sanz, M.T.; Cardeira, M.; Rodrigues, L.; Fernández, N.; Matias, A.A.; Bronze, M.R.; Marques, M.; Paiva, A.; et al. Supercritical $CO_2$ and subcritical water technologies for the production of bioactive extracts from sardine (*Sardina pilchardus*) waste. *J. Supercrit. Fluids* **2020**, *164*, 104943. [CrossRef]
56. Cho, Y.J.; Getachew, A.T.; Park, J.S.; Lim, C.T.; Lee, H.J.; Chun, B.S. Influence of temperature on decomposition reaction of compressed hot water to valorize *Achatina fulica* as a functional material. *Food Bioprod. Process* **2020**, *122*, 89–97. [CrossRef]
57. Sereewatthanawut, I.; Prapintip, S.; Watchiraruji, K.; Goto, M.; Sasaki, M.; Shotipruk, A. Extraction of protein and amino acids from deoiled rice bran by subcritical water hydrolysis. *Bioresour. Technol.* **2008**, *99*, 555–561. [CrossRef]
58. Watchararuji, K.; Goto, M.; Sasaki, M.; Shotipruk, A. Value-added subcritical water hydrolysate from rice bran and soybean meal. *Bioresour. Technol.* **2008**, *99*, 6207–6213. [CrossRef]
59. Enteshari, M.; Martínez-Monteagudo, S.I. Subcritical hydrolysis of ice-cream wastewater: Modeling and functional properties of hydrolysate. *Food Bioprod. Process* **2018**, *111*, 104–113. [CrossRef]
60. Park, J.S.; Jeong, Y.R.; Chun, B.S. Physiological activities and bioactive compound from laver (*Pyropia yezoensis*) hydrolysates by using subcritical water hydrolysis. *J. Supercrit. Fluids* **2019**, *148*, 130–136. [CrossRef]
61. Espinoza, A.D.; Morawicki, R.O.; Hager, T. Hydrolysis of whey protein isolate using subcritical water. *J. Food Sci.* **2012**, *77*, 20–26. [CrossRef]
62. Mikhaylin, V.; Nikonenko, G.; Pourcelly, G.; Bazinet, L. Hybrid bipolar membrane electrodialysis/ultrafiltration technology assisted by a pulsed electric field for casein production. *Green Chem.* **2016**, *18*, 307. [CrossRef]
63. Siow, H.-L.; Gan, C.-Y. Extraction of antioxidative and antihypertensive bioactive peptides from *Parkia speciosa* seeds. *Food Chem.* **2013**, *141*, 3435–3442. [CrossRef]
64. Agyei, D.; Pan, S.; Acquah, C.; Bekhit, A.E.D.A.; Danquah, M.K. Structure-informed detection, and quantification of peptides in food and biological fluids. *J. Food Biochem.* **2019**, *43*, e12482. [CrossRef]
65. Kašička, V. Recent developments in capillary and microchip electroseparations of peptides (2017–mid 2019). *Electrophoresis* **2020**, *41*, 10–35. [CrossRef] [PubMed]
66. Kašička, V. Recent developments in capillary and microchip electroseparations of peptides (2019–mid 2021). *Electrophoresis* **2022**, *43*, 82–108. [CrossRef] [PubMed]
67. Lu, Z.; Sun, N.; Dong, L.; Gao, Y.; Lin, S. Production of Bioactive Peptides from Sea Cucumber and Its Potential Health Benefits: A Comprehensive Review. *J. Agric. Food Chem.* **2022**, *70*, 7607–7625. [CrossRef]
68. Ho, C.S.; Lam, C.; Chan, M.; Cheung, R.; Law, L.; Lit, L.; Ng, K.; Suen, M.; Tai, H. Electrospray ionisation mass spectrometry: Principles and clinical applications. *Clin. Biochem. Rev.* **2003**, *24*, 3. [PubMed]
69. Worsztynowicz, P.; Białas, W.; Grajek, W. Integrated approach for obtaining bioactive peptides from whey proteins hydrolysed using a new proteolytic lactic acid bacteria. *Food Chem.* **2020**, *312*, 126035. [CrossRef]
70. Zanoni, C.; Aiello, G.; Arnoldi, A.; Lammi, C. Hempseed Peptides Exert Hypocholesterolemic Effects with a Statin-Like Mechanism. *J. Agric. Food Chem.* **2017**, *65*, 8829–8838. [CrossRef] [PubMed]
71. Soboleva, A.; Schmidt, R.; Vikhnina, M.; Grishina, T.; Frolov, A. Maillard Proteomics: Opening new pages. *Int. J. Mol. Sci.* **2017**, *18*, 2677. [CrossRef] [PubMed]
72. Ayala-Niño, A.; Rodríguez-Serrano, G.M.; González-Olivares, L.G.; Contreras-López, E.; Regal-López, P.; Cepeda-Saez, A. Sequence Identification of bioactive peptides from amaranth seed proteins (*Amaranthus hypochondriacus* spp.). *Molecules* **2019**, *24*, 3033. [CrossRef]
73. Smolikova, G.; Gorbach, D.; Lukasheva, E.; Mavropolo-Stolyarenko, G.; Bilova, T.; Soboleva, A.; Tsarev, A.; Romanovskaya, E.; Podolskaya, E.; Zhukov, V.; et al. Bringing New Methods to the Seed Proteomics Platform: Challenges and Perspectives. *Int. J. Mol. Sci.* **2020**, *21*, 9162. [CrossRef] [PubMed]
74. Cakir, B.; Gulseren, I. Identification of Novel Proteins from Black Cumin Seed Meals Based on 2D Gel Electrophoresis and MALDI-TOF/TOF-MS Analysis. *Plant Foods Hum. Nutr.* **2019**, *74*, 414–420. [CrossRef] [PubMed]
75. Lafarga, T.; Acién-Fernándeza, F.G.; Garcia-Vaquero, M. Bioactive peptides and carbohydrates from seaweed for food applications: Natural occurrence, isolation, purification, and identification. *Algal. Res.* **2020**, *48*, 101909. [CrossRef]
76. Youssef, F.S.; Ashour, M.L.; Singab, A.N.B.; Wink, M. A Comprehensive Review of Bioactive Peptides from Marine Fungi and Their Biological Significance. *Mar. Drugs* **2019**, *17*, 559. [CrossRef] [PubMed]
77. Barbieri, J.S.; Wanat, K.; Seykora, J. *Skin: Basic Structure and Function*; Elsevier: Amsterdam, The Netherlands, 2014; ISBN 9780123864567.
78. Bonté, F.; Girard, D.; Archambault, J.C.; Desmoulière, A. Skin Changes during Ageing. *Subcell Biochem.* **2019**, *91*, 249–280.
79. Jensen, J.-M.; Förl, M.; Winoto-Morbach, S.; Seite, S.; Schunck, M.; Proksch, E.; Schütze, S. Acid and neutral sphingomyelinase, ceramide synthase, and acid ceramidase activities in cutaneous aging. *Exp. Dermatol.* **2005**, *14*, 609–618. [CrossRef]
80. Kaya, G.; Tran, C.; Sorg, O.; Hotz, R.; Grand, D.; Carraux, P.; Didierjean, L.; Stamenkovic, I.; Saurat, J.-H. Hyaluronate Fragments Reverse Skin Atrophy by a CD44-Dependent Mechanism. *PLoS Med.* **2006**, *3*, e493. [CrossRef]
81. Makrantonaki, E.; Zouboulis, C.C.; William, J. Cunliffe Scientific Awards. Characteristics and pathomechanisms of endogenously aged skin. *Dermatology* **2007**, *214*, 352–360. [CrossRef]
82. López-Otín, C.; Blasco, M.A.; Partridge, L.; Serrano, M.; Kroemer, G. The hallmarks of aging. *Cell* **2013**, *153*, 1194–1217. [CrossRef]
83. Lovell, C.R.; Smolenski, K.A.; Duance, V.C.; Light, N.D.; Young, S.; Dyson, M. Type I and III collagen content and fibre distribution in normal human skin during ageing. *Br. J. Dermatol.* **1987**, *117*, 419–428. [CrossRef]

84. Autio, P.; Risteli, J.; Haukipuro, K.; Risteli, L.; Oikarinen, A. Collagen synthesis in human skin in vivo: Modulation by aging, ultraviolet B irradiation and localization. *Photodermatol. Photoimmunol. Photomed.* **1994**, *10*, 212–216. [PubMed]
85. Csekes, E.; Račková, L. Skin Aging, Cellular Senescence and Natural Polyphenols. *Int. J. Mol. Sci.* **2021**, *22*, 12641. [CrossRef] [PubMed]
86. Wang, A.S.; Dreesen, O. Biomarkers of cellular senescence and skin aging. *Front. Genet.* **2018**, *9*, 247. [CrossRef] [PubMed]
87. Krutmann, J.; Bouloc, A.; Sore, G.; Bernard, B.A.; Passeron, T. The skin aging exposome. *J. Dermatol. Sci.* **2017**, *85*, 152–161. [CrossRef]
88. Gu, Y.; Han, J.; Jiang, C.; Zhang, Y. Biomarkers, oxidative stress and autophagy in skin aging. *Ageing Res. Rev.* **2020**, *59*, 101036. [CrossRef]
89. Dąbrowska, A.K.; Spano, F.; Derler, S.; Adlhart, C.; Spencer, N.D.; Rossi, R.M. The relationship between skin function, barrier properties, and body-dependent factors. *Ski. Res. Technol.* **2018**, *24*, 165–174. [CrossRef]
90. Law, M.H.; Medland, S.E.; Zhu, G.; Yazar, S.; Viñuela, A.; Wallace, L.; Shekar, S.N.; Duffy, D.L.; Bataille, V.; Glass, D.; et al. Genome-wide association shows that pigmentation genes play a role in skin aging. *J. Investig. Dermatol.* **2017**, *137*, 1887–1894. [CrossRef]
91. Krutmann, J.; Schikowski, T.; Morita, A.; Berneburg, M. Environmentally-Induced (Extrinsic) Skin Aging: Exposomal Factors and Underlying Mechanisms. *J. Investig. Dermatol.* **2021**, *141*, 1096–1103. [CrossRef]
92. Blume-Peytavi, U.; Kottner, J.; Sterry, W.; Hodin, M.W.; Griffiths, T.W.; Watson, R.E.B.; Hay, R.J.; Griffiths, C.E.M. Age-Associated Skin Conditions and Diseases: Current Perspectives and Future Options. *Gerontologist* **2016**, *56*, S230–S242. [CrossRef]
93. Kurban, R.S.; Kurban, A.K. Common skin disorders of aging: Diagnosis and treatment. *Geriatrics* **1993**, *48*, 30–31, 35–36, 39–42.
94. Aguilar-Toalá, J.; Santiago-López, L.; Peres, C.; Garcia, H.; Vallejo-Cordoba, B.; González-Córdova, A.; Hernández-Mendoza, A. Assessment of multifunctional activity of bioactive peptides derived from fermented milk by specific *Lactobacillus plantarum* strains. *J. Dairy Sci.* **2017**, *100*, 65–75. [CrossRef] [PubMed]
95. Amigo, L.; Hernández-Ledesma, B. Current Evidence on the Bioavailability of Food Bioactive Peptides. *Molecules* **2020**, *25*, 4479. [CrossRef] [PubMed]
96. Sedighi, M.; Shrestha, N.; Mahmoudi, Z.; Khademi, Z.; Ghasempour, A.; Dehghan, H.; Talebi, S.F.; Toolabi, M.; Préat, V.; Chen, B.; et al. Multifunctional Self-Assembled Peptide Hydrogels for Biomedical Applications. *Polymers* **2023**, *15*, 1160. [CrossRef] [PubMed]
97. Udenigwe, C.C.; Fogliano, V. Food matrix interaction and bioavailability of bioactive peptides: Two faces of the same coin? *J. Funct. Foods* **2017**, *35*, 9–12. [CrossRef]
98. Li-Chan, E.C.Y. Bioactive peptides and protein hydrolysates may pass into the systemic circulation: Research trends and challenges for application as nutraceuticals and functional food ingredients. *Curr. Opin. Food Sci.* **2015**, *1*, 28–37. [CrossRef]
99. Ying, X.; Agyei, D.; Udenigwe, C.; Adhikari, B.; Wang, B. Manufacturing of Plant-Based Bioactive Peptides Using Enzymatic Methods to Meet Health and Sustainability Targets of the Sustainable Development Goals. *Front. Sustain. Food Syst.* **2021**, *5*, 769028. [CrossRef]
100. Manzanares, P.; Gandía, M.; Garrigues, S.; Marcos, J.F. Improving Health-Promoting Effects of Food-Derived Bioactive Peptides through Rational Design and Oral Delivery Strategies. *Nutrients* **2019**, *11*, 2545. [CrossRef]
101. McClements, D.J. Encapsulation, protection, and delivery of bioactive proteins and peptides using nanoparticle and microparticle systems: A review. *Adv. Colloid Interfac. Sci.* **2018**, *253*, 1–22. [CrossRef]
102. Patil, P.J.; Usman, M.; Zhang, C.N.; Mehmood, A.; Zhou, M.C.; Teng, C.; Li, X.T. An updated review on food-derived bioactive peptides: Focus on the regulatory requirements, safety, and bioavailability. *Compr. Rev. Food Sci. Food Saf.* **2022**, *21*, 1732–1776. [CrossRef]
103. Apone, F.; Barbulova, A.; Colucci, M.G. Plant and Microalgae Derived Peptides Are Advantageously Employed as Bioactive Compounds in Cosmetics. *Front. Plant Sci.* **2019**, *10*, 756. [CrossRef]
104. Shin, J.-W.; Kwon, S.-H.; Choi, J.-Y.; Na, J.-I.; Huh, C.-H.; Choi, H.-R.; Park, K.-C. Molecular Mechanisms of Dermal Aging and Antiaging Approaches. *Int. J. Mol. Sci.* **2019**, *20*, 2126. [CrossRef] [PubMed]
105. Castro-Jacome, T.P.; Alcantara-Quintana, L.E.; Montalvo-Gonzalez, E.; Chacon-Lopez, A.; Kalixto-Sanchez, M.A.; Rivera, M.D.; Lopez-Garcia, U.M.; Tovar-Perez, E.G. Skin-protective properties of peptide extracts produced from white sorghum grain kafirins. *Ind. Crops Prod.* **2021**, *167*, 113551. [CrossRef]
106. Zhmak, M.N.; Utkin, Y.N.; Andreeva, T.V.; Kudryavtsev, D.S.; Kryudova, E.V.; Tsetlin, V.I.; Shelukhina, I.V.E. Peptide Inhibitors of Nicotinic Acetylcholine Receptor. US Patent US 20,150,361,137 A1, 17 December 2015.
107. Gruber, J.V.; Smith, D.; Bouldin, L. Personal Care Composition Containing Yeast Extract and Hexapeptide. U.S. Patent Application No. 13/247,689, 12, 03, 2014.
108. Hama, R.; Ulziibayar, A.; Reinhardt, J.W.; Watanabe, T.; Kelly, J.; Shinoka, T. Recent Developments in Biopolymer-Based Hydrogels for Tissue Engineering Applications. *Biomolecules* **2023**, *13*, 280. [CrossRef]
109. Kisling, A.; Lust, R.M.; Katwa, L.C. What is the role of peptide fragments of collagen I and IV in health and disease? *Life Sci.* **2019**, *228*, 30–34. [CrossRef] [PubMed]
110. Sionkowska, A.; Adamiak, K.; Musiał, K.; Gadomska, M. Collagen Based Materials in Cosmetic Applications: A Review. *Materials* **2020**, *13*, 4217. [CrossRef]
111. Kwatra, B. Collagen Supplementation: Therapy for Skin Disorders: A Review. *World J. Pharm. Res.* **2020**, *9*, 2504–2518.

112. Zhao, X.; Zhang, X.; Liu, D. Collagen peptides and the related synthetic peptides: A review on improving skin health. *J. Funct. Foods* **2021**, *86*, 104680. [CrossRef]
113. Ahmed, M.; Verma, A.K.; Patel, R. Collagen extraction and recent biological activities of collagen peptides derived from sea-food waste: A review. *Sustain. Chem. Pharm.* **2020**, *18*, 100315. [CrossRef]
114. Kim, D.-U.; Chung, H.-C.; Choi, J.; Sakai, Y.; Lee, B.-Y. Oral intake of low-molecular-weight collagen peptide improves hydration, elasticity, and wrinkling in human skin: A randomized, double-blind, placebo-controlled study. *Nutrients* **2018**, *10*, 826. [CrossRef]
115. Addor, F.A.S.; Cotta Vieira, J.; Abreu Melo, C.S. Improvement of dermal parameters in aged skin after oral use of a nutrient supplement. *Clin. Cosmet. Investig. Dermatol.* **2018**, *11*, 195–201. [CrossRef]
116. Czajka, A.; Kania, E.M.; Genovese, L.; Corbo, A.; Merone, G.; Luci, C.; Sibilla, S. Daily oral supplementation with collagen peptides combined with vitamins and other bioactive compounds improves skin elasticity and has a beneficial effect on joint and general wellbeing. *Nutr. Res.* **2018**, *57*, 97–108. [CrossRef] [PubMed]
117. Maia Campos, P.M.; Melo, M.O.; Siqueira César, F.C. Topical application and oral supplementation of peptides in the improvement of skin viscoelasticity and density. *J. Cosmet. Dermatol.* **2019**, *18*, 1693–1699. [CrossRef]
118. Song, H.; Zhang, S.; Zhang, L.; Li, B. Effect of Orally Administered Collagen Peptides from Bovine Bone on Skin Aging in Chronologically Aged Mice. *Nutrients* **2017**, *9*, 1209. [CrossRef] [PubMed]
119. Yazaki, M.; Ito, Y.; Yamada, M.; Goulas, S.; Teramoto, S.; Nakaya, M.-a.; Ohno, S.; Yamaguchi, K. Oral ingestion of collagen hydrolysate leads to the transportation of highly concentrated Gly-Pro-Hyp and its hydrolyzed form of Pro-Hyp into the bloodstream and skin. *J. Agric. Food Chem.* **2017**, *65*, 2315–2322. [CrossRef]
120. Offengenden, M.; Chakrabarti, S.; Wu, J. Chicken collagen hydrolysates differentially mediate antiinflammatory activity and type I collagen synthesis on human dermal fibroblasts. *Food Sci. Hum. Wellness* **2018**, *7*, 138–147. [CrossRef]
121. Xiao, Z.; Liang, P.; Chen, J.; Chen, M.F.; Gong, F.; Li, C.; Zhou, C.; Hong, P.; Yang, P.; Qian, Z.J. A Peptide YGDEY from Tilapia gelatin hydrolysates inhibits UVB-mediated skin photoaging by regulating MMP-1 and MMP-9 expression in HaCaT cells. *Photochem Photobiol.* **2019**, *95*, 1424–1432. [CrossRef] [PubMed]
122. Tkaczewska, J.; Bukowski, M.; Mak, P. Identification of antioxidant peptides in enzymatic hydrolysates of carp (*Cyprinus carpio*) skin gelatin. *Molecules* **2019**, *24*, 97. [CrossRef]
123. Sanchez, A.; Blanco, M.; Correa, B.; Perez-Martin, R.I.; Sotelo, C.G. Effect of Fish Collagen Hydrolysates on Type I Collagen mRNA Levels of Human Dermal Fibroblast Culture. *Mar. Drugs* **2018**, *16*, 144. [CrossRef] [PubMed]
124. Liu, C.; Ma, X.; Che, S.; Wang, C.; Li, B. The effect of hydrolysis with neutrase on molecular weight, functional properties, and antioxidant activities of Alaska pollock protein isolate. *J. Ocean. Univ. China* **2018**, *17*, 1423–1431. [CrossRef]
125. Zhao, Y.; Wang, Z.; Zhang, J.; Su, T. Extraction and characterization of collagen hydrolysates from the skin of *Rana chensinensis*. *3 Biotech.* **2018**, *8*, 181. [CrossRef]
126. Montalvo, G.E.B.; Thomaz-Soccol, V.; Vandenberghe, L.P.S.; Carvalho, J.C.; Faulds, C.B.; Bertrand, E.; Prado, M.R.M.; Bonatto, S.J.R.; Soccol, C.R. *Arthrospira maxima* OF15 biomass cultivation at laboratory and pilot scale from sugarcane vinasse for potential biological new peptides production. *Bioresour. Technol.* **2019**, *273*, 103–113. [CrossRef]
127. Norzagaray-Valenzuela, C.D.; Valdez-Ortiz, A.; Shelton, L.M.; Jiménez-Edeza, M.; Rivera-López, J.; Valdez-Flores, M.A.; Germán-Báez, L.J. Residual biomasses and protein hydrolysates of three green microalgae species exhibit antioxidant and antiaging activity. *J. Appl. Phycol.* **2016**, *29*, 189–198. [CrossRef]
128. Huang, G.; Chen, J. Preparation and applications of hyaluronic acid and its derivatives. *Int. J. Biol. Macromol.* **2019**, *125*, 478–484. [CrossRef] [PubMed]
129. Garg, C.; Khurana, P.; Garg, M. Molecular mechanisms of skin photoaging and plant inhibitors. *Inter. J. Green Pharm.* **2017**, *11*, 217–232.
130. Saranraj, P.; Naidu, M.A. Hyaluronic acid production and its applications-a review. *Int. J. Pharm. Biol. Arch.* **2013**, *4*, 853–859.
131. Cui, Y.; Hu, Y.-H.; Yu, F.; Zheng, J.; Chen, L.-S.; Chen, Q.-X.; Wang, Q. Inhibition kinetics and molecular simulation of p-substituted cinnamic acid derivatives on tyrosinase. *Int. J. Biol. Macromol.* **2017**, *95*, 1289–1297. [CrossRef] [PubMed]
132. Aguilar-Toalá, J.; Hernández-Mendoza, A.; González-Córdova, A.; Vallejo-Cordoba, B.; Liceaga, A.M. Potential role of natural bioactive peptides for development of cosmeceutical skin products. *Peptides* **2019**, *122*, 170170. [CrossRef]
133. Kubglomsong, S.; Theerakulkait, C.; Reed, R.L.; Yang, L.; Maier, C.S.; Stevens, J.F. Isolation and Identification of Tyrosinase-Inhibitory and Copper-Chelating Peptides from Hydrolyzed Rice-Bran-Derived Albumin. *J. Agric. Food Chem.* **2018**, *66*, 8346–8354. [CrossRef]
134. Ochiai, A.; Tanaka, S.; Tanaka, T.; Taniguchi, M. Rice Bran Protein as a Potent Source of Antimelanogenic Peptides with Tyrosinase Inhibitory Activity. *J. Nat. Prod.* **2016**, *79*, 2545–2551. [CrossRef]
135. Nakchum, L.; Kim, S.M. Preparation of squid skin collagen hydrolysate as an antihyaluronidase, antityrosinase, and antioxidant agent. *Prep. Biochem. Biotechnol.* **2016**, *46*, 123–130. [CrossRef]
136. Karkouch, I.O.; Tabbene, D.; Gharbi, M.A.; Ben Mlouka, S.; Elkahoui, C.; Rihouey, L.; Coquet, P.; Cosette Jouenne, T.; Limam, F. Antioxidant, antityrosinase and antibiofilm activities of synthesized peptides derived from vicia faba protein hydrolysate: A powerful agents in cosmetic application. *Ind. Crops Prod.* **2017**, *109*, 310–319. [CrossRef]
137. Chen, Y.M.; Shih, T.W.; Chiu, C.P.; Pan, T.M.; Tsai, T.Y. Effects of lactic acid bacteria-fermented soy milk on melanogenesis in B16F0 melanocytes. *J. Funct. Foods* **2013**, *5*, 395–405. [CrossRef]
138. Qin, Z. Soluble elastin peptides in cardiovascular homeostasis: Foe or ally. *Peptides* **2015**, *67*, 64–73. [CrossRef] [PubMed]

139. Cao, M.; Shen, Y.; Wang, Y.; Wang, X.; Li, D. Self-assembly of short elastin-like amphiphilic peptides: Effects of temperature, molecular hydrophobicity and charge distribution. *Molecules* **2019**, *24*, 202. [CrossRef] [PubMed]
140. Leiros, G.J.; Kusinsky, A.G.; Balana, M.E.; Hagelin, K. Triolein reduces MMP-1 upregulation in dermal fibroblasts generated by ROS production in UVB-irradiated keratinocytes. *J. Dermatol. Sci.* **2017**, *85*, 124–130. [CrossRef] [PubMed]
141. Liu, Y.; Su, G.; Zhou, F.; Zhang, J.; Zheng, L.; Zhao, M. Protective effect of bovine elastin peptides against photoaging in mice and identification of novel antiphotoaging peptides. *J. Agric. Food Chem.* **2018**, *66*, 10760–10768. [CrossRef] [PubMed]
142. Le Page, A.; Khalil, A.; Vermette, P.; Frost, E.H.; Larbi, A.; Witkowski, J.M.; Fulop, T. The role of elastin-derived peptides in human physiology and diseases. *Matrix Biol.* **2019**, *84*, 81–96. [CrossRef] [PubMed]
143. Amakye, W.K.; Yang, L.; Yao, M.; Yuan, E.; Ren, R.; Ren, J. Skipjack (*Katsuwonus pelamis*) elastin hydrolysate-derived peptides attenuate UVA irradiation-induced cell damage in human HaCaT keratinocytes. *Food Frontiers* **2021**, *2*, 184–194. [CrossRef]
144. Xiong, Y.; Peng, P.; Chen, S.J.; Chang, M.; Wang, Q.; Yin, S.N.; Ren, D.F. Preparation, identification, and molecular docking of novel elastase inhibitory peptide from walnut (*Juglans regia* L.) meal. *Food Chem. Mol. Sci.* **2022**, *5*, 100139. [CrossRef]
145. Martemucci, G.; Costagliola, C.; Mariano, M.; D'andrea, L.; Napolitano, P.; D'Alessandro, A.G. Free Radical Properties, Source and Targets, Antioxidant Consumption and Health. *Oxygen* **2022**, *2*, 48–78. [CrossRef]
146. Dini, I. Bio Discarded from Waste to Resource. *Foods* **2021**, *10*, 2652. [CrossRef] [PubMed]
147. Dini, I.; Grumetto, L. Recent Advances in Natural Polyphenol Research. *Molecules* **2022**, *27*, 8777. [CrossRef]
148. Wang, L.; Wang, X.; Bai, F.; Fang, Y.; Wang, J.; Gao, R. The anti-skin-aging effect of oral administration of gelatin from the swim bladder of Amur sturgeon (*Acipenser schrenckii*). *Food Funct.* **2019**, *10*, 3890–3897. [CrossRef]
149. Dini, I. The Potential of Dietary Antioxidants. *Antioxidants* **2021**, *10*, 1752. [CrossRef] [PubMed]
150. Dini, I.; Laneri, S. Spices, Condiments, Extra Virgin Olive Oil and Aromas as Not Only Flavorings, but Precious Allies for Our Wellbeing. *Antioxidants* **2021**, *10*, 868. [CrossRef] [PubMed]
151. Dini, I.; De Biasi, M.-G.; Mancusi, A. An Overview of the Potentialities of Antimicrobial Peptides Derived from Natural Sources. *Antibiotics* **2022**, *11*, 1483. [CrossRef]
152. Stefanucci, A.; Mollica, A.; Macedonio, G.; Dall'Acqua, S.; Ahmed, A.A.; Novellino, E. Exogenous opioid peptides derived from food proteins and their possible uses as dietary supplements: A critical review. *Food Rev. Int.* **2016**, *34*, 70–86. [CrossRef]
153. Chai, T.-T.; Law, Y.-C.; Wong, F.-C.; Kim, S.-K. Enzyme-Assisted Discovery of Antioxidant Peptides from Edible Marine Invertebrates: A Review. *Mar. Drugs* **2017**, *15*, 42. [CrossRef]
154. Lorenzo, J.M.; Munekata, P.E.S.; Gómez, B.; Barba, F.J.; Mora, L.; Pérez-Santaescolástica, C.; Toldrá, F. Bioactive peptides as natural antioxidants in food products—A review. *Trends Food Sci. Technol.* **2018**, *79*, 136–147. [CrossRef]
155. Yang, X.; Zhang, L.; Ding, D.; Chi, C.; Wang, B.; Huo, J.C. Preparation, identification, and activity evaluation of eight antioxidant peptides from protein hydrolysate of hairtail (*Trichiurus japonicas*) muscle. *Mar. Drugs* **2019**, *17*, 23. [CrossRef]
156. Ketnawa, S.; Wickramathilaka, M.; Liceaga, A.M. Changes on antioxidant activity of microwave-treated protein hydrolysates after simulated gastrointestinal digestion: Purification and identification. *Food Chem.* **2018**, *254*, 36–46. [CrossRef]
157. Abd-Talib, N.; Yaji, E.L.A.; Wahab, N.S.A.; Razali, N.; Len, K.Y.T.; Roslan, J.; Saari, N.; Pa'ee, K.F. Bioactive peptides and its alternative processes: A review. *Biotechnol. Bioprocess Eng.* **2022**, *27*, 306–335. [CrossRef]
158. Pouzo, L.B.; Descalzo, A.M.; Zaritzky, N.E.; Rossetti, L.; Pavan, E. Antioxidant status, lipid and color stability of aged beef from grazing steers supplemented with corn grain and increasing levels of flaxseed. *Meat Sci.* **2016**, *111*, 1–8. [CrossRef] [PubMed]
159. Tironi, V.A.; Anon, M.C. Amaranth proteins as a source of antioxidant peptides: Effect of proteolysis. *Food Res. Int.* **2010**, *43*, 315–322. [CrossRef]
160. Chen, H.; Zhao, M.; Lin, L.; Wang, J.; Sun-Waterhouse, D.; Dong, Y.; Zhuang, M.; Su, G. Identification of antioxidant peptides from defatted walnut meal hydrolysate with potential for improving learning and memory. *Food Res. Int.* **2015**, *78*, 216–233. [CrossRef]
161. Eftekharzadeh, B.; Khodagholi, F.; Abdi, A.; Maghsoudi, N. Alginate protects NT2 neurons against $H_2O_2$-induced neurotoxicity. *Carbohydr. Polym.* **2010**, *79*, 1063–1072. [CrossRef]
162. Cai, L.Y.; Wu, X.S.; Zhang, Y.H.; Li, X.X.; Ma, S.; Li, J.R. Purification and characterization of three antioxidant peptides from protein hydrolysate of grass carp (*Ctenopharyngodon idella*) skin. *J. Funct. Foods* **2015**, *16*, 234. [CrossRef]
163. Chi, C.F.; Hu, F.Y.; Wang, B.; Li, T.; Ding, G.F. Antioxidant and anticancer peptides from the protein hydrolysate of blood clam (*Tegillarca granosa*) muscle. *J. Funct. Foods* **2015**, *15*, 301–313. [CrossRef]
164. Vásquez-Villanueva, R.; Marina, M.L.M.; Garcia, C. Identification by hydrophilic interaction and reversed-phase liquid chromatography-tandem mass spectrometry of peptides with antioxidant capacity in food residues. *J. Chromatogr. A* **2015**, *1428*, 185–192. [CrossRef]
165. Song, R.; Wei, R.B.; Ruan, G.Q.; Luo, H.Y. Isolation and identification of antioxidant peptides from peptic hydrolysates of half-fin anchovy (*Setipinna taty*). *LWT-Food Sci. Technol.* **2015**, *60*, 221–229. [CrossRef]
166. Yan, Q.J.; Huang, L.H.; Sun, Q.; Jiang, Z.Q.; Wu, X. Isolation, identification and synthesis of four novel antioxidant peptides from rice residue protein hydrolyzed by multiple proteases. *Food Chem.* **2015**, *179*, 290–295. [CrossRef] [PubMed]
167. Montone, A.M.I.; Capuano, F.; Mancusi, A.; Di Maro, O.; Peruzy, M.F.; Proroga, Y.T.R.; Cristiano, D. Exposure to *Bacillus cereus* in Water Buffalo Mozzarella Cheese. *Foods* **2020**, *9*, 1899. [CrossRef]
168. Hsieh, C.C.; Hernández-Ledesma, B.; Jeong, H.J.; Park, J.H.; de Lumen, B.O. Complementary roles in cancer prevention: Protease inhibitor makes the cancer preventive peptide lunasin bioavailable. *PLoS ONE* **2010**, *5*, e8890. [CrossRef] [PubMed]

169. Pihlanto, A. Lactic fermentation and bioactive peptides. In *Lactic Acid Bacteria R & R for Food, Health and Livestock Purposes*; IntechOpen: Marcelino, Kongo, 2013.
170. Rizzello, C.G.; Tagliazucchi, D.; Babini, E.; Rutella, G.S.; Saa, D.L.T.; Gianotti, A. Bioactive peptides from vegetable food matrices: Research trends and novel biotechnologies for synthesis and recovery. *J. Funct. Foods* **2016**, *27*, 549–569. [CrossRef]
171. Malaguti, M.; Dinelli, G.; Leoncini, E.; Bregola, V.; Bosi, S.; Cicero, A.F.G.; Hrelia, S. Bioactive peptides in cereals and legumes: Agronomical, biochemical and clinical aspects. *Int. J. Mol. Sci.* **2014**, *15*, 21120–21135. [CrossRef]
172. Aluko, R. Amino acids, peptides, and proteins as antioxidants for food preservation. In *Handbook of Antioxidants for Food Preservation*; Elsevier: Amsterdam, The Netherlands, 2015; pp. 105–140.
173. Power, O.; Jakeman, P.; FitzGerald, R.J. Antioxidative peptides: Enzymatic production, in vitro and in vivo antioxidant activity and potential applications of milk-derived antioxidative peptides. *Amino Acids* **2013**, *44*, 797–820. [CrossRef]
174. Brandelli, A.; Daroit, D.J.; Corrêa, A.P.F. Whey as a source of peptides with remarkable biological activities. *Food Res. Int.* **2015**, *73*, 149–161. [CrossRef]
175. El-Salam, M.A.; El-Shibiny, S. Bioactive peptides of buffalo, camel, goat, sheep, mare, and yak milks and milk products. *Food Rev. Int.* **2013**, *29*, 1–23. [CrossRef]
176. Sila, A.; Bougatef, A. Antioxidant peptides from marine by-products: Isolation, identification and application in food systems. A review. *J. Funct. Foods* **2016**, *21*, 10–26. [CrossRef]
177. Samaranayaka, A.G.; Li-Chan, E.C. Food-derived peptidic antioxidants: A review of their production, assessment, and potential applications. *J. Funct. Foods* **2011**, *3*, 229–254. [CrossRef]
178. Wu, R.B.; Wu, C.L.; Liu, D.; Yang, X.H.; Huang, J.F.; Zhang, J.; Liao, B.Q.; He, H.L.; Li, H. Overview of antioxidant peptides derived from marine resources: The Sources, characteristic, purification, and evaluation methods. *Appl. Biochem. Biotechnol.* **2015**, *176*, 1815–1833. [CrossRef] [PubMed]
179. Nimalaratne, C.; Wu, J. Hen Egg as an Antioxidant Food Commodity: A Review. *Nutrients* **2015**, *7*, 8274–8293. [CrossRef]
180. Mora, L.; Reig, M.; Toldrá, F. Bioactive peptides generated from meat industry by-products. *Food Res. Int.* **2014**, *65*, 344–349. [CrossRef]
181. Zhang, M.; Mu, T.H.; Sun, M.J. Purification and identification of antioxidant peptides from sweet potato protein hydrolysates by Alcalase. *J. Funct. Foods* **2014**, *7*, 191–200. [CrossRef]
182. Huang, D.; Ou, B.; Prior, R.L. The chemistry behind antioxidant capacity assay. *J. Agric. Food Chem.* **2005**, *53*, 1841–1856. [CrossRef]
183. Zheng, L.; Zhao, M.; Xiao, C.; Zhao, Q.; Su, G. Practical problems when using ABTS assay to assess the radical-scavenging activity of peptides: Importance of controlling reaction pH and time. *Food Chem.* **2016**, *192*, 288–294. [CrossRef] [PubMed]
184. Kobayashi, T.; Nagao, K. "Deepening" Insight on Skin Aging and Antimicrobial Immunity. *Cell Metab.* **2019**, *29*, 515–517. [CrossRef] [PubMed]
185. Wang, J.; Kadyan, S.; Ukhanov, V.; Cheng, J.; Nagpal, R.; Cui, L. Recent advances in the health benefits of pea protein (*Pisum sativum*): Bioactive peptides and the interaction with the gut microbiome. *Curr. Opin. Food Sci.* **2022**, *48*, 100944. [CrossRef]
186. Niyonsaba, F.; Kiatsurayanon, C.; Chieosilapatham, P.; Ogawa, H. Friends or Foes? Host defense (antimicrobial) peptides and proteins in human skin diseases. *Exp. Dermatol.* **2017**, *26*, 989–998. [CrossRef]
187. Iinuma, K.; Noguchi, N.; Nakaminami, H.; Sasatsu, M.; Nishijima, S.; Tsuboi, I. Susceptibility of Propionibacterium acnes isolated from patients with acne vulgaris to zinc ascorbate and antibiotics. *Clin. Cosmet. Investig. Dermatol.* **2011**, *4*, 161–165.
188. Pfalzgraff, A.; Brandenburg, K.; Weindl, G. Antimicrobial Peptides and Their Therapeutic Potential for Bacterial Skin Infections and Wounds. *Front. Pharmacol.* **2018**, *9*, 281. [CrossRef]
189. Le, C.F.; Fang, C.M.; Sekaran, S.D. Intracellular Targeting Mechanisms by Antimicrobial Peptides. *Antimicrob. Agents Chemother.* **2017**, *61*, e02340-16. [CrossRef]
190. Taniguchi, M.; Kameda, M.; Namae, T.; Ochiai, A.; Saitoh, E.; Tanaka, T. Identification and characterization of multifunctional cationic peptides derived from peptic hydrolysates of rice bran protein. *J. Funct. Foods* **2017**, *34*, 287–296. [CrossRef]
191. Beaulieu, L.; Bondu, S.; Doiron, K.; Rioux, L.-E.; Turgeon, S.L. Characterization of antibacterial activity from protein hydrolysates of the macroalga *Saccharina longicruris* and identification of peptides implied in bioactivity. *J. Funct. Foods* **2015**, *17*, 685–697. [CrossRef]
192. Kobbi, S.; Nedjar, N.; Chihib, N.; Balti, R.; Chevalier, M.; Silvain, A.; Chaabouni, S.; Dhulster, P.; Bougatef, A. Synthesis and antibacterial activity of new peptides from Alfalfa RuBisCO protein hydrolysates and mode of action via a membrane damage mechanism against *Listeria innocua*. *Microb. Pathog.* **2018**, *115*, 41–49. [CrossRef]
193. Lueangsakulthai, J.; Jangpromma, N.; Temsiripong, T.; McKendrick, J.E.; Khunkitti, W.; Maddocks, S.E.; Klaynongsruang, S. A novel antibacterial peptide derived from *Crocodylus siamensis* haemoglobin hydrolysate induces membrane permeabilization causing iron dysregulation, oxidative stress and bacterial death. *J. Appl. Microbiol.* **2017**, *123*, 819–831. [CrossRef] [PubMed]
194. Zanutto-Elgui, M.R.; Vieira, J.C.S.; do Prado, D.Z.; Buzalaf, M.A.R.; de Magalhães Padilha, P.; de Oliveira, D.E.; Fleuri, L.F. Production of milk peptides with antimicrobial and antioxidant properties through fungal proteases. *Food Chem.* **2019**, *278*, 823–831. [CrossRef]
195. Liu, W.; Chen, X.; Li, H.; Zhang, J.; An, J.; Liu, X. Antiinflammatory Function of Plant-Derived Bioactive Peptides: A Review. *Foods* **2022**, *11*, 2361. [CrossRef]
196. Dragoș, D.; Petran, M.; Gradinaru, T.-C.; Gilca, M. Phytochemicals and Inflammation: Is Bitter Better? *Plants* **2022**, *11*, 2991. [CrossRef]

197. Zlobin, A.; Bloodworth, J.C.; Osipo, C. Mitogen-activated protein kinase (MAPK) signaling. In *Predictive Biomarkers in Oncology*; Springer: Berlin/Heidelberg, Germany, 2019; pp. 213–221.
198. Alharbi, K.S.; Fuloria, N.K.; Fuloria, S.; Rahman, S.B.; Al-Malki, W.H.; Shaikh, M.A.J.; Thangavelu, L.; Singh, S.K.; Raju, V.S.R.; Jha, N.K. Nuclear factor-kappa B and its role in inflammatory lung disease. *Chem. Biol. Interact.* **2021**, *345*, 109568. [CrossRef]
199. Xu, Z.; Shang, W.; Zhao, Z.; Zhang, B.; Liu, C.; Cai, H. Curcumin alleviates rheumatoid arthritis progression through the phosphatidylinositol 3-kinase/protein kinase B pathway: An in vitro and in vivo study. *Bioengineered* **2022**, *13*, 12899–12911. [CrossRef] [PubMed]
200. Yu, W.; Field, C.J.; Wu, J. Purification and identification of antiinflammatory peptides from spent hen muscle proteins hydrolysate. *Food Chem.* **2018**, *253*, 101–107. [CrossRef] [PubMed]
201. Meram, C.; Wu, J. Antiinflammatory effects of egg yolk livetins (α, β, and γ-livetin) fraction and its enzymatic hydrolysates in lipopolysaccharide-induced RAW 264.7 macrophages. *Food Res. Int.* **2017**, *100*, 449–459. [CrossRef] [PubMed]
202. Craik, D.J.; Fairlie, D.P.; Liras, S.; Price, D. The future of peptide-based drugs. *Chem. Biol. Drug Des.* **2013**, *81*, 136–147. [CrossRef]
203. Zhang, S.H.; Ren, M.; Zeng, X.F.; He, P.L.; Ma, X.; Qiao, S.Y. Leucine stimulates ASCT2 amino acid transporter expression in porcine jejunal epithelial cell line (IPEC-J2) through PI3K/Akt/mTOR and ERK signaling pathways. *Amino Acids* **2014**, *46*, 2633–2642. [CrossRef]
204. Bonvini, A.; Rogero, M.M.; Coqueiro, A.Y.; Raizel, R.; Bella, L.M.; Fock, R.A.; Borelli, P.; Tirapegui, J. Effects of different branched-chain amino acids supplementation protocols on the inflammatory response of LPS-stimulated RAW 264.7 macrophages. *Amino Acids* **2021**, *53*, 597–607. [CrossRef]
205. Singh, S.; Datta, A.; Schmidtchen, A.; Bhunia, A.; Malmsten, M. Tryptophan end-tagging for promoted lipopolysaccharide interactions and antiinflammatory effects. *Sci. Rep.* **2017**, *7*, 212. [CrossRef]
206. Ding, L.; Wang, L.Y.; Yu, Z.P.; Ma, S.T.; Du, Z.Y.; Zhang, T.; Liu, J.B. Importance of terminal amino acid residues to the transport of oligopeptides across the Caco-2 cell monolayer. *J. Agric. Food Chem.* **2017**, *65*, 7705–7712. [CrossRef]
207. Han, H.; Yin, J.; Wang, B.; Huang, X.G.; Yao, J.M.; Zheng, J.; Fan, W.J.; Li, T.J.; Yin, Y.L. Effects of dietary lysine restriction on inflammatory responses in piglets. *Sci. Rep.* **2018**, *8*, 2451. [CrossRef]
208. Lan, J.; Dou, X.J.; Li, J.W.; Yang, Y.; Xue, C.Y.; Wang, C.X.; Gao, N.; Shan, A. L-arginine ameliorates lipopolysaccharide-induced intestinal inflammation through inhibiting the TLR4/NF-κB and MAPK pathways and stimulating β-defensins expression in vivo and in vitro. *J. Agric. Food Chem.* **2020**, *68*, 2648–2663. [CrossRef]
209. Tang, N.; Skibsted, L.H. Calcium binding to amino acids and small glycine peptides in aqueous solution: Toward peptide design for better calcium bioavailability. *J. Agric. Food Chem.* **2016**, *64*, 4376–4389. [CrossRef] [PubMed]
210. Li, X.G.; Bradford, B.U.; Wheeler, M.D.; Stimpson, S.A.; Pink, H.M.; Brodie, T.A.; Schwab, J.H.; Thurman, R.G. Dietary glycine prevents peptidoglycan polysaccharide-induced reactive arthritis in the rat: Role for glycine-gated chloride channel. *Infect. Immun.* **2001**, *69*, 5883–5891. [CrossRef] [PubMed]
211. Huang, J.; Liu, J.; Chang, G.J.; Wang, Y.; Ma, N.; Roy, A.C.; Shen, X.Z. Glutamine supplementation attenuates the inflammation caused by LPS-induced acute lung injury in mice by regulating the TLR4/MAPK signaling pathway. *Inflammation* **2021**, *44*, 2180–2192. [CrossRef] [PubMed]
212. Vo, T.; Ryu, B.M.; Kim, S.K. Purification of novel anti-inflammatory peptides from enzymatic hydrolysate of the edible microalgal *Spirulina maxima*. *J. Funct. Food* **2013**, *5*, 1336–1346. [CrossRef]
213. He, R.; Liu, M.T.; Zou, Z.P.; Wang, M.J.; Wang, Z.G.; Ju, X.R.; Hao, G.F. Anti-inflammatory activity of peptides derived from millet bran in vitro and in vivo. *Food Funct.* **2022**, *13*, 1881–1889. [CrossRef] [PubMed]
214. Suttisuwan, R.; Phunpruch, S.; Saisavoey, T.; Sangtanoo, P.; Karnchanatat, A. Isolation and characterization of anti-inflammatory peptides derived from trypsin hydrolysis of microalgae protein (*Synechococcus* sp. VDW). *Food Biotech.* **2019**, *33*, 303–324. [CrossRef]
215. Velliquette, R.A.; Fast, D.J.; Maly, E.R.; Alashi, A.M.; Aluko, R.E. Enzymatically derived sunflower protein hydrolysate and peptides inhibit NF-κB and promote monocyte differentiation to a dendritic cell phenotype. *Food Chem.* **2020**, *319*, 126563. [CrossRef]
216. Saisavoey, T.; Sangtanoo, P.; Chanchao, C.; Reamtong, O.; Karnchanatat, A. Identification of novel anti-inflammatory peptides from bee pollen (*Apis mellifera*) hydrolysate in lipopolysaccharide-stimulated RAW264.7 macrophages. *J. Apic. Res.* **2020**, *60*, 280–289. [CrossRef]
217. Swain, S.; Mondal, D.; Beg, S.; Niranjan Patra, C.; Chandra Dinda, S.; Sruti, J.; Eswara Bhanoji Rao, M. Stabilization and delivery approaches for protein and peptide pharmaceuticals: An extensive review of patents. *Recent Pat. Biotechnol.* **2013**, *7*, 28–46. [CrossRef]
218. Barua, S.; Mitragotri, S. Challenges associated with penetration of nanoparticles across cell and tissue barriers: A review of current status and future prospects. *Nano Today* **2014**, *9*, 223–243. [CrossRef]
219. Liu, X.; Wang, P.; Zou, Y.X.; Luo, Z.G.; Tamer, T.M. Co-encapsulation of Vitamin C and beta-Carotene in liposomes: Storage stability, antioxidant activity, and in vitro gastrointestinal digestion. *Food Res. Int.* **2020**, *136*, 109587. [CrossRef]
220. Ge, X.; Wei, M.; He, S.; Yuan, W.-E. Advances of Non-Ionic Surfactant Vesicles (Niosomes) and Their Application in Drug Delivery. *Pharmaceutics* **2019**, *11*, 55. [CrossRef] [PubMed]
221. Dubey, S.K.; Dey, A.; Singhvi, G.; Pandey, M.M.; Singh, V.; Kesharwani, P. Emerging trends of nanotechnology in advanced cosmetics. *Colloids Surf. B Biointerfaces* **2022**, *214*, 112440. [CrossRef] [PubMed]

222. Yasamineh, S.; Yasamineh, P.; Ghafouri Kalajahi, H.; Gholizadeh, O.; Yekanipour, Z.; Afkhami, H.; Eslami, M.; Hossein Kheirkhah, A.; Taghizadeh, M.; Yazdani, Y.; et al. A State-of-the-Art Review on the Recent Advances of Niosomes as a Targeted Drug Delivery System. *Int. J. Pharm.* **2022**, *624*, 121878. [CrossRef]
223. Kaul, S.; Gulati, N.; Verma, D.; Mukherjee, S.; Nagaich, U. Role of nanotechnology in cosmeceuticals: A review of recent advances. *J. Pharm.* **2018**, *2018*, 3420204. [CrossRef]
224. Trucillo, P.; Campardelli, R.; Reverchon, E. Liposomes: From Bangham to Supercritical Fluids. *Processes* **2020**, *8*, 1022. [CrossRef]
225. Sguizzato, M.; Pepe, A.; Baldisserotto, A.; Barbari, R.; Montesi, L.; Drechsler, M.; Mariani, P.; Cortesi, R. Niosomes for Topical Application of Antioxidant Molecules: Design and In Vitro Behavior. *Gels* **2023**, *9*, 107. [CrossRef] [PubMed]
226. Kaur, I.P.; Agrawal, R. Nanotechnology: A new paradigm in cosmeceuticals. *Recent Pat. Drug Deliv. Formul.* **2007**, *1*, 171–182. [CrossRef]
227. Mosallam, S.; Ragaie, M.H.; Moftah, N.H.; Elshafeey, A.H.; Abdelbary, A.A. Use of novasomes as a vesicular carrier for improving the topical delivery of terconazole: In vitro characterization, in vivo assessment and exploratory clinical experimentation. *Int. J. Nanomed.* **2021**, *16*, 119–132. [CrossRef]
228. Opatha, S.A.T.; Titapiwatanakun, V.; Chutoprapat, R. Transfersomes: A promising nanoencapsulation technique for transdermal drug delivery. *Pharmaceutics* **2020**, *12*, 855. [CrossRef]
229. Saraf, G.J.S. Topical Delivery of Curcuma longa Extract Loaded Nanosized Ethosomes to Combat Facial Wrinkles Research Article. *J. Pharm. Drug Deliv. Res.* **2014**, *3*, 1.
230. Gupta, V.; Mohapatra, S.; Mishra, H.; Farooq, U.; Kumar, K.; Ansari, M.J.; Aldawsari, M.F.; Alalaiwe, A.S.; Mirza, M.A.; Iqbal, Z. Nanotechnology in Cosmetics and Cosmeceuticals—A Review of Latest Advancements. *Gels* **2022**, *8*, 173. [CrossRef]
231. Wu, P.-S.; Li, Y.-S.; Kuo, Y.-C.; Tsai, S.-J.J.; Lin, C.-C. Preparation and Evaluation of Novel Transfersomes Combined with the Natural Antioxidant Resveratrol. *Molecules* **2019**, *24*, 600. [CrossRef]
232. Yang, J.; Kim, B. Synthesis and characterization of ethosomal carriers containing cosmetic ingredients for enhanced transdermal delivery of cosmetic ingredients. *Korean J. Chem. Eng.* **2018**, *35*, 792–797. [CrossRef]
233. Yaghmur, A.; Mu, H. Recent advances in drug delivery applications of cubosomes, hexosomes, and solid lipid nanoparticles. *Acta Pharm. Sin. B* **2021**, *11*, 871–885. [CrossRef]
234. Fytianos, G.; Rahdar, A.; Kyzas, G.Z. Nanomaterials in Cosmetics: Recent Updates. *Nanomaterials* **2020**, *10*, 979. [CrossRef]
235. Kim, B.; Cho, H.-E.; Moon, S.H.; Ahn, H.-J.; Bae, S.; Cho, H.-D.; An, S. Transdermal delivery systems in cosmetics. *Biomed. Dermatol.* **2020**, *4*, 10. [CrossRef]
236. Decome, L.; De Méo, M.; Geffard, A.; Doucet, O.; Duménil, G.; Botta, A. Evaluation of photolyase (Photosome®) repair activity in human keratinocytes after a single dose of ultraviolet B irradiation using the comet assay. *J. Photochem. Photobiol. B Biol.* **2005**, *79*, 101–108. [CrossRef] [PubMed]
237. Patravale, V.B.; Mandawgade, S.D. Novel cosmetic delivery systems: An application update. *Int. J. Cosmet. Sci.* **2008**, *30*, 19–33. [CrossRef] [PubMed]
238. Zhang, X.-Y.; Zhang, P.-Y. Polymersomes in nanomedicine—A review. *Curr. Nanosci.* **2017**, *13*, 124–129. [CrossRef]
239. Souto, E.B.; Fernandes, A.R.; Martins-Gomes, C.; Coutinho, T.E.; Durazzo, A.; Lucarini, M.; Souto, S.B.; Silva, A.M.; Santini, A. Nanomaterials for Skin Delivery of Cosmeceuticals and Pharmaceuticals. *Appl. Sci.* **2020**, *10*, 1594. [CrossRef]
240. Bermudez, H.; Brannan, A.K.; Hammer, D.A.; Bates, F.S.; Discher, D.E. Molecular weight dependence of polymersome membrane structure, elasticity, and stability. *Macromolecules* **2002**, *35*, 8203–8208. [CrossRef]
241. Joye, I.J.; McClements, D.J. Biopolymer-based nanoparticles and microparticles: Fabrication, characterization, and application. *Curr. Opin. Colloid Interface Sci.* **2014**, *19*, 417–427. [CrossRef]
242. Allamraju, K.V. Green Hydrogels. *Green Compos.* **2021**, 225–249. [CrossRef]
243. Pramanik, B.; Ahmed, S. Peptide-Based Low Molecular Weight Photosensitive Supramolecular Gelators. *Gels* **2022**, *8*, 533. [CrossRef]
244. Perry, S.L.; McClements, D.J. Recent Advances in Encapsulation, Protection, and Oral Delivery of Bioactive Proteins and Peptides using Colloidal Systems. *Molecules* **2020**, *25*, 1161. [CrossRef]
245. Nair, A.; Mallya, R.; Suvarna, V.; Khan, T.A.; Momin, M.; Omri, A. Nanoparticles—Attractive Carriers of Antimicrobial Essential Oils. *Antibiotics* **2022**, *11*, 108. [CrossRef]
246. Rasouli, R.; Barhoum, A.; Bechelany, M.; Dufresne, A. Nanofibers for Biomedical and Healthcare Applications. *Macromol. Biosci.* **2018**, *19*, e1800256. [CrossRef]
247. Haddada, M.B.; Gerometta, E.; Chawech, R.; Sorres, J.; Bialecki, A.; Pesnel, S.; Morel, A.L. Assessment of antioxidant and dermoprotective activities of gold nanoparticles as safe cosmetic ingredient. *Colloids Surf. B Biointerfaces* **2020**, *189*, 110855. [CrossRef]
248. Mondéjar-López, M.; López-Jiménez, A.J.; Abad-Jordá, M.; Rubio-Moraga, A.; Ahraz, O.; Gómez-Gómez, L.; Niza, E. Biogenic Silver Nanoparticles from *Iris tuberosa* as Potential Preservative in Cosmetic Products. *Molecules* **2021**, *26*, 4696. [CrossRef]
249. Lee, C.-C.; Lin, Y.-H.; Hou, W.-C.; Li, M.-H.; Chang, J.-W. Exposure to $ZnO/TiO_2$ Nanoparticles Affects Health Outcomes in Cosmetics Salesclerks. *Int. J. Environ. Res. Public Health* **2020**, *17*, 6088. [CrossRef] [PubMed]
250. Sánchez-López, E.; Gomes, D.; Esteruelas, G.; Bonilla, L.; Lopez-Machado, A.L.; Galindo, R.; Cano, A.; Espina, M.; Ettcheto, M.; Camins, A.; et al. Metal-Based Nanoparticles as Antimicrobial Agents: An Overview. *Nanomaterials* **2020**, *10*, 292. [CrossRef] [PubMed]

251. Walters, K.A.; Lane, M.E. Dermal and Transdermal Drug Delivery Systems. In *Dermal Drug Delivery*, 1st ed.; Ghosh, T.K., Ed.; CRC Press: Boca Raton, FL, USA, 2020; pp. 1–60.
252. Gugleva, V.; Ivanova, N.; Sotirova, Y.; Andonova, V. Dermal Drug Delivery of Phytochemicals with Phenolic Structure via Lipid-Based Nanotechnologies. *Pharmaceuticals* **2021**, *14*, 837. [CrossRef] [PubMed]
253. Morais, R.P.; Hochheim, S.; de Oliveira, C.C.; Riegel-Vidotti, I.C.; Marino, C.E.B. Skin interaction, permeation, and toxicity of silica nanoparticles: Challenges and recent therapeutic and cosmetic advances. *Int. J. Pharm.* **2022**, *614*, 121439. [CrossRef] [PubMed]
254. Bilal, M.; Iqbal, H.M.N. New Insights on Unique Features and Role of Nanostructured Materials in Cosmetics. *Cosmetics* **2020**, *7*, 24. [CrossRef]
255. Pentek, T.; Newenhouse, E.; O'Brien, B.; Singh Chauhan, A. Development of a Topical Resveratrol Formulation for Commercial Applications Using Dendrimer Nanotechnology. *Molecules* **2017**, *22*, 137. [CrossRef] [PubMed]
256. Alkilani, A.Z.; Nasereddin, J.; Hamed, R.; Nimrawi, S.; Hussein, G.; Abo-Zour, H.; Donnelly, R.F. Beneath the Skin: A Review of Current Trends and Future Prospects of Transdermal Drug Delivery Systems. *Pharmaceutics* **2022**, *14*, 1152. [CrossRef] [PubMed]
257. Zhou, H.; Luo, D.; Chen, D.; Tan, X.; Bai, X.; Liu, Z.; Yang, X.; Liu, W. Current Advances of Nanocarrier Technology-Based Active Cosmetic Ingredients for Beauty Applications. *Clin. Cosmet. Investig. Dermatol.* **2021**, *14*, 867–887. [CrossRef]
258. Eckert, R.W.; Wiemann, S.; Keck, C.M. Improved dermal and transdermal delivery of curcumin with smartfilms and nanocrystals. *Molecules* **2021**, *26*, 1633. [CrossRef]
259. Wadhawan, J.; Parmar, P.K.; Bansal, A.K. Nanocrystals for improved topical delivery of medium soluble drug: A case study of acyclovir. *J. Drug Deliv. Sci. Technol.* **2021**, *65*, 102662. [CrossRef]
260. Kazemzadeh, H.; Mozafari, M. Fullerene-based delivery systems. *Drug Discov. Today* **2019**, *24*, 898–905. [CrossRef] [PubMed]
261. De Jong, W.H.; Borm, P.J. Drug delivery and nanoparticles: Applications and hazards. *Int. J. Nanomed.* **2008**, *3*, 133. [CrossRef]
262. Miljkovic, S.; Jeftic, B.; Stankovic, I.; Stojiljkovic, N.; Koruga, D. Mechanisms of skin moisturization with hyperharmonized hydroxyl modified fullerene substance. *J. Cosmet. Dermatol.* **2021**, *20*, 3018–3025. [CrossRef]
263. Krabicová, I.; Appleton, S.L.; Tannous, M.; Hoti, G.; Caldera, F.; Rubin Pedrazzo, A.; Cecone, C.; Cavalli, R.; Trotta, F. History of Cyclodextrin Nanosponges. *Polymers* **2020**, *12*, 1122. [CrossRef]
264. Torne, S.J.; Ansari, K.A.; Vavia, P.R.; Trotta, F.; Cavalli, R. Enhanced oral paclitaxel bioavailability after administration of paclitaxel-loaded nanosponges. *Drug Deliv.* **2010**, *17*, 419–425. [CrossRef] [PubMed]
265. Trotta, F.; Cavalli, R.; Martina, K.; Biasizzo, M.; Vitillo, J.; Bordiga, S.; Vavia, P.; Ansari, K. Cyclodextrin nanosponges as effective gas carriers. *J. Incl. Phenom. Macrocycl. Chem.* **2011**, *71*, 189–194. [CrossRef]
266. Mortazavi, S.M.; Moghimi, H.R. Skin permeability, a dismissed necessity for anti-wrinkle peptide performance. *Int. J. Cosmet. Sci.* **2022**, *44*, 232–248. [CrossRef] [PubMed]
267. Engelskirchen, S.; Maurer, R.; Levy, T.; Berghaus, R.; Auweter, H.; Glatter, O. Highly concentrated emulsified microemulsions as solvent-free plant protection formulations. *J. Colloid Interface Sci.* **2012**, *388*, 151–161. [CrossRef]
268. Yaghmur, A.; de Campo, L.; Sagalowicz, L.; Leser, M.E.; Glatter, O. Emulsified microemulsions and oil-containing liquid crystalline phases. *Langmuir* **2005**, *21*, 569–577. [CrossRef]
269. Dini, I. Contribution of Nanoscience Research in Antioxidants Delivery Used in Nutricosmetic Sector. *Antioxidants* **2022**, *11*, 563. [CrossRef]
270. Sabri, F.; Raphael, W.; Berthomier, K.; Fradette, L.; Tavares, J.R.; Virgilio, N. One-Step Processing of Highly Viscous Multiple Pickering Emulsions. *J. Colloid Interface Sci.* **2020**, *560*, 536–545. [CrossRef] [PubMed]
271. Lin, C.; Debeli, D.K.; Gan, L.; Deng, J.; Hu, L.; Shan, G. Polyether-modified siloxane stabilized dispersion system on the physical stability and control release of double (W/O/W) emulsions. *Food Chem.* **2020**, *332*, 127381. [CrossRef] [PubMed]
272. Ying, X.; Gao, J.; Lu, J.; Ma, C.; Lv, J.; Adhikari, B.; Wang, B. Preparation and drying of water-in-oil-in-water (W/O/W) double emulsion to encapsulate soy peptides. *Food Res. Int.* **2021**, *141*, 110148. [CrossRef] [PubMed]
273. Giroux, H.J.; Shea, R.; Sabik, H.; Fustier, P.; Robitaille, G.; Britten, M. Effect of oil phase properties on peptide release from water-in-oil-in-water emulsions in gastrointestinal conditions. *LWT* **2019**, *109*, 429–435. [CrossRef]

*antioxidants*

MDPI

Article

# Fucosterol Isolated from Dietary Brown Alga *Sargassum horneri* Protects TNF-α/IFN-γ-Stimulated Human Dermal Fibroblasts Via Regulating Nrf2/HO-1 and NF-κB/MAPK Pathways

**Kirinde Gedara Isuru Sandanuwan Kirindage [1], Arachchige Maheshika Kumari Jayasinghe [1], Eui-Jeong Han [1], Youngheun Jee [2], Hyun-Jin Kim [3], Sun Gil Do [3], Ilekuttige Priyan Shanura Fernando [4,*] and Ginnae Ahn [1,*]**

1 Department of Food Technology and Nutrition, Chonnam National University, Yeosu 59626, Korea; 218388@jnu.ac.kr (K.G.I.S.K.); 218385@jnu.ac.kr (A.M.K.J.); iosu5772@jnu.ac.kr (E.-J.H.)

2 Department of Veterinary Medicine and Veterinary Medical Research Institute, Jeju National University, Jeju 63243, Korea; yhjee@jejunu.ac.kr

3 Research and Development Center, Naturetch Co., Ltd., Cheonnam-si 31257, Korea; hjmari@naturetech.co.kr (H.-J.K.); sgildo@naturetech.co.kr (S.G.D.)

4 Department of Marine Bio-Food Sciences, Chonnam National University, Yeosu 59626, Korea

* Correspondence: shanura@chonnam.ac.kr (I.P.S.F.); gnahn@jnu.ac.kr (G.A.)

**Abstract:** *Sargassum horneri* is a well-known edible brown alga that is widely abundant in the sea near China, Korea, and Japan and has a wide range of bioactive compounds. Fucosterol (FST), which is a renowned secondary metabolite in brown algae, was extracted from *S. horneri* to 70% ethanol, isolated via high-performance liquid chromatography (HPLC), followed by the immiscible liquid-liquid separation, and its structure was confirmed by NMR spectroscopy. The present study was undertaken to investigate the effects of FST against oxidative stress, inflammation, and its mechanism of action in tumor necrosis factor (TNF)-α/interferon (IFN)-γ-stimulated human dermal fibroblast (HDF). FST was biocompatible with HDF cells up to the 120 μM dosage. TNF-α/IFN-γ stimulation significantly decreased HDF viability by notably increasing reactive oxygen species (ROS) production. FST dose-dependently decreased the intracellular ROS production in HDFs. Western blot analysis confirmed a significant increment of nuclear factor erythroid 2-related factor 2 (Nrf2)/ heme oxygenase-1 (HO-1) involvement in FST-treated HDF cells. In addition, the downregulation of inflammatory mediators, molecules related to connective tissue degradation, and tissue inhibitors of metalloproteinases were identified. TNF-α/IFN-γ stimulation in HDF cells increased the phosphorylation of nuclear factor-κB (NF-κB) and mitogen-activated protein kinase (MAPK) mediators, and its phosphorylation was reduced with the treatment of FST in a dose-dependent manner. Results obtained from western blot analysis of the NF-κB nuclear translocation were supported by immunocytochemistry results. Collectively, the outcomes suggested that FST significantly upregulates the Nrf2/HO-1 signaling and regulates NF-κB/MAPK signaling pathways to minimize the inflammatory responses in TNF-α/IFN-γ-stimulated HDF cells.

**Keywords:** *Sargassum horneri*; fucosterol; human dermal fibroblasts; Nrf2/HO-1; MAPK; NF-κB

**Citation:** Kirindage, K.G.I.S.; Jayasinghe, A.M.K.; Han, E.-J.; Jee, Y.; Kim, H.-J.; Do, S.G.; Fernando, I.P.S.; Ahn, G. Fucosterol Isolated from Dietary Brown Alga *Sargassum horneri* Protects TNF-α/IFN-γ- Stimulated Human Dermal Fibroblasts Via Regulating Nrf2/HO-1 and NF-κB/MAPK Pathways. *Antioxidants* **2022**, *11*, 1429. https://doi.org/10.3390/antiox11081429

Academic Editors: Irene Dini and Sonia Laneri

Received: 13 July 2022

Accepted: 22 July 2022

Published: 23 July 2022



## 1. Introduction

Since ancient times, humans have used certain species of marine algae as culinary ingredients. *Sargassum horneri* is a type of edible marine brown algae that is abundant in the sea nearby China, Korea, and Japan. Apart from its dietary uses, *S. horneri* has gained increased attention towards research on antioxidants, and anti-inflammatory agents due to its high content of bioactive compounds, such as sulfated polysaccharides, sargachromenol, phlorotannins, phenolic substances, and proteoglycans [1–3]. Although *S. horneri* was first reported as a naturally available source in 1962 in the Manual of Chinese Economic Seaweeds, it is more suitable in modern medicine due to its promising protective role against bacteria, anti-proliferative effects on cancer cells, and anti-inflammatory and antioxidant activity. The huge boom of future drug developments focuses more on synergetic pharmacological effects rather than the "one-target, one-drug" approach [4]. Several studies indicated that lipid compounds, such as carotenoids and sterols, are equally beneficial to human health while these substances are prevalent in diverse brown seaweeds [5]. Prior investigation revealed the presence of fucosterol (FST) as a principal sterol in brown seaweeds [6,7]. Moreover, FST is relatively abundant in *Sargassum* spp. [8]. In particular, the protective effect of FST against acute liver injury [9], urban particulate matter-induced injury and inflammation in human lung epithelial cells [10], and particulate matter-induced skin lesions [11], as well as antioxidative and anti-inflammatory effects on macrophages [12] have been reported. However, there is no evidence that FST has an anti-inflammatory effect on human dermal fibroblasts (HDF) stimulated by tumor necrosis factor (TNF)-α/interferon (IFN)-γ.

Inflammation is a complex biological response that takes place in the body for maintaining homeostasis alongside internal or external stimuli. Under specific circumstances, inflammation can increase the risk of numerous diseases reaching a chronic level. Inflammatory responses in the skin are accompanied by the occurrence of oxidative stress from intracellular reactive oxygen species (ROS) that can damage tissues and cause chronic inflammatory illnesses [13]. TNF-α and IFN-γ are inflammatory mediators that have the ability to promote oxidative stress and inflammatory reactions in cells. It causes abnormal expression of inflammatory cytokines, chemokines, and inflammatory mediated signaling pathways [14,15]. The nuclear transcription factor kappa B (NF-κB) pathway is activated by dysregulated ROS generation in cells, which promotes phosphorylation of IκBα and NF-κB and enables the NF-κB p65 subunit to enter the nucleus from the cytosol. Moreover, the mitogen-activated protein kinase (MAPK) pathway is activated by the phosphorylation of ERK, JNK, and p38. Activation of the NF-κB and MAPK pathways collectively stimulates inflammatory responses by producing inflammatory cytokines and chemokines. The nuclear factor erythroid 2-related factor 2 (Nrf2)/heme oxygenase-1 (HO-1) signaling is considered a key pathway, which protects cells by boosting the production of antioxidant enzyme genes which is essential for defending cells from inflammatory responses [13,16,17]. Keratinocytes are the outermost live cells of the skin that are directly exposed to external stimuli which may occur inflammatory reactions and release inflammatory mediators [13]. It can affect underneath cells such as HDFs to cause inflammatory reactions. In addition, HDF cells themselves can indicate inflammation, and the production of matrix metalloproteinases (MMPs), which are involved in the degradation of connective tissue components in the skin by increasing oxidative stress [18]. Moreover, external stimuli can increase oxidative stress in HDFs, activate protective as well as apoptotic pathways, and ultimately cause abnormal inflammatory reactions.

Skincare and related nutricosmetic products which contain biologically active natural ingredients are seen as the next big innovation in the cosmetic industry [19]. The present study aims to isolate bioactive compounds by using the immiscible liquid–liquid separation of 70% ethanol extract of *S. horneri*. High-performance liquid chromatography (HPLC) was implicated to identify prominent UV absorbing peaks assuming the hexane fraction contains a significant number of sterols. The collected fraction from HPLC was identified as FST by subjecting nuclear magnetic resonance (NMR) spectroscopy. Further experiments were designed, hypothesizing that FST isolated from 70% ethanol extract of *S. horneri* possesses potent antioxidant and anti-inflammatory effects on TNF-α/IFN-γ-stimulated HDF cells by regulating the Nrf2/ HO-1 signaling pathway, thereby regulating the NF-κB, MAPK signaling, and production of inflammatory cytokines and chemokines.

## 2. Materials and Methods

### 2.1. Materials

*S. horneri* samples were collected from the western shores of Jeju Island, South Korea. Recombinant TNF-α and IFN-γ were purchased from R&D Systems (Minneapolis, MN, USA). Dulbecco's modified eagle medium (DMEM), and a mixture of streptomycin and penicillin (P/S) as antibiotics were purchased from GibcoBRL (Grand Island, NY, USA), and Fetal bovine serum (FBS) was purchased from Welgene (Gyeongsangbuk-do, South Korea). 2,2-azino-bis (3-ethylbenzothiazoline-6-sulfonic acid) diammonium salt (ABTS), 2,2-diphenyl-1-picrylhydrazyl (DPPH), 2′7′-dichlorodihydrofluorescein diacetate (DCFH-DA), 3-(4,5-dimethylthiazol-2-yl)-2,5-diphenyltetrazolium bromide (MTT), Dimethyl sulfoxide (DMSO), bovine serum albumin (BSA), Folin and Ciocalteu's phenol reagent, ethidium bromide, and agarose were bought from Sigma-Aldrich (St. Louis, MO, USA). D-glucose was purchased from Junsei Chemical Co., Ltd. (Tokyo, Japan). BCA protein assay kit, NE-PER® nuclear and cytoplasmic extraction kit, 1-Step transfer buffer, Pierce™ RIPA buffer, protein ladder, and SuperSignal™ West Femto Maximum Sensitivity Substrate were purchased from Thermo Fisher Scientific (Rockford, IL, USA). Antibodies needed for the western blot analysis were purchased from Santa Cruz Biotechnology Inc. (Dallas, TX, USA) and Cell Signaling Technology Inc. (Beverly, MA, USA). Skim milk powder was obtained from BD Difco™ (Sparks, MD, USA). Normal goat serum, Prolong® Gold antifade reagent with DAPI reagent, and DyLihgtTM 554 Phalloidin were purchased from Cell Signaling Technology (Danvers, MA, USA). The primers for the reverse transcription-polymerase chain reaction (RT-PCR) were purchased from Bioneer Co. (Deadeock-gu, Daejeon, Korea). The remaining chemicals and reagents used were of analytical grade.

### 2.2. Sample Collection and Extraction

Samples were collected from the western shores of Jeju Island, South Korea. *S. horneri* was identified by the Biodiversity Research Institute in Jeju, South Korea (voucher specimen (SH2017J005), and was kept in the laboratory of Marine Bioresource Technology at Jeju National University, South Korea). Sands and impurities were removed by washing with running water, air-dried at room temperature, and pulverized into powder by using IKA MF10 laboratory pulverizer (Staufen, Germany). The powder was extracted into 70% ethanol for 12 h and repeated three times. After centrifugation and filtration, the crude extract was obtained by concentrating the extract in a rotary evaporator.

### 2.3. Compositional Analysis of Crude Extract

The contents of total polyphenol, total protein, and carbohydrate were determined according to the method mentioned in one of the previous studies [20]. In brief, the dried crude extract was dissolved and incubated in the dark after mixing with Folin–Ciocalteu regent. Total polyphenolic compounds were measured by comparing the absorbance values with the gallic acid standard. Total protein content was measured by using the Lowry method. For that, the dissolved sample was incubated in the dark after mixing with Folin–Ciocalteu regent. The total protein content was calculated by using bovine serum albumin as the standard. Carbohydrate content was measured by using phenol–sulphuric method. For that sample, it was dissolved and then mixed with phenol and sulphuric acid. Then, absorbance values were obtained after incubating in the dark. Carbohydrate content was calculated by comparing it with the standard glucose series.

### 2.4. High-Performance Liquid Chromatography (HPLC) Analysis and Structural Identification of Compounds

The crude extract was dissolved in water and partitioned with hexane by thoroughly mixing and equilibrating in a separatory funnel. The resulted hexane fraction was dried by removing the solvents using a rotary evaporator and resolved by HPLC system equipped with an 'Ultimate 3000'-variable wavelength detector. Among the four different fractions, the concentrated hexane fraction was resolved in a BioBasic SEC-60 (300 $\times$ 7.3 mm) PREP column with the use of gradient acetonitrile: water (49:1) solvent system. Proton ($^1$H) NMR data were used to identify and elucidate the structure of the isolated compound. For that, the compound was dissolved in $CDCl_3$ and analyzed by an AVANCE III HD 400 spectrometer (400 MHz) (Bruker, Mundelein, IL, USA).

### 2.5. Cell Culture

HDF (KCLB, Seoul, Korea) were cultured and maintained in DMEM media supplemented with 25% F-12, 10% FBS, and 1% penicillin/streptomycin mixture at 37 °C with 5% of $CO_2$ in a humidified atmosphere. Cells were sub-cultured once every 5 days while cell culture media was replaced once every 2 days with fresh media. When cells reached exponential growth during the passages of 3–6, they were accordingly seeded in multi-well plates, chamber slides, or cell culture dishes for further experiments.

### 2.6. Cell Viability and ROS Production Analysis

HDF cells seeded in a 96-well plate were treated with a series of FST concentrations and incubated for 1 h. Then, 10 μL of TNF-α/IFN-γ mixture in 1:1 ratio was added and incubated at 37 °C for 24 h. Then, MTT assay was conducted to investigate the cell viability as described in one of our previous publications [13]. In brief, the absorbance of formazan crystals dissolved in DMSO was measured by using a SpectraMax M2 microplate reader (Molecular Devices, Sunnyvale, CA, USA) at 570 nm. The effect of FST on intracellular ROS levels in HDF cells stimulated with TNF-α/IFN-γ was measured by 2′,7′-dichlorofluorescein diacetate (DCF-DA) assay, as described in one of the previous publications [20].

### 2.7. Western Blot Analysis

HDF cells were seeded at $2 \times 10^5$ cells/mL of cell counts in 10 cm culture dishes for 24 h. Cells were stimulated with 10 ng/mL TNF-α/IFN-γ after being treated with a series of FST concentrations for 2 h. Then, cells were harvested and lysed using the NE-PER® nuclear and cytoplasmic extraction kit (Thermo Scientific, Rockford, IL, USA). A BCA protein assay kit (Thermo Scientific, Rockford, IL, USA) was used to estimate the protein concentrations in cell lysate, and 30 μg of protein of each lysate were subjected to electrophoresis on 10% polyacrylamide gels. Then, we followed the same procedure as described in our previous study [20].

### 2.8. Immunofluorescence Analysis

Immunostaining was conducted according to the method described in the previous study with slight modifications [21]. In brief, a density of $1 \times 10^4$ cells/chamber was used for cell culture in chamber slides, and samples were treated accordingly after 24 h of incubation in a humidified atmosphere. Then, wells were rinsed with PBS after 30 min of TNF-α/IFN-γ exposure, fixed, and incubated in blocking buffer for 1 h before being incubated overnight with primary antibodies (anti-NF-κB p65). The cells were then rinsed in PBS, and treated with Alexa Fluor® 488 conjugated Anti-Mouse IgG for 2 h. Slides were covered with coverslips with Prolong® Gold antifade reagent containing DAPI, after being washed in PBS. Then an EVOS M5000 (Thermo Fisher Scientific, Waltham, MA, USA) Imaging microscope was used to visualize the cells.

### 2.9. Reverse Transcription-Polymerase Chain Reaction (RT-PCR)

RT-PCR was carried out to investigate the expression of mRNA of inflammatory cytokines. The experiment was conducted following the method described by Jayasinghe et al. [13]. In brief, HDF cells were seeded in 6 cm culture dishes and incubated for 24 h at 37 °C in a humidified atmosphere. Then cells were treated with a series of concentrations of FST prior to the stimulation of TNF-α/IFN-γ. RNA was collected from the harvested cells and cDNAs were synthesized from 17.5 μL of RNA solution (2 μg/μL) in each by using TaKaRa PCR Thermal Cycler (TaKaRa Bio Inc., Otsu, Japan). PCR for corresponding cytokines was carried out for 35 cycles in a TaKaRa PCR Thermal Cycler. Then, the relevant bands of ethidium bromide-stained PCR products were visualized after agarose gel electrophoresis by using Wisd WUV-L20 UV transilluminator (Daihan Scientific Co., Gang-won-Do, Korea). The final analysis was conducted using NIH Image J software (Version No. 1, US National Institutes of Health, Bethesda, MD, USA).

### 2.10. Statistical Analysis

All statistical analyses of the study were performed using the SPSS software (Version 24.0, Chicago, IL, USA). One-way analysis of variance (ANOVA) followed by Duncan's multiple range tests was used to evaluate the significant variations among data sets, and data were presented as the mean ± standard error of the mean (SEM). In this study, $p < 0.05$ was considered statistically significant.

## 3. Results

### 3.1. Extraction of S. horneri, Isolation of Fucosterol by HPLC, and Structural Elucidation

Dried *S. horneri* was ground to a powder and extracted to 70% ethanol for 24 h. The extraction yield of the *S. horneri* 70% ethanol extract (SHE) was 8.12 ± 0.26% from the initial dry weight. Of that, 3.92 ± 0.19% were carbohydrates and 1.09 ± 0.07% were proteins; meanwhile, polyphenols indicated the highest among all measured compositions, which were 14.82 ± 0.68% (Table 1). The results of this analysis do not differ significantly from previous research findings [22].

Based on the results of preliminary assessments, crude ethanol extract of *S. horneri* was further separated by the immiscible liquid–liquid separation method indicated in Figure 1A. Among all resulted fractions, the subsequent hexane fraction was subjected to HPLC based on the potent bioactivities identified through bioactivity evaluation. Seven HPLC fractions were obtained from the hexane fraction following the chromatogram presented in Figure 1B. The prominent fractions were initially screened for potent antioxidant and anti-inflammatory activities. The seventh fraction was recognized showing prominent bioactivities. HPLC was used to further purify a single peak in the chromatogram from the seventh fraction (Figure 1C). The analysis of $^1$H NMR, along with a comparison with the literature data, allowed its structure to be elucidated as FST [12]. Further, numerical values of spectral data of the $^1$H NMR, which are presented in Table 2, confirm the structure of the substance was FST.

**Table 1.** Composition of SHE.

| SHE | Composition % |
|---|---|
| Yield | 8.12 ± 0.26 |
| Carbohydrates | 3.92 ± 0.19 |
| Protein | 1.09 ± 0.07 |
| Total polyphenols | 14.82 ± 0.68 |

Mean ± SEM (all experiments were performed in triplicate ($n = 3$) to determine the repeatability).

**Table 2.** Numerical values of spectra of the $^1$H-NMR values of FST.

| No. | $^1$H-NMR Value |
|---|---|
| 1 | 1.81 (1H, m), 1.10 (1H, m) |
| 2 | 1.52 (1H, m), 1.38 (1H, m) |
| 3 | 3.51 (1H, m) |
| 4 | 2.17 (1H, m), 2.07 (1H, m) |
| 6 | 5.34 (1H, d) |
| 7 | 1.89 (1H, m), 1.60 (1H, m) |
| 8 | 1.43 (1H, m) |
| 9 | 0.91 (1H, m) |
| 11 | 1.50 (1H, m), 1.43 (1H, m) |
| 12 | 1.99 (1H, m, 1.16 (1H, m) |
| 14 | 1.01 (1H, m) |
| 15 | 1.58 (1H, m), 1.07 (1H, m) |
| 16 | 1.82 (1H, m), 1.26 (1H, m) |
| 17 | 1.16 (1H, m) |
| 18 | 0.69 (3H, s) |
| 19 | 0.99 (3H, s) |
| 20 | 1.40 (1H, m) |

**Table 2.** *Cont.*

| No. | $^1$H-NMR Value |
|---|---|
| 21 | 1.00 (3H, d) |
| 22 | 1.41 (1H, m), 1.09 (1H, m) |
| 23 | 2.03 (1H, m), 1.90 (1H, m) |
| 25 | 2.20 (1H, m) |
| 26 | 0.97 (3H, s) |
| 27 | 0.97 (3H, s) |
| 28 | 5.17 (1H, dd) |
| 29 | 1.56 (3H, s) |

All spectra were recorded in $CDCL_3$ at 400 MHz.

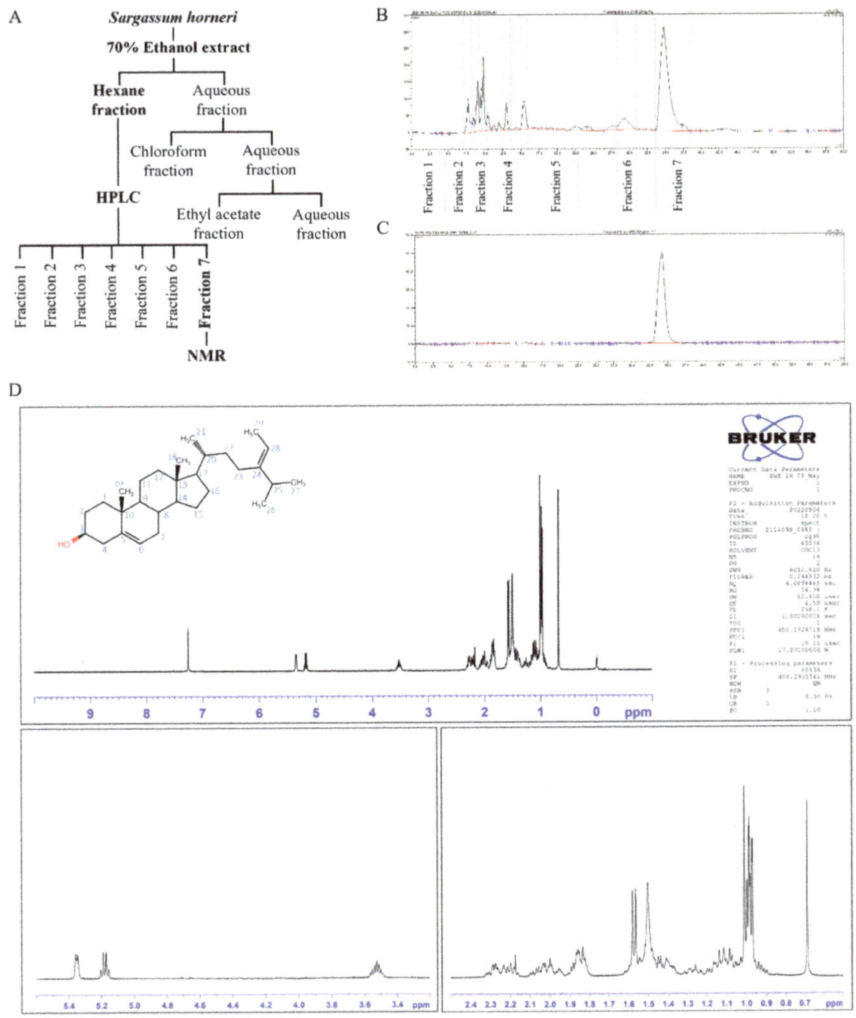


**Figure 1.** (**A**) Flow diagram representing the extraction and fractionation of *S. horneri* 70% ethanol extract. (**B**) HPLC chromatogram of hexane fraction. (**C**) HPLC chromatogram of prominent peak collected from hexane fraction and, (**D**) pure compound was characterized using $^1$H NMR analysis. All experiments were performed in triplicate ($n$ = 3) to determine repeatability.

### 3.2. Effect of Fucosterol on Cell Viability and Intracellular ROS Production

Significant cytotoxicity of FST on HDFs was not observed up to the concentration of 120 μM (Figure 2A). As shown in Figure 2B, TNF-α/IFN-γ-stimulation significantly decreased the cell viability of HDFs. Treatment of FST on stimulated HDFs significantly and dose-dependently increased the cell viability up to 120 μM. Henceforward, 30, 60, and 120 μM concentrations of FST were used throughout the study. TNF-α/IFN-γ stimulation increased intracellular ROS in HDFs, while treatment of FST significantly and dose-dependently reduced the ROS production (Figure 2C). This result was strengthened by the findings of DCF-DA flow cytometric analysis and DCF-DA fluorescence imaging which are illustrated in Figure 2D,E, respectively. FACs analysis with the use of fluoroprobe is a reliable method of cell sorting due to its ability to omit the error caused by cell death. Rightward shifting of the peaks on FITC-A axis is reduced with the increase in FST concentrations. Meanwhile, increased green fluorescence with TNF-α/IFN-γ-stimulation compared to the control cells was gradually reduced in the FST-pre-treated HDF cells. DCF-DA flow cytometric analysis, as well as fluorescence microscopy imaging, indicated the dose-dependent effect of FST on intracellular ROS production in TNF-α/IFN-γ-stimulated HDF cells. The bioactivity of FST concentrations (30, 60, and 120 μM) was compared with the positive control Indomethacin (50 μM).

### 3.3. Fucosterol Regulated the Nrf2/HO-1 Signaling

Activation of the Nrf2/ HO-1 signaling is a key pathway in the reduction of intracellular ROS generation and contributes to the regulation of the inflammatory responses and apoptosis in cells [23]. Result of the western blot analysis of Nrf2, HO-1, and NQO1, the pretreatment of FST boosted the nuclear translocation of Nrf2 in TNF-α/IFN-γ-stimulated HDFs and increased the levels of cytosolic HO-1 and NQO1 in a dose-dependent manner (Figure 3).

### 3.4. Fucosterol Downregulated Inflammatory Mediators, MMP and Tissue Inhibitors of Metalloproteinases (TIMP)

The effect of FST on the expression of inflammatory cytokines and chemokines was evaluated by using RT-PCR by examining the key inflammatory mediators. TNF-α/IFN-γ stimulation upregulated the mRNA expression levels of key inflammatory mediators compared to the control group and along with that, FST treatments downregulated the mRNA expressions of inflammatory cytokines (IL-6, IL-8, IL-13, IL-33, IL-1β, TNF-α, and IFN-γ) in HDF cells which were upregulated by the TNF-α/IFN-γ stimulation (Figure 4A,C).

Meanwhile, the expression of mRNA related to connective tissue degradation (MMP1, MMP2, MMP3, MMP8, MMP9, and MMP13) was evaluated by using RT-PCR. As denoted in Figure 4B,D, expression of the above-mentioned molecules was upregulated by the TNF-α/IFN-γ stimulation in HDF cells compared to control, and dose-dependently downregulated by the treatment of FST. TIMPs are vital for the regulation of MMP activities in fibroblasts, and the physiological functions of ECM are significantly reliant on the balance of TIMPs and MMPs [18,24]. TNF-α/IFN-γ stimulation increased the expression of TIMP1 and TIMP2 in HDF cells compared to the control, and particularly downregulated TIMP expression in a dose-dependent manner.

### 3.5. Fucosterol Regulated the NF-κB and MAPK Signaling

Activated canonical NF-κB signaling in cells are responding to the inflammation, immune response, cell proliferation, and differentiation stimulated by external stimuli [25]. Moreover, it is a well-known fact that NF-κB and MAPK are important upstream pathways that bear a major role in inflammatory responses [11]. As illustrated in Figure 5A, western blot analysis revealed that TNF-α/IFN-γ stimulation in HDF cells increase the phosphorylation of NF-κB mediators; cytosolic IκBα, NF-κB p65, and its phosphorylation is reduced with the treatment of FST in a dose-dependent manner. Besides, western blot analysis indicated that nuclear translocation of NF-κB p65 is reduced in the same manner which increased with the TNF-α/IFN-γ stimulation. Similarly, immunofluorescence analysis indicated that the FST dose-dependently reduced NF-κB p65 nuclear translocation in TNF-α/IFN-γ-stimulated HDF cells. The intense green fluorescence in immunostaining implied an increased NF-κB p65 nuclear translocation in stimulated cells (Figure 5B). In a similar manner, phosphorylated p38, ERK, and JNK expressions in TNF-α/IFN-γ stimulated HDF cells were dose-dependently downregulated with the FST (Figure 5C).

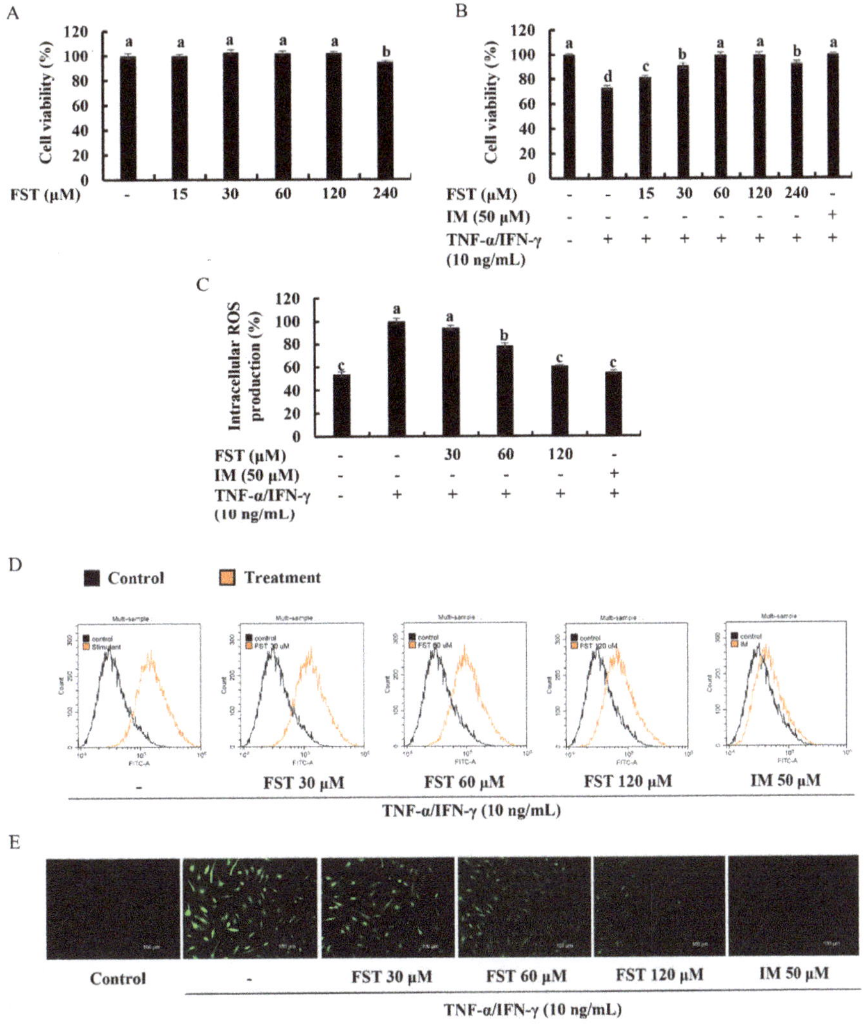


**Figure 2.** Cytoprotective effect of FST on human dermal fibroblasts (HDF). (**A**) Cytotoxicity, (**B**) cell viability (%), and analysis of intracellular ROS generation through (**C**) fluorometry, (**D**) flow cytometry analysis, and (**E**) fluorescence microscopy of FST-pre-treated TNF-α/IFN-γ-stimulated HDF cells. Indomethacin (IM, 50 μM) was used as a positive control. All experiments were performed in triplicate ($n = 3$) to determine the repeatability and lettered error bars were significantly different ($p < 0.05$).

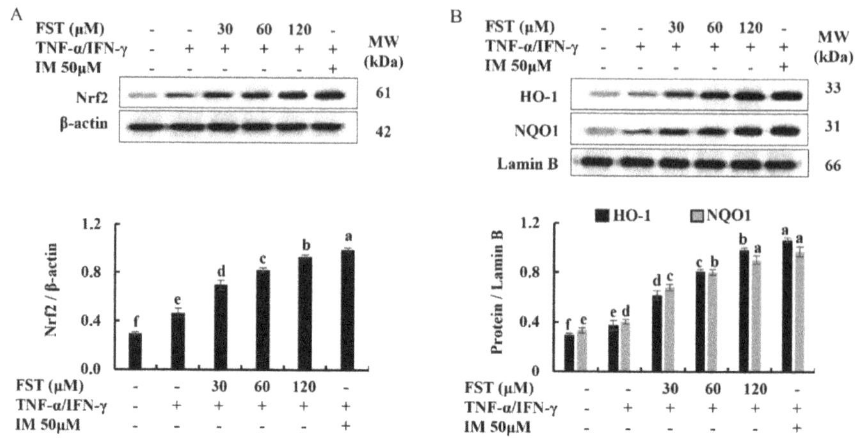


**Figure 3.** Dose-dependent effect of FST on Nrf2-mediated activation of HO-1 and NQO1 in TNF-α/IFN-γ-stimulated HDF cells. (**A**) Nrf2 expression, and (**B**) HO-1 and NQO1 expression. All experiments were performed in triplicate ($n = 3$) to determine the repeatability and lettered error bars were significantly different within the same molecule ($p < 0.05$).

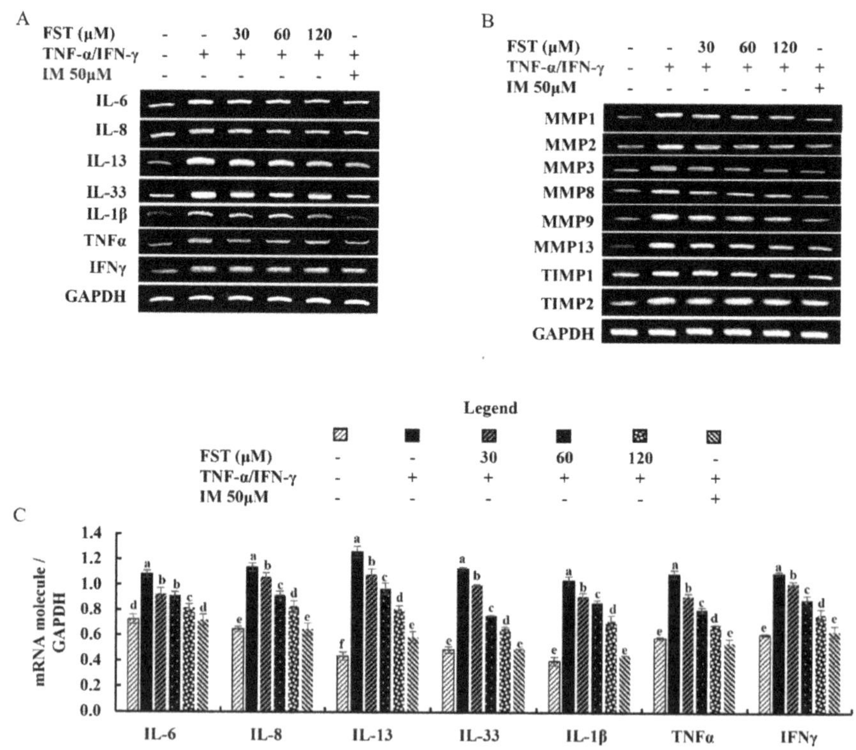


**Figure 4.** *Cont.*

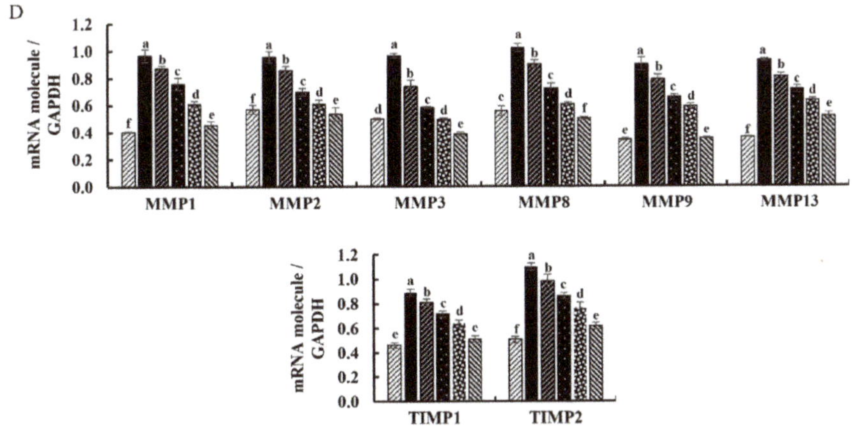


**Figure 4.** Inhibitory effect of FST on mRNA expression of inflammatory mediators, MMPs, and TIMP in TNF-α/IFN-γ-stimulated HDF cells. Effect of FST on (**A**) inflammatory cytokine expression, and (**B**) MMPs and TIMPs expression in TNF-α/IFN-γ-stimulated HDF cells evaluated by RT-PCR analysis. Densitometric analysis of (**C**) inflammatory cytokine expression, and (**D**) MMPs and TIMPs expression in TNF-α/IFN-γ-stimulated HDF cells. All experiments were performed in triplicate ($n = 3$) to determine the repeatability and lettered error bars were significantly different within the same molecule ($p < 0.05$).

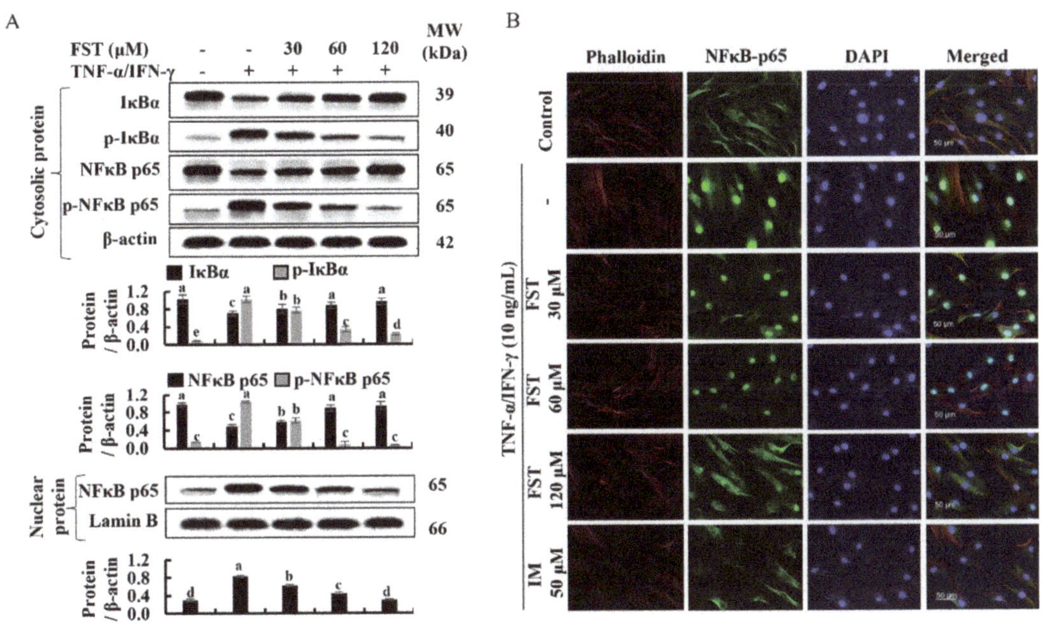


**Figure 5.** *Cont.*

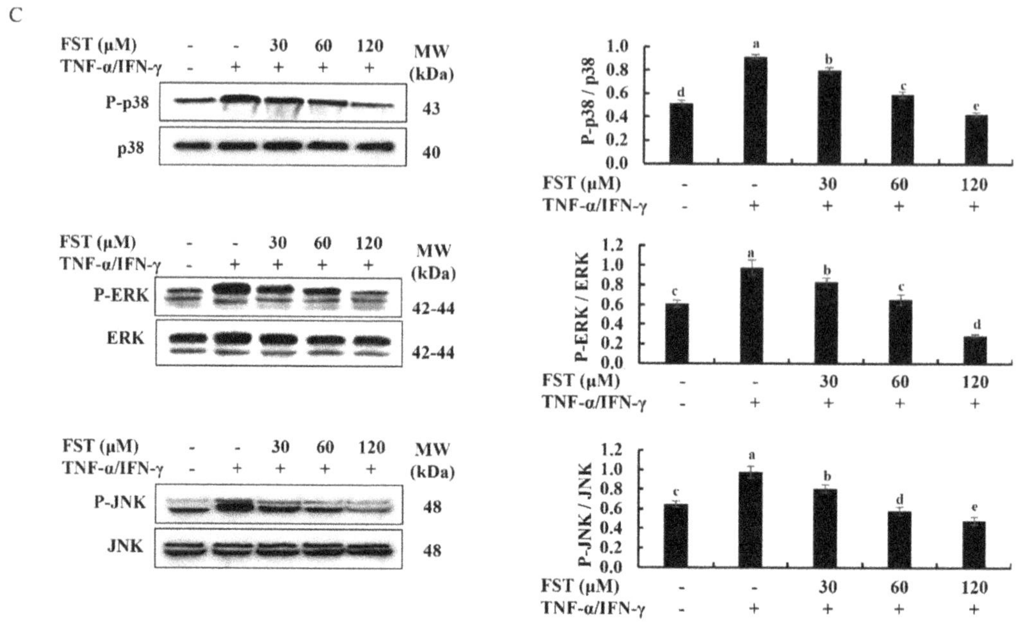


**Figure 5.** Protective effect of FST on NF-κB and MAPK inflammatory mediators in TNF-α/IFN-γ-stimulated HDF cells. Levels of molecular mediators were assessed by (**A**) western blot analysis for NF-κB, (**B**) evaluation of NF-κB p65 nuclear translocation by immunofluorescence analysis, and (**C**) western blot analysis of MAPK signaling. All experiments were performed in triplicate ($n = 3$) to determine the repeatability, and lettered error bars were significantly different ($p < 0.05$).

## 4. Discussion

During the last few decades, inflammation and related complications are a focus of extensive research. It has emerged as one of the hottest topics in medical science due to its impact on the human body at chronic levels, as well as the swift growth of skin health-related concerns in the public. The inflammatory response, as a whole, is considered a primary protective mechanism that takes place as a consequence of alterations in tissue homeostasis caused by a range of stimuli, such as pathogens, tissue damage, or pollutants, and includes the activation of innate and adaptive immunity in the human body [23,26]. Nonetheless, when inflammatory reactions go beyond the controllable limit and become recurring in the body, they can cause chronic inflammatory diseases, such as atopic dermatitis, cardiovascular diseases, cancer, bronchitis, and asthma, and it can be life-threatening [13]. Inflammatory and phagocytic cells produce a variety of chemical mediators and signaling molecules, such as histamine, serotonin, leukotrienes, and prostaglandins, which contribute to the onset of inflammation [27]. Scientific evidence revealed that exacerbation of inflammatory skin diseases has occurred with the excessive levels of ROS accumulation in the cellular environment. It leads to oxidative modifications and biomolecular damage, while triggering the inflammatory signaling cascades [28]. Implementing naturally available bioactive metabolites to treat numerous diseases, including skin inflammation-related disorders, has gained attention in the current context of pharmaceutical developments [10,29].

Brown seaweeds, such as *S. horneri*, are well-known sources of sterols, and many studies highlighted that investigating the isolation of bioactive compounds and their therapeutic potential is crucial [30,31]. Previous studies have revealed that FST isolated from edible brown alga, such as *Eisenia bicyclia* and *Undaria pinnatifida*, have potent anti-inflammatory effects on RAW 264.7 macrophages, which stimulate to produce nitric oxide inside cells by lipopolysaccharide [16,32]. FST isolated from 70% ethanol extract of *S. horneri* was used in the present study. Immiscible liquid–liquid separation of crude ethanol extract was implemented to separate compounds in the crude extract by using their solubilities in hexane, chloroform, ethyl acetate, and water. The most effective fraction was non-cytotoxic, and had a good effect of inhibition of ROS production in TNF-α/IFN-γ-stimulated HDF cells, was continued for further investigations. According to that, the hexane fraction was chosen and the most prominent peak on the HPLC chromatograph was collected. The collected fraction was dried by evaporating the solvent and dry powder was identified as FST by using $^{1}$H NMR analysis.

Then, a range of experiments was carried out to investigate the anti-inflammatory effect of FST isolated from *S. horneri* in TNF-α/IFN-γ-stimulated HDF cells. Many studies revealed that TNF-α/IFN-γ triggers ROS and pro-inflammatory cytokine production in vitro [13,14,33]. Dysregulated ROS production in cells leads to inflammatory gene expression by mediating redox-based activation of the NF-κB signaling pathway [16]. As indicated in the Figure 2, DCF-DA fluorometry, fluorescence microscopic imaging, as well as FACs analysis, confirmed that TNF-α/IFN-γ stimulation promptly augmented the intracellular ROS production in HDF cells. In particular, fluorometry represents the level of ROS in cells quantitatively, and flow cytometry analysis of DCF-DA is considered a trustworthy approach, whereas it measures the fluorescence intensity of each cell [18]. Different concentrations of FST (30, 60, and 120 μM) decreased the ROS levels in TNF-α/IFN-γ-stimulated HDF cells in a dose-dependent manner.

Accumulated shreds of evidence suggest that the significance of the Nrf2/HO-1 signaling is stimuli-specific and cell type-specific [17]. Nrf2 is a redox-sensitive transcription factor that resides in the cytoplasm as an inactive complex with Kelch-like ECH-associated protein 1 (Keap1). It can regulate the production of numerous antioxidant enzymes, including HO-1, and enters into the nucleus and binds to the antioxidant response element site. Nrf2 protects diverse cells from oxidative stress by boosting the production of antioxidant enzyme genes and proteins when stimuli activate the associated pathways [34]. FST enhanced nuclear translocation of Nrf2 in a dose-dependent manner in stimulated HDFs, as expected. At the same time, the levels of HO-1 and NQO1 change in the same way. Moreover, several studies have revealed that HO-1 and its metabolites have anti-inflammatory actions that are mediated via Nrf2 [35]. According to the best understanding over the years, numerous inflammatory cytokines are overproduced when NF-κB is activated by oxidative stress. In the meantime, the elevation of HO-1 expression, which is mediated by activated Nrf2, leads to the inhibition of NF-κB signaling [23].

One of the adaptive immune responses to inflammatory stimuli is the release of pro-inflammatory cytokines and chemokines [22]. Inflammatory mediators in HDFs, in particular, play a critical role in modulating the structural integrity of the extracellular matrix of the skin by regulating MMP, collagenase, and elastase transcription [36]. A recent study conducted by Jayasinghe et al. contained evidence of the exposure of the epidermal cells to TNF-α/IFN-γ stimulating abnormal expression of cytokines, chemokines, and inflammatory mediated signaling pathways [13]. Expression levels of IL-6, IL-8, IL-13, IL-33, IL-1β, TNF-α, and IFN-γ are significantly increased in TNF-α/IFN-γ-stimulated HDF cells, while dose-dependently downregulated by the FST treatment. Aside from that, the expression of chosen MMPs was investigated further to determine the influence of FST on skin inflammation. Even though MMPs are involved in the regeneration of normal tissues, an aberrant increase in expressions of some MMPs, such as MMP1, MMP8, and MMP13, in dermal fibroblasts is associated with type I and type III collagen (ECM) degeneration [37]. Numerous studies linked IL-6 and TNF-α as implicated in the regulation of TIMP1, MMP1, and MMP9s' mRNA expression. Furthermore, it regulates activator protein-1 (AP-1) activation, NF-κB mediators, and the expression of MMP1, MMP3, and MMP9 in HDFs [24,38,39]. RT-PCR results revealed that increased levels of MMP1, MMP2, MMP3, MMP8, MMP9, MMP13, TIMP1, and TIMP2 due to TNF-α/IFN-γ stimulation in HDFs are down-regulated with the FST treatments. Results imply that TNF-α/IFN-γ stimulation in HDFs has a higher probability of inducing skin inflammation in underlying mechanisms, while FST treatment is noteworthy in downregulating.

Furthermore, TNF-α/IFN-γ stimulation of HDFs significantly activated crucial intracellular upstream signaling molecules, such as NF-κB and MAPK. As reported, MAPK proteins are also important for ECM breakdown [18]. According to the findings from western blot analysis, the expression levels of cytosolic p-IκBα, p-NF-κB p65, and nucleic NF-κB p65 were downregulated along with the FST doses in stimulated HDFs. Results conclude that FST positively regulates the nuclear translocation of NF-κB p65, which is strengthened by the findings of immunocytochemistry analysis of NF-κB p65 nuclear translocation. TNF-α/IFN-γ stimulation of HDFs significantly increases the phosphorylation of MAPK mediators which are JNK, ERK, and p38, while FST lowers phosphorylation in a dose-dependent manner. Thus, the findings of this study reveal that FST effectively regulates the Nrf2/HO-1 signaling, NF-κB, and MAPK pathways, and, hence, the regulation of inflammatory cytokines in TNF-α/IFN-γ-stimulated HDFs.

## 5. Conclusions

Based on the present study, hexane fraction collected from the 70% ethanol extract of *S. horneri* in an immiscible liquid–liquid separation system contains FST, which shows a prominent peak at 220 nm of UV absorbance in HPLC analysis. FST isolated from the HPLC possesses potent anti-inflammatory activities on HDFs that are stimulated with TNF-α/IFN-γ in vitro. Results show that the regulation of Nrf2/HO-1, as well as NF-κB and MAPK signaling collectively, contribute to the anti-inflammatory activity of FST in stimulated HDFs.

**Author Contributions:** Conceptualization, K.G.I.S.K., I.P.S.F., Y.J. and G.A.; methodology, K.G.I.S.K. and I.P.S.F.; validation, I.P.S.F., Y.J. and E.-J.H.; formal analysis, K.G.I.S.K., A.M.K.J.; investigation, K.G.I.S.K., A.M.K.J.; resources, G.A., Y.J., H.-J.K., S.G.D.; data curation, K.G.I.S.K., A.M.K.J. and E.-J.H.; writing—original draft preparation, K.G.I.S.K.; writing—review and editing, I.P.S.F. and A.M.K.J.; visualization, K.G.I.S.K., A.M.K.J. and E.-J.H.; supervision, G.A.; project administration, G.A., H.-J.K. and S.G.D.; funding acquisition, G.A., H.-J.K. and S.G.D.; All authors have read and agreed to the published version of the manuscript.

**Funding:** This research was funded by the Ministry of Oceans and Fisheries, Korea (Grants No. M01201920150306), as a part of the project titled "Development of functional materials derived from marine living resources for regulation of immune diseases".

**Institutional Review Board Statement:** Not applicable.

**Informed Consent Statement:** Not applicable.

**Data Availability Statement:** Data are included within the article.

**Conflicts of Interest:** The authors declare no conflict of interest.

## References

1. Fernando, I.S.; Jayawardena, T.U.; Sanjeewa, K.A.; Wang, L.; Jeon, Y.-J.; Lee, W.W. Anti-inflammatory potential of alginic acid from *Sargassum horneri* against urban aerosol-induced inflammatory responses in keratinocytes and macrophages. *Ecotoxicol. Environ. Saf.* **2018**, *160*, 24–31. [CrossRef] [PubMed]
2. Wen, Z.-S.; Xiang, X.-W.; Jin, H.-X.; Guo, X.-Y.; Liu, L.-J.; Huang, Y.-N.; OuYang, X.-K.; Qu, Y.-L. Composition and anti-inflammatory effect of polysaccharides from Sargassum horneri in RAW264. 7 macrophages. *Int. J. Biol. Macromol.* **2016**, *88*, 403–413. [CrossRef] [PubMed]
3. Sanjeewa, K.K.A.; Fernando, I.P.S.; Kim, E.-A.; Ahn, G.; Jee, Y.; Jeon, Y.-J. Anti-inflammatory activity of a sulfated polysaccharide isolated from an enzymatic digest of brown seaweed Sargassum horneri in RAW 264.7 cells. *Nutr. Res. Pract.* **2017**, *11*, 3–10. [CrossRef] [PubMed]
4. Casas, A.I.; Hassan, A.A.; Larsen, S.J.; Gomez-Rangel, V.; Elbatreek, M.; Kleikers, P.W.; Guney, E.; Egea, J.; López, M.G.; Baumbach, J. From single drug targets to synergistic network pharmacology in ischemic stroke. *Proc. Natl. Acad. Sci. USA* **2019**, *116*, 7129–7136. [CrossRef] [PubMed]

5. Terasaki, M.; Kawagoe, C.; Ito, A.; Kumon, H.; Narayan, B.; Hosokawa, M.; Miyashita, K. Spatial and seasonal variations in the biofunctional lipid substances (fucoxanthin and fucosterol) of the laboratory-grown edible Japanese seaweed (*Sargassum horneri* Turner) cultured in the open sea. *Saudi J. Biol. Sci.* **2017**, *24*, 1475–1482. [CrossRef] [PubMed]
6. Kalsait, R.P.; Khedekar, P.B.; Saoji, A.N.; Bhusari, K.P. Isolation of phytosterols and antihyperlipidemic activity of Lagenaria siceraria. *Arch. Pharmacal Res.* **2011**, *34*, 1599–1604. [CrossRef]
7. Juárez-Portilla, C.; Olivares-Bañuelos, T.; Molina-Jiménez, T.; Sánchez-Salcedo, J.A.; Del Moral, D.I.; Meza-Menchaca, T.; Flores-Muñoz, M.; López-Franco, Ó.; Roldán-Roldán, G.; Ortega, A. Seaweeds-derived compounds modulating effects on signal transduction pathways: A systematic review. *Phytomedicine* **2019**, *63*, 153016. [CrossRef] [PubMed]
8. Rosa, G.P.; Tavares, W.R.; Sousa, P.M.; Pagès, A.K.; Seca, A.M.; Pinto, D.C. Seaweed secondary metabolites with beneficial health effects: An overview of successes in in vivo studies and clinical trials. *Mar. Drugs* **2019**, *18*, 8. [CrossRef]
9. Mo, W.; Wang, C.; Li, J.; Chen, K.; Xia, Y.; Li, S.; Xu, L.; Lu, X.; Wang, W.; Guo, C. Fucosterol protects against concanavalin A-induced acute liver injury: Focus on P38 MAPK/NF-κB pathway activity. *Gastroenterol. Res. Pract.* **2018**, *2018*, 2824139. [CrossRef] [PubMed]
10. Fernando, I.S.; Jayawardena, T.U.; Kim, H.-S.; Lee, W.W.; Vaas, A.; De Silva, H.; Abayaweera, G.; Nanayakkara, C.; Abeytunga, D.; Lee, D.-S. Beijing urban particulate matter-induced injury and inflammation in human lung epithelial cells and the protective effects of fucosterol from Sargassum binderi (Sonder ex J. Agardh). *Environ. Res.* **2019**, *172*, 150–158. [CrossRef] [PubMed]
11. Fernando, I.S.; Jayawardena, T.U.; Kim, H.-S.; Vaas, A.; De Silva, H.; Nanayakkara, C.; Abeytunga, D.; Lee, W.; Ahn, G.; Lee, D.-S. A keratinocyte and integrated fibroblast culture model for studying particulate matter-induced skin lesions and therapeutic intervention of fucosterol. *Life Sci.* **2019**, *233*, 116714. [CrossRef] [PubMed]
12. Jayawardena, T.U.; Sanjeewa, K.K.A.; Lee, H.-G.; Nagahawatta, D.P.; Yang, H.-W.; Kang, M.-C.; Jeon, Y.-J. Particulate Matter-Induced Inflammation/Oxidative Stress in Macrophages: Fucosterol from Padina boryana as a Potent Protector, Activated via NF-κB/MAPK Pathways and Nrf2/HO-1 Involvement. *Mar. Drugs* **2020**, *18*, 628. [CrossRef]
13. Jayasinghe, A.M.K.; Kirindage, K.G.I.S.; Fernando, I.P.S.; Han, E.J.; Oh, G.-W.; Jung, W.-K.; Ahn, G. Fucoidan Isolated from Sargassum confusum Suppresses Inflammatory Responses and Oxidative Stress in TNF-α/IFN-γ-Stimulated HaCaT Keratinocytes by Activating Nrf2/HO-1 Signaling Pathway. *Mar. Drugs* **2022**, *20*, 117. [CrossRef] [PubMed]
14. Han, E.-J.; Fernando, I.P.S.; Kim, H.-S.; Lee, D.-S.; Kim, A.; Je, J.-G.; Seo, M.-J.; Jee, Y.-H.; Jeon, Y.-J.; Kim, S.-Y.; et al. (−)-Loliolide Isolated from Sargassum horneri Suppressed Oxidative Stress and Inflammation by Activating Nrf2/HO-1 Signaling in IFN-γ/TNF-α-Stimulated HaCaT Keratinocytes. *Antioxidants* **2021**, *10*, 856. [CrossRef] [PubMed]
15. Ju, S.M.; Song, H.Y.; Lee, S.J.; Seo, W.Y.; Sin, D.H.; Goh, A.R.; Kang, Y.-H.; Kang, I.-J.; Won, M.-H.; Yi, J.-S. Suppression of thymus-and activation-regulated chemokine (TARC/CCL17) production by 1, 2, 3, 4, 6-penta-O-galloyl-β-d-glucose via blockade of NF-κB and STAT1 activation in the HaCaT cells. *Biochem. Biophys. Res. Commun.* **2009**, *387*, 115–120. [CrossRef]
16. Jung, H.A.; Jin, S.E.; Ahn, B.R.; Lee, C.M.; Choi, J.S. Anti-inflammatory activity of edible brown alga Eisenia bicyclis and its constituents fucosterol and phlorotannins in LPS-stimulated RAW264. 7 macrophages. *Food Chem. Toxicol.* **2013**, *59*, 199–206. [CrossRef] [PubMed]
17. Lee, J.; Kim, S. Upregulation of heme oxygenase-1 expression by dehydrodiconiferyl alcohol (DHCA) through the AMPK–Nrf2 dependent pathway. *Toxicol. Appl. Pharmacol.* **2014**, *281*, 87–100. [CrossRef]
18. Fernando, I.P.S.; Dias, M.K.H.M.; Madusanka, D.M.D.; Kim, H.-S.; Han, E.-J.; Kim, M.-J.; Seo, M.-J.; Ahn, G. Effects of (−)-Loliolide against Fine Dust Preconditioned Keratinocyte Media-Induced Dermal Fibroblast Inflammation. *Antioxidants* **2021**, *10*, 675. [CrossRef] [PubMed]
19. Priyan Shanura Fernando, I.; Kim, K.-N.; Kim, D.; Jeon, Y.-J. Algal polysaccharides: Potential bioactive substances for cosmeceutical applications. *Crit. Rev. Biotechnol.* **2019**, *39*, 99–113. [CrossRef] [PubMed]
20. Kirindage, K.G.I.S.; Fernando, I.P.S.; Jayasinghe, A.M.K.; Han, E.-J.; Dias, M.K.H.M.; Kang, K.-P.; Moon, S.-I.; Shin, T.-S.; Ma, A.; Ahn, G. Moringa oleifera Hot Water Extract Protects Vero Cells from Hydrogen Peroxide-Induced Oxidative Stress by Regulating Mitochondria-Mediated Apoptotic Pathway and Nrf2/HO-1 Signaling. *Foods* **2022**, *11*, 420. [CrossRef]
21. Fernando, I.P.S.; Dias, M.K.H.M.; Madusanka, D.M.D.; Han, E.J.; Kim, M.J.; Jeon, Y.-J.; Ahn, G. Fucoidan refined by Sargassum confusum indicate protective effects suppressing photo-oxidative stress and skin barrier perturbation in UVB-induced human keratinocytes. *Int. J. Biol. Macromol.* **2020**, *164*, 149–161. [CrossRef] [PubMed]
22. Dias, M.K.H.M.; Madusanka, D.M.D.; Han, E.J.; Kim, H.-S.; Jeon, Y.-J.; Jee, Y.; Kim, K.-N.; Lee, K.; Fernando, I.P.S.; Ahn, G. Sargassum horneri (Turner) C. Agardh ethanol extract attenuates fine dust-induced inflammatory responses and impaired skin barrier functions in HaCaT keratinocytes. *J. Ethnopharmacol.* **2021**, *273*, 114003. [CrossRef]
23. Saha, S.; Buttari, B.; Panieri, E.; Profumo, E.; Saso, L. An overview of Nrf2 signaling pathway and its role in inflammation. *Molecules* **2020**, *25*, 5474. [CrossRef] [PubMed]
24. Dasu, M.R.; Barrow, R.E.; Spies, M.; Herndon, D.N. Matrix metalloproteinase expression in cytokine stimulated human dermal fibroblasts. *Burns* **2003**, *29*, 527–531. [CrossRef]
25. Yu, H.; Lin, L.; Zhang, Z.; Zhang, H.; Hu, H. Targeting NF-κB pathway for the therapy of diseases: Mechanism and clinical study. *Signal Transduct. Target. Ther.* **2020**, *5*, 209. [CrossRef]
26. Jayawardena, T.U.; Fernando, I.S.; Lee, W.W.; Sanjeewa, K.A.; Kim, H.-S.; Lee, D.-S.; Jeon, Y.-J. Isolation and purification of fucoidan fraction in Turbinaria ornata from the Maldives; Inflammation inhibitory potential under LPS stimulated conditions in in-vitro and in-vivo models. *Int. J. Biol. Macromol.* **2019**, *131*, 614–623. [CrossRef] [PubMed]

27. Vijayalakshmi, T.; Muthulakshmi, V.; Sachdanandam, P. Salubrious effect of Semecarpus anacardium against lipid peroxidative changes in adjuvant arthritis studied in rats. *Mol. Cell Biochem.* **1997**, *175*, 65–69. [CrossRef] [PubMed]
28. Xu, H.; Zheng, Y.-W.; Liu, Q.; Liu, L.-P.; Luo, F.-L.; Zhou, H.-C.; Isoda, H.; Ohkohchi, N.; Li, Y.-M. Reactive oxygen species in skin repair, regeneration, aging, and inflammation. *React. Oxyg. Species (ROS) Living Cells* **2018**, *8*, 69–87.
29. Sanjeewa, K.A.; Lee, W.; Jeon, Y.-J. Nutrients and bioactive potentials of edible green and red seaweed in Korea. *Fish. Aquat. Sci.* **2018**, *21*, 19. [CrossRef]
30. Xia, M.; Liu, C.; Gao, L.; Lu, Y. One-step preparative separation of phytosterols from edible brown seaweed Sargassum horneri by high-speed countercurrent chromatography. *Mar. Drugs* **2019**, *17*, 691. [CrossRef]
31. Ko, W.; Lee, H.; Kim, N.; Jo, H.G.; Woo, E.-R.; Lee, K.; Han, Y.S.; Park, S.R.; Ahn, G.; Cheong, S.H. The anti-oxidative and anti-neuroinflammatory effects of sargassum horneri by heme oxygenase-1 induction in BV2 and HT22 cells. *Antioxidants* **2021**, *10*, 859. [CrossRef] [PubMed]
32. Yoo, M.-S.; Shin, J.-S.; Choi, H.-E.; Cho, Y.-W.; Bang, M.-H.; Baek, N.-I.; Lee, K.-T. Fucosterol isolated from Undaria pinnatifida inhibits lipopolysaccharide-induced production of nitric oxide and pro-inflammatory cytokines via the inactivation of nuclear factor-κB and p38 mitogen-activated protein kinase in RAW264. 7 macrophages. *Food Chem.* **2012**, *135*, 967–975. [CrossRef] [PubMed]
33. Phung, H.M.; Lee, S.; Hong, S.; Lee, S.; Jung, K.; Kang, K.S. Protective Effect of Polymethoxyflavones Isolated from Kaempferia parviflora against TNF-α-Induced Human Dermal Fibroblast Damage. *Antioxidants* **2021**, *10*, 1609. [CrossRef] [PubMed]
34. Ko, W.; Yoon, C.-S.; Kim, K.-W.; Lee, H.; Kim, N.; Woo, E.-R.; Kim, Y.-C.; Kang, D.G.; Lee, H.S.; Oh, H. Neuroprotective and anti-inflammatory effects of Kuwanon C from Cudrania tricuspidata are mediated by heme oxygenase-1 in HT22 hippocampal cells, RAW264. 7 macrophage, and BV2 microglia. *Int. J. Mol. Sci.* **2020**, *21*, 4839. [CrossRef] [PubMed]
35. Ahmed, S.M.U.; Luo, L.; Namani, A.; Wang, X.J.; Tang, X. Nrf2 signaling pathway: Pivotal roles in inflammation. *Biochim. Biophys. Acta (BBA)-Mol. Basis Dis.* **2017**, *1863*, 585–597. [CrossRef] [PubMed]
36. Russo, B.; Brembilla, N.C.; Chizzolini, C. Interplay between keratinocytes and fibroblasts: A systematic review providing a new angle for understanding skin fibrotic disorders. *Front. Immunol.* **2020**, *11*, 648. [CrossRef]
37. Wang, L.; Kim, H.S.; Oh, J.Y.; Je, J.G.; Jeon, Y.-J.; Ryu, B. Protective effect of diphlorethohydroxycarmalol isolated from Ishige okamurae against UVB-induced damage in vitro in human dermal fibroblasts and in vivo in zebrafish. *Food Chem. Toxicol.* **2020**, *136*, 110963. [CrossRef] [PubMed]
38. Youn, U.J.; Nam, K.-W.; Kim, H.-S.; Choi, G.; Jeong, W.S.; Lee, M.Y.; Chae, S. 3-Deoxysappanchalcone inhibits tumor necrosis factor-α-induced matrix metalloproteinase-9 expression in human keratinocytes through activated protein-1 inhibition and nuclear factor-kappa B DNA binding activity. *Biol. Pharm. Bull.* **2011**, *34*, 890–893. [CrossRef] [PubMed]
39. Oh, J.H.; Joo, Y.H.; Karadeniz, F.; Ko, J.; Kong, C.-S. Syringaresinol inhibits UVA-induced MMP-1 expression by suppression of MAPK/AP-1 signaling in HaCaT keratinocytes and human dermal fibroblasts. *Int. J. Mol. Sci.* **2020**, *21*, 3981. [CrossRef]

antioxidants

Article

# Chestnut Wood Mud as a Source of Ellagic Acid for Dermo-Cosmetic Applications

**Federica Moccia [1], Davide Liberti [1], Samuele Giovando [2], Carla Caddeo [3], Daria Maria Monti [1], Lucia Panzella [1,*] and Alessandra Napolitano [1]**

1 Department of Chemical Sciences, University of Naples "Federico II", Via Cintia 4, I-80126 Naples, Italy
2 Centro Ricerche per la Chimica Fine Srl for Silvateam Spa, Via Torre 7, I-12080 San Michele Mondovì, Italy
3 Department of Scienze della Vita e dell'Ambiente, Sezione di Scienze del Farmaco, University of Cagliari, Via Ospedale 72, I-09124 Cagliari, Italy
* Correspondence: panzella@unina.it; Tel.: +39-081674131

**Abstract:** Ellagic acid (EA) has long been recognized as a very active antioxidant, anti-inflammatory, and antimicrobial agent. However, its low bioavailability has often hampered its applications in health-related fields. Here, we report a phospholipid vesicle-based controlled release system for EA, involving the exploitation of chestnut wood mud (CWM), an industrial by-product from chestnut tannin production, as a largely available and low-cost source of this compound. Two kinds of CWM with different particle size distributions, indicated as CWM-A and CWM-B (<100 and 32 μm, respectively), containing 5 ± 1% *w*/*w* EA, were incorporated into transfersomes. The latter were small in size (~100 nm), homogeneously dispersed, and negatively charged. 2,2-Diphenyl-1-picrylhydrazyl (DPPH) and ferric reducing/antioxidant power (FRAP) assays indicated up to three-fold improvement in the antioxidant properties of CWM upon incorporation into transfersomes. The kinetics of EA released under simulated physiological conditions were evaluated by UV-Vis spectroscopy and HPLC analysis. The best results were obtained with CWM-B (100% of EA gradually released after 37 days at pH 7.4). A stepwise increase in the antioxidant properties of the released material was also observed. Cell-based experiments confirmed the efficacy of CWM-B transfersomes as antioxidant agents in contrasting photodamage.

**Keywords:** ellagic acid; chestnut wood; antioxidant; controlled release; transfersomes; HaCaT; 2,2-diphenyl-1-picrylhydrazyl (DPPH) assay; ferric reducing/antioxidant power (FRAP) assay; UVA; reactive oxygen species

**Citation:** Moccia, F.; Liberti, D.; Giovando, S.; Caddeo, C.; Monti, D.M.; Panzella, L.; Napolitano, A. Chestnut Wood Mud as a Source of Ellagic Acid for Dermo-Cosmetic Applications. *Antioxidants* **2022**, *11*, 1681. https://doi.org/10.3390/antiox11091681

Academic Editor: Stanley Omaye

Received: 22 July 2022
Accepted: 25 August 2022
Published: 28 August 2022



## 1. Introduction

Ellagic acid (EA) is a phenolic compound naturally present in many red fruits and berries. Apart from being the main product of ellagitannin hydrolysis, it is endowed with remarkable biological properties, including antioxidant [1–3], anti-inflammatory [4], antimicrobial [5], antidiabetic [6], antiviral [7], antidegenerative [8], and anticancer activities [9]. In addition to systemic uses, topical applications of EA have been widely described [10]. Several studies have reported the potential use of EA for the prevention or treatment of skin disorders. For example, EA was found to be effective against skin tumors [11], contact dermatitis [12], or cutaneous leishmaniasis [13]. It can be used in wound bandaging [14], or as a photoprotective [15] and antiaging agent [16]. Furthermore, EA is considered a useful compound in the treatment of skin pigmentation disorders, such as hyperpigmentation, melasma, and other dyschromia [17].

Despite its remarkable properties, the wide application of EA is limited by its low permeability and low solubility in aqueous solvents. To overcome these drawbacks, several approaches have been proposed, involving modulation of EA solubility properties through encapsulation or chemical derivatization [18–23], and different type of formulations based on e.g., pectins [24,25], chitosan [25–27], chitin [28], zein [5], cellulose [29], cyclodextrins [30–32], poly(lactide-co-glycolide) (PLGA) [33], graphene oxide [34], alginate [35], and microalgae [36] have been designed for the controlled release of this compound.

In this context, liposomes (spherical vesicles composed of one or more bilayers formed by dispersion of phospholipids in aqueous medium) have been largely utilized as a drug delivery vehicle for administration of nutrients and pharmaceutical drugs in biomedical, food, and agricultural industries, and have also been exploited for enhancing the biological effects [37–39], improving skin permeation [40], and guaranteeing a sustained release [41] of EA. In particular, over the last few years, liposomes have been the target of reformulating studies aimed at producing vesicles capable of delivering active compounds to the deeper skin layers. A number of additives have been explored in combination with conventional components of liposomes, producing new classes of vesicles, such as transfersomes. Transfersomes are composed of phospholipids and an edge activator, which is a membrane-softening agent (e.g., Tween 80, Span 80, or sodium cholate) that makes the vesicle ultra-deformable and capable of penetrating the skin more efficiently than conventional liposomes [42–45].

In addition to the development of novel formulations to improve its bioavailability, another primary aim of the recent scientific research on EA is the discovery of sustainable, low cost and easily available sources of this compound, prompted by the global increasing demand for green products and processes. Among these sources, a prominent role is occupied by agri-food by-products such as pomegranate peel [46–50], although other ellagitannin-rich wastes have recently emerged as possible sources of EA. A noticeable example is represented by chestnut shell [51–53] as well as chestnut wood fiber, which is the residual exhausted material from chestnut tannin industrial production [54,55].

Within this scenario, we report herein the exploitation of chestnut wood mud (CWM) as an easily available source of EA for dermo-cosmetic applications upon incorporation into transfersomes. CWM is an industrial by-product of the chestnut tannin production, deriving from exhausted chestnut wood subjected to a natural fermentation process. The antioxidant properties of the samples were investigated by chemical assays and the protective effects on UVA-induced oxidative photodamage were evaluated on immortalized human keratinocytes (HaCaT). Finally, the controlled-release profile of EA under simulated physiological conditions was investigated by UV-Vis spectroscopy and HPLC.

## 2. Materials and Methods

### 2.1. Materials

CWM was provided by Silvateam (S. Michele Mondovì, Cuneo, Italy). CWM was first dried in an oven at 35 °C for one week, then ground in a common blender and finally passed through sieves to obtain two fractions with particle sizes lower than 100 and 32 μm, indicated as CWM-A and CWM-B, respectively.

Lipoid S75 (S75), a mixture of soybean phospholipids (70% phosphatidylcholine, 9% phosphatidylethanolamine and 3% lysophosphatidylcholine), triglycerides and fatty acids, was purchased from Lipoid GmbH (Ludwigshafen, Germany). Tween 80 (polysorbate 80, polyoxyethylene sorbitan monooleate; non-ionic hydrophilic surfactant, HLB 15) was supplied by Galeno (Carmignano, Prato, Italy).

2,2-Diphenyl-1-picrylhydrazyl (DPPH), iron(III) chloride (97%), phosphate buffer saline (PBS) 10×, 2,4,6-tris(2-pirydyl)-s-triazine (TPTZ) (≥98%), and (±)-6-hydroxy-2,5,7,8-tetramethylchromane-2-carboxylic acid (Trolox) (97%) were obtained from Sigma-Aldrich (Milan, Italy).

### 2.2. Methods

UV–Vis spectra were recorded on a Jasco (Lecco, Italy) V-730 Spectrophotometer.

HPLC analysis was performed with an Agilent (Cernusco sul Naviglio, Milan, Italy) instrument equipped with a UV-Vis detector; a Phenomenex (Castel Maggiore, Bologna, Italy)

Sphereclone ODS column (250 × 4.60 mm, 5 μm) was used at a flow rate of 1.0 mL/min. A gradient elution using 0.1% formic acid in water (solvent A) and methanol (solvent B) was performed as follows: 5% B, 0–10 min; from 5 to 80% B, 10–57.5 min. The detection wavelength was set at 254 nm.

### 2.3. Preparation and Characterization of Transfersomes

CWM (A or B) was weighed in a glass vial along with S75; thereafter, Tween 80 and water were added (Table 1). To obtain the transfersomes, the dispersion was sonicated (5 s on and 2 s off, 10 cycles; 13 microns of probe amplitude) with an ultrasonic disintegrator (Soniprep 150 plus; MSE Crowley, London, UK).

For comparative purposes, empty transfersomes (i.e., those without CWM) were also prepared under the same conditions as CWM transfersomes (Table 1).

The mean diameter, polydispersity index, and zeta potential of the transfersomes were determined by dynamic and electrophoretic light scattering using a Zetasizer nano-ZS (Malvern Panalytical, Worcestershire, UK). Samples ($n > 10$) were diluted with water (1:100) and analyzed at 25 °C.

The above three parameters were monitored for 90 days to assess the long-term stability of the formulations.

**Table 1.** Composition of the transfersome formulations.

| Formulation | S75 | CWM | Tween 80 | $H_2O$ |
|---|---|---|---|---|
| Empty transfersomes | 120 mg | - | 0.05 mL | 0.95 mL |
| CWM-A transfersomes | 120 mg | 2 mg | 0.05 mL | 0.95 mL |
| CWM-B transfersomes | 120 mg | 2 mg | 0.05 mL | 0.95 mL |

### 2.4. Antioxidant Properties of CWM Samples

#### 2.4.1. DPPH Assay

CWM or CWM transfersomes (0.02–0.15 mg/mL final dose) (concentrations are referred to as CWM content in the formulations) were added to 3 mL of a 0.2 mM ethanolic solution of DPPH [56], and after 10 min under stirring at room temperature, the absorbance at 515 nm was measured. Experiments were run in triplicate.

#### 2.4.2. Ferric Reducing/Antioxidant Power (FRAP) Assay

CWM and CWM transfersomes were added (0.001–0.1 mg/mL final dose) (concentrations are referred to as CWM content in the formulations) to 3 mL of 0.3 M acetate buffer (pH 3.6) containing 1.7 mM $FeCl_3$ and 0.83 mM TPTZ [57], and after 10 min of stirring at room temperature, the absorbance of the solutions at 593 nm was measured. Results were expressed as Trolox equivalents (eqs). Experiments were run in triplicate.

### 2.5. Release Experiments from CWM Transfersomes

Each CWM transfersome formulation (3 g) was placed in a dialysis membrane (MWCO 100–500 Da) and dialyzed against 30 mL of PBS 1×. The samples were kept at 37 °C in a water bath. Next, 0.5 mL of release medium was periodically withdrawn and replaced with an equal volume of corresponding fresh medium and analyzed using UV-Vis spectroscopy or HPLC. Each experiment was run in triplicate.

### 2.6. Antioxidant Properties of Released Fractions from CWM Transfersomes

Aliquots (150 μL) of the released fractions from CWM transfersomes were added to 2 mL of FRAP reagent prepared as described in Section 2.4.2. After 10 min under stirring, the mixtures were centrifuged (3 min at 5000 rpm) and the absorbance of the supernatants at 593 nm was measured.

### 2.7. Analysis of Cell Viability

Immortalized human keratinocytes (HaCaT, Innoprot, Derio, Spain) were cultured in 10% fetal bovine serum in Dulbecco's Modified Eagle's Medium, in the presence of 1% antibiotics and 2 mM L-glutamine, in a 5% $CO_2$ humidified atmosphere at 37 °C. To verify the biocompatibility of each sample, cells were seeded in 96-well plates at a density of $2 \times 10^3/cm^2$ and 24 h after seeding. Cells were incubated in the presence of increasing concentration of EA (up to 10 μM) or transfersome samples (up to 25 μL/mL) for 24 and 48 h. At the end of incubation, cell viability was assessed by the 3-(4,5-dimethylthiazol-2-yl)-2,5-diphenyltetrazoliumbromide (MTT) assay. Cell survival was expressed as the percentage of viable cells in the presence of each sample and compared with control cells (represented by the average obtained between untreated cells and cells supplemented with the highest concentration of buffer). Each sample was tested in three independent analyses, each carried out in triplicate.

### 2.8. UVA Irradiation and Dichlorofluorescein Diacetate (DCFDA) Assay

The protective effect of each sample was measured by determining the intracellular reactive oxygen species (ROS) levels. A previously reported protocol [58] was followed, with some modifications. Briefly, HaCaT cells were preliminarily exposed for 2, 6, and 16 h to 10 μM EA to define the proper incubation time. After that, cells were incubated with the samples (10 μM EA or 25 μL/mL transfersomes, providing a 10 μM EA concentration) for 6 h in the absence or presence of 10 min UVA irradiation (100 J/cm$^2$). At the end of the irradiation, $H_2$-DCFDA was added to measure intracellular ROS level. Fluorescence intensity of the probe was measured at an emission wavelength of 525 nm and an excitation wavelength of 488 nm using a Perkin-Elmer (Milan, Italy) LS50 spectrofluorometer. Emission spectra were acquired at a scanning speed of 300 nm/min, with 5 slit widths for both excitation and emission. ROS production was expressed as percentage of DCF fluorescence intensity of the sample under test, compared to the untreated sample. Results are presented as mean of results obtained after three independent experiments (mean $\pm$ SD) and compared by one-way ANOVA according to Bonferroni's method (post hoc) using Graphpad Prism for Windows, version 6.01.

## 3. Results and Discussion

### 3.1. Determination of the EA Content in CWM Samples

To gain information about the amount of EA contained in the two CWM samples, DMSO solutions of CWM-A and CWM-B were prepared and then analyzed using UV-Vis spectroscopy and HPLC after proper dilution in methanol. DMSO was chosen as the solvent based on its ability to dissolve a wide range of most polar and non-polar natural phenolic compounds, including EA [50,54,59].

As an example, the UV-Vis spectrum and elutographic profile of CWM-A are reported in Figure 1. The UV-Vis spectrum was characterized by absorption maxima at around 280 and 360 nm, as expected based on the presence of EA [60]. In agreement with this observation, HPLC analysis showed the presence of a single chromatographable compound eluted at ca. 38 min, identified as EA by comparison with an authentic standard. Quantitative analysis indicated content of EA of 5 $\pm$ 1% *w*/*w* for both CWM-A and CWM-B.

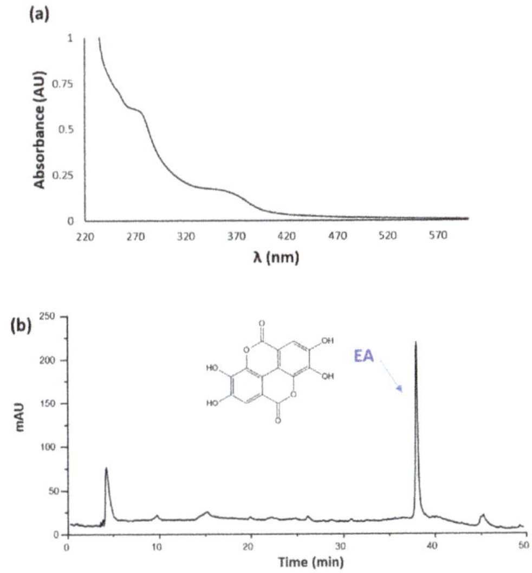


**Figure 1.** (**a**) UV-Vis spectrum (recorded at 0.02 mg/mL) and (**b**) HPLC profile (recorded at 1 mg/mL) of CWM-A.

### *3.2. Incorporation of CWM Samples into Transfersomes*

Transfersomes—that is, phospholipid vesicles modified with Tween 80 to promote skin penetration—containing CWM were produced and characterized in terms of size, homogeneity, and surface charge. To evaluate the CWM effect on the vesicles, the CWM transfersomes were compared with the empty transfersomes.

The light scattering results, as reported in Table 2, showed that the empty transfersomes had a mean diameter of 106 nm, and were homogeneously dispersed (polydispersity index 0.27) and highly negatively charged (−71 mV). CMW-A incorporation significantly increased the mean diameter of the vesicles, although they remained small (around 120 nm); the polydispersity index was unaltered, and the zeta potential value became less negative (Table 2), but it was still high enough to allow particle repulsion and prevent aggregation. On the other hand, CMW-B incorporation did not affect the vesicle size, nor the homogeneity of the dispersion, but produced less negative surface charge, as much as CMW-A.

The stability of the transfersome formulations was evaluated by monitoring the mean diameter, the polydispersity index, and the zeta potential during a 90-day storage period at 4 °C. No significant alterations (<10%) were detected.

**Table 2.** Characteristics of empty and CWM transfersomes: mean diameter (MD), polydispersity index (PI), and zeta potential (ZP). Each value represents the mean ± SD ($n > 10$). * values statistically different ($p < 0.05$) with respect to empty transfersomes.

| Formulation | MD (nm) | PI | ZP (mV) |
|---|---|---|---|
| Empty transfersomes | 106 ± 3.1 | 0.27 ± 0.01 | −71 ± 5.8 |
| CWM-A transfersomes | * 121 ± 7.8 | 0.27 ± 0.01 | * −56 ± 5.7 |
| CWM-B transfersomes | 105 ± 2.9 | 0.27 ± 0.03 | * −58 ± 9.4 |

The physicochemical characteristics of the herein described transfersomes are in line with those reported in literature for other EA-incorporating nanosystems [20]. As an example, Tween 80-coated chitosan-based nanoformulations exhibited an average hydrodynamic diameter of 155 nm and a PI of 0.37, although a lower ZP (−9.7 mV) compared to CWM transfersomes was determined. These nanoformulations led to a sustained release of EA

(47% after 24 h) at pH 7.4 and exhibited more efficient anticancer effects in tumor-bearing mice compared to EA alone [27]. EA-loaded schizophyllan and chitin nanoparticles showed size distributions of 217.8 and 39.82 nm, and ZP of +27 and −9.14 mV, respectively. The chitin nanoparticles in particular led to a rapid release of EA (ca. 50%) after 8 h at pH 7.4, followed by a gradual release (up to 63%) that continued up to 50 h. MTT assay indicated that both nanoformulations effectively inhibited the growth of breast cancer cell lines, with $IC_{50}$ values of 60 and 115 μg/mL, respectively [28]. Zein nanoparticles containing EA showed a mean size between 260 and 370 nm and a PI lower than 0.3. These formulations were found to be positively charged, with ZP ranging from +24 to + 37 mV, and showed inhibitory and bactericide activity against *S. aureus* and *P. aeruginosa* (MIC <72 μg/mL) [5]. Finally, poly(ε-caprolactone)-based EA nanoparticles formulated by applying various stabilizing agents exhibited average diameters ranging from 193 to 1252 nm, PI of 0.36–0.98 and ZP of −25–+62 mV. A fast release followed by a linear release period with a slower rate was observed at pH 7.4, with a cumulative release ranging from 25% to 48% after 8 days. These nanoparticles enhanced the cytotoxicity of EA up to 6.9-fold against colon adenocarcinoma cells, as well as the absorption extent of orally taken EA in rabbits [61].

### 3.3. Antioxidant Properties of CWM Transfersomes

The antioxidant properties of the CWM transfersomes were initially investigated with respect to the starting CWM samples by widely used chemical assays; that is, the DPPH and FRAP assays. Standard EA was also tested for comparison. The results are shown in Table 3. Both CWM-A and CWM-B exhibited antioxidant properties in line with what was expected based on a 5% *w/w* EA content. Notably, incorporation into transfersomes induced an about 2.5-fold decrease in the $EC_{50}$ values determined in the DPPH assay for the CWM samples, and an even higher improvement in the reducing properties was observed in the FRAP assay. Since empty transfersomes were not found to exhibit significant antioxidant properties, these results clearly suggest a larger availability of the antioxidant compound EA following incorporation into the vesicles.

**Table 3.** Antioxidant properties of CWM samples. Reported are the mean ± SD values of at least three experiments. Data for CWM transfersomes have been normalized based on the CWM content in the formulation.

| | **DPPH Assay $EC_{50}$ (mg/mL)** | **FRAP Assay (mg of Trolox/mg of Sample)** |
|---|---|---|
| CWM-A transfersomes | 0.0389 ± 0.0005 | 0.36 ± 0.06 |
| CWM-B transfersomes | 0.0375 ± 0.0004 | 0.39 ± 0.04 |
| Empty transfersomes | - | 0.00015 ± 0.00002 |
| CWM-A | 0.103 ± 0.001 | 0.047 ± 0.002 |
| CWM-B | 0.106 ± 0.001 | 0.050 ± 0.001 |
| EA | 0.0051 ± 0.0004 | 1.04 ± 0.02 |

### 3.4. Release of EA from CWM Transfersomes and Antioxidant Properties of the Released Fractions under Simulated Physiological Conditions

The release of EA from the CWM transfersomes in PBS at 37 °C was followed by UV-Vis spectroscopy and HPLC over 5 weeks. No significant release of EA was observed in the case of CWM-A, probably as a result of the higher particle size of the sample, whereas very promising results were obtained with the CWM-B transfersomes. Indeed, the UV-Vis spectra of the released fractions from the latter exhibited absorption maxima at ca. 280 and 360 nm, which linearly increased over time (Figure 2a). HPLC analysis confirmed a controlled release of EA, which was complete after 30 days, reaching a concentration of ca. 56 μM (Figure 2b).

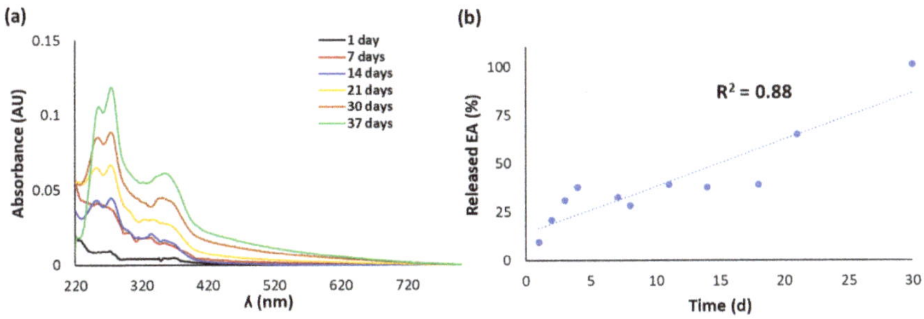


**Figure 2.** (**a**) UV-Vis spectra of fractions released over time from CWM-B transfersomes in PBS at 37 °C. (**b**) Kinetics of release of EA, determined by HPLC analysis. Reported are the mean values of at least three experiments (SD $\leq$ 10%).

The released fractions from CWM-B transfersomes were also evaluated for their antioxidant properties by chemical assays. Actually, it was not possible to perform the DPPH assay due to interference of the released material with the assay medium. On the other hand, the reducing properties evaluated by the FRAP assay (Figure 3a) linearly increased over time on account of the progressive release of EA from the transfersomes. A good linear correlation ($R^2$ = 0.91) of the antioxidant properties with the amount of total released EA was indeed observed (Figure 3b).

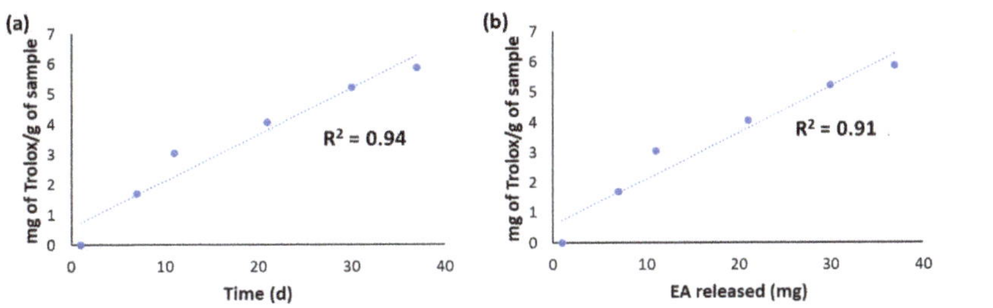


**Figure 3.** (**a**) Results of the FRAP assay on the fractions released over time from CWM-B transfersomes in PBS at 37 °C. (**b**) Correlation between the $Fe^{3+}$-reducing properties of the released fractions and the amount of released EA.

### 3.5. Cell Viability of CWM Transfersomes

Based on the encouraging results of the release experiments, and with the aim of further probing the potential of CWM transfersomes for dermo-cosmetic applications, in subsequent experiments the sample biocompatibility was evaluated on HaCaT, since these cells are normally present in the outermost layer of the skin. EA was also tested for comparison in a range of concentrations corresponding to those provided by the CWM transfersome samples. MTT assay (not shown) showed that both EA (up to 10 µM) and the transfersomes (up to 25 µL/mL) were biocompatible under all the experimental conditions.

### 3.6. Protective Effect of CWM Transfersomes on Photoinduced Oxidative Stress

The antioxidant cytoprotective properties of CWM-A and CWM-B transfersomes were evaluated on UVA-irradiated HaCaT. Preliminary experiments (data not shown) were performed to define the optimal time (2, 6, or 16 h) for cell preincubation with 10 µM EA (corresponding to a non-cytotoxic concentration of 25 µL/mL CWM transfersomes), and 6 h incubation was chosen for further experiments. As shown in Figure 4, UVA irradiation induced a significant increase in intracellular ROS levels (150–200%) with respect to untreated cells. When cells were pretreated with 10 µM EA (Figure 4a, gray bars) prior to UVA exposure, a significant lowering of intracellular ROS levels was observed. As expected, empty transfersomes did not exert any protective effect against oxidative stress. Interestingly, when cells were treated with 25 μL/mL CWM-B transfersomes (providing an EA concentration of 10 μM) (Figure 4b, gray bars) prior to UVA exposure, a significant reduction ($p \leq 0.05$) in intracellular ROS levels, compared to untreated UVA-exposed cells, was observed. On the other hand, CWM-A transfersomes (Figure 4b, white bars) were unable to protect cells from UVA-induced oxidative stress injury. Thus, these results, combined with those from the release experiments, suggest that the particle size of the CWM incorporated into transfersomes is fundamental to allow EA to be active as an antioxidant in cellular models.

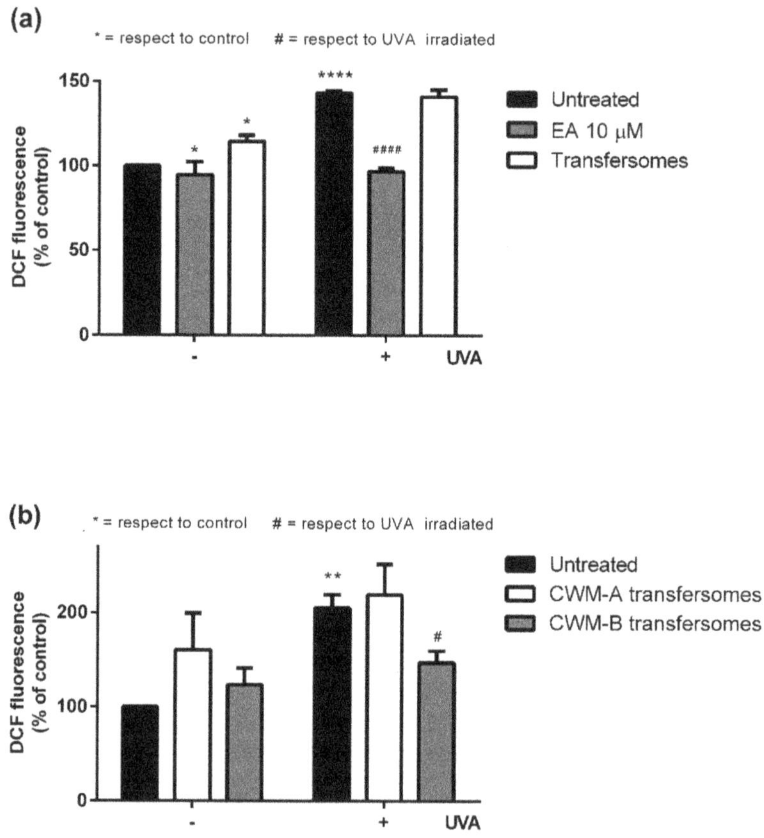


**Figure 4.** Protective effects of EA and transfersome samples on UVA-stressed HaCaT cells. Intracellular ROS levels were determined by DCFDA assay. Cells were preincubated with (**a**) 10 μM EA (gray bars) or 25 μL/mL empty transfersomes (white bars), or (**b**) 25 μL/mL of CWM-A (white bars) or CWM-B (gray bars) transfersomes (both providing a 10 μM EA concentration). Black bars refer to untreated cells in the absence (-) or in the presence (+) of UVA stress. Values are expressed as percentage with respect to untreated cells. Data shown are means ± SD of three independent experiment. *, # indicate $p < 0.05$; ** indicates $p < 0.01$; ****, #### indicate $p < 0.001$ with respect to control cells and UVA treated cells, respectively.

## 4. Conclusions

In conclusion, the present work reports the efficacy of transfersomes as carriers for the controlled release of the biologically active compound EA from CWM, an industrial by-product deriving from tannin extraction. The incorporation into the transfersomes induced a significant improvement of the antioxidant properties of CWM, likely as a result of the larger availability of EA. Moreover, the transfersomal CWM-B was found to be able to decrease ROS production in UVA-irradiated keratinocytes and to provide a complete and controlled release of EA in pseudophysiological conditions at pH 7.4, a result of interest in dermo-cosmetic applications. For example, open wounds are characterized by a neutral or alkaline pH ranging from 6.5 to 8.5, whereas chronic wounds exhibit a pH in the range of 7.5–8.5 [62,63]. All together, these results highlight nanoformulated CWM of proper particle size as an easily accessible and biocompatible material that could warrant a sustained release of the water-insoluble bioactive EA under physiologically relevant conditions; for example for the treatment and protection of damaged skin.

**Author Contributions:** Conceptualization, C.C., D.M.M. and L.P.; Data curation, F.M., D.L., C.C., D.M.M. and L.P.; Investigation, F.M., D.L. and C.C.; Methodology, F.M., D.L., C.C., D.M.M. and L.P.; Resources, S.G.; Supervision, D.M.M., L.P. and A.N.; Validation, C.C., D.M.M., L.P. and A.N.; Writing—original draft, F.M., C.C. and D.M.M.; Writing—review and editing, L.P. and A.N. All authors have read and agreed to the published version of the manuscript.

**Funding:** This research received no external funding.

**Data Availability Statement:** The data are contained within the article.

**Acknowledgments:** The authors wish to thank the European Union (FSE, PON Ricerca e Innovazione 2014–2020, Azione I.1 "Dottorati Innovativi con caratterizzazione Industriale") for funding a PhD grant to Federica Moccia. Financial support to Lucia Panzella by Italian MIUR, PRIN 2017 2017YJMPZN is acknowledged.

**Conflicts of Interest:** The authors declare no conflict of interest.

## References

1. Tošović, J.; Bren, U. Antioxidative action of ellagic acid—A kinetic DFT study. *Antioxidants* **2020**, *9*, 587. [CrossRef]
2. Xiao, Y.; Huang, R.; Wang, N.; Deng, Y.; Tan, B.; Yin, Y.; Qi, M.; Wang, J. Ellagic acid alleviates oxidative stress by mediating Nrf2 signaling pathways and protects against paraquat-induced intestinal injury in piglets. *Antioxidants* **2022**, *11*, 252. [CrossRef]
3. Kim, D.H.; Sim, Y.; Hwang, J.H.; Kwun, I.S.; Lim, J.H.; Kim, J.; Kim, J.I.; Baek, M.C.; Akbar, M.; Seo, W.; et al. Ellagic acid prevents binge alcohol-induced leaky gut and liver injury through inhibiting gut dysbiosis and oxidative stress. *Antioxidants* **2021**, *10*, 1386. [CrossRef]
4. Romeo, I.; Vallarino, G.; Turrini, F.; Roggeri, A.; Olivero, G.; Boggia, R.; Alcaro, S.; Costa, G.; Pittaluga, A. Presynaptic release-regulating alpha2 autoreceptors: Potential molecular target for ellagic acid nutraceutical properties. *Antioxidants* **2021**, *10*, 1759. [CrossRef]
5. de Souza Tavares, W.; Pena, G.R.; Martin-Pastor, M.; de Sousa, F.F.O. Design and characterization of ellagic acid-loaded zein nanoparticles and their effect on the antioxidant and antibacterial activities. *J. Mol. Liq.* **2021**, *341*, 116915. [CrossRef]
6. Amor, A.J.; Gómez-Guerrero, C.; Ortega, E.; Sala-Vila, A.; Lázaro, I. Ellagic acid as a tool to limit the diabetes burden: Updated evidence. *Antioxidants* **2020**, *9*, 1226. [CrossRef]
7. Umar, A.K.; Zothantluanga, J.H.; Aswin, K.; Maulana, S.; Sulaiman Zubair, M.; Lalhlenmawia, H.; Rudrapal, M.; Chetia, D. Antiviral phytocompounds "ellagic acid" and "(+)-sesamin" of *Bridelia retusa* identified as potential inhibitors of SARS-CoV-2 3CL pro using extensive molecular docking, molecular dynamics simulation studies, binding free energy calculations, and bioactivity prediction. *Struct. Chem.* **2022**, *10*, 1–21. [CrossRef]
8. Javaid, N.; Shah, M.A.; Rasul, A.; Chauhdary, Z.; Saleem, U.; Khan, H.; Ahmed, N.; Uddin, M.S.; Mathew, B.; Behl, T.; et al. Neuroprotective effects of ellagic acid in Alzheimer's disease: Focus on underlying molecular mechanisms of therapeutic potential. *Curr. Pharm. Des.* **2021**, *27*, 3591–3601. [CrossRef]
9. Mohammadinejad, A.; Mohajeri, T.; Aleyaghoob, G.; Heidarian, F.; Oskuee, R.K. Ellagic acid as a potent anticancer drug: A comprehensive review on in vitro, in vivo, in silico, and drug delivery studies. *Biotechnol. Appl. Biochem.* 2021; *in press*. [CrossRef]
10. Ríos, J.L.; Giner, R.M.; Marín, M.; Recio, M.C. A pharmacological update of ellagic acid. *Planta Med.* **2018**, *84*, 1068–1093. [CrossRef]
11. Korkina, L.G.; Pastore, S.; Dellambra, E.; De Luca, C. New molecular and cellular targets for chemoprevention and treatment of skin tumors by plant polyphenols: A critical review. *Curr. Med. Chem.* **2013**, *20*, 852–868. [CrossRef]
12. Mo, J.; Panichayupakaranant, P.; Kaewnopparat, N.; Songkro, S.; Reanmongkol, W. Topical anti-inflammatory potential of standardized pomegranate rind extract and ellagic acid in contact dermatitis. *Phyther. Res.* **2014**, *28*, 629–632. [CrossRef]
13. De Moraes Alves, M.M.; Arcanjo, D.D.R.; Figueiredo, K.A.; De Sousa Macêdo Oliveira, J.S.; Viana, F.J.C.; De Sousa Coelho, E.; Lopes, G.L.N.; Gonçalves, J.C.R.; Carvalho, A.L.M.; Dos Santos Rizzo, M.; et al. Gallic and ellagic acids are promising adjuvants to conventional amphotericin B for the treatment of cutaneous leishmaniasis. *Antimicrob. Agents Chemother.* **2020**, *64*, e00807–e00820. [CrossRef]

14. Tavares, W.S.; Tavares-Júnior, A.G.; Otero-Espinar, F.J.; Martín-Pastor, M.; Sousa, F.F.O. Design of ellagic acid-loaded chitosan/zein films for wound bandaging. *J. Drug Deliv. Sci. Technol.* **2020**, *59*, 101903. [CrossRef]
15. Lembo, S.; Balato, A.; Di Caprio, R.; Cirillo, T.; Giannini, V.; Gasparri, F.; Monfrecola, G. The modulatory effect of ellagic acid and rosmarinic acid on ultraviolet-B-induced cytokine/chemokine gene expression in skin keratinocyte (HaCaT) cells. *Biomed Res. Int.* **2014**, *2014*, 346793. [CrossRef] [PubMed]
16. Moon, N.R.; Kang, S.; Park, S. Consumption of ellagic acid and dihydromyricetin synergistically protects against UV-B induced photoaging, possibly by activating both TGF-β1 and wnt signaling pathways. *J. Photochem. Photobiol. B Biol.* **2018**, *178*, 92–100. [CrossRef]
17. Panzella, L.; Napolitano, A. Natural and bioinspired phenolic compounds as tyrosinase inhibitors for the treatment of skin hyperpigmentation: Recent advances. *Cosmetics* **2019**, *6*, 57. [CrossRef]
18. Muráth, S.; Szerlauth, A.; Sebok, D.; Szilágyi, I. Layered double hydroxide nanoparticles to overcome the hydrophobicity of ellagic acid: An antioxidant hybrid material. *Antioxidants* **2020**, *9*, 153. [CrossRef]
19. Li, Y.; Mei, L.; Guan, X.; Hu, Y. Ellagic acid solid dispersion: Characterization and bioactivity in the hydroxyl radical oxidation system. *Food Res. Int.* **2021**, *142*, 110184. [CrossRef]
20. Ceci, C.; Graziani, G.; Faraoni, I.; Cacciotti, I. Strategies to improve ellagic acid bioavailability: From natural or semisynthetic derivatives to nanotechnological approaches based on innovative carriers. *Nanotechnology* **2020**, *31*, 382001. [CrossRef]
21. Nyamba, I.; Lechanteur, A.; Semdé, R.; Evrard, B. Physical formulation approaches for improving aqueous solubility and bioavailability of ellagic acid: A review. *Eur. J. Pharm. Biopharm.* **2021**, *159*, 198–210. [CrossRef]
22. Elbehairi, S.E.I.; Alfaifi, M.Y.; Shati, A.A.; Fahmy, U.A.; Gorain, B.; Md, S. Encapsulation of ellagic acid in di-block copolymeric micelle for non-small cell lung cancer therapy. *Sci. Adv. Mater.* **2021**, *13*, 66–72. [CrossRef]
23. Villalgordo, J.M.; Trulli, L.; García-Villalba, R.; García, V.; Althobaiti, Y.; Tomás-Barberán, F.A. Novel regioselective synthesis of urolithin glucuronides–human gut microbiota cometabolites of ellagitannins and ellagic acid. *J. Agric. Food Chem.* **2022**, *70*, 5819–5828. [CrossRef] [PubMed]
24. Ortenzi, M.A.; Antenucci, S.; Marzorati, S.; Panzella, L.; Molino, S.; Rufián-Henares, J.Á.; Napolitano, A.; Verotta, L. Pectin-based formulations for controlled release of an ellagic acid salt with high solubility profile in physiological media. *Molecules* **2021**, *26*, 433. [CrossRef]
25. Karakas, C.Y.; Ordu, H.R.; Bozkurt, F.; Karadag, A. Electrosprayed chitosan-coated alginate–pectin beads as potential system for colon-targeted delivery of ellagic acid. *J. Sci. Food Agric.* **2022**, *102*, 965–975. [CrossRef] [PubMed]
26. Ahlawat, J.; Neupane, R.; Deemer, E.; Sreenivasan, S.T.; Narayan, M. Chitosan-ellagic acid nanohybrid for mitigating rotenone-induced oxidative stress. *ACS Appl. Mater. Interfaces* **2020**, *12*, 18964–18977. [CrossRef]
27. Kaur, H.; Ghosh, S.; Kumar, P.; Basu, B.; Nagpal, K. Ellagic acid-loaded, tween 80-coated, chitosan nanoparticles as a promising therapeutic approach against breast cancer: In-vitro and in-vivo study. *Life Sci.* **2021**, *284*, 119927. [CrossRef]
28. Pirzadeh-Naeeni, S.; Mozdianfard, M.R.; Shojaosadati, S.A.; Khorasani, A.C.; Saleh, T. A comparative study on schizophyllan and chitin nanoparticles for ellagic acid delivery in treating breast cancer. *Int. J. Biol. Macromol.* **2020**, *144*, 380–388. [CrossRef]
29. Li, B.; Harich, K.; Wegiel, L.; Taylor, L.S.; Edgar, K.J. Stability and solubility enhancement of ellagic acid in cellulose ester solid dispersions. *Carbohydr. Polym.* **2013**, *92*, 1443–1450. [CrossRef]
30. Savic, I.M.; Jocic, E.; Nikolic, V.D.; Popsavin, M.M.; Rakic, S.J.; Savic-Gajic, I.M. The effect of complexation with cyclodextrins on the antioxidant and antimicrobial activity of ellagic acid. *Pharm. Dev. Technol.* **2019**, *24*, 410–418. [CrossRef]
31. Sharma, K.; Kadian, V.; Kumar, A.; Mahant, S.; Rao, R. Evaluation of solubility, photostability and antioxidant activity of ellagic acid cyclodextrin nanosponges fabricated by melt method and microwave-assisted synthesis. *J. Food Sci. Technol.* **2022**, *59*, 898–908. [CrossRef]
32. Gontijo, A.V.L.; G Sampaio, A.D.; Koga-Ito, C.Y.; Salvador, M.J. Biopharmaceutical and antifungal properties of ellagic acid-cyclodextrin using an *in vitro* model of invasive candidiasis. *Future Microbiol.* **2019**, *14*, 957–967. [CrossRef] [PubMed]
33. Das, J.; Debbarma, A.; Lalhlenmawia, H. Formulation and in vitro evaluation of poly-(D,L-lactide-co-glycolide) (PLGA) nanoparticles of ellagic acid and its effect on human breast cancer, MCF-7 cell line. *Int. J. Curr. Pharm. Res.* **2021**, *13*, 56–62. [CrossRef]
34. Kakran, M.; Sahoo, N.G.; Bao, H.; Pan, Y.; Li, L. Functionalized graphene oxide as nanocarrier for loading and delivery of ellagic acid. *Curr. Med. Chem.* **2011**, *18*, 4503–4512. [CrossRef] [PubMed]
35. Ichiura, H.; Konishi, T.; Morikawa, M. Alginate film prepared on polyethylene nonwoven sheet and its function for ellagic acid release in response to sodium ions. *J. Mater. Sci.* **2009**, *44*, 992–997. [CrossRef]
36. Yağmur, N.; Şahin, S.; Korkmaz, E. Microencapsulation of ellagic acid extracted from pomegranate peel onto Spirulina: Characterization, loading, and storage stability properties. *J. Food Process. Preserv.* **2021**, *45*, e15086. [CrossRef]
37. Wei, Y.; Wang, Y.; Xia, D.; Guo, S.; Wang, F.; Zhang, X.; Gan, Y. Thermosensitive liposomal codelivery of HSA-paclitaxel and HSA-ellagic acid complexes for enhanced drug perfusion and efficacy against pancreatic cancer. *ACS Appl. Mater. Interfaces* **2017**, *9*, 25138–25151. [CrossRef]
38. Najafi, A.; Taheri, R.A.; Mehdipour, M.; Martinez-Pastor, F.; Rouhollahi, A.A.; Nourani, M.R. Improvement of post-thawed sperm quality in broiler breeder roosters by ellagic acid-loaded liposomes. *Poult. Sci.* **2019**, *98*, 440–446. [CrossRef]
39. Stojiljković, N.; Ilić, S.; Stojanović, N.; Janković-Veličković, L.; Stojnev, S.; Kocić, G.; Radenković, G.; Arsić, I.; Stojanović, M.; Petković, M. Nanoliposome-encapsulated ellagic acid prevents cyclophosphamide-induced rat liver damage. *Mol. Cell. Biochem.* **2019**, *458*, 185–195. [CrossRef]

40. Gonçalves, M.M.; Carneiro, J.; Döll-Boscardin, P.M.; Justus, B.; Budel, J.M.; Farago, P.V.; de Paula, J.P. Preparation of ellagic acid-loaded vesicles and method validation to quantify encapsulation efficiency. *Lat. Am. J. Pharm.* **2018**, *37*, 1000–1004.
41. Madrigal-Carballo, S.; Lim, S.; Rodriguez, G.; Vila, A.O.; Krueger, C.G.; Gunasekaran, S.; Reed, J.D. Biopolymer coating of soybean lecithin liposomes via layer-by-layer self-assembly as novel delivery system for ellagic acid. *J. Funct. Foods* **2010**, *2*, 99–106. [CrossRef]
42. Rai, S.; Pandey, V.; Rai, G. Transfersomes as versatile and flexible nano-vesicular carriers in skin cancer therapy: The state of the art. *Nano Rev. Exp.* **2017**, *8*, 1325708. [CrossRef] [PubMed]
43. Cevc, G. Material transport across permeability barriers by means of lipid vesicles. In *Handbook of Biological Physics*; Lipowsky, R., Sackmann, E., Eds.; Elsevier B.V.: Amsterdam, The Netherlands, 1995; pp. 465–490.
44. Sguizzato, M.; Ferrara, F.; Hallan, S.S.; Baldisserotto, A.; Drechsler, M.; Malatesta, M.; Costanzo, M.; Cortesi, R.; Puglia, C.; Valacchi, G.; et al. Ethosomes and transethosomes for mangiferin transdermal delivery. *Antioxidants* **2021**, *10*, 768. [CrossRef] [PubMed]
45. Kocbek, P.; Baumgartner, S.; Kristl, J. Preparation and evaluation of nanosuspensions for enhancing the dissolution of poorly soluble drugs. *Int. J. Pharm.* **2006**, *312*, 179–186. [CrossRef]
46. Magangana, T.P.; Makunga, N.P.; Fawole, O.A.; Opara, U.L. Processing factors affecting the phytochemical and nutritional properties of pomegranate (*Punica granatum* L.) peel waste: A review. *Molecules* **2020**, *25*, 4690. [CrossRef] [PubMed]
47. Verotta, L.; Panzella, L.; Antenucci, S.; Calvenzani, V.; Tomay, F.; Petroni, K.; Caneva, E.; Napolitano, A. Fermented pomegranate wastes as sustainable source of ellagic acid: Antioxidant properties, anti-inflammatory action, and controlled release under simulated digestion conditions. *Food Chem.* **2018**, *246*, 129–136. [CrossRef]
48. Panzella, L.; Moccia, F.; Nasti, R.; Marzorati, S.; Verotta, L.; Napolitano, A. Bioactive phenolic compounds from agri-food wastes: An update on green and sustainable extraction methodologies. *Front. Nutr.* **2020**, *7*, 60. [CrossRef] [PubMed]
49. Moccia, F.; Flores-Gallegos, A.C.; Chávez-González, M.L.; Sepúlveda, L.; Marzorati, S.; Verotta, L.; Panzella, L.; Ascacio-Valdes, J.A.; Aguilar, C.N.; Napolitano, A. Ellagic acid recovery by solid state fermentation of pomegranate wastes by *Aspergillus Niger* and *Saccharomyces cerevisiae*: A comparison. *Molecules* **2019**, *24*, 3689. [CrossRef]
50. Moccia, F.; Agustin-Salazar, S.; Verotta, L.; Caneva, E.; Giovando, S.; D'Errico, G.; Panzella, L.; d'Ischia, M.; Napolitano, A. Antioxidant properties of agri-food byproducts and specific boosting effects of hydrolytic treatments. *Antioxidants* **2020**, *9*, 438. [CrossRef]
51. An, J.Y.; Wang, L.T.; Lv, M.J.; Wang, J.D.; Cai, Z.H.; Wang, Y.Q.; Zhang, S.; Yang, Q.; Fu, Y.J. An efficiency strategy for extraction and recovery of ellagic acid from waste chestnut shell and its biological activity evaluation. *Microchem. J.* **2021**, *160*, 105616. [CrossRef]
52. Husanu, E.; Mero, A.; Rivera, J.G.; Mezzetta, A.; Ruiz, J.C.; D'Andrea, F.; Pomelli, C.S.; Guazzelli, L. Exploiting deep eutectic solvents and ionic liquids for the valorization of chestnut shell waste. *ACS Sustain. Chem. Eng.* **2020**, *8*, 18386–18399. [CrossRef]
53. Lameirão, F.; Pinto, D.; Vieira, E.F.; Peixoto, A.F.; Freire, C.; Sut, S.; Dall'Acqua, S.; Costa, P.; Delerue-Matos, C.; Rodrigues, F. Green-sustainable recovery of phenolic and antioxidant compounds from industrial chestnut shells using ultrasound-assisted extraction: Optimization and evaluation of biological activities in vitro. *Antioxidants* **2020**, *9*, 267. [CrossRef] [PubMed]
54. Moccia, F.; Gallucci, N.; Giovando, S.; Zuorro, A.; Lavecchia, R.; D'Errico, G.; Panzella, L.; Napolitano, A. A tunable deep eutectic solvent-based processing for valorization of chestnut wood fiber as a source of ellagic acid and lignin. *J. Environ. Chem. Eng.* **2022**, *10*, 107773. [CrossRef]
55. Panzella, L.; Moccia, F.; Toscanesi, M.; Trifuoggi, M.; Giovando, S.; Napolitano, A. Exhausted woods from tannin extraction as an unexplored waste biomass: Evaluation of the antioxidant and pollutant adsorption properties and activating effects of hydrolytic treatments. *Antioxidants* **2019**, *8*, 84. [CrossRef] [PubMed]
56. Goupy, P.; Dufour, C.; Loonis, M.; Dangles, O. Quantitative kinetic analysis of hydrogen transfer reactions from dietary polyphenols to the DPPH radical. *J. Agric. Food Chem.* **2003**, *51*, 615–622. [CrossRef] [PubMed]
57. Benzie, I.F.F.; Strain, J.J. The ferric reducing ability of plasma (FRAP) as a measure of '"antioxidant power"': The FRAP assay. *Anal. Biochem.* **1996**, *239*, 70–76. [CrossRef] [PubMed]
58. Imbimbo, P.; Romanucci, V.; Pollio, A.; Fontanarosa, C.; Amoresano, A.; Zarrelli, A.; Olivieri, G.; Monti, D.M. A cascade extraction of active phycocyanin and fatty acids from *Galdieria phlegrea*. *Appl. Microbiol. Biotechnol.* **2019**, *103*, 9455–9464. [CrossRef]
59. Garcia-Villalba, R.; Espín, J.C.; Kroon, P.A.; Alasalvar, C.; Heinonen, M.; Voorspoels, S.; Tomas-Barberan, F. A validated method for the characterization and quantification of extractable and non-extractable ellagitannins after acid hydrolysis in pomegranate fruits, juices, and extracts. *J. Agric. Food Chem.* **2015**, *63*, 6555–6566. [CrossRef]
60. Tokutomi, H.; Takeda, T.; Hoshino, N.; Akutagawa, T. Molecular structure of the photo-oxidation product of ellagic acid in solution. *ACS Omega* **2018**, *3*, 11179–11183. [CrossRef]
61. Mady, F.M.; Shaker, M.A. Enhanced anticancer activity and oral bioavailability of ellagic acid through encapsulation in biodegradable polymeric nanoparticles. *Int. J. Nanomed.* **2017**, *12*, 7405–7417. [CrossRef]
62. Dissemond, J.; Witthoff, M.; Brauns, T.C.; Haberer, D.; Goos, M. pH values in chronic wounds. Evaluation during modern wound therapy. *Der Hautarzt* **2003**, *54*, 959–965. [CrossRef]
63. Jones, E.M.; Cochrane, C.A.; Percival, S.L. The effect of pH on the extracellular matrix and biofilms. *Adv. Wound Care* **2015**, *4*, 431–439. [CrossRef]

antioxidants

Article

# Response of *Thymus lotocephalus* In Vitro Cultures to Drought Stress and Role of Green Extracts in Cosmetics

**Inês Mansinhos [1], Sandra Gonçalves [1,*], Raquel Rodríguez-Solana [1,2], Hugo Duarte [1], José Luis Ordóñez-Díaz [2], José Manuel Moreno-Rojas [2] and Anabela Romano [1,*]**

1 MED–Mediterranean Institute for Agriculture, Environment and Development & CHANGE–Global Change and Sustainability Institute, Faculdade de Ciências e Tecnologia, Universidade do Algarve, Campus de Gambelas, 8005-139 Faro, Portugal; ifmansinhos@ualg.pt (I.M.); raquel.rodriguez.solana@juntadeandalucia.es (R.R.-S.); hmduarte@ualg.pt (H.D.)

2 Department of Agroindustry and Food Quality, Andalusian Institute of Agricultural and Fisheries Research and Training (IFAPA), Avenida Menendez-Pidal, SN, 14004 Córdoba, Spain; josel.ordonez@juntadeandalucia.es (J.L.O.-D.); josem.moreno.rojas@juntadeandalucia.es (J.M.M.-R.)

* Correspondence: smgoncalves@ualg.pt (S.G.); aromano@ualg.pt (A.R.); Tel.: +351-289800900 (S.G.); +351-289800910 (A.R.)

**Abstract:** The impact of drought stress induced by polyethylene glycol (PEG) on morphological, physiological, (bio)chemical, and biological characteristics of *Thymus lotocephalus* López and Morales shoot cultures have been investigated, as well as the potential of iron oxide nanoparticles, salicylic acid, and methyl jasmonate (MeJA) as alleviating drought stress agents. Results showed that PEG caused oxidative stress in a dose-dependent manner, raising $H_2O_2$ levels and reducing shoots' growth, photosynthetic pigment contents, and phenolic compounds production, especially phenolic acids, including the major compound rosmarinic acid. Moreover, Fourier Transform Infrared Spectra analysis revealed that PEG treatment caused changes in shoots' composition, enhancing terpenoids biosynthesis. PEG also decreased the biological activities (antioxidant, anti-tyrosinase, and photo-protective) of the eco-friendly extracts obtained with a Natural Deep Eutectic Solvent. MeJA was the most efficient agent in protecting cells from oxidative damage caused by drought, by improving the biosynthesis of phenolics, like methyl 6-O-galloyl-β-D-glucopyranoside and salvianolic acids, as well as improving the extracts' antioxidant activity. Altogether, the obtained results demonstrated a negative impact of PEG on *T. lotocephalus* shoots and an effective role of MeJA as a mitigating agent of drought stress. Additionally, extracts showed a good potential to be used in the cosmetics industry as skincare products.

**Keywords:** aromatic plant; abiotic stress; green extract; phenolics; tyrosinase inhibition; UV protecting extracts

**Citation:** Mansinhos, I.; Gonçalves, S.; Rodríguez-Solana, R.; Duarte, H.; Ordóñez-Díaz, J.L.; Moreno-Rojas, J.M.; Romano, A. Response of *Thymus lotocephalus* In Vitro Cultures to Drought Stress and Role of Green Extracts in Cosmetics. *Antioxidants* **2022**, *11*, 1475. https://doi.org/10.3390/antiox11081475

Academic Editors: Irene Dini and Sonia Laneri

Received: 1 July 2022
Accepted: 25 July 2022
Published: 28 July 2022



## 1. Introduction

Plants face abiotic stresses every day, being affected at several levels, such as morphological, physiological, biochemical, and metabolic. According to climate change models, the Mediterranean basin is one of the regions across the globe that will be strongly affected by this situation, especially by drought events [1]. Drought is one of the biggest concerns that delimits plant development, causing severe or lethal consequences. Water comprises 80–90% of the biomass of plants and is implicated in many critical physiological functions, such as growth and photosynthesis [2,3]. Usually, drought stress triggers the accumulation of reactive oxygen species (ROS), which act as signaling molecules, leading to changes in several stress-responsive genes expression [4] and in the biosynthesis of secondary metabolites [5]. Furthermore, the accumulation of osmolytes, such as proline, soluble sugars, and betaines is crucial to maintain the cell water status and alleviate the oxidative damage caused by water deficiency [4]. Typically, tolerant plants exhibit a significantly lesser number of metabolites accumulation than sensitive plants, though demonstrating an increment in osmolytes production [2]. Nevertheless, the plant's capacity to survive under stressed conditions depends on several factors, such as plant species, growth stage, and extent and intensity of the stress conditions. In vitro simulation of abiotic stresses is an advantageous approach to investigating plants' response to those factors. Besides being performed under controlled conditions and free of pathogens, in vitro culture allow the production of plant material on a large scale, without compromising species conservation. Moreover, in vitro culture allows the application of several strategies to change the production of secondary metabolites (e.g., adding precursors or elicitors to the culture media) [6]. Among osmotic agents that can be applied to mimic drought conditions in vitro, polyethylene glycol (PEG) is the most commonly used. PEG increases the solute potential of the culture media, blocking the absorption of water by the plant [7]. Moreover, owing to its high molecular weight, this osmotic agent can decrease the water potential without being phytotoxic or taken up by the plant [4]. Different agents, such as iron oxide ($Fe_3O_4$) nanoparticles (NPs), salicylic acid (SA), and methyl jasmonate (MeJA) were tested for their capacity to alleviate the deleterious impact of drought stress. Included in abiotic nano compounds, $Fe_3O_4$ NPs have been shown to stimulate several mechanisms in stressed plants, as well as to improve the biosynthesis of secondary metabolites [8]. SA and MeJA are two important endogenous plant growth regulators involved in stress tolerance in several plant species [9], which, when applied exogenously, have a great capability as elicitors [10].

The genus *Thymus* is the eighth-most abundant genus of the Lamiaceae (Labiatae) family [11], comprising about 350 worldwide species of perennial, aromatic herbs, and subshrubs [7,12]. These perennial herbs have been known since ancient times in virtue of their medicinal and aromatic attributes [2,11], being widely used as flavoring agents, culinary herbs, and ornamentals [6]. More recently, *Thymus* have been demonstrated to exhibit a broad range of functional opportunities for food, pharmaceutical, and the cosmetic industries [13], due to their main groups of secondary metabolites–volatile and phenolic compounds–being responsible for a great diversity of biological actions [12]. *Thymus lotocephalus* López and Morales is a Mediterranean species endemic to the Algarve, a southern region of Portugal, that is listed as Near Threatened in The IUCN Red List of Threatened Species [14]. Previous studies indicated that this species produces phenolic (e.g., rosmarinic acid, salvianolic acids, luteolin, caffeic acid) and volatile (e.g., linalool, caryophyllene oxide, camphor, borneol) compounds with biological effects, and that in vitro culture is a good alternative for the production of bioactive compounds from this species, without compromising natural populations [6,15–17]. Phenolic compounds, which may be categorized as phenolic acids, flavonoids, tannins, stilbenes, lignans, and coumarins, present a great potential for therapeutic applications to numerous human diseases, such as diabetes, cancer, neurodegenerative and cardiovascular pathologies, due to their multifunctional biological activities, particularly antioxidant [18].

The first decade of the 21st century was marked by the significant growth in sales of cosmetic products, representing 23% of the total market share of consumer products. The consumer demand for innovative and natural products to treat skin is continuously increasing. The harmful effects on human health of excessive exposure to UV (ultraviolet) photons (UV-A and UV-B) are well known and, under climate change context, the correct use of products protecting the skin is even more essential. UV filters can be classified into organic (natural or artificial) and inorganic substances. In general, while inorganic compounds reflect and scatter UV radiation, organic sunscreens absorb it [19]. The safety and efficacy of most artificial sunscreens ingredients are hampered by their photostability, toxicity, and damage to aquatic ecosystems [20], so new strategies are being looked for and developed. One increasingly appreciated strategy consists of the application of plant extracts (or their isolated bioactive compounds) as ingredients, instead of synthetic chemicals. Phytochemicals exhibit antioxidant activity, which is very important to fight against environmental free radicals caused, for example, by UV radiation, which trigger skin-damaging conditions. In addition, antioxidants are essential in several cosmetic formulations, such as facial anti-aging, antioxidant-based, anti-allergy, and dry skin hydrating products [21]. Besides the other recognized biological properties of phytochemicals, which are extremely valuable to cosmetic products, such as antioxidant, anti-inflammatory, anti-aging, and antimicrobial, they also present the ability to absorb the energy of photons, making them potential UV-protecting agents [22,23]. In this scenario, plant extracts, or their isolated bioactive compounds, can be used in sunscreen formulations. Aging signs (e.g., wrinkles, hyperpigmentation, flaccidity, texture changes) are another concern of modern society and therefore the demand for anti-aging products is, likewise, increasing. Melanogenesis is the process responsible for melanin production as a defensive mechanism against UV radiation, in which tyrosinase is a key enzyme. The excessive accumulation of melanin in the skin can cause hyperpigmentation, so the inhibition of tyrosinase is a target in the cosmetic industry [24].

The solvent used to extract plant bioactive compounds for cosmetic applications is of utmost importance. Ethanol and glycols are two conventional solvents largely employed to extract polar or semi-polar compounds, but they present some drawbacks. In addition to ethanol requiring specific safety procedures due to its volatility and flammability nature, it can also disturb the skin hydrolipidic film and irritate sensitive skins due to its drying nature. Glycols are usually associated with petrochemicals and, for this reason, consumers often try to avoid glycol-containing cosmetic formulations. Water is also used to extract polar compounds; however, microbiological safety is its major limitation. Recently, Natural Deep Eutectic Solvents (NADES) have appeared as new eco-friendly alternatives to conventional solvents, which are recognized to be dangerous to human health and the environment. NADES are composed of at least two components, a hydrogen bond acceptor (HBA) and a hydrogen bond donor (HBD), having a much lower melting point than that of their individual components [18]. Due to their natural origin, biodegradable and non-toxic nature, NADES represent an unexplored opportunity to develop innovative extracts with unique phytochemical footprints and biological properties, making it possible to be securely employed in the pharmaceutical, food, and cosmetic industries [18,25]. In the previous study conducted by our group, NADES were demonstrated to be more effective than conventional solvents (methanol, ethanol 80%, and water) to extract phenolic compounds from a Lamiaceae species [18].We hypothesize that, as in other Mediterranean species, drought stress might have harmful effects on *T. lotocephalus* and that the use of some agents may alleviate these negative effects. In this context, the key goal of this work is to investigate how drought induced by PEG (2, 5, and 7%) influences some morphological, physiological, and (bio)chemical traits *of T. lotocephalus* shoot cultures, as well the biological activities (antioxidant, anti-tyrosinase, and photoprotective) of extracts obtained with NADES. The potential of $Fe_3O_4$ NPs, SA, and MeJA in alleviating drought stress effects were also investigated. As far as we know, this is the first report investigating the response of *T. lotocephalus* to drought stress, analyzing the chemical structures of this species using FTIR spectroscopy and using NADES for the extraction of their bioactive compounds. In a climate change scenario, these studies are important as a first approach to understanding the response of *T. lotocephalus* to drought stress, and how these conditions can affect the potential applications of this aromatic species. Results from this study can also be useful to predict the impact of drought stress conditions on other Mediterranean aromatic species with medicinal/industrial properties.

## 2. Materials and Methods

### 2.1. Chemicals and Reagents

L-Proline (>99%), PEG 6000, ninhydrin, 2,2′-azobis(2 methylpropionamidine)dihydrochloride (AAPH), (±)-6-hydroxy-2,5,7,8-tetramethylchromane-2-carboxylic acid (Trolox) and potassium ferricyanide ($K_3[Fe(CN)_6]$) were acquired from Acros Organics (Geel, Germany). Lactic acid, and fluorescein were purchased from Panreac (Barcelona, Spain). Sodium carbonate anhydrous ($Na_2CO_3$), ferric chloride ($FeCl_3$), Folin-Ciocalteu's reagent, and gallic acid were acquired from VWR (Leuven, Belgium). Ascorbic acid, acetic acid, and potassium iodide were acquired from Merck (Darmstadt, Germany). Salicylic acid (SA), methyl jasmonate (MeJA), $Fe_3O_4$ nanoparticles (50–100 nm), hydrogen peroxide ($H_2O_2$), trichloroacetic acid (TCA), L- proline ($\geq$99%), 2,2-diphenyl-1-picrylhydrazyl (DPPH), 3,4-dihydroxy-L-phenylalanine (L-DOPA), 2,2′-azino-bis (3-ethylbenzothiazoline-6-sulfonic acid) diammonium salt tablets (ABTS), mushroom tyrosinase (EC 1.14.18.1), kojic acid, potassium persulfate ($K_2S_2O_8$), potassium bromide, HPLC-MS-grade water, HPLC-MS-grade acetonitrile, luteolin, epigallocatechin gallate, protocatechuic acid, and formic acid were obtained from Sigma–Aldrich (Steinheim, Germany). Rosmarinic acid and quercetin were supplied by Extrasynthese (Genay, France), and ρ-coumaric, caffeic acid, and catechin were provided by AASC Ltd. (Southhampton, UK).

### 2.2. Plant Material, In Vitro Culture Conditions, and Experiments

Shoots of *T. lotocephalus* were proliferated in vitro as described by Coelho et al. [26] in MS medium [27], containing 2% (*w*/*v*) sucrose and 0.7% (*w*/*v*) agar, and pH was corrected to 5.75 ± 0.05. Culture media were autoclaved for 20 min at 121 °C. To induce drought, different concentrations of PEG 6000 [2, 5, or 7% (*w*/*v*)] were added to the culture media, according to the diffusion-based method reported by Girma and Krieg [28]. PEG was spread on the MS solidified medium surface for 24 h. The concentration of PEG that induced higher drought effects, 7%, was selected to evaluate the potential of $Fe_3O_4$ NPs, SA, and MeJA in alleviating drought stress effects. Thus, these agents were added to PEG-free and 7% PEG culture media. $Fe_3O_4$ NPs (10 mg/L) were added before autoclaving, and SA (50 µM) and MeJA (50 µM) solutions were sterilized through 0.2 µm microfilters before addition to the autoclaved media. Multiplication medium was used as control. Cultures were incubated at 25 ± 1 °C, with 16 h light (40 µmol $m^{-2}$ $s^{-1}$, cool white fluorescent lamps) and 8 h dark cycle, for 7 weeks. Twelve Erlenmeyer flasks with seven shoots each were tested for each treatment.

### 2.3. Morphological Traits

After culture for 7 weeks, the biometric parameters (total number of shoots, the longest shoot length, and the fresh and dry weight of the biomass) were registered. Shoots were dried at 40 °C until constant weight to determine the dry weight.

### 2.4. Physiological and Biochemical Measurements

Physiological attributes such as the chlorophyll and carotenoids, hydrogen peroxide ($H_2O_2$), and proline contents were assessed using fresh material. Chlorophylls and carotenoids were extracted, according to Lichtenthaler [29], using pure acetone and 25 mg of plant material. The absorbance was measured at 661.6, 644.8, and 470 nm using UV–Vis spectrophotometer (T70+ UV/Vis Spectrophotometer, PG Instruments Ltd., Leicestershire, UK).

The $H_2O_2$ content was determined according to a method from Loreto and Velikova [30] with slight alterations. A hundred milligrams of plant material were ground in ice with 0.1% TCA. The same volume of the supernatant obtained after centrifugation (15 min at 12,000× *g*) and 10 mM potassium phosphate buffer were mixed, before adding the double 1 M KI solution. The absorbance was measured at 390 nm (Tecan Infinite M200 microplate reader, Männedorf, Switzerland) after 30 min in darkness. For proline estimation, plant material was extracted three times using 80% (*v*/*v*) ethanol at 80 °C for 30 min [31]. Then, the extract was incubated for 1 h at 100 °C with 1% (*w*/*v*) ninhydrin reagent prepared with 60% (*v*/*v*) acetic acid. After cooling, and the addition of toluene, the absorbance of the organic phase was recorded at 520 nm.

### 2.5. Fourier Transform Infrared Spectra (FTIR) Spectroscopy

Functional group distribution of the main compounds present in samples was followed by FTIR spectroscopy (Bruker Tensor 27, Billerica, MA, USA). For that, the samples were dried at 60 °C, mixed with potassium bromide, pressed into KBr pellets, and measured in a

4000–600 $cm^{-1}$ range. Each FTIR spectra was evaluated by the data analysis and graphing software OriginPro, version 2022 (OriginLab Corporation, Northampton, MA, USA) and compared with those of other reports [32–36].

### *2.6. Extraction of Phenolic Compounds*

The plant material was dried at 40 °C until constant weight and powdered to <2 mm. A green extraction was performed using proline: lactic acid (1:1) mixture containing 30% (*w*/*w*) water as NADES according to Mansinhos et al. [18], using a plant/solvent ratio of 2.5:100 (*w*/*v*). Extraction was performed in an ultrasound bath (Elmasonic S 100 (H, Elma Hans Schmidbauer GmbH & Co. KG, Singen, Germany) using a frequency of 37 kHz for 30 min at 50 °C. After being filtered using a Whatman n°. 1 filter paper (Whatman Int. Ltd., Maidstone, UK), the extracts were kept at −20 °C up to usage.

### *2.7. Spectrophotometric and Chromatographic Analysis for Phenolic Compounds*

#### 2.7.1. Spectrophotometric Measurement of Total Phenolic Content (TPC)

The content of total phenolics in the extracts was determined using Folin-Ciocalteu (F-C) reagent, as described by Ainsworth and Gillespie [37]. A mixture containing 200 µL 10% (*v*/*v*) F-C reagent, 100 µL plant extracts diluted in phosphate buffer (75 mM, pH 7.0), and 800 µL $Na_2CO_3$ (700 mM) was incubated for 2 h. The standard used was gallic acid (0.004–0.5 mM) and the absorbance was recorded at 765 nm. The results were calculated as gallic acid equivalents per gram of dry weight ($mg_{GAE}/g_{DW}$).

#### 2.7.2. Analysis of Individual Phenolic Compounds by HPLC-HRMS

The plant extracts were examined utilizing a Dionex Ultimate 3000 HPLC system, with an HPLC pump and an autosampler operating at 10 °C (Thermo Fisher Scientific, San Jose, CA, USA). The sample separation was performed on a 150 × 4.6 mm i.d. 5 µm 100 A C18 Kinetex column (Phenomenex, UK) with 1 mL/min of flow rate and with 40 °C of column temperature. The chromatographic conditions were performed according to Gonçalves et al. and Mansinhos et al. [17,18]. The solvent system was composed of solvent A (distilled water) and solvent B (acetonitrile), both with 0.1% formic acid. Gradient mode was 0 min—90% A; 10 min—74% A; 22 min—35% A; 30 min—5% A; 40 min—5% A; 40.1 min—90% A; and 45 min—90% A. The column flow rate was 0.2 mL/min directed to an Exactive Orbitrap mass spectrometer (Thermo Fisher Scientific, San Jose, CA, USA) fitted with a heated electrospray ionization probe (HESI). Negative ions were analyzed at scan mode of auto MS/MS, at the range of 100–1000 $m/z$. The analyses were also based on in-source collision-induced dissociation scans at 25 eV. The source condition was the spray voltage of 4000 V, the capillary temperature of 320 °C, heater temperature of 150 °C, and the sheath gas and auxiliary gas flow rate of 25 and 5 units, respectively. Identification of compounds was based on the retention time and the exact mass in conjunction with standards. When standards were unavailable, it was compared with the theoretical exact mass of the molecular ion with the determined accurate mass of the molecular ion, to tentatively identify the compound and was looked for in several metabolite databases (PubChem, Metlin, ChemSpider, Phenol Explorer). In addition, the biocompound's identification was performed following the MSI MS levels [38]. Supplementary Table S1 summarizes the chemical formula, theoretical exact mass, delta ppm, retention time (RT), and MSI MI level of the compounds. To quantify the compounds, the theoretical exact mass of the molecular ion was selected by standard curves, or by the calibration curve of a close parent metabolite based on the structure. Limits of detection (LOD) and quantification (LOQ) were determined from the standard deviation of ten blank determinations, ranging LOD and LOQ from 0.10 to 228.13 µg/L and 0.33 to 760.43 µg/L, respectively. The criteria used in the quantification of phenolics are outlined in Supplementary Table S2. The results were expressed in milligrams per kilogram of extract.

### 2.8. Assessment of the Biological Properties of the Extracts for Dermo-Cosmetic Application

#### 2.8.1. Antioxidant Activity

The antioxidant activity of the extracts was assessed using different assays with two mechanisms [atom hydrogen transfer-based method (ORAC), single electron transfer-based method (FRAP), and mixed methods making use of hydrogen-atom transfer and single-electron transfer (DPPH and ABTS)].

DPPH Free Radical Scavenging

Applying the method described by Soler-Rivas et al. [39], the capacity of the plant samples to scavenge the free radical DPPH$^{\bullet}$ was analyzed. For that, before 30 min of incubation, 30 µL extract was mixed with 300 µL DPPH methanolic solution (90 µM) and methanol 80% until 900 µL. The absorbance was measured at 515 nm, using Trolox (0.025–0.3 mM) as standard, and the results were expressed as milligrams of Trolox equivalents per gram of dry weight ($mg_{TE}/g_{DW}$).

ABTS Free Radical Scavenging

The ABTS$^{\bullet}$ free radical scavenging activity of the extracts was analyzed according to Re et al. [40]. Using $K_2S_2O_8$, the ABTS stock solution (7 mM) was prepared and stored for 12–16 h. After this period, the stock solution was diluted with $H_2O$ until obtaining an absorbance (734 nm) of 0.700 ± 0.02. The samples (10 µL) were added to the diluted ABTS solution (190 µL) and the absorbance was read, using Trolox as standard. The results were expressed as milligrams of Trolox equivalents per gram of dry weight ($mg_{TE}/g_{DW}$).

Ferric Reducing Antioxidant Power (FRAP)

FRAP assay comprises the reduction of Fe (III) to Fe (II) in the presence of an antioxidant. Based on the method described by Yen and Chen [41], the reaction involving the plant extracts (100 µL), 1% $K_3[Fe\ (CN)_6]$ (250 µL), and potassium phosphate buffer (200 mM, pH 6.6) (250 µL) was incubated for 20 min at 50 °C. After it was added 10% TCA (250 µL) to the reaction and centrifuged. $FeCl_3$ (80 µL) at 0.1% was mixed with the same amount of supernatant and water (400 µL). The absorbance was measured at 700 nm, using ascorbic acid as standard, and the results were expressed as milligrams of ascorbic acid equivalents per gram of dry weight ($mg_{AAE}/g_{DW}$).

Oxygen Radical Absorbance Capacity (ORAC)

According to Gillespie et al. [42], plant extracts (25 µL) were incubated with fluorescein (0.2 µM) for 10 min at 37 °C. Then, 150 mM AAPH (25 µL) was added and the fluorescence was read every 5 min (90 min), up to value zero at 530 nm emission and 485 nm excitation. Applying the differences in areas under the fluorescein decay curve between the blank and the samples, the results were calculated. Trolox was used as standard and the results were expressed as milligrams of Trolox equivalents per gram of dry weight ($mg_{TE}/g_{DW}$).

#### 2.8.2. Inhibition Effects against Melanogenesis Key Enzyme

The tyrosinase (Tyr) inhibitory assay was carried out according to Masuda et al. [43]. The extracts (50 µL) were mixed with mushroom Tyr (50 µL, 46 U/mL) and 20 mM sodium phosphate buffer (80 µL, pH 6.8), and incubated for 10 min at room temperature. After the addition of 80 µL of the substrate (L-DOPA, 2.5 mM) and an incubation period of 10 min at room temperature, the absorbance was recorded at 475 nm (microplate reader). The results were calculated as kojic acid equivalents ($mg_{KAE}/g_{DW}$).

#### 2.8.3. Photoprotective Properties

The capacity of the plant extracts as a natural filter protecting from ultraviolet (UV) radiation was examined using UV–Vis spectrophotometer. UV spectra of the extracts at 250 µL/mL were measured in the range from 250 to 400 nm with an interval of 2 nm. The extraction solvent [proline: lactic acid (1:1) with 30% (*w*/*w*) water] was used as a blank.

2.8.4. Determination of Sun Protection Factor (SPF)

The photoprotective activity of the extracts was evaluated by measuring the SPF values, which are frequently utilized to evaluate the efficacy of sunscreen against UV radiation. Extracts were diluted in the extraction solvent to obtain different concentrations (50–2500 μL/mL), and the present solvent served as a blank. Spectrophotometric scanning was accomplished at wavelengths in the range of 290–320 nm, with intervals of 5 nm with UV–Vis spectrophotometer. SPF values were obtained following the equation developed by Mansur et al. [44]:

$$SPF = CF \times \sum_{290}^{320} EE(\lambda) \times I(\lambda) \times Abs(\lambda)$$

where CF is the correction factor (=10); EE(λ) is the erythemal effect spectrum; I(λ) is the solar intensity spectrum; Abs(λ) is the absorbance. The values of EE(λ) × I(λ), described by Sayre et al. [45] are presented in Table S3.

### 2.9. Statistical Analysis

Data are presented as mean ± standard error for the total number of the results and analyzed by one-way analysis of variance (ANOVA), and Tukey's New Multiple Range Test ($p < 0.05$). Correlations were determined using Pearson's test. Statistical analyses were performed by IBM SPSS Statistics for Windows (version 26.0, Armonk, NY, USA: IBM Corporation). Hierarchical cluster analysis, K-means cluster analysis, and Principal Component Analysis (PCA) were analyzed by the software OriginPro, version 2022 (OriginLab Corporation, Northampton, MA, USA).

## 3. Results and Discussion

### 3.1. Evaluation of Shoots Growth

Physiological mechanisms of plants, especially growth and development, are sensitive to water limitations. The visual appearance of *T. lotocephalus* in vitro-regenerated shoots cultured in different media is shown in Figure 1. Results show that the increasing PEG concentration gradually reduced the mean shoot length. The highest PEG concentration (7%) caused the biggest decrease in shoot length (11.5 ± 0.43 mm), which was statistically different from that of the control (48.3 ± 2.68 mm) and the PEG concentration 2% (Table 1). Similar results were obtained in in vitro cultures of *Thymus vulgaris* [4], *Thymus citriodorus* [7], *Salvia leriifolia* [46], *Stevia rebaudiana* [3], and *Amsonia orientalis* [47], under drought stress. When plants are subjected to drought stress, the water movement through the xylem decreases, and to maintain turgor status plant cells have to reduce their osmotic potential [47]. The decrease in shoot length could be explained by the restricted water absorption and the turgescence pressure for cell enlargement, which disrupt cell division and elongation [46]. Curiously, in comparison with the control, the shoot number and the biomass produced were higher in PEG-containing media, especially with 2%. Even so, shoots obtained in the control group demonstrated normal aspect (Figure 1), while those produced in media with PEG (2, 5, and 7%) showed symptoms of hyperhydricity (a plant disorder in which shoots presents a rigid translucent aspect with little internodes). Usually, this disorder appears in in vitro cultures and is caused by different factors (e.g., water availability, growth regulators) [6]. This phenomenon can explain the highest biomass produced when cultures were submitted to drought stress. Similar signs of hyperhydricity were evidenced in *S. lerrifolia* cultures exposed to PEG stress [46].

Among the three levels of PEG tested, 7% induced the highest negative effects in *T. lotocephalus* cultures, therefore this concentration was selected to study the potential of three different agents ($Fe_3O_4$ NPs, SA, and MeJA) in alleviating drought stress. Nevertheless, as observed in Table 1, none of the agents tested were capable of reversing the negative impacts of PEG on the shoot's length, with MeJA still potentiating the negative effects in this parameter (6.67 ± 0.47 mm). Similar results were obtained in *Oryza sativa* shoots stressed with 3% PEG 6000, in which MeJA (5 mM) reduced the length of the shoots. Nevertheless, the three agents tested reduced the signs of hyperhydricity of the shoots cultured in media with 7% PEG (Figure 1). $Fe_3O_4$ NPs, SA, and MeJA were also added to PEG-free culture media to investigate their effects in non-stressful conditions. Results indicated that $Fe_3O_4$ NPs improved shoot number and biomass production, in comparison to the control. It is known that iron has a great influence on the growth and development of plants, and its nanoscale application in in vitro cultures facilitates its absorption and plant nutritional balance [17,48,49]. Furthermore, the application of these NPs has been demonstrated to improve the uptake of important macronutrients in some plants [49,50]. The results showed that SA decreased 2.4-fold *T. lotocephalus* shoot fresh weight (FW). Distinct findings were obtained by Mozafari et al. [48] who compared the influence of SA (0.05 mM) and $Fe_3O_4$ NPs (0.08 ppm) on the biomass produced by in vitro cultures of strawberry. In that case, the authors demonstrated more beneficial effects of SA than $Fe_3O_4$ NPs, although they used a much lower concentration of iron NPs (0.08 mg/L) than that used in this study (10 mg/L). Using concentrations higher than those tested in this study, Karamian et al. [9] observed that 100 µM of SA significantly improved the biomass of *Verbascum sinuatum* L. shoots; however, 3 mM did not affect the growth of *Impatiens walleriana* L. [51]. Similar to SA, MeJA seems to exert an inhibitory effect on *T. lotocephalus* shoots's length (11.7 $\pm$ 0.56 mm), in accordance with previous results observed in *V. sinuatum* [9]. Interestingly, enhanced shoot length was evidenced in *Glycyrrhiza glabra* cultures by MeJA (0.1–2 mM) after 24 h of exposure, but after 48 h it decreased significantly [52]. These distinct outcomes prove that the efficacy of the tested agents is affected by several factors, such as plant species, culture type, time of exposure, and concentration.

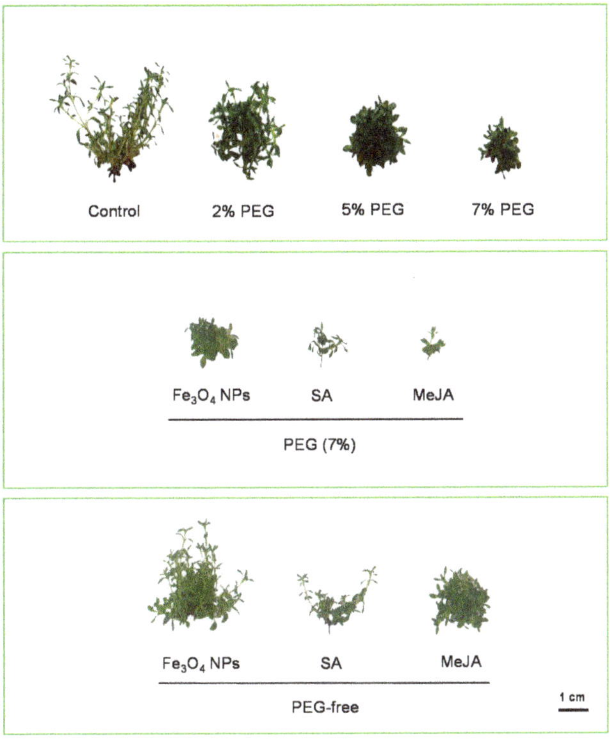


**Figure 1.** The aspect of in vitro shoots of *Thymus lotocephalus* grown in media with 0 (control), 2, 5, or 7% of PEG, $Fe_3O_4$ nanoparticles (NPs), salicylic acid (SA), methyl jasmonate (MeJA), 7% PEG + $Fe_3O_4$ NPs, 7% PEG + SA, or 7% PEG + MeJA after 7 weeks of culture. The scale bar represents 1 cm.

**Table 1.** Shoot growth and total chlorophyll (Cltotal), carotenoids (Crt), hydrogen peroxide ($H_2O_2$), and proline contents of *Thymus lotocephalus* López and Morales shoots cultured in media with 0 (control), 2, 5, or 7% of PEG, 7% PEG + $Fe_3O_4$ nanoparticles (NPs), 7% PEG+ salicylic acid (SA), or 7% PEG+ methyl jasmonate (MeJA), $Fe_3O_4$ NPs, SA, or MeJA.

| Treatment | No. Shoots | Length of the Longest Shoot (mm) | Fresh Weight (g) | Dry Weight (mg) | Cltotal (mg/gFW) | Crt (mg/gFW) | $H_2O_2$ (μmol/$g_{FW}$) | Proline (μmol/$g_{FW}$) |
|---|---|---|---|---|---|---|---|---|
| Control | 7.94 ± 1.70 $^{c}$ | 48.3 ± 2.68 $^{a}$ | 1.40 ± 0.383 $^{cd}$ | 144 ± 26.6 $^{c}$ | 0.74 ± 0.02 $^{a}$ | 0.16 ± 0.01 $^{ab}$ | 0.63 ± 0.00 $^{de}$ | 0.59 ± 0.01 $^{d}$ |
| PEG | | | | | | | | |
| 2% | 33.2 ± 1.77 $^{a}$ | 22.9 ± 0.89 $^{c}$ | 3.69 ± 0.20 $^{a}$ | 417 ± 13.9 $^{a}$ | 0.59 ± 0.03 $^{bc}$ | 0.14 ± 0.00 $^{abc}$ | 0.43 ± 0.03 $^{e}$ | 1.39 ± 0.14 $^{bcd}$ |
| 5% | 21.9 ± 1.72 $^{b}$ | 15.7 ± 0.80 $^{d}$ | 2.06 ± 0.11 $^{bc}$ | 295 ± 17.1 $^{b}$ | 0.50 ± 0.01 $^{cdef}$ | 0.12 ± 0.01 $^{bcd}$ | 0.79 ± 0.07 $^{cde}$ | 1.53 ± 0.16 $^{bc}$ |
| 7% | 18.4 ± 2.32 $^{b}$ | 11.5 ± 0.43 $^{de}$ | 1.33 ± 0.10 $^{cd}$ | 318 ± 33.5 $^{ab}$ | 0.38 ± 0.02 $^{f}$ | 0.09 ± 0.01 $^{d}$ | 1.28 ± 0.13 $^{bc}$ | 4.55 ± 0.33 $^{a}$ |
| 7% PEG | | | | | | | | |
| $Fe_3O_4$ NPs | 4.71 ± 0.47 $^{c}$ | 10.3 ± 0.39 $^{de}$ | 0.59 ± 0.12 $^{de}$ | 93.3 ± 18.2 $^{c}$ | 0.45 ± 0.04 $^{def}$ | 0.11 ± 0.01 $^{cd}$ | 1.47 ± 0.08 $^{b}$ | 1.57 ± 0.13 $^{bc}$ |
| SA | 4.32 ± 0.93 $^{c}$ | 10.4 ± 0.95 $^{de}$ | 0.66 ± 0.14 $^{de}$ | 102 ± 13.4 $^{c}$ | 0.41 ± 0.02 $^{ef}$ | 0.10 ± 0.00 $^{cd}$ | 1.37 ± 0.00 $^{b}$ | 2.27 ± 0.25 $^{b}$ |
| MeJA | 3.44 ± 0.67 $^{c}$ | 6.67 ± 0.47 $^{e}$ | 0.21 ± 0.08 $^{e}$ | 49.1 ± 14.6 $^{c}$ | 0.51 ± 0.03 $^{cde}$ | 0.12 ± 0.00 $^{cd}$ | 0.94 ± 0.11 $^{bcde}$ | 5.37 ± 0.25 $^{a}$ |
| PEG-free | | | | | | | | |
| $Fe_3O_4$ NPs | 25.4 ± 2.34 $^{ab}$ | 29.6 ± 1.10 $^{b}$ | 3.10 ± 0.35 $^{ab}$ | 320 ± 23.8 $^{ab}$ | 0.68 ± 0.01 $^{ab}$ | 0.17 ± 0.01 $^{a}$ | 1.16 ± 0.12 $^{bcd}$ | 0.76 ± 0.10 $^{cd}$ |
| SA | 8.11 ± 1.34 $^{c}$ | 13.8 ± 0.95 $^{d}$ | 0.59 ± 0.15 $^{de}$ | 89.0 ± 14.3 $^{c}$ | 0.57 ± 0.01 $^{bcd}$ | 0.17 ± 0.01 $^{a}$ | 2.45 ± 0.17 $^{a}$ | 1.12 ± 0.07 $^{cd}$ |
| MeJA | 7.83 ± 0.79 $^{c}$ | 11.7 ± 0.56 $^{de}$ | 1.12 ± 0.05 $^{cde}$ | 106 ± 2.70 $^{c}$ | 0.43 ± 0.04 $^{ef}$ | 0.10 ± 0.01 $^{cd}$ | 0.60 ± 0.10 $^{e}$ | 0.77 ± 0.06 $^{cd}$ |

Values are expressed as mean ± SE. For each variable, the values followed by different letters (a–f) are significantly different at $p < 0.05$ (Tukey's New Multiple Range Test).

### 3.2. Physiological and Biochemical Traits

Photosynthesis is a crucial process for ideal plant development, metabolism, and biomass production, which depends directly on plant pigment levels. The obtained results showed that drought stress induced by PEG had damaging effects on photosynthetic pigment contents (Table 1). The highest contents were observed in the control, and the increasing drought stress progressively reduced total chlorophylls and carotenoid contents, with 7% PEG inducing a reduction of 51.35% and 56.25%, respectively. A decline in photosynthetic pigment levels under PEG stress was also observed in other species [4,46,47,51,53]. MeJA caused a reversion of the 7% PEG effects on the levels of total chlorophylls (from 0.38 ± 0.02 to 0.51 ± 0.03 mg/$g_{FW}$). In contrast, in *V. sinuatum* cultures treated with PEG, the addition of 200 μM MeJA decreased the pigments' contents [9]. The reduced amount of pigment under abiotic stress may be attributed to the high amount of ROS, which results in damage to the plant cells. ROS, which includes hydrogen peroxide ($H_2O_2$), hydroxyl radical ($OH^-$), superoxide anions ($O_2^-$), and singlet oxygen ($^1O_2$) [54], act as signaling molecules to induce the expression of several genes and pathways [55].

As shown in Table 1, there were no significant differences between the $H_2O_2$ levels of the control and the media containing 2% and 5% PEG, but the $H_2O_2$ accumulation was significantly increased by 7% PEG (from 0.63 ± 0.00 to 1.28 ± 0.13 μmol/$g_{FW}$). Likewise, cellular damage marked by the elevated levels of $H_2O_2$ was reported in several species under PEG-induced drought stress [46,47,56,57]. MeJA was able to reduce the high $H_2O_2$ accumulation induced by 7% PEG in *T. lotocephalus* shoots (from 1.28 ± 0.13 to 0.94 ± 0.11 μmol/$g_{FW}$). Contrary to these results, MeJA negatively affected the redox status of *V. sinuatum*, increasing the content of $H_2O_2$ and being not able to alleviate the harmful consequences of drought stress [9]. The addition of $Fe_3O_4$ NPs and SA, to PEG-free and 7% PEG-containing media, also induced a significant accumulation of $H_2O_2$ in comparison with the control in *T. lotocephalus* shoots. In agreement with our results, it has been previously reported that SA significantly increases the levels of $H_2O_2$ [56]. A significantly high accumulation of $H_2O_2$ was reported in *Artemisia aucheri* treated with PEG 4% and 0.1 mM SA [56]. Moreover, SA treatments showed a positive effect in decreasing $H_2O_2$ contents in *I. walleriana* under 1–4% PEG [51]. Excessive oxidative stress was also observed in other species subjected to different metal oxide NPs [9,17,58,59], which denotes the ROS effects induced by metals binding with proteins presenting SH bonds, which are found in plants.

Under drought stress, to regulate cellular redox status due to excessive ROS production and cellular osmotic adjustment, as well as to stabilize membranes and proteins, plants accumulate several osmoprotectants, such as proline [54]. Thus, proline accumulation under drought conditions is an adequate marker of stress tolerance [53,60]. In this investigation, 5% and 7% PEG caused a significant rise in proline accumulation, compared to the control. The values ranged from $0.59 \pm 0.01$ μmol/$g_{FW}$ in the control treatment to $4.55 \pm 0.33$ μmol/$g_{FW}$ in the greatest percentage of PEG treatment (Table 1). Similar results were achieved in other species [46,47,56,57]. Interestingly, in *T. vulgaris* proline accumulation increased until 6% PEG but decreased significantly at 8%, probably due to the greater proline utilization outpacing their biosynthesis [4]. In *T. lotocephalus*, proline accumulation triggered by drought stress was accompanied by $H_2O_2$ accumulation in a dose-dependent manner, which demonstrates the protective role of proline against ROS production. This evidence is consistent with that observed in in vitro cultures of *A. orientalis* [47], *S. leriifolia* [46], and *I. walleriana* [51]. In higher plants, this osmoprotectant derives from glutamate, and its biosynthesis occurs in chloroplasts or cytosol [61]. Since the photosynthetic pigments' degradation increases with increasing PEG concentration (which may be due to chloroplast damages), the majority of the proline biosynthesis in *T. lotocephalus* under drought stress probably occurs in the cytosol. Although $Fe_3O_4$ NPs, SA, and MeJA did not influence proline content per si, MeJA increased proline accumulation in PEG-stressed cultures ($5.37 \pm 0.25$ μmol/$g_{FW}$), suggesting the protective role of this signaling molecule in improving drought tolerance. Similar results were obtained in two cultivars of *O. sativa* seedlings under PEG stress, in which the priming with 2.5 and 5 mM MeJA significantly enhanced the proline accumulation [60].

### *3.3. Chemical Composition of the Shoots by Fourier-Transform Infrared (FTIR) Spectroscopy*

FTIR is a powerful spectroscopic tool used to obtain detailed information about the major functional groups in the chemical composition of a sample. In plants, the initial responses to biotic or abiotic stresses can be assessed by changes in functional groups, which present characteristic frequencies in the infrared spectrum [62]. In this way, to investigate the molecular structural changes of the *T. lotocephalus* cells caused by PEG stress and the mitigating agents ($Fe_3O_4$ NPs, SA, and MeJA), FTIR analysis was performed. The FTIR spectra (4000 to 600 $cm^{-1}$) are shown in Figure 2A,B and the wavenumbers of characteristic bands and corresponding functional groups are listed in Table 2.

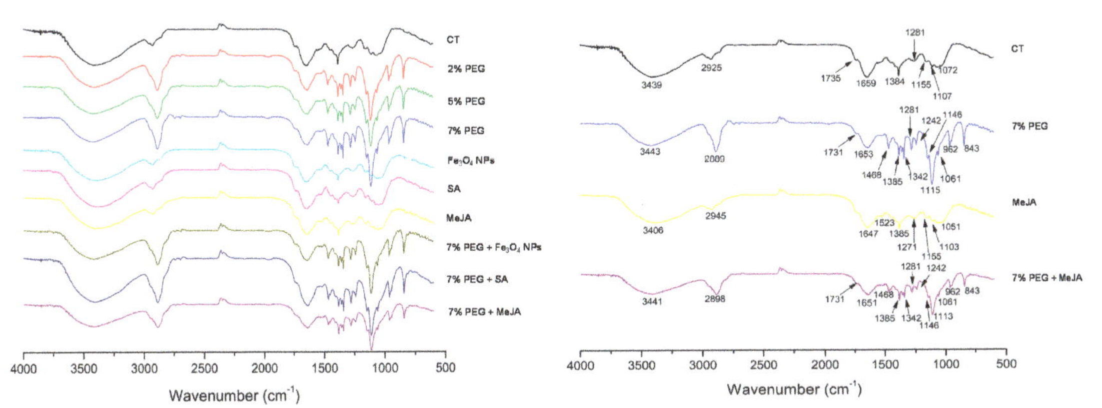


(A) (B)

**Figure 2.** (**A**) FTIR spectra (600–4000 $cm^{-1}$) of *Thymus lotocephalus* shoots cultured in media with 0 (control, CT), 2, 5, or 7% PEG, $Fe_3O_4$ nanoparticles (NPs), salicylic acid (SA), or methyl jasmonate (MeJA) and a combination of PEG (7%) with $Fe_3O_4$ NPs, SA and MeJA; (**B**) FTIR spectra of *T. lotocephalus* shoots cultured in media with 0 (control, CT), 7% PEG, MeJA or 7% PEG + MeJA with the identification of the bands (the corresponding bonds and functional groups are presented in Table 2).

**Table 2.** FTIR spectral bands and functional groups of *Thymus lotocephalus* shoots cultured in media with 0 (control) or 7% PEG, methyl jasmonate (MeJA), or 7% PEG + MeJA reported in Figure 2B.

| **Wavenumber Range** | | **Bond** | **Functional Group** | **References** |
|---|---|---|---|---|
| **Present Study ($cm^{-1}$)** | **Reference ($cm^{-1}$)** | | | |
| 3443–3406 | 3415–3369 | O–H stretching | Alcohols | [32,33] |
| 2945–2889 | 2920–2800 | C–H stretching | Lipids, carbohydrates | [32–34,36] |
| 1735–1647 | 1723–1607 | C=O stretching, C–N stretching, COO– antisymmetric stretching | Proteins, lipids, carbohydrates | [32,34–36] |
| 1468–1342 | 1454–1366 | C–O stretching, C–C stretching, COO symmetric stretching, $CH_2$ bending | Phenyl groups of aromatic compounds | [33,34,36] |
| 1281–1271 | 1300–1260 | C–O stretching | Hydroxyflavonoids | [34,36] |
| 1242–1155 | 1270–1150 | C–O stretching, C–N stretching | Acid or ester | [33,34] |
| 1115–1051 | 1170–950 | C–O and C–C stretching | Carbohydrates, flavonoids | [34–36] |
| 962–843 | 980–960 | C=H bending, C–H out-of-plane bending | Terpenoids | [33–35] |

FTIR spectra showed some chemical differences between the control and the shoots growing in all media containing PEG (with or without mitigating agents) (Figure 2A). PEG stress showed additional characteristic bands, namely at 1468, 1342, 1242, 1115, 962, and 843 $cm^{-1}$ (Figure 2B), that are associated with phenyl groups or terpenoids (Table 2). According to Schulz and Baranska [35], apparently, drought stress affects positively the biosynthesis of some specific volatile compounds, namely 1,8-cineol (peak 843 $cm^{-1}$), lutein (peak 962 $cm^{-1}$), and especially citronellal (peak 1115 $cm^{-1}$) compounds with characteristic bands that appeared in our samples. These results were also consistent with the results of Sevindik et al. [63] who observed that FTIR analysis of *Ocimum basilicum* (Lamiaceae) irrigated with PEG 6000 exhibited major alterations in the functional groups corresponding to bands below 1400 $cm^{-1}$. FTIR spectra from shoots cultivated in the control medium and PEG-free media containing $Fe_3O_4$ NPs, SA, or MeJA were similar (Figure 2A,B), indicating that these agents do not induce appreciable chemical modifications evaluated with this technique. Similar results were obtained in *Rosmarinus officinalis*, another Lamiaceae species [34].

### *3.4. Phenolics Biosynthesis and Biological Activity of the Extracts*

#### 3.4.1. Total Phenolic Contents by F-C Method

Many abiotic stresses, including drought induced by PEG, frequently affect the production of phenolic compounds as a reaction to oxidative injury [53,64,65]. In this work, all PEG concentrations significantly reduced the total phenolic content (TPC) in the extracts obtained by F-C method (Table 3). According to Moradi et al. [2], when sensitive plants are exposed to stress conditions they usually present a higher accumulation of metabolites than tolerant plants, which can evidence a certain drought tolerance of *T. lotocephalus* cultures. In accordance with these results, PEG also showed a negative impact on phenolics accumulation in *Taxus baccata* (1%, 2%, and 3% PEG) [57] and *V. sinuatum* (−0.6 MPa osmotic potential) [9]. However, in *T. vulgaris* grown in vitro [4] or ex vitro [12,55] conditions, drought significantly improved TPC compared to the control treatment. The exposure of stressed plants (7% PEG) to SA ($53.8 \pm 3.07$ $mg_{GAE}/g_{DW}$) and MeJA ($57.4 \pm 2.34$ $mg_{GAE}/g_{DW}$) caused a significant rise in TPC in comparison with the 7% PEG treatment ($46.8 \pm 0.42$ $mg_{GAE}/g_{DW}$). Other authors also reported the beneficial effects of MeJA on the phenolic compound accumulation in plants under drought stress [9,66]. TPC significantly increased in non-stressed cultures treated with $Fe_3O_4$ NPs, SA, and MeJA. Similar findings were achieved by other authors [6,9,66], reinforcing the role of these agents as elicitors. MeJA was shown to be the best elicitor for *T. lotocephalus* shoot cultures, enhancing phenolics accumulation in 38.8%. The same was obtained in *Rubus idaeus* [67], *Mentha* × *piperita* [66], and *Brassica rapa* L. ssp. *chinensis* [68] treated with MeJA. Recently, Kianersi et al. [69] tested distinct concentrations of MeJA (10, 100, 150, and 200 μM) in different *Thymus* species (*T. vulgaris*, *T. migricus*, and *T. daenensis*) and observed that the maximum phenolics accumulation was achieved using 100 μM MeJA in those species.

**Table 3.** Total phenolic content (TPC) and tyrosinase inhibitory capacity of extracts from *Thymus lotocephalus* shoots cultured in media with 0 (control), 2, 5, or 7% of PEG, 7% PEG + $Fe_3O_4$ nanoparticles (NPs), 7% PEG+ salicylic acid (SA), or 7% PEG + methyl jasmonate (MeJA), $Fe_3O_4$ NPs, SA, or MeJA.

| Treatment | TPC ($mg_{GAE}/g_{DW}$) | Tyrosinase Inhibition ($mg_{KAE}/g_{DW}$) |
|---|---|---|
| **Control** | 62.2 ± 1.19 $^{cd}$ | 21.72 ± 1.25 $^{b}$ |
| PEG | | |
| 2% PEG | 45.1 ± 2.61 $^{ef}$ | 9.59 ± 0.10 $^{c}$ |
| 5% PEG | 45.0 ± 3.39 $^{ef}$ | 5.98 ± 1.09 $^{c}$ |
| 7% PEG | 46.8 ± 0.42 $^{ef}$ | 9.03 ± 1.61 $^{c}$ |
| **7% PEG** | | |
| $Fe_3O_4$ NPs | 43.4 ± 2.35 $^{f}$ | 10.8 ± 1.89 $^{c}$ |
| SA | 53.8 ± 3.07 $^{de}$ | 10.2 ± 0.60 $^{c}$ |
| MeJA | 57.4 ± 2.34 $^{d}$ | 8.46 ± 1.50 $^{c}$ |
| **PEG-free** | | |
| $Fe_3O_4$ NPs | 68.8 ± 0.70 $^{bc}$ | 28.2 ± 1.27 $^{ab}$ |
| SA | 76.6 ± 0.32 $^{ab}$ | 29.2 ± 0.48 $^{a}$ |
| MeJA | 86.4 ± 1.09 $^{a}$ | 34.2 ± 0.31 $^{a}$ |

Values are expressed as mean ± standard error. In each column values followed by different letters (a–f) are significantly different at $p < 0.05$ (Tukey's New Multiple Range Test).

### 3.4.2. Phenolic Profile Analysis by HPLC-HRMS

Secondary metabolites are substances produced by plants that make them competitive in their environment and are essential in responding to biotic and abiotic stresses. Phenolic compounds are one of the main classes of secondary metabolites with important biological properties in plants [5]. In the present study, the phenolic profile of *T. lotocephalus* extracts obtained from shoots cultivated in media with PEG and/or with different mitigating agents, obtained for the first time using NADES, was analyzed by HPLC-HRMS. A total of twenty-six phenolics (19 phenolic acids, five flavonoids, a coumarin derivative, and a hydroxybenzaldehyde) were identified and quantified in *T. lotocephalus* extracts (Table 4 and Supplementary Tables S1 and S2).

**Table 4.** Qualitative and quantitative [mg/kg or g/kg (marked with *) of extract, mean ± SE] analysis by HPLC-HRMS of the phenolic profile from *Thymus lotocephalus* shoots cultured in media with 0 (control), 2, 5, or 7% of PEG, 7% PEG + $Fe_3O_4$ nanoparticles (NPs), 7% PEG + salicylic acid (SA), or 7% PEG + methyl jasmonate (MeJA), $Fe_3O_4$ NPs, SA, or MeJA.

| Compound | Treatment | | | | | | | | | |
|---|---|---|---|---|---|---|---|---|---|---|
| | Control | PEG | | | 7% PEG | | | PEG-Free | | |
| | | 2% | 5% | 7% | $Fe_3O_4$ NPs | SA | MeJA | $Fe_3O_4$ NPs | SA | MeJA |
| **Phenolic acids** | | | | | | | | | | |
| Salvianolic acid A isomer I | 468 ± 1 [a] | 137 ± 6 [bc] | 146 ± 6 [bc] | 150 ± 8 [bc] | 107 ± 16 [c] | 288 ± 38 [abc] | 172 ± 10 [bc] | 339 ± 20 [ab] | 323 ± 103 [ab] | 172 ± 41 [bc] |
| Salvianolic acid A isomer II | <LOQ | <LOQ | <LOQ | <LOQ | <LOQ | <LOQ | 133 ± 6 | <LOQ | <LOQ | 227 ± 64 |
| Salvianolic acid A isomer IV | <LOQ | <LOQ | <LOQ | <LOQ | <LOQ | <LOQ | <LOQ | <LOQ | <LOQ | 149 ± 28 |
| Salvianolic acid B/Salvianolic acid L isomer I | <LOD | <LOD | <LOD | <LOD | n.d. | <LOD | <LOD | <LOD | <LOD | <LOD |
| Salvianolic acid B/Salvianolic acid L isomer II | <LOQ | <LOQ | <LOQ | <LOQ | <LOQ | <LOQ | <LOQ | <LOQ | <LOQ | <LOQ |
| Salvianolic acid B/Salvianolic acid L isomer III | <LOQ | <LOQ | <LOQ | <LOQ | <LOQ | 174 ± 8 | 127 ± 10 | <LOQ | <LOQ | <LOQ |
| Salvianolic acid B/Salvianolic acid L isomer IV | <LOQ | <LOQ | <LOQ | <LOQ | <LOQ | <LOQ | 126 ± 10 [b] | <LOQ | 180 ± 39 [b] | 391 ± 46 [a] |
| Salvianolic acid C | <LOQ | <LOQ | <LOQ | <LOQ | <LOQ | <LOQ | <LOQ | <LOQ | 143 ± 8 | 199 ± 62 |
| Salvianolic acid F isomer I | <LOQ | n.d. | n.d. | n.d. | n.d. | <LOQ | n.d. | <LOQ | <LOQ | <LOQ |
| Salvianolic acid F isomer II | 173 ± 23 [a] | 113 ± 20 [a] | <LOQ | <LOQ | <LOQ | <LOQ | <LOQ | 118 ± 20 [a] | 199 ± 2 [a] | 204 ± 3 [a] |
| Salvianolic acid I/Melitric acid A | 252 ± 8 [b] | 224 ± 12 [b] | 176 ± 9 [b] | 178 ± 12 [b] | 295 ± 17 [ab] | 199 ± 31 [b] | 307 ± 7 [ab] | 178 ± 6 [b] | 192 ± 56 [b] | 591 ± 158 [a] |
| **Total salvianolic acids *** | 0.89 ± 0.03 [b] | 0.47 ± 0.00 [b] | 0.32 ± 0.02 [b] | 0.33 ± 0.02 [b] | 0.40 ± 0.03 [b] | 0.66 ± 0.06 [b] | 0.86 ± 0.02 [b] | 0.64 ± 0.05 [b] | 1.04 ± 0.21 [b] | 1.93 ± 0.40 [a] |
| Caffeic acid | 46 ± 16 [cd] | 30 ± 8 [cd] | 20 ± 3 [cd] | 23 ± 3 [cd] | 8 ± 2 [d] | 35 ± 4 [cd] | 30 ± 1 [cd] | 54 ± 7 [bc] | 90 ± 3 [ab] | 112 ± 10 [a] |
| Melitric acid B | <LOD | <LOD | <LOD | <LOD | <LOD | <LOD | <LOD | <LOD | <LOD | <LOD |
| Methyl 6-O-galloyl-β-D-glucopyranoside * | 6.00 ± 1.25 [a] | 4.71 ± 0.56 [a] | 3.61 ± 0.19 [a] | 2.46 ± 0.24 [a] | 4.30 ± 0.43 [a] | 4.20 ± 0.30 [a] | 5.01 ± 0.08 [a] | 6.02 ± 0.26 [a] | 4.39 ± 1.29 [a] | 6.15 ± 1.11 [a] |
| Methylrosmarinic acid isomer I | 94 ± 4 [a] | <LOQ | <LOQ | 83 ± 2 [a] | <LOQ | <LOQ | <LOQ | 84 ± 4 [a] | 75 ± 18 [a] | 78 ± 7 [a] |
| Methylrosmarinic acid isomer II | 651 ± 19 [a] | 342 ± 30 [ab] | 272 ± 38 [b] | 420 ± 0 [ab] | 265 ± 16 [b] | 384 ± 11 [ab] | 204 ± 38 [b] | 394 ± 91 [ab] | 616 ± 150 [a] | 632 ± 38 [a] |
| Rosmarinic acid * | 53.0 ± 3.57 [a] | 16.8 ± 1.25 [bc] | 15.9 ± 1.04 [bc] | 23.9 ± 0.55 [b] | 23.5 ± 0.78 [b] | 20.3 ± 1.98 [bc] | 12.5 ± 0.16 [c] | 48.2 ± 1.95 [a] | 54.9 ± 2.05 [a] | 58.2 ± 2.36 [a] |
| Sagerinic acid * | 2.88 ± 0.03 [ab] | 0.85 ± 0.03 [c] | 0.72 ± 0.01 [c] | 0.97 ± 0.05 [c] | 0.63 ± 0.17 [c] | 0.84 ± 0.08 [c] | 0.55 ± 0.04 [c] | 1.98 ± 0.12 [bc] | 3.70 ± 0.70 [a] | 2.73 ± 0.38 [ab] |
| Salviaflaside | <LOQ | <LOQ | <LOQ | <LOQ | <LOQ | <LOQ | <LOQ | <LOQ | <LOQ | <LOQ |
| **Total phenolic acids *** | 63.5 ± 2.36 [ab] | 23.3 ± 0.75 [cd] | 20.8 ± 0.85 [cd] | 28.2 ± 0.87 [cd] | 29.1 ± 1.43 [c] | 26.4 ± 2.12 [cd] | 19.2 ± 0.26 [d] | 57.4 ± 2.28 [b] | 64.8 ± 2.59 [ab] | 69.8 ± 1.30 [a] |
| **Flavonoids** | | | | | | | | | | |
| Dihydromorelloflavone | 218 ± 34 [a] | 216 ± 18 [a] | 231 ± 19 [a] | 217 ± 7 [a] | 183 ± 4 [a] | 190 ± 4 [a] | 157 ± 9 [a] | 194 ± 6 [a] | 208 ± 34 [a] | 175 ± 9 [a] |
| Epigallocatechin gallate * | 6.60 ± 1.69 [a] | 4.51 ± 0.29 [a] | 7.59 ± 0.95 [a] | 6.60 ± 0.70 [a] | 4.22 ± 0.57 [a] | 4.56 ± 1.23 [a] | 4.15 ± 0.00 [a] | 5.65 ± 0.26 [a] | 6.13 ± 1.50 [a] | 3.96 ± 0.35 [a] |
| Luteolin | 12 ± 0 [a] | <LOQ | <LOQ | <LOQ | 12 ± 0 [a] | <LOQ | <LOD | <LOQ | 13 ± 1 [a] | <LOQ |
| Luteolin-7-O-glucuronide | 157 ± 6 [a] | 68 ± 3 [b] | 44 ± 2 [b] | 48 ± 3 [b] | 39 ± 6 [b] | 48 ± 4 [b] | 20 ± 3 [b] | 156 ± 23 [a] | 154 ± 16 [a] | 52 ± 0 [b] |
| Theaflavic acid * | 1.22 ± 0.00 [a] | 1.06 ± 0.02 [a] | 0.956 ± 0.07 [a] | 0.94 ± 0.03 [a] | 0.97 ± 0.00 [a] | 0.93 ± 0.00 [a] | 0.78 ± 0.00 [a] | 1.15 ± 0.09 [a] | 1.20 ± 0.26 [a] | 1.21 ± 0.12 [a] |
| **Total flavonoids *** | 8.21 ± 1.64 [a] | 5.85 ± 0.28 [a] | 8.82 ± 0.86 [a] | 7.80 ± 0.71 [a] | 5.42 ± 0.60 [a] | 5.73 ± 1.22 [a] | 5.11 ± 0.00 [a] | 7.14 ± 0.37 [a] | 7.70 ± 1.19 [a] | 5.40 ± 0.24 [a] |
| **Coumarin derivative** | | | | | | | | | | |
| Herniarin | <LOD | <LOD | <LOD | <LOD | <LOD | <LOD | <LOD | <LOD | <LOD | <LOD |
| **Hydroxybenzaldehyde** | | | | | | | | | | |
| Protocatechuic aldehyde | 66 ± 6 [bcd] | 49 ± 4 [cd] | 29 ± 5 [d] | 23 ± 3 [d] | 21 ± 0 [d] | 32 ± 3 [d] | 25 ± 13 [d] | 82 ± 16 [bc] | 112 ± 17 [ab] | 135 ± 0 [a] |
| **Total phenolic compounds *** | 71.8 ± 4.00 [a] | 29.2 ± 0.47 [bc] | 29.6 ± 0.00 [bc] | 36.0 ± 1.57 [b] | 34.6 ± 0.86 [bc] | 32.1 ± 0.91 [bc] | 24.3 ± 0.28 [c] | 64.6 ± 2.26 [a] | 72.6 ± 3.80 [a] | 75.3 ± 1.06 [a] |

Notes: n.d.—not detected; LOD—limit of detection; LOQ—limit of quantification. The results were analyzed using a one-way analysis of variance (ANOVA) followed by Tukey's New Multiple Range Test. Different letters (a–d) in each row and for each phenolic compound mean significant differences ($p < 0.05$) among extracts.

To the best of our knowledge, this is the first time that epigallocatechin gallate, methyl 6-*O*-galloyl-β-D-glucopyranoside, theaflavic acid, methylrosmarinic acid (I and II), dihydromorelloflavone, protocatechuic aldehyde, melitric acid B and salviaflaside are identified in *Thymus* genera. Nevertheless, melitric acid B, methylrosmarinic acid, salviaflaside and protocatechuic aldehyde, were previously identified in other genera belonging to the Lamiaceae family, namely *Melissa* and *Salvia* [70–72]. Salvianolic acid F and sagerinic acid, although being identified for the first time in *T. lotocephalus*, were detected in other *Thymus* species (*T. zygis*, *T. pulegioides*, *T. fragrantissimus*, and *T. herba-barona*) [73]. As reported in previous studies with this species [6,15,17], rosmarinic acid was the major biocompound identified in the extracts (with a range of 12.5 g/kg in 7% PEG + MeJA to 58.2 g/kg in MeJA) (Table 4). Rosmarinic acid has been demonstrated to have important biological properties, such as antioxidant, anti-inflammatory, antitumor, neuroprotective, and antimicrobial [74,75]. This phenolic acid is utilized as a food and cosmetic ingredient and many pharmaceutical applications have also been reported [76]. The second most abundant compounds were epigallocatechin gallate (EGCG) (flavonoid) and methyl 6-*O*-galloyl-β-D-glucopyranoside (phenolic acid), although both eight times less abundant than rosmarinic acid in the control. Although the greatest total phenolic content (and rosmarinic acid) was achieved in the MeJA extract, it was not significantly different from the other agents (SA and $Fe_3O_4$ NPs) and control extracts, as observed when TPCs were determined by F-C method. Even though a high correlation ($p < 0.01$) between TPC by F-C assay and HPLC exists, there are some differences in the statistical analysis, which can be explained by the poor specificity of F-C reagent that can oxidize other substances, overestimating the results [77].

Regarding the effect of drought on phenolics biosynthesis, as noted for morphological and physiological traits and in agreement with F-C results, PEG significantly reduced the content of almost all phenolic compounds (especially rosmarinic acid), compared to the control. Likewise, rosmarinic acid content decreased by water stress in other Lamiaceae species, such as *O. basilicum* [78], *Melissa officinalis* [79,80], and *Salvia miltiorrhiza* [81]. On the other hand, the production of flavonoids (epigallocatechin gallate, dihydromorelloflavone, and theaflavic acid) was not affected by drought. It was observed that MeJA showed beneficial effects on phenolic acids production in stressed plants, improving the synthesis of methyl 6-O-galloyl-β-D-glucopyranoside (two-fold) and salvianolic acids (A and B). Under non-stress conditions, MeJA enhanced two-fold the biosynthesis of salvianolic acids, caffeic acid, and protocatechuic aldehyde. This agent also significantly improved the production of salvianolic and caffeic acids in *Salvia* species [82,83].

#### 3.4.3. Potential Properties of the Extracts for Skincare Products and Correlation with Phenolic Composition

Antioxidant Activity

Since plant extracts are extraordinarily complex multicomponent mixtures, the antioxidant capacity of *T. lotocephalus* extract was evaluated using more than one assay (ORAC, FRAP, DPPH, and ABTS) (Figure 3). The ORAC measures the capacity of an antioxidant to break hydrophilic and lipophilic chains in the presence of peroxyl radicals. FRAP assay determines the capacity of the extracts to reduce ferric ion ($Fe^{3+}$) to ferrous ion ($Fe^{2+}$). Finally, DPPH and ABTS quantify the scavenging capability against stable free radicals.

Similar to TPCs, in all assays, the antioxidant activity of the *T. lotocephalus* extracts were significantly reduced by all concentrations of PEG compared to the control, except in the ORAC assay, in which no significant differences were observed. These results confirm those previously observed for *T. vulgaris*, in which antioxidant activity accessed by FRAP decreased to half at 8% PEG [4]. Otherwise, an increase in antioxidant activity was reported in *S. rebaudiana* under rising concentrations of PEG (0.5, 1, 2, 4%) [64], as well as in *T. vulgaris* [12] and *Mentha piperita* [66] grown under water deficit. MeJA increased the antioxidant capacity of the cultures subjected to PEG stress in almost all the assays. Similar results were obtained in infusions prepared from peppermint plants grown under water deficit, in which MeJA (50 μM) revealed the greatest capability to scavenge $ABTS^{+}$ and $DPPH^{\bullet}$ radicals [66]. Regarding the impact of the tested agents under non-stress conditions, similar to TPC by F-C assay, the antioxidant activity was significantly higher in MeJA treatment in all the assays, except in ORAC in which SA, $Fe_3O_4$ NPs and control demonstrated a similar capacity to quench free radicals by hydrogen donation. This exception can be justified by the distinct mechanisms implicated in the various methods. The increase in antioxidant activity was observed in extracts from other plants grown with MeJA [67,68]. The extracts obtained from the cultures treated with $Fe_3O_4$ NPs and SA showed higher antioxidant capacities (ABTS and ORAC) compared to the control. A similar trend was obtained in our previous works, in which $Fe_3O_4$ NPs [17] and SA [6] improved the antioxidant capability of *T. lotocephalus* methanolic extracts. Since this is the first study investigating the impact of MeJA, there are no reports in the literature to compare the obtained results. A strong correlation was established between all antioxidant results (DPPH, ABTS, FRAP, and ORAC) and TPC by HPLC ($p < 0.01$) (Figure 4) showing that phenolic compounds, especially phenolic acids, are the major contributors to *T. lotocephalus* antioxidant activities. The correlation between antioxidant activity and this class of bioactive compounds has been recently reported in *Thymus* species [6,17,84,85]. Although rosmarinic acid possesses a great significance to the antioxidant activity of the extracts because it is the major compound, the strong correlations ($p < 0.01$) between other minor compounds and antioxidant activity assays, such as sagerinic acid, protocatechuic aldehyde, and caffeic acid, equally demonstrate some influence of these compounds to this activity.

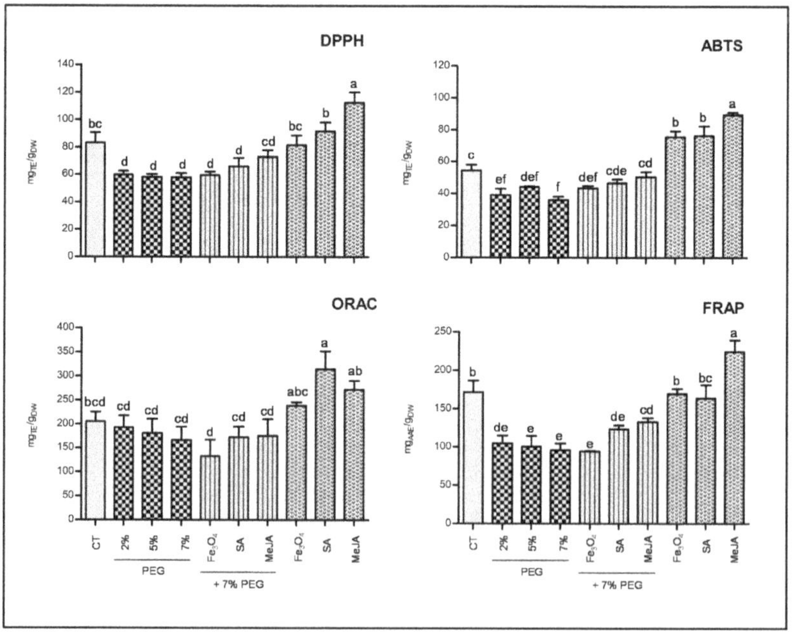


**Figure 3.** Antioxidant activity evaluated by 2,2-diphenyl-1- picrylhydrazyl (DPPH), 2,2-azino-bis(3-ethylbenzothiazoline-6-sulfonic acid) (ABTS), oxygen radical absorbance capacity (ORAC), and ferric reducing antioxidant power (FRAP) methods of the extracts from *T. lotocephalus* shoots cultured in media with 0 (control, CT), 2, 5, or 7% of PEG, 7% PEG + $Fe_3O_4$ nanoparticles (NPs), 7% PEG + salicylic acid (SA), or 7% PEG + methyl jasmonate (MeJA), $Fe_3O_4$ NPs, SA, or MeJA. Values are presented as mean ± SE. Different letters (a–f) in each graph bars specify significant differences ($p < 0.05$, Tukey's new multiple test).

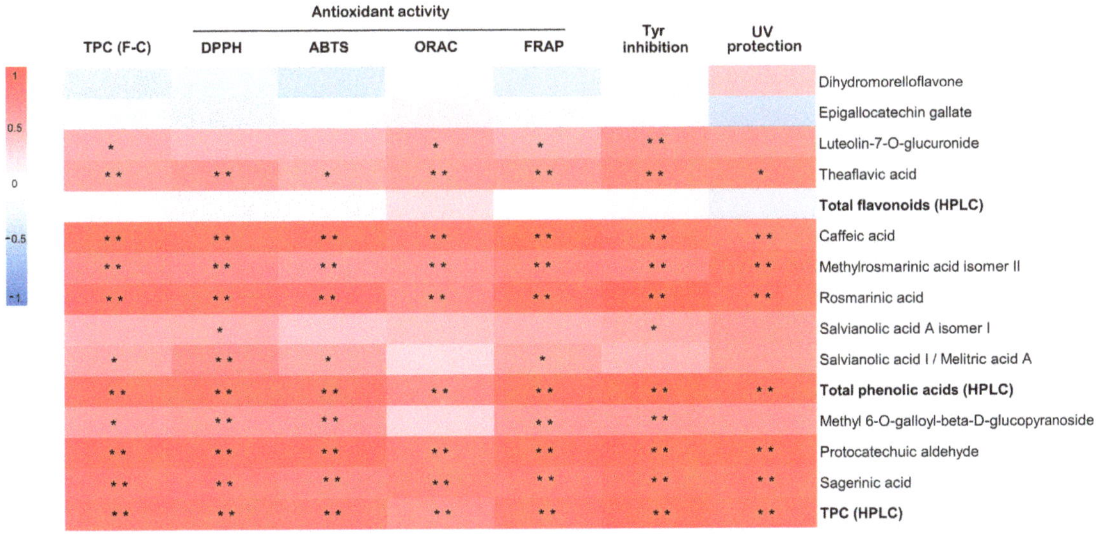


**Figure 4.** Heat map corresponding to Pearson's correlation coefficients between antioxidant activity measured DPPH, ABTS, ORAC, and FRAP, tyrosinase (Tyr) inhibition capacity, UV protection ability, total phenolic contents measured by F-C and HPLC, total flavonoids (HPLC), total phenolic acids (HPLC) and individual phenolic compounds (HPLC). ** Correlation is significant ($p < 0.01$). * Correlation is significant ($p < 0.05$).

Extracts as Anti-Depigmentation Agents

The depigmentation or whitening effects of a product are related to its capacity to inhibit Tyr enzyme activity. Thus, the potential of *T. lotocephalus* extracts to inhibit this enzyme were evaluated in this study (Table 3). The Tyr inhibition activity varied from $5.98 \pm 1.09$ to $34.1 \pm 0.31$ $mg_{KAE}/g_{DW}$. Extracts from cultures exposed to PEG demonstrated the lowest Tyr inhibition. The three agents tested in this work under non-stressed conditions, particularly MeJA, significantly improved the anti-Tyr activity in *T. lotocephalus*. A strong correlation ($p < 0.01$) was observed between Tyr inhibition and the total phenolic compounds (HPLC) (Figure 4), which were similarly reported in other studies [17,18]. Rosmarinic acid was the phenolic compound that demonstrated the highest correlation with Tyr inhibition (r = 0.943). This phenolic acid was previously defined as a good inhibitor of this enzyme [86]. Although present in smaller amounts, protocatechuic aldehyde, caffeic acid, sagerinic acid, theaflavic acid, and methylrosmarinic acid strongly contributed ($p < 0.01$) to inhibiting Tyr. The inhibitory capacity of protocatechuic aldehyde and caffeic acid against Tyr was also previously reported by Ko and Lee [87] and Crespo et al. [88], respectively.

Extracts as Ultraviolet (UV) Protecting Agents

In cosmetics, the capacity to absorb UV-A and UV-B radiation is an outstanding starting point to consider the extract as a natural sun protective agent [22]. To assess the potential of *T. lotocephalus* extracts as UV-protecting agents, different concentrations of the extracts were tested and their capacity to absorb UV-A (315–400 nm) and UV-B (200–280 nm) radiation was evaluated spectrophotometrically. The UV–vis spectra of the extracts at 250 μg/mL are shown in Figure 5A.

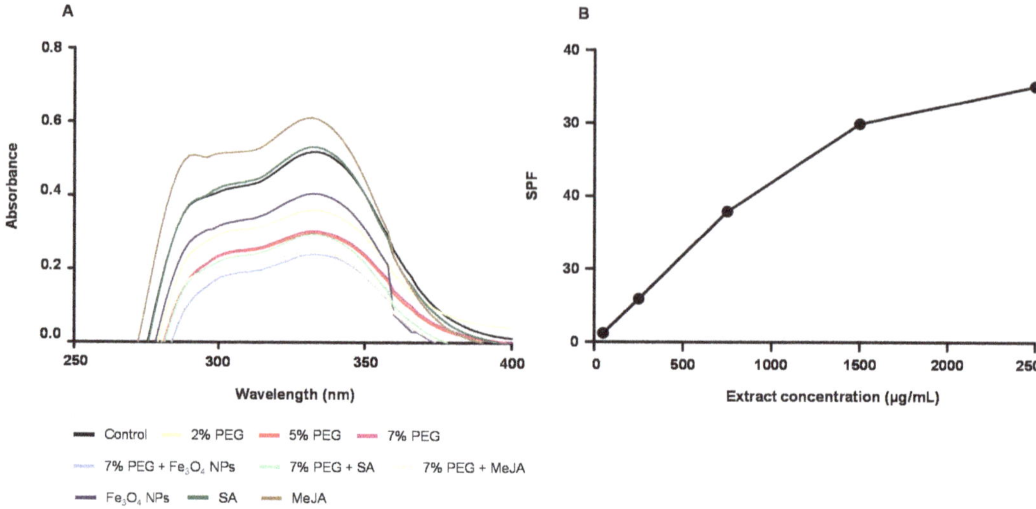


**Figure 5.** (**A**) UV-Vis absorption spectra of the extracts (250 μg/mL) from *T. lotocephalus* shoots cultured in media with 0 (control), 2, 5, or 7% of PEG, $Fe_3O_4$ nanoparticles (NPs), salicylic acid (SA), methyl jasmonate (MeJA), 7% PEG + $Fe_3O_4$ NPs, 7% PEG + SA, or 7% PEG + MeJA; (**B**) Sun Protection Factor (SPF) of the extract from *Thymus lotocephalus* shoots cultured in media with MeJA (50 μM).

All *T. lotocephalus* extracts exhibited the capacity to absorb UV radiation, although the extract from cultures grown with MeJA showed the highest capacity with a band at the wavelength of 332 nm (Abs = 0.611). Even though the greatest ability of this extract is to absorb UV-A rays (55.79%), it also demonstrated a good capacity to inhibit UV-B radiation (36.86%). Similar to antioxidant and anti-melanogenic activities results, all cultures exposed to PEG demonstrated the worst photoprotective capacity. A strong correlation was found between several phenolic compounds and UV protecting activity ($p < 0.01$) (Figure 4), such as protocatechuic aldehyde, and rosmarinic, methylrosmarinic, sagerinic, and caffeic acids. Rosmarinic and caffeic acids were previously reported as photoprotective agents against UV and other ionizing radiations [89,90].

Since MeJA extract showed the highest ability to absorb UV radiation, the sun protection factor (SPF) was only calculated for this extract at different concentrations (50, 250, 750, 1500, and 2500 μg/mL). Compared to the in vivo models for the determination of SPF, the in vitro models present several advantages since they are simple, reproducible, fast, and avoid subject UV exposure [91]. The UV radiation in the range of 290–320 nm retains the highest biological activity to induce skin damage, such as burning, photoaging, and cancer, so the SPF of a product is evaluated in these wavelengths. According to Yakoubi et al. [92] the rating of sun-protective activity of sunscreens can be determined according to SPF values in minimum (2–12), moderate (12–30), and high (≥30). The SPF values of the *T. lotocephalus* extract increased in a concentration-dependent manner (Figure 5B). Fifty μg/mL (SPF = 1.20) and 250 μg/mL (SPF = 5.93) of *T. lotocephalus* extract resulted in minimum, 750 μg/mL (SPF = 17.92) in moderate, and finally, 1500 μg/mL (SPF = 29.92) and 2500 μg/mL (SPF = 35.13) in high photoprotective activities. Other plant extracts have been reported in the literature as good agents to absorb UV radiation, such as ethanolic and aqueous-glycerin extracts of *Plantago lanceolata* [22] and polyol extracts of *Camellia oleifera* [93]. With regards to *Thymus* extracts, as far as is known, there are no previous reports describing their capacity to absorb UV radiation and its SPF evaluation. Although NADES are considered green solvents, only a few can be applied in cosmetics because of safety or regulatory reasons, especially in Europe [25]. This is the case of the mixture used in the present study (proline: lactic acid, 1:1), which is accepted by the European Cosmetic Regulation EC No.1223/2009 to be used in cosmetic products formulations.

3.4.4. Multivariate Analyses: Cluster Analysis and Principal Component Analysis (PCA)

Principal Component Analysis (PCA) is used to decrease the dimensionality of a multivariate data set in a few principal components (PC), which summarize the predominant patterns in the data. PCA was based on the individual and total phenolic content (F-C and HPLC) and the biological properties (photoprotective, anti-melanogenic, and antioxidant) of the extracts. The analysis can be observed in Figure 6 in a PCA biplot [PCA score plot (green) + loading plot (purple)] form. Each point on the score plot characterizes the different extracts, and each point on the loading plot indicates the contribution of each variable to the score. The first two PC described 85.47% of the total variation in the dataset, explaining the first principal component (PC1) 68.24% of the data variability, and the second principal component (PC2) 17.23%. The score plot of the first two principal components demonstrated a clear grouping of the extracts by PEG influence. It is possible to observe a clear separation between samples grown with PEG (negative PC1 values, second and third quadrants) and samples grown without PEG (positive PC1 values, first and fourth quadrants). Similarly, PCA analysis of *S. rebaudiana* in vitro plants subjected to PEG 6000 (2.5, 5, 7, and 10%) revealed a pronounced separation between the controls and the treatment groups [94].The results indicate that the different agents ($Fe_3O_4$ NPs, SA, and MeJA) and PEG concentrations (2, 5, and 7%) added to the culture media present a different ability to induce the production of distinct classes of phenolics. In general, the major number of phenolic compounds and consequent biological properties were produced in higher amounts in the media containing the different mitigating agents ($Fe_3O_4$ NPs, SA, and MeJA), or in the control. Positioned at higher positive values of PC1, MeJA, SA, and PC2, control presented the highest contents in rosmarinic acid and its derivatives methyl rosmarinic (isomer II) and sagerinic acids. The control extract stands out for its content in the flavonoid luteolin-7-O-glucuronide and the rosmarinic derivative salvianolic acid A (isomer I), while SA and MeJA showed the highest contents in protocatechuic aldehyde and caffeic acid. Furthermore, the MeJA sample, situated in the fourth quadrant and with the greatest influence on PC1 and PC2, presented the highest difference among samples, and was shown to be the best agent to promote the production of salvianolic acid I/melitric acid A, exhibiting the highest abilities as UV protection, Tyr inhibition and with the highest TPC (F-C method) and antioxidant capacity by FRAP, DPPH, and ABTS.

Hierarchical cluster analysis and K-means cluster analysis are two multivariate analyses to recognize the clustering pattern and group objects according to the similarities among samples. A cluster analysis from data of total phenolic content (F-C) and biological properties (photoprotective, anti-melanogenic, and antioxidant) of the extracts provided good separation among the samples (Figure 7A,B). The dendrogram (Figure 7A) grouped the samples into two principal clusters: cluster 1 (C1) included control and $Fe_3O_4$ NPs, SA, and MeJA, and cluster 2 (C2) comprised all PEG-treated samples (2, 5, 7%, and 7% PEG + agents). Both hierarchical (Figure 7A) and K-means (Figure 7B) clusters showed close relationships in the two sub-groups of the first cluster, namely between the control and $Fe_3O_4$ NPs, and between the signaling molecules SA and MeJA. Regarding PEG treatments, both analyses grouped all samples from PEG media (with or without agents) into one cluster, showing the strong role of this osmotic agent in inducing drought stress in *T. lotocephalus* shoot cultures, and its huge influence on the biological activities of the extracts. Moreover, it suggests that the agents tested in this study and/or their concentrations were not efficient enough to mitigate the stress produced in the plant by the use of PEG.

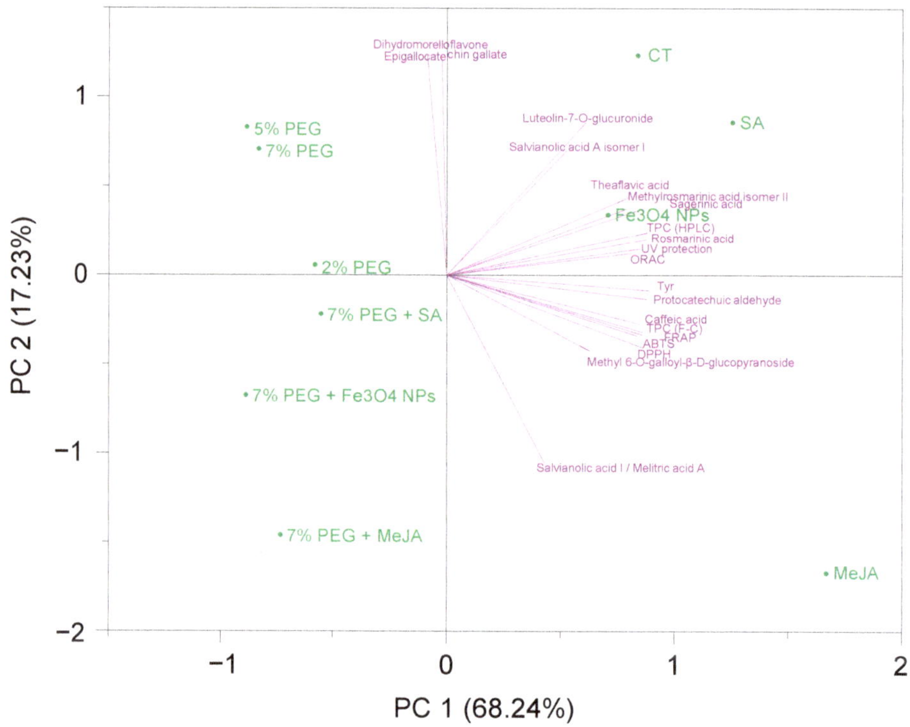


**Figure 6.** Principal component analysis (PCA) biplot of the different polyphenols and bioactivities studied in extracts from *T. lotocephalus* shoots cultured in media with 0 (control, CT), 2, 5, or 7% of PEG, $Fe_3O_4$ nanoparticles (NPs), salicylic acid (SA), methyl jasmonate (MeJA), 7% PEG + $Fe_3O_4$ NPs, 7% PEG + SA, or 7% PEG + MeJA.

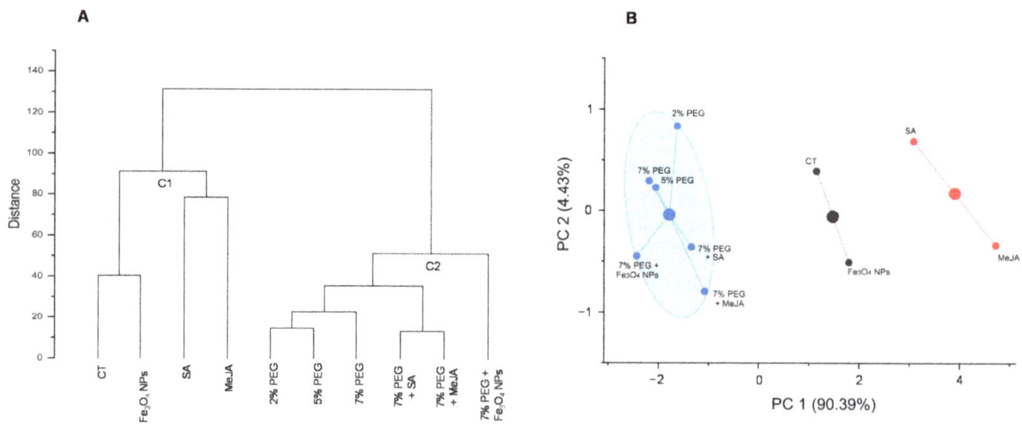


**Figure 7.** (**A**) Hierarchical cluster analysis based on the biological properties (photoprotective, anti-melanogenic, and antioxidant activities) of the extracts from *T. lotocephalus* shoots cultured in media with 0 (control, CT), 2, 5, or 7% of PEG, $Fe_3O_4$ nanoparticles (NPs), salicylic acid (SA), methyl jasmonate (MeJA), 7% PEG + $Fe_3O_4$ NPs, 7% PEG + SA, or 7% PEG + MeJA; (**B**) K-means cluster analysis according to the biological properties (photoprotective, anti-melanogenic, and antioxidant activities) of the *T. lotocephalus* extracts.

## 4. Conclusions

The results obtained suggest that PEG added to culture media has a considerable impact on *T. lotocephalus* in vitro cultures, namely increasing the oxidative stress and causing direct damage on shoots growth, photosynthetic pigments, and bioactive compounds synthesis and bioactivity (Figure 8). Among the different agents tested, the addition of MeJA to the culture media played a prominent role in the mitigation/reduction of some drought harm (Figure 8). MeJA was also shown to be useful as an elicitor, improving the biosynthesis of some phenolic compounds and consequent biological properties of the extracts. Moreover, this study showed that *T. lotocephalus* extracts prepared with Natural Deep Eutectic Solvents (NADES) have the potential to be used in cosmetics due to their ability to absorb UV radiation, as well as antioxidant and depigmentation properties. Nevertheless, more detailed parameters must be carefully investigated, such as concentration, stability, compatibility, and toxicity of the extracts. Overall, although our findings indicate that in vitro culture proves to be an adequate tool for a first estimation of the effect of drought in this species, it will be important, in future works, to evaluate the response of plants to drought stress under ex vitro conditions, and to understand if PEG stress indeed alters the production of terpenoids, as FTIR studies suggest.

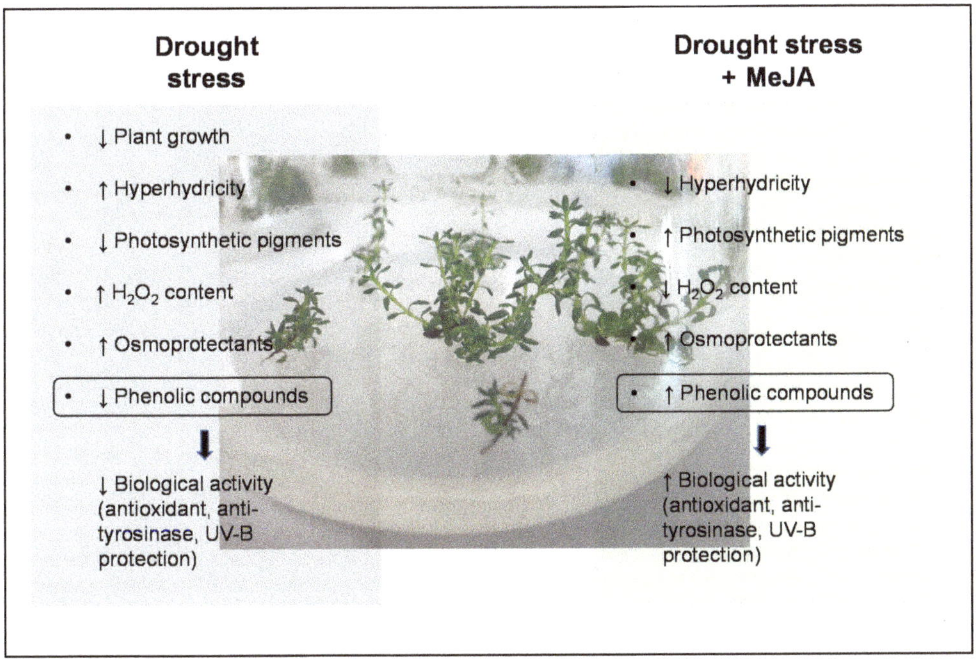


**Figure 8.** The overall response of *Thymus lotocephalus* in vitro cultures to drought stress induced by PEG and the positive effects of methyl jasmonate (MeJA) in relieving damage caused by stress.

**Supplementary Materials:** The following are available online at https://www.mdpi.com/article/10.3390/antiox11081475/s1, Table S1: HPLC-HRMS data of identified phenolics in *Thymus lotocephalus* López and Morales extracts, Table S2: Summary of HPLC-HRMS criterion for quantification of phenolics in *Thymus lotocephalus* López and Morales extracts, Table S3: EE(λ) × I(λ) constants used in the determination of SPF for each wavelength [45].

**Author Contributions:** Conceptualization, I.M., S.G. and A.R.; methodology, I.M., S.G., R.R.-S. and H.D.; software, S.G., I.M. and R.R.-S.; validation, I.M., S.G., R.R.-S., H.D., J.L.O.-D. and A.R.; formal analysis, I.M., S.G. and R.R.-S.; investigation, I.M., S.G., R.R.-S. and A.R.; resources, J.M.M.-R. and A.R.; data curation, S.G., I.M. and R.R.-S.; writing—original draft preparation, S.G., I.M. and R.R.-S.; writing—review and editing, S.G., I.M., R.R.-S., H.D., J.L.O.-D., J.M.M.-R. and A.R.; supervision, S.G. and A.R.; project administration, S.G. and A.R.; funding acquisition, J.M.M.-R. and A.R. All authors have read and agreed to the published version of the manuscript.

**Funding:** This research was funded by National Funds through FCT-Foundation for Science and Technology under the Project UIDB/05183/2020. Inês Mansinhos (Grant SFRH/BD/145243/2019), Sandra Gonçalves (under the Norma Transitória–DL 57/2016/CP1361/CT0022) and Hugo Duarte (PTDC/ASP-SIL/30619/2017) are funded by National Funds through FCT. Raquel Rodríguez Solana was supported by a Juan de la Cierva–Incorporation contract from the Spanish Ministry of Science, Innovation, and Universities (IJC2018-036207-I).

**Institutional Review Board Statement:** Not applicable.

**Informed Consent Statement:** Not applicable.

**Data Availability Statement:** The data presented are included within the article and in the Supplementary Materials.

**Conflicts of Interest:** The authors declare no conflict of interest. The funders had no role in the design of the study; in the collection, analyses, or interpretation of data; in the writing of the manuscript, or in the decision to publish the results.

## References

1. Tramblay, Y.; Koutroulis, A.; Samaniego, L.; Vicente-Serrano, S.M.; Volaire, F.; Boone, A.; Le Page, M.; Llasat, M.C.; Albergel, C.; Burak, S.; et al. Challenges for drought assessment in the Mediterranean region under future climate scenarios. *Earth-Sci. Rev.* **2020**, *210*, 103348. [CrossRef]
2. Moradi, P.; Mahdavi, A.; Khoshkam, M.; Iriti, M. Lipidomics Unravels the Role of Leaf Lipids in Thyme Plant Response to Drought Stress. *Int. J. Mol. Sci.* **2017**, *18*, 2067. [CrossRef] [PubMed]
3. Pradhan, N.; Singh, P.; Dwivedi, P.; Pandey, D.K. Evaluation of sodium nitroprusside and putrescine on polyethylene glycol induced drought stress in *Stevia rebaudiana* Bertoni under in vitro condition. *Ind. Crops Prod.* **2020**, *154*, 112754. [CrossRef]
4. Razavizadeh, R.; Farahzadianpoor, F.; Adabavazeh, F.; Komatsu, S. Physiological and morphological analyses of *Thymus vulgaris* L. in vitro cultures under polyethylene glycol (PEG)-induced osmotic stress. *Vitr. Cell. Dev. Biol.-Plant* **2019**, *55*, 342–357. [CrossRef]
5. Isah, T. Stress and defense responses in plant secondary metabolites production. *Biol. Res.* **2019**, *52*, 39. [CrossRef]
6. Gonçalves, S.; Mansinhos, I.; Rodríguez-Solana, R.; Pérez-Santín, E.; Coelho, N.; Romano, A. Elicitation improves rosmarinic acid content and antioxidant activity in *Thymus lotocephalus* shoot cultures. *Ind. Crops Prod.* **2019**, *137*, 214–220. [CrossRef]
7. Tátrai, Z.A.; Sanoubar, R.; Pluhár, Z.; Mancarella, S.; Orsini, F.; Gianquinto, G. Morphological and Physiological Plant Responses to Drought Stress in Thymus citriodorus. *Int. J. Agron.* **2016**, *2016*, e4165750. [CrossRef]
8. Niazian, M.; Sabbatini, P. Traditional in vitro strategies for sustainable production of bioactive compounds and manipulation of metabolomic profile in medicinal, aromatic and ornamental plants. *Planta* **2021**, *254*, 111. [CrossRef]
9. Karamian, R.; Ghasemlou, F.; Amiri, H. Physiological evaluation of drought stress tolerance and recovery in Verbascum sinuatum plants treated with methyl jasmonate, salicylic acid and titanium dioxide nanoparticles. *Plant Biosyst. -Int. J. Deal. All Asp. Plant Biol.* **2020**, *154*, 277–287. [CrossRef]
10. Thakur, M.; Bhattacharya, S.; Khosla, P.K.; Puri, S. Improving production of plant secondary metabolites through biotic and abiotic elicitation. *J. Appl. Res. Med. Aromat. Plants* **2019**, *12*, 1–12. [CrossRef]
11. Ashrafi, M.; Azimi-Moqadam, M.-R.; Moradi, P.; MohseniFard, E.; Shekari, F.; Kompany-Zareh, M. Effect of drought stress on metabolite adjustments in drought tolerant and sensitive thyme. *Plant Physiol. Biochem.* **2018**, *132*, 391–399. [CrossRef] [PubMed]
12. Khalil, N.; Fekry, M.; Bishr, M.; El-Zalabani, S.; Salama, O. Foliar spraying of salicylic acid induced accumulation of phenolics, increased radical scavenging activity and modified the composition of the essential oil of water stressed *Thymus vulgaris* L. *Plant Physiol. Biochem.* **2018**, *123*, 65–74. [CrossRef] [PubMed]
13. Salehi, B.; Mishra, A.P.; Shukla, I.; Sharifi-Rad, M.; del Contreras, M.M.; Segura-Carretero, A.; Fathi, H.; Nasrabadi, N.N.; Kobarfard, F.; Sharifi-Rad, J. Thymol, thyme, and other plant sources: Health and potential uses. *Phytother. Res.* **2018**, *32*, 1688–1706. [CrossRef] [PubMed]
14. Caldas, F.B. 2011. *Thymus lotocephalus*. The IUCN Red List of Threatened Species 2011:e.T161974A5522381. Available online: https://www.iucnredlist.org/species/161974/5522381 (accessed on 25 July 2022).
15. Costa, P.; Gonçalves, S.; Valentão, P.; Andrade, P.B.; Coelho, N.; Romano, A. Thymus lotocephalus wild plants and in vitro cultures produce different profiles of phenolic compounds with antioxidant activity. *Food Chem.* **2012**, *135*, 1253–1260. [CrossRef] [PubMed]

16. Costa, P.; Goncalves, S.; Grosso, C.; Andrade, P.B.; Valentao, P.; Gabriela Bernardo-Gil, M.; Romano, A. Chemical profiling and biological screening of Thymus lotocephalus extracts obtained by supercritical fluid extraction and hydrodistillation. *Ind. Crops Prod.* **2012**, *36*, 246–256. [CrossRef]
17. Gonçalves, S.; Mansinhos, I.; Rodríguez-Solana, R.; Pereira-Caro, G.; Moreno-Rojas, J.M.; Romano, A. Impact of Metallic Nanoparticles on In Vitro Culture, Phenolic Profile and Biological Activity of Two Mediterranean Lamiaceae Species: *Lavandula viridis* L'Hér and *Thymus lotocephalus* G. López and R. Morales. *Molecules* **2021**, *26*, 6427. [CrossRef] [PubMed]
18. Mansinhos, I.; Gonçalves, S.; Rodríguez-Solana, R.; Ordóñez-Díaz, J.L.; Moreno-Rojas, J.M.; Romano, A. Ultrasonic-Assisted Extraction and Natural Deep Eutectic Solvents Combination: A Green Strategy to Improve the Recovery of Phenolic Compounds from *Lavandula pedunculata* subsp. lusitanica (Chaytor) Franco. *Antioxidants* **2021**, *10*, 582. [CrossRef]
19. Krishnan, S.K.; Salian, A.; Dutta, S.; Mandal, S. A roadmap to UV-protective natural resources: Classification, characteristics, and applications. *Mater. Chem. Front.* **2021**, *5*, 7696–7723. [CrossRef]
20. He, H.; Li, A.; Li, S.; Tang, J.; Li, L.; Xiong, L. Natural components in sunscreens: Topical formulations with sun protection factor (SPF). *Biomed. Pharmacother.* **2021**, *134*, 111161. [CrossRef] [PubMed]
21. Mapoung, S.; Semmarath, W.; Arjsri, P.; Umsumarng, S.; Srisawad, K.; Thippraphan, P.; Yodkeeree, S.; Dejkriengkraikul, P.L. Determination of Phenolic Content, Antioxidant Activity, and Tyrosinase Inhibitory Effects of Functional Cosmetic Creams Available on the Thailand Market. *Plants* **2021**, *10*, 1383. [CrossRef]
22. Nizioł-Łukaszewska, Z.; Gaweł-Bęben, K.; Rybczyńska-Tkaczyk, K.; Jakubczyk, A.; Karaś, M.; Bujak, T. Biochemical properties, UV-protecting and fibroblast growth-stimulating activity of *Plantago lanceolata* L. extracts. *Ind. Crops Prod.* **2019**, *138*, 111453. [CrossRef]
23. Stevanato, R.; Bertelle, M.; Fabris, S. Photoprotective characteristics of natural antioxidant polyphenols. *Regul. Toxicol. Pharmacol.* **2014**, *69*, 71–77. [CrossRef] [PubMed]
24. Lee, B.; Moon, K.M.; Lim, J.S.; Park, Y.; Kim, D.H.; Son, S.; Jeong, H.O.; Kim, D.H.; Lee, E.K.; Chung, K.W.; et al. 2-(3,4-dihydroxybenzylidene) malononitrile as a novel anti-melanogenic compound. *Oncotarget* **2017**, *8*, 91481–91493. [CrossRef] [PubMed]
25. Benoit, C.; Virginie, C.; Boris, V. The use of NADES to support innovation in the cosmetic industry. In *Eutectic Solvents and Stress in Plants*; Verpoorte, R., Witkamp, G.J., Choi, Y.H., Eds.; Academic Press Ltd-Elsevier Science Ltd.: London, UK, 2021; Volume 97, pp. 309–332, ISBN 978-0-12-821691-0.
26. Coelho, N.; Goncalves, S.; Elena Gonzalez-Benito, M.; Romano, A. Establishment of an in vitro propagation protocol for *Thymus lotocephalus*, a rare aromatic species of the Algarve (Portugal). *Plant Growth Regul.* **2012**, *66*, 69–74. [CrossRef]
27. Murashige, T.; Skoog, F. A Revised Medium for Rapid Growth and Bio Assays with Tobacco Tissue Cultures. *Physiol. Plant.* **1962**, *15*, 473–497. [CrossRef]
28. Girma, F.S.; Krieg, D.R. Osmotic adjustment in sorghum: I. Mechanisms of diurnal osmotic potential changes. *Plant Physiol.* **1992**, *99*, 577–582. [CrossRef] [PubMed]
29. Lichtenthaler, H.K. Chlorophylls and carotenoids: Pigments of photosynthetic biomembranes. In *Methods in Enzymology*; Plant Cell Membranes; Academic Press: Cambridge, MA, USA, 1987; Volume 148, pp. 350–382.
30. Loreto, F.; Velikova, V. Isoprene Produced by Leaves Protects the Photosynthetic Apparatus against Ozone Damage, Quenches Ozone Products, and Reduces Lipid Peroxidation of Cellular Membranes. *Plant Physiol.* **2001**, *127*, 1781–1787. [CrossRef]
31. Martins, N.; Gonçalves, S.; Palma, T.; Romano, A. The influence of low pH on in vitro growth and biochemical parameters of Plantago almogravensis and P. algarbiensis. *Plant Cell Tissue Organ. Cult.* **2011**, *107*, 113–121. [CrossRef]
32. Cao, Z.; Wang, Z.; Shang, Z.; Zhao, J. Classification and identification of *Rhodobryum roseum* Limpr. and its adulterants based on fourier-transform infrared spectroscopy (FTIR) and chemometrics. *PLoS ONE* **2017**, *12*, e0172359. [CrossRef]
33. Caunii, A.; Pribac, G.; Grozea, I.; Gaitin, D.; Samfira, I. Design of optimal solvent for extraction of bio–active ingredients from six varieties of Medicago sativa. *Chem. Cent. J.* **2012**, *6*, 123. [CrossRef]
34. Rezanejad, R.; Ojagh, S.M.; Heidarieh, M.; Raeisi, M.; Rafiee, G.; Alishahi, A. Characterization of Gamma-Irradiated *Rosmarinus officinalis* L. (Rosemary). *Turk. J. Pharm. Sci.* **2019**, *16*, 43–47. [CrossRef] [PubMed]
35. Schulz, H.; Baranska, M. Identification and quantification of valuable plant substances by IR and Raman spectroscopy. *Vib. Spectrosc.* **2007**, *43*, 13–25. [CrossRef]
36. Sinanoglou, V.J.; Zoumpoulakis, P.; Fotakis, C.; Kalogeropoulos, N.; Sakellari, A.; Karavoltsos, S.; Strati, I.F. On the Characterization and Correlation of Compositional, Antioxidant and Colour Profile of Common and Balsamic Vinegars. *Antioxidants* **2018**, *7*, 139. [CrossRef]
37. Ainsworth, E.A.; Gillespie, K.M. Estimation of total phenolic content and other oxidation substrates in plant tissues using Folin–Ciocalteu reagent. *Nat. Protoc.* **2007**, *2*, 875–877. [CrossRef] [PubMed]
38. Sumner, L.W.; Amberg, A.; Barrett, D.; Beale, M.H.; Beger, R.; Daykin, C.A.; Fan, T.W.-M.; Fiehn, O.; Goodacre, R.; Griffin, J.L.; et al. Proposed minimum reporting standards for chemical analysis Chemical Analysis Working Group (CAWG) Metabolomics Standards Initiative (MSI). *Metabolomics* **2007**, *3*, 211–221. [CrossRef] [PubMed]
39. Soler-Rivas, C.; Espín, J.C.; Wichers, H.J. An easy and fast test to compare total free radical scavenger capacity of foodstuffs. *Phytochem. Anal.* **2000**, *11*, 330–338. [CrossRef]
40. Re, R.; Pellegrini, N.; Proteggente, A.; Pannala, A.; Yang, M.; Rice-Evans, C. Antioxidant activity applying an improved ABTS radical cation decolorization assay. *Free Radic. Biol. Med.* **1999**, *26*, 1231–1237. [CrossRef]

41. Yen, G.-C.; Chen, H.-Y. Antioxidant Activity of Various Tea Extracts in Relation to Their Antimutagenicity. *J. Agric. Food Chem.* **1995**, *43*, 27–32. [CrossRef]
42. Gillespie, K.M.; Chae, J.M.; Ainsworth, E.A. Rapid measurement of total antioxidant capacity in plants. *Nat. Protoc.* **2007**, *2*, 867–870. [CrossRef]
43. Masuda, T.; Yamashita, D.; Takeda, Y.; Yonemori, S. Screening for tyrosinase inhibitors among extracts of seashore plants and identification of potent inhibitors from *Garcinia subelliptica*. *Biosci. Biotechnol. Biochem.* **2005**, *69*, 197–201. [CrossRef]
44. De Mansur, J.S.; Breder, M.N.R.; Mansur, M.C.; Azulay, R.D. Determination of sun protection factor by spectrophotometry. *An. Bras. Derm.* **1986**, 121–124.
45. Sayre, R.M.; Agin, P.P.; LeVee, G.J.; Marlowe, E. A Comparison of in Vivo and in Vitro Testing of Sunscreening Formulas. *Photochem. Photobiol.* **1979**, *29*, 559–566. [CrossRef] [PubMed]
46. Hosseini, N.; Hagh, Z.G.; Khoshghalb, H. Morphological, antioxidant enzyme activity and secondary metabolites accumulation in response of polyethylene glycol-induced osmotic stress in embryo-derived plantlets and callus cultures of Salvia leriifolia. *Plant Cell Tissue Organ Cult.* **2019**, *140*, 143–155. [CrossRef]
47. Acemi, A.; Duman, Y.A.; Karakus, Y.Y.; Özen, F. Developmental and biochemical analyses of in vitro drought stress response in ornamental European Bluestar (*Amsonia orientalis* Decne.). *Folia Hortic.* **2018**, *3*, 357–366. [CrossRef]
48. Akbar Mozafari, A.; Havas, F.; Ghaderi, N. Application of iron nanoparticles and salicylic acid in in vitro culture of strawberries (Fragaria × ananassa Duch.) to cope with drought stress. *Plant Cell Tissue Organ Cult.* **2018**, *132*, 511–523. [CrossRef]
49. Ngan, H.T.M.; Tung, H.T.; Le, B.V.; Nhut, D.T. Evaluation of root growth, antioxidant enzyme activity and mineral absorbability of carnation (*Dianthus caryophyllus* "Express golem") plantlets cultured in two culture systems supplemented with iron nanoparticles. *Sci. Hortic.* **2020**, *272*, 109612. [CrossRef]
50. Joseph, S.; Anawar, H.M.; Storer, P.; Blackwell, P.; Chia, C.; Lin, Y.; Munroe, P.; Donne, S.; Horvat, J.; Wang, J.; et al. Effects of Enriched Biochars Containing Magnetic Iron Nanoparticles on Mycorrhizal Colonisation, Plant Growth, Nutrient Uptake and Soil Quality Improvement. *Pedosphere* **2015**, *25*, 749–760. [CrossRef]
51. Antonić, D.; Milošević, S.; Cingel, A.; Lojić, M.; Trifunović-Momčilov, M.; Petrić, M.; Subotić, A.; Simonović, A. Effects of exogenous salicylic acid on *Impatiens walleriana* L. grown in vitro under polyethylene glycol-imposed drought. *S. Afr. J. Bot.* **2016**, *105*, 226–233. [CrossRef]
52. Shabani, L.; Ehsanpour, A.A.; Asghari, G.; Emami, J. Glycyrrhizin production by in vitro cultured *Glycyrrhiza glabra* elicited by methyl Jasmonate and salicylic acid. *Russ. J. Plant Physiol.* **2009**, *56*, 621–626. [CrossRef]
53. Goswami, B.; Rankawat, R.; Gadi, B. Physiological and antioxidative responses associated with drought tolerance of *Lasiurus sindicus* Henr. endemic to Thar desert, India. *Braz. J. Bot.* **2020**, *43*, 761–773. [CrossRef]
54. Mellacheruvu, S.; Talakayala, A.; Garladinne, M. Chapter 7—Crop Improvement of Cereals through Manipulation of Signaling Pathways in Response to Drought Stress. In *Plant Signaling Molecules*; Khan, M.I.R., Reddy, P.S., Ferrante, A., Khan, N.A., Eds.; Woodhead Publishing: Sawston, UK, 2019; pp. 125–139, ISBN 978-0-12-816451-8.
55. Kleinwächter, M.; Paulsen, J.; Bloem, E.; Schnug, E.; Selmar, D. Moderate drought and signal transducer induced biosynthesis of relevant secondary metabolites in thyme (*Thymus vulgaris*), greater celandine (*Chelidonium majus*) and parsley (*Petroselinum crispum*). *Ind. Crops Prod.* **2015**, *64*, 158–166. [CrossRef]
56. Abbaspour, J.; Ehsanpour, A.A. Physiological targets of salicylic acid on Artemisia aucheri BOISS as a medicinal and aromatic plant grown under in vitro drought stress. *Bot. Stud.* **2016**, *57*, 39. [CrossRef] [PubMed]
57. Sarmadi, M.; Karimi, N.; Palazón, J.; Ghassempour, A.; Mirjalili, M.H. Improved effects of polyethylene glycol on the growth, antioxidative enzymes activity and taxanes production in a *Taxus baccata* L. callus culture. *Plant Cell Tissue Organ Cult.* **2019**, *137*, 319–328. [CrossRef]
58. Ali, M.A.; Mosa, K.A.; El-Keblawy, A.; Alawadhi, H. Exogenous Production of Silver Nanoparticles by *Tephrosia apollinea* Living Plants under Drought Stress and Their Antimicrobial Activities. *Nanomaterials* **2019**, *9*, 1716. [CrossRef]
59. Baskar, V.; Safia, N.; Sree Preethy, K.; Dhivya, S.; Thiruvengadam, M.; Sathishkumar, R. A comparative study of phytotoxic effects of metal oxide (CuO, ZnO and NiO) nanoparticles on in-vitro grown *Abelmoschus esculentus*. *Plant Biosyst.-Int. J. Deal. All Asp. Plant Biol.* **2021**, *155*, 374–383. [CrossRef]
60. Sheteiwy, M.S.; Gong, D.; Gao, Y.; Pan, R.; Hu, J.; Guan, Y. Priming with methyl jasmonate alleviates polyethylene glycol-induced osmotic stress in rice seeds by regulating the seed metabolic profile. *Environ. Exp. Bot.* **2018**, *153*, 236–248. [CrossRef]
61. Zhang, L.; Becker, D. Connecting proline metabolism and signaling pathways in plant senescence. *Front. Plant Sci.* **2015**, *6*, 552. [CrossRef]
62. Phansak, P.; Siriwong, S.; Kanawapee, N.; Thumanu, K.; Gunnula, W.; Buensanteai, N. Drought Response of Rice in Northeastern Thailand Assessed via Fourier Transform Infrared Spectroscopy. *Acta Agrobot.* **2021**, *74*, 7421. [CrossRef]
63. Sevindik, B.; Sevindik, O.; Selli, S. Effect of drought stress induced by PEG 6000 on *Ocimum basilicum* L. aroma profile. *J. Food Process. Preserv.* **2021**, *46*, e15948. [CrossRef]
64. Ahmad, M.A.; Javed, R.; Adeel, M.; Rizwan, M.; Yang, Y. PEG 6000-Stimulated Drought Stress Improves the Attributes of In Vitro Growth, Steviol Glycosides Production, and Antioxidant Activities in Stevia rebaudiana Bertoni. *Plants* **2020**, *9*, 1552. [CrossRef]
65. Razavizadeh, R.; Karami, M. Antioxidant capacity and chemical composition of Carum copticum under PEG treatment. *Iran. J. Plant Physiol.* **2018**, *8*, 2321–2331. [CrossRef]

66. Abdi, G.; Shokrpour, M.; Karami, L.; Salami, S.A. Prolonged Water Deficit Stress and Methyl Jasmonate-Mediated Changes in Metabolite Profile, Flavonoid Concentrations and Antioxidant Activity in Peppermint (*Mentha × piperita* L.). *Not. Bot. Horti. Agrobot. Cluj-Napoca* **2019**, *47*, 70–80. [CrossRef]
67. Chanjirakul, K.; Wang, S.Y.; Wang, C.Y.; Siriphanich, J. Effect of natural volatile compounds on antioxidant capacity and antioxidant enzymes in raspberries. *Postharvest. Biol. Technol.* **2006**, *40*, 106–115. [CrossRef]
68. Baek, M.W.; Choi, H.R.; Solomon, T.; Jeong, C.S.; Lee, O.-H.; Tilahun, S. Preharvest Methyl Jasmonate Treatment Increased the Antioxidant Activity and Glucosinolate Contents of Hydroponically Grown Pak Choi. *Antioxidants* **2021**, *10*, 131. [CrossRef] [PubMed]
69. Kianersi, F.; Pour-Aboughadareh, A.; Majdi, M.; Poczai, P. Effect of Methyl Jasmonate on Thymol, Carvacrol, Phytochemical Accumulation, and Expression of Key Genes Involved in Thymol/Carvacrol Biosynthetic Pathway in Some Iranian Thyme Species. *Int. J. Mol. Sci.* **2021**, *22*, 11124. [CrossRef]
70. Agata, I.; Kusakabe, H.; Hatano, T.; Nishibe, S.; Okuda, T. Melitric Acid-a and Acid-B, New Trimeric Caffeic Acid-Derivatives from Melissa-Officinalis. *Chem. Pharm. Bull.* **1993**, *41*, 1608–1611. [CrossRef]
71. Pereira, O.R.; Catarino, M.D.; Afonso, A.F.; Silva, A.M.S.; Cardoso, S.M. *Salvia elegans*, *Salvia greggii* and *Salvia officinalis* Decoctions: Antioxidant Activities and Inhibition of Carbohydrate and Lipid Metabolic Enzymes. *Molecules* **2018**, *23*, 3169. [CrossRef]
72. Zeng, H.; Su, S.; Xiang, X.; Sha, X.; Zhu, Z.; Wang, Y.; Guo, S.; Yan, H.; Qian, D.; Duan, J. Comparative Analysis of the Major Chemical Constituents in Salvia miltiorrhiza Roots, Stems, Leaves and Flowers during Different Growth Periods by UPLC-TQ-MS/MS and HPLC-ELSD Methods. *Molecules* **2017**, *22*, 771. [CrossRef]
73. Afonso, A.F.; Pereira, O.R.; Válega, M.; Silva, A.M.S.; Cardoso, S.M. Metabolites and Biological Activities of *Thymus zygis*, *Thymus pulegioides*, and *Thymus fragrantissimus* Grown under Organic Cultivation. *Molecules* **2018**, *23*, 1514. [CrossRef]
74. Gonçalves, S.; Romano, A. Rosmarinic Acid: Sources, Properties, Applications and Biotechnological Production. In *Medicinal Plants*; CRC Press: Boca Raton, FL, USA, 2019; ISBN 978-0-429-25996-8.
75. Marchev, A.S.; Vasileva, L.V.; Amirova, K.M.; Savova, M.S.; Koycheva, I.K.; Balcheva-Sivenova, Z.P.; Vasileva, S.M.; Georgiev, M.I. Rosmarinic acid—From bench to valuable applications in food industry. *Trends Food Sci. Technol.* **2021**, *117*, 182–193. [CrossRef]
76. Babaei, M.; Borja Zamfir, G.M.; Chen, X.; Christensen, H.B.; Kristensen, M.; Nielsen, J.; Borodina, I. Metabolic Engineering of Saccharomyces cerevisiae for Rosmarinic Acid Production. *ACS Synth. Biol.* **2020**, *9*, 1978–1988. [CrossRef]
77. Xiaowei, H.; Zhihua, L.; Tahir, H.E.; Xiaobo, Z.; Jiyong, S.; Yiwei, X.; Xiaodong, Z. Chapter 13—Conventional and rapid methods for measurement of total bioactive components and antioxidant activity in Hibiscus sabdariffa. In *Roselle (Hibiscus sabdariffa)*; Mariod, A.A., Tahir, H.E., Mahunu, G.K., Eds.; Academic Press: Cambridge, MA, USA, 2021; pp. 199–214, ISBN 978-0-12-822100-6.
78. Celikcan, F.; Kocak, M.Z.; Kulak, M. Vermicompost applications on growth, nutrition uptake and secondary metabolites of *Ocimum basilicum* L. under water stress: A comprehensive analysis. *Ind. Crops Prod.* **2021**, *171*, 113973. [CrossRef]
79. Ahmadi, T.; Shabani, L.; Sabzalian, M.R. Improvement in drought tolerance of lemon balm, *Melissa officinalis* L. under the pre-treatment of LED lighting. *Plant Physiol. Biochem.* **2019**, *139*, 548–557. [CrossRef]
80. Szabó, K.; Radácsi, P.; Rajhárt, P.; Ladányi, M.; Németh, É. Stress-induced changes of growth, yield and bioactive compounds in lemon balm cultivars. *Plant Physiol. Biochem.* **2017**, *119*, 170–177. [CrossRef] [PubMed]
81. Liu, H.; Wang, X.; Wang, D.; Zou, Z.; Liang, Z. Effect of drought stress on growth and accumulation of active constituents in Salvia miltiorrhiza Bunge. *Ind. Crops Prod.* **2011**, *33*, 84–88. [CrossRef]
82. Li, J.; Li, B.; Luo, L.; Cao, F.; Yang, B.; Gao, J.; Yan, Y.; Zhang, G.; Peng, L.; Hu, B. Increased phenolic acid and tanshinone production and transcriptional responses of biosynthetic genes in hairy root cultures of *Salvia przewalskii* Maxim. treated with methyl jasmonate and salicylic acid. *Mol. Biol. Rep.* **2020**, *47*, 8565–8578. [CrossRef] [PubMed]
83. Pei, T.; Ma, P.; Ding, K.; Liu, S.; Jia, Y.; Ru, M.; Dong, J.; Liang, Z. SmJAZ8 acts as a core repressor regulating JA-induced biosynthesis of salvianolic acids and tanshinones in *Salvia miltiorrhiza* hairy roots. *J. Exp. Bot.* **2018**, *69*, 1663–1678. [CrossRef] [PubMed]
84. Jaouadi, R.; Silva, A.M.S.; Boussaid, M.; Yahia, I.B.H.; Cardoso, S.M.; Zaouali, Y. Differentiation of Phenolic Composition among Tunisian Thymus algeriensis Boiss. et Reut. (Lamiaceae) Populations: Correlation to Bioactive Activities. *Antioxidants* **2019**, *8*, 515. [CrossRef] [PubMed]
85. Taghouti, M.; Martins-Gomes, C.; Felix, L.M.; Schaefer, J.; Santos, J.A.; Bunzel, M.; Nunes, F.M.; Silva, A.M. Polyphenol composition and biological activity of Thymus citriodorus and Thymus vulgaris: Comparison with endemic Iberian Thymus species. *Food Chem.* **2020**, *331*, 127362. [CrossRef]
86. Kang, H.S.; Kim, H.R.; Byun, D.S.; Park, H.J.; Choi, J.S. Rosmarinic Acid as a Tyrosinase Inhibitors from Salvia miltiorrhiza. *Nat. Prod. Sci.* **2004**, *10*, 80–84.
87. Ko, S.-C.; Lee, S.-H. Protocatechuic Aldehyde Inhibits α-MSH-Induced Melanogenesis in B16F10 Melanoma Cells via PKA/CREB-Associated MITF Downregulation. *Int. J. Mol. Sci.* **2021**, *22*, 3861. [CrossRef]
88. Crespo, M.I.; Chabán, M.F.; Lanza, P.A.; Joray, M.B.; Palacios, S.M.; Vera, D.M.A.; Carpinella, M.C. Inhibitory effects of compounds isolated from *Lepechinia meyenii* on tyrosinase. *Food Chem. Toxicol.* **2019**, *125*, 383–391. [CrossRef]
89. Saija, A.; Tomaino, A.; Trombetta, D.; De Pasquale, A.; Uccella, N.; Barbuzzi, T.; Paolino, D.; Bonina, F. In vitro and in vivo evaluation of caffeic and ferulic acids as topical photoprotective agents. *Int. J. Pharm.* **2000**, *199*, 39–47. [CrossRef]

90. Sanchez-Campillo, M.; Gabaldon, J.A.; Castillo, J.; Benavente-Garcia, O.; Del Bano, M.J.; Alcaraz, M.; Vicente, V.; Alvarez, N.; Lozano, J.A. Rosmarinic acid, a photo-protective agent against UV and other ionizing radiations. *Food Chem. Toxicol.* **2009**, *47*, 386–392. [CrossRef]
91. Chen, J.; Ran, M.; Wang, M.; Liu, X.; Liu, S.; Ruan, Z.; Jin, N. Evaluation of antityrosinase activity and mechanism, antioxidation, and UV filter properties of theaflavin. *Biotechnol. Appl. Biochem.* **2021**, *69*, 3. [CrossRef]
92. Yakoubi, R.; Megateli, S.; Hadj Sadok, T.; Gali, L. Photoprotective, antioxidant, anticholinesterase activities and phenolic contents of different Algerian *Mentha pulegium* extracts. *Biocatal. Agric. Biotechnol.* **2021**, *34*, 102038. [CrossRef]
93. Tsai, C.-E.; Lin, L.-H. The liquid polyol extracts of camellia seed dregs used in sunscreen cosmetics. *Chem. Pap.* **2018**, *73*, 501–508. [CrossRef]
94. Magangana, T.P.; Stander, M.A.; Masondo, N.A.; Makunga, N.P. Steviol glycoside content and essential oil profiles of Stevia rebaudiana Bertoni in response to NaCl and polyethylene glycol as inducers of salinity and drought stress in vitro. *Plant Cell Tissue Organ Cult.* **2021**, *145*, 1–18. [CrossRef]

antioxidants

MDPI

*Article*

# Chemical, Antioxidant, and Antimicrobial Properties of the Peel and Male Flower By-Products of Four Varieties of *Punica granatum* L. Cultivated in the Marche Region for Their Use in Cosmetic Products

**Maria Rosa Gigliobianco [1], Manuela Cortese [2], Samanta Nannini [3], Lucrezia Di Nicolantonio [2,4], Dolores Vargas Peregrina [4], Giulio Lupidi [2], Luca Agostino Vitali [2], Elena Bocchietto [3], Piera Di Martino [5,*] and Roberta Censi [2,4]**

1 Percuros B.V., Zernikedreef 8, 2333 CL Leiden, The Netherlands; m.gigliobianco@percuros.nl
2 CHiP Research Center, School of Pharmacy, University of Camerino, Via Madonna delle Carceri, 62032 Camerino, Italy; manuela.cortese@unicam.it (M.C.); lucrezia.dinicolantonio@unicam.it (L.D.N.); giulio.lupidi@unicam.it (G.L.); luca.vitali@unicam.it (L.A.V.); roberta.censi@unicam.it (R.C.)
3 ABICH Srl, Biological and Chemical Toxicology Research Laboratory, 28924 Verbania, Italy; samanta.nannini@hotmail.it (S.N.); elena.bocchietto@abich.it (E.B.)
4 Recusol Srl, 62032 Camerino, Italy; dolores.vargas@unicam.it
5 Department of Pharmacy, Università "G. D'Annunzio" Chieti e Pescara, Via dei Vestini, 1, 66100 Chieti, Italy
* Correspondence: piera.dimartino@unich.it

**Abstract:** We are now seeing an increase in the production of agri-food waste, which is an essential resource for the recovery of bioactive compounds that may be employed as innovative natural ingredients in cosmetics. To date, the approach to cosmetics preservation has seen a significant shift in the search for biological components that give healthier alternatives for customers and help businesses operate in an environmentally friendly manner. To achieve this goal, we studied pomegranate extracts using the peel and, for the first time, extracts from the male flowers of a wide pomegranate variety cultivated in the Marche region, specifically, the Wonderful, Mollar de Elche, Parfianka, and less-studied G1 varieties. We studied the phenol compounds profile, antioxidant capacity, antimicrobial activity, and cell viability of the obtained pomegranate extracts. The identification and quantification of phenol compounds belonging to different classes, such as hydrolysable tannins, hydroxybenzoic acid, hydroxycinnamic acid, dihydroflavonol, gallocatechin, and anthocyanins, were performed using UPLC-ESI-MS/MS. Punicalagin isomers and punicalin resulted in the most abundant polyphenols found in the peel and male flower extracts. Mollar de Elche 2020 peel extract revealed a high concentration of punicalagin A and B (7206.4 mg/kg and 5812.9), while the content of gallic acid revealed high results in the G1 and Parfianka varieties. All extracts were spectrophotometrically analysed to determine their total phenol content (TPC) using the Folin–Ciocalteu method and their antioxidant capacity (AC). In terms of the total phenol obtained by the Folin–Ciocalteu colorimetric method, Mollar de Elche 2020 extracts reported the highest TPC content of 12.341 μmol GAE/g. Results revealed that the Mollar de Elche and Wonderful 2020 peel extracts demonstrated the highest TPC and AC. Furthermore, AC results indicated that the peel extracts displayed higher AC than the male flower extract due to the high punicalagin content detected by UPLC analysis. The antimicrobial activity testing revealed that the Wonderful and G1 2020 peel extracts resulted active against *Escherichia coli*, while all extracts exhibited promising anticandidal activity. Additionally, the cytocompatibility was evaluated in keratinocytes HaCaT cells by testing concentrations of pomegranate extracts ranging from 0.15 to 5.00 mg/mL. Extracts were non-toxic for the cells in the tested concentration range. The acquired results may help exploit pomegranate agri-food waste products provided by the Marche region's short supply chain for their use as an antimicrobial and antioxidant booster in the formulation of cosmetic products.

**Citation:** Gigliobianco, M.R.; Cortese, M.; Nannini, S.; Di Nicolantonio, L.; Peregrina, D.V.; Lupidi, G.; Vitali, L.A.; Bocchietto, E.; Di Martino, P.; Censi, R. Chemical, Antioxidant, and Antimicrobial Properties of the Peel and Male Flower By-Products of Four Varieties of *Punica granatum* L. Cultivated in the Marche Region for Their Use in Cosmetic Products. *Antioxidants* **2022**, *12*, 768. https://doi.org/10.3390/antiox11040768

Academic Editors: Irene Dini and Sonia Laneri

Received: 25 March 2022
Accepted: 10 April 2022
Published: 12 April 2022

**Keywords:** agri-food waste; pomegranate wastes; pomegranate's flower by-products; green extraction; antioxidant capacity; UPLC-ESI-MS/MS analysis; antimicrobial activity

## 1. Introduction

In recent years, the cosmetics sector's new focus has been on two crucial challenges: ecology and sustainability. Given this sustainable approach, companies are currently using natural ingredients to enhance the quality of the cosmetics, keep them free of microorganisms contamination, enhance their shelf-life, obtain safer products, and avoid the use of synthetic preservatives.

Furthermore, the typical approach to cosmetic preservation is to employ the most significant number of conventional preservatives allowed (refer to Annex V of the European Commission's Regulation (EC) No 1223/2009 for a list of authorized preservatives [1]). Conventional preservatives have shown many secondary effects and risks on human health. Recent studies have also reported side effects from preservatives on the skin microbiota [2,3]. The use of cosmetic products has recently been shown to change the equilibrium of the skin microbiota, which can cause it to lose its healthy status. This effect can be linked to various causes, including the preservatives' residual impact on the skin.

Against this background, the approach to cosmetics preservation can be managed with several strategies to prevent microbial contamination. Various methods can be used to achieve these goals. One solution is based on the concept of "Hurdle Technology", which encompasses all intelligent ways of preventing and reducing microbe proliferation throughout the formulation production process. As a result of this approach, several steps must be taken to avoid microbial contamination, ranging from strict adherence to GMP, raw material microbial control, and water treatment using a different strategy to reduce water activity, as water is one of the most abundant ingredients in many cosmetic formulations. In adherence to GMP, qualified employees can manage cross-contamination, disinfection of equipment, and proper packaging. (e.g., airless dispensers, pumps) are essential [4]. Any of these measures, when combined, can help lower any microbial contamination.

On the other hand, using novel raw materials obtained from agri-food waste is one way to avoid using conventional preservatives and it depends more and more on innovative ingredients. This strategy provides healthier options for customers and assists businesses in working in an ecologically sustainable manner. This method is also feasible since it is widely recognised that some natural ingredients may play many roles in cosmetic formulation, such as absorbed water making a formulation alkaline or acid. They can also be applied as a natural antibacterial. Herbal extracts and essential oils have shown evidence of microbial activity despite not being preservatives but also exhibiting antimicrobial activity with double safety rules: safety of the formula, where we do not want microorganisms, and protection of the skin, where we do not want side effects [5]. Currently, a feasible strategy for reintroducing agri-food waste into natural cosmetics manufacturing is the circular economy [6–8]. Natural compounds obtained from agri-food waste products can be used successfully as a natural preservative in cosmetic compositions [9,10]. One of the most interesting by-products that may be utilised for this purpose is pomegranate by-products, which are becoming recognised as having numerous bioactive compounds with considerable antioxidant activity and antibacterial properties [11].

Pomegranate, or *Punica Granatum* L., one of the most ancient fruits, is a fruit that belongs to the family Punica and has many benefits to human health. It is grown worldwide with approximately 1500 ha of cultivars in Italy, mostly in Sicily, Puglia, Calabria, Campania, and Lazio [12]. Still, many of these cultivations are also present in the Marche region. In the last decade, the interest in the cultivation of *Punica Granatum* L. also highlights the varieties that have encouraged the revival of pomegranates in Italy as an income crop. Therefore, many entrepreneurs, driven by the need to diversify and expand their business, have identified a viable alternative to traditional orchards. Consequently, this situation has led to a significant increase in waste. Usually, pomegranates generate 669 kg of waste materials for every 1 ton of fresh fruit, with 78% composed of peel and 22% seed [13,14]. Different studies report pomegranate peel as an interesting by-product [12,15,16], but we also identified another part of the pomegranate tree that provides an abundant waste material. During the flowering period, from May to July, the pomegranate tree develops two principal types of flowers: male flowers and hermaphrodite flowers. The first one has a poorly designed or no pistil and atrophied ovaries with few ovules. It is infertile and drops down without fruit set, representing considerable waste material. In contrast, the hermaphrodite flowers are fertile, with an ovary producing fruit. The primary by-products of the pomegranate flower, the male ones, have not been investigated singularly as potential sources of bioactive polyphenols. To the best of our knowledge, no data are available.

Pomegranate peel and flowers represent an enormous waste of raw materials. As reported by the literature, pomegranates contain a high number of phenol ingredients in their pulp, seed, and juice [17–20]. Polyphenols represent the predominant phytochemicals of the pomegranate peel and are mainly composed of hydrolysable tannins (pedunculagin, punicalin, punicalagin, and ellagic acids). Among them, the predominant hydrolysable tannin of pomegranates is known as the punicalagin [21]. In addition, there are also other flavonoids such as quercetin, kaempferol, luteolin glycosides, catechins, anthocyanins, and complex flavonoids [20,22]. These bioactive compounds are responsible for many pharmacological properties, notably their significant antioxidant activity and antibacterial characteristics [18,23]. Total phenolic content, antioxidant capacity, and antibacterial activity against various microbes have been highly correlated [24]. Pomegranate peel extracts and other fruit by-products, such as juice or seeds, have exhibited remarkable antioxidant and antibacterial properties, with a high scavenging capacity, reducing the microbiological proliferation of several bacteria, among other therapeutic benefits to the human body [25].

Due to their high content of ellagitannins, such as punicalagin, *Punica Granatum* L. peel extracts show a remarkable antimicrobial capacity influenced by the different types of cultivars, which has been demonstrated against pathogenic such as *Staphylococcus aureus* [26]. The antimicrobial activity was associated with a high concentration of punicalagin and an ellagic acid concentration in the pomegranate peel extract, demonstrating that the ellagic acid content substantially impacts the antimicrobial activity [27–29].

Based on the potential benefits of pomegranate extract and the need for research exploring natural ingredients that may be used as alternative preservatives in cosmetic formulations, our study aims to evaluate the antioxidant capacity and anti-microbial activities of extracts of the peel and male flower by-products of four pomegranate varieties cultivated in the Marche region. The US green extraction processes were applied to efficiently recover the polyphenols from the pomegranate peel and male flower extract. The main polyphenol compounds present in the peel and male flower extracts were identified and quantified by UPLC-ESI-MS/MS. The majority of the data on pomegranate extracts in the literature came from one of the most common varieties. As a result, our research presents a chemical and antibacterial analysis of pomegranate varieties that have received less attention in the past, such as G1 and Parfianka. To our knowledge, the characterisation of polyphenols and antibacterial activity for the combination of pomegranate and other fruits, such as the G1 cultivar, has never been studied. As a result, this study will provide a scientific foundation for supporting the valorisation of pomegranate peels and male flower by-products received through a short supply chain as a substantial source of natural preservatives used in the production of cosmetics with a reduced rate of conventional ones. To achieve this goal, pomegranate extracts might be used as a preservative booster with antimicrobial properties, decreasing the need for synthetic preservatives in cosmetics and as a booster fulfilling other specific purposes, such as moisturiser or antioxidants.

## 2. Materials and Methods

### 2.1. Chemicals

The chemicals 1,1-Diphenyl-2-picrylhydrazyl (DPPH), 2,4,6-Tris(2-pyridyl)-s-triazine (TPTZ), (±)6-Hydroxy-2,5,7,8-tetramethylchromane-2-carboxylic acid (TROLOX), 2,2′-Azino-bis(3-ethylbenzothiazoline-6-sulfonic acid), diammonium salt (98%TLC), ABTS, gallic acid, sodium carbonate monohydrate ACS reagents, and ethanol (final ethanol grade) were purchased from Sigma-Aldrich (Steinheim, Germany); Manganese (IV) oxidise activated (>90%), and Folin–Ciocalteu's phenol reagents were purchased from Fluka (Buchs, Switzerland). Analytical grade solvents and reagents were used for all of the work. Ellagic acid, gallic acid, punicalagin A and B, punicalin, cyanidin 3-glucoside, cyanidin 3,5-diglucoside, pelargonidin 3,5-diglucoside, and pelargonidin 3-glucoside were obtained from Sigma-Aldrich (Stenheim, Germany). Formic acid and acetonitrile for LC/MS were purchased from Carlo Erba Reagents (Cornaredo, MI, Italy). Sigma Aldrich provided all the other chemicals (Steinheim, Germany). The ultrapure water was produced from the Millipore system (Millipore Sigma, Darmstadt, Germany), and filtered with a 0.20 µm Sartolon polyamide filter (Sartorius Stedim Biotech, Göttingen, Germany).

### 2.2. Pomegranate Samples

In this study, Wonderful, Mollar de Elche, Parfianka, and G1 pomegranate cultivars, grown in the Marche region, were harvested in October 2019 and 2020 at commercial maturity at the agricultural enterprise "Onori Maria Rosaria", Sant'Elpidio a Mare (Fermo, Marche). Immediately after the fruits were collected, the pomegranate peels were manually removed to accurately separate the exocarp (rind) used for the extraction process from the mesocarp (white spongy tissue). The exocarp was dried at 37 °C in a ventilated oven (VEC2103/8, Everest, Rimini, Italy). The male flowers, and more specifically sepals and stamen, of Wonderful and Mollar de Elche were picked during the flowering period, in June 2020, from the same agricultural enterprise from which we obtained all of the materials for our investigation. The waste male flowers were promptly dried in a ventilated oven at 37 °C until they reached a consistent weight and were then utilised for extraction. For each sample, three extractions were performed.

### 2.3. Extraction of the Phenol Compounds from Pomegranate Peel and Male Flowers

Extractions from the pomegranate peel and male flowers were carried out in an ultrasonic bath (FALC-LABSONIC LBS2, Treviglio, BG, Italy), with a maximum capacity of 10 L. Extractions were performed using a water and ethanol at 70:30 (*v*/*v*) mixture as extraction solvent. The plants (1 g) were dispersed in 100 mL of solvent and then immersed in an ultrasonic bath (50 °C, 59.2 W/cm$^2$ for 15 min). The extracts were lyophilised at −53 °C for 24 h at 0.03 millibar (FreeZone, 1 L, LABCONCO, Kansas City, MO, USA). Each sample was kept at −20 °C in 50 mL screw-top cap polyethylene vials (BD Falcon TMBD Biosciences, Bedford, MA, USA). The lyophilised solids were reconstituted with appropriate solvents for subsequent investigations.

### 2.4. Ultra-Performance Liquid Chromatography Mass Spectrometry Analysis

UPLC (Agilent 1290 Infinity Technologies UPLC, Santa Clara, CA, USA) with a diode array detector and a triple quadrupole mass spectrometer (MS-QQQ), combined with an electrospray ionization source (ESI), was used to identify and quantify polyphenols in pomegranate extracts. The UPLC had a binary pump and an auto-sampler. We used a Luna (C18 1504.6 mm) column (Phenomenex, Castel Maggiore, BO, Italy) at 25 °C for chromatographic separation. The gradient elution used water/0.1% formic acid (solvent A) and acetonitrile/0.1% formic acid (solvent B) at a 1 mL/min flow rate. The optimised gradient was 0–3 min, 7% B, 3–30 min, 27% B, 30–60 min, and 100% B, and 5 µL was injected. The mass spectrometer detection in both negative and positive ionisation modes was utilised to identify the different signals. However, in the negative mode, a significant group of compounds matching deprotonate molecular ions of diverse chemicals were identified.

In contrast, in the positive mode, the characteristics of anthocyanins of pomegranate were detected. The peak characterisation was performed based on their retention time and accurate molecular mass in the MS and $MS^2$ experiment. The ESI source worked at 350 °C, 12 L/min, and a nebulizer pressure of 55 psi. Samples were diluted 1:2 with mobile phase, sonicated for 5 min, then centrifuged at 12,000 rpm for 10 min (Scilogex D3024R High-Speed Refrigerated Micro-Centrifuge, Rocky Hill, CT, USA). Next, a 0.20 µm filter (Captiva Econofilter, PTFE) was used to filter the samples. A full scan mass spectrometer analysis was performed in *m/z* 150–1500 (negative) and 50–1500 (positive). Following that, several production experiments were conducted on the corresponding precursor ion. The quantification of polyphenols was carried out for individual phenol compounds using a calibration curve of the respective reference external standards. To this end, the mixed standard was prepared at 1000 mg/L. The linearity, sensitivity, accuracy, and precision of the developed method were verified according to the Food and Drug Administration Guidelines (FDA), as already mentioned in our previous studies [30]. The linearity was explored in the concentration range of 1–50 mg/L for ellagic acid, gallic acid, punicalagin A and B, and punicalin. External anthocyanins standards as cyanidin 3-glucoside, cyanidin 3,5-glucoside, pelargonidin 3,5-glucoside, and pelargonidin 3-glucoside were used with a concentration range of 1–20 mg/L, obtaining the corresponding calibration curve for further calculations (Table 1). The performances of this method were also tested for accuracy, precision, and sensitivity (Table 1). The sensitivity was evaluated in terms of the limit of detection (LOD) and limit of quantification (LOQ) by triple injections of a standard mixture at concentrations responsible for a signal-to-noise ratio equal to 3 and 10 for LOD and LOQ, respectively. To evaluate the method's accuracy, we chose a concentration of 10 mg/L for all standards. They were injected in triplicate among the unknown samples in the daily sequence. The accuracy was expressed as standard deviation percentage and reported in the Supplementary Materials. The overall method precision was evaluated by analysing nine extracts of the same sample. Each section was then injected individually as an unknown sample. The accuracy was calculated for each compound and expressed as a relative standard deviation percentage.

**Table 1.** Linearity, sensitivity, accuracy, and precision of the developed UPLC-ESI-MS/MS method for pomegranate peel and male flower by-products extracts.

| Analytes | RT (min.) | Linearity [a] Regression Curves | $r^2$ | Sensibility LOD [b] | LOQ [c] | Accuracy [d] | Precision [e] |
|---|---|---|---|---|---|---|---|
| Gallic acid | 3.83 | y = 109,036x | 0.973 | 0.07 | 0.24 | 14.7 | 0.4–16.8 |
| Punicalin | 3.75 | y = 115,291x | 0.992 | 0.04 | 0.14 | 11.9 | 0.5–18.8 |
| Punicalagin A | 9.25 | y = 139,898x | 0.997 | 0.04 | 0.14 | 6.8 | 0.4–6.7 |
| Punicalagin B | 11.71 | y = 149,562x | 0.998 | 0.04 | 0.14 | 4.7 | 0.4–3.8 |
| Ellagic Acid | 21.84 | y = 233,729x + 84,858 | 0.994 | 0.03 | 0.11 | 7.8 | 0.2–19.7 |
| Cyanidin 3,5-diglucoside | 6.2 | y = 491,074x | 0.998 | 0.08 | 0.24 | 3.0 | 3.6–14.4 |
| Pelargonidin 3,5-diglucoside | 8.4 | y = 557,447x | 0.999 | 0.05 | 0.14 | 1.6 | 0.5–13.2 |
| Cyanidin 3-glucoside | 10.3 | y = 1,073,535.261 | 0.998 | 0.03 | 0.11 | 2.8 | 9.4–15.7 |
| Pelargonidin 3-glucoside | 12.0 | y = 1,095,405x | 0.993 | 0.01 | 0.03 | 1.7 | 0.2–18.6 |

[a] explored in the concentration range of 1–50 mg/L for ellagic acid, gallic acid, punicalagin A and B, and punicalin; and in a concentration range of 1–20 mg/L in the case of cyanidin 3-glucoside, cyanidin 3,5-diglucoside, pelargonidin 3,5-diglucoside, and pelargonidin 3-glucoside, obtaining the corresponding calibration curve for further calculations. [b] Limit of detection. [c] Limit of quantification. [d] Accuracy was expressed as standard deviation %. [e] The precision was calculated for each compound, and it was expressed as relative standard deviation percentage. A range of SD% was reported in the table.

The matrix effect (ME) for the matrices used in this study was also evaluated. The Matrix Effect percentage (ME%) was calculated using this Equation (1):

$$ME\% = \frac{B}{A} \times 100 \quad (1)$$

where $B$ is the difference between the areas of the fortified sample and non-fortified one, and $A$ is the area of the standard solution in the pure solvent. Thus, it was possible to evaluate if the matrix effect induced an ion suppression (values < 100%) or an ion enhancement (values > 100%). A ME value equal to 100% means that no matrix effect occurs; deals over 100% reveal a signal enhancement and matters less than 100% reveal a signal suppression [31].

### 2.5. Determination of the Total Phenol Content

Total Phenol Content (TPC) was determined by using the Folin–Ciocalteu spectrophotometric method according to Gigliobianco et al. [30,32,33]. The results were expressed as milligrams of gallic acid equivalents (GAE) per grams of by-product (mg GAE/g).

### 2.6. Antioxidant Capacity (AC) Evaluation

Three methods were adopted in our approach to measure antioxidant activity: DPPH• radical scavenging activity, 2,2′-and-bis (3-ethylbenzothiazoline-6-sulphonic acid) (ABTS•+) radical cation scavenging capacity, and Ferric Reducing Antioxidant capacity (FRAP) [22,23]. Trolox, which is 6-hydroxy-2,5,7,8-tetramethylchroman-2-carboxylic acid, was used as a standard. Results were shown in terms of IC50, which is the concentration of the test sample that would reduce the concentration of DPPH, ABTS by 50%, as well as the amount of Trolox equivalent (TE)g-1 in the sample.

The DPPH free radical scavenging activity was assessed using previously published methods with slight modifications [34,35]. In a 96-well microliter plate, 100 μL of the sample (10 mg $mL^{-1}$) and standard were added to 150 μL of DPPH in 100 ethanol. At 517 nm, the absorbance of each well was measured after 20 min of incubation at 37 °C.

The ABTS analysis was carried out as described in [36] using a 96-well microliter plate [35]. We made the ABTS•+ solution by oxidizing it with $MnO_2$ in water for 30 min. The ABTS•+ solution (150 μL) was added to a 50 μL aliquot of the different concentrations of the sample (concentration 10 mg $mL^{-1}$) and standard (Trolox). After 10 min of incubation at room temperature, the absorbance of each well was measured at 734 nm.

The FRAP values of the extracts were calculated using a previously described technique [37], with minimal modifications [38]. The FRAP reagent was produced by dissolving the following three solutions: 50 mL 0.3 M acetate buffer pH 3.6 (1.23 g sodium acetate in 50 mL water acidified with acetic acid); 5 mL stock solution of 5 mM TPTZ (2,4,6-tripyridyl-s-triazine) (15.6 mg) in 40 mM HCl; 5 mL stock solution of 5 mM FeCl3 6H2O (16.2 mg) in 40 mM HCl. The FRAP solution was warmed to 37 °C. Aliquots of 50 μL sample solution (10 mg $mL^{-1}$) were applied in triplicate to the 96-well plate (BD FalconTM). Each well was initiated with 175 μL of FRAP reagent. The plate was immediately shaken for 30 s in a FLUOstar Omega plate reader, and the reaction was allowed to run for 10 min before being read on a plate reader (593 nm). Meanwhile, a reference solution of Trolox was run to create the calibration curve by linear regression with the linearity of $R^2$ = 0.9996. The data were presented in μM Trolox equivalent (TE) g-1 sample.

### 2.7. In Vitro Antimicrobial Activity Assay

Extracts were tested against the Gram-negative bacterial species Pseudomonas aeruginosa (ATCC 9027) and Escherichia coli (ATCC 8739), the Gram-positive Staphylococcus aureus (ATCC 6538), and the yeast Candida albicans (ATCC 10231). These microorganisms were from the culture collections maintained by the In Vitro Testing Laboratory of Abich s.r.l. (Verbania, VB, Italy).

The raw materials of the pomegranate peel and male flower extracts were initially evaluated for their TVC (total viable counts) according to ISO 21149 for bacteria and ISO 16212 for yeast and moulds. The bacterial TVC were < 10 UFC/g for all analysed pomegranate cultivars. Only male flowers had a bacterial TVC > 10 UFC/g, probably because the flowers were collected directly from the field after they had fallen from the tree. No contamination was detected in the yeast and moulds except for the Mollar de Elche peel (2019 and 2020). Furthermore, we also found no microbiological contamination in pomegranate extracts. The antimicrobial activity of extracts was evaluated by following the ISO 20645 based on an agar-disk diffusion method with some modifications for *C. albicans* and the filter paper size used. Petri dishes were prepared with a bottom layer of about 10 mL of a non-inoculated agar medium, which was Tryptone Soy Agar (TSA) (VWR, Milano, Italia) for bacteria and Sabouraud Dextrose Agar (SDA) for Candida. Then a top agar inoculated with $10^6$ CFU/mL of cells was poured onto the bottom layer. After solidification, a sterile filter paper disc (about 6 mm in diameter) containing the test compound was placed on the surface of the inoculated top agar. The disks were prepared by soaking up 100 μL of extract and then placing them still wet on top of the agar layer. Negative control disks were soaked in the solvent used for the extraction. Plates were incubated at 37 °C for 24 h. To assess the antimicrobial activity of the extracts, bacterial and yeast growth inhibition under and around the disk was evaluated, and the diameter of the inhibition zone was measured and given in millimetres. The growth inhibition was calculated on a scale of 0% to 100%, with 0% indicating no inhibition and 100% indicating the complete absence of growth below the disk. Each experiment was repeated three times.

Effect of Pomegranate Extracts on Bacterial Growth

The susceptibility of the four microorganisms to different concentrations of pomegranate extracts was determined by the broth microdilution method, and the inhibition of growth was assessed by a turbidimetric assay. A serial dilution of the extracts (highest concentration: 50 μg/μL) was prepared in a growth medium and inoculated with bacteria at 37 °C for 24 h. A calibration curve correlated the turbidimetric values with bacterial cell counts. Fertility and sterility tests were run in parallel with ranging concentrations of Tryptone soy broth (TSB) (VWR, Milano, Italia) and extracts, respectively. After the incubation at 37 °C for 24 h, the optical density at $A_{600nm}$ was determined, and the percentage of inhibition was obtained by the following Equation (2):

$$Inibition(\%) = 100 - \frac{O.D./mL_{\ extract}}{O.D./mL_{\ no\ extract}} \times 100 \quad (2)$$

*2.8. Cell Viability Evaluation of Pomegranate Extracts in Keratinocyte Cells*

The human keratinocyte cell line HaCaT was cultured at 37 °C in a humidified incubator with 5% $CO_2$ in Dulbecco's Medium (DMEM) supplemented with 10% foetal bovine serum, two mM L-glutamine, 50 U/mL penicillin, and 50 g/mL streptomycin. To assess the intracellular antioxidant activity of HaCaT cells, they were seeded at a density of 3104 cells per well in 96-well plates. The studies were conducted 24 h after incubation at 37 °C in 5% $CO_2$.

Cell viability was determined by decreasing MTT as described in our previous work [30]. At 37 °C in 5% $CO_2$, HaCaT cells were treated for 24 h with various concentrations of extracts (0.160–5 mg/mL). The treatment medium was changed to MTT in HBSS (0.5 mg/mL) for two hours at 37 °C in 5% $CO_2$. Formazan crystals were dissolved in isopropanol after being washed with HBSS. The formazan concentration was determined (570 nm, reference filter 690 nm) using the VICTOR$^{TM}$X3 multilabel plate reader (PerkinElmer, Waltham, MA, USA). The viability of the cells was expressed as a percentage of total vitality.

### 2.9. Statistical Analysis

As appropriate, we used one-way ANOVA together with Dunnett or Bonferroni post hoc tests to conduct our statistical study. On a Windows platform, analyses and Pearson correlation analysis were performed using GraphPad PRISM software (version 5.0; GraphPad Software, La Jolla, CA, USA).

## 3. Results and Discussion

### 3.1. Qualitative Polyphenols Identification

We obtained the mass data in the full-scan analysis, and product ion mass data were acquired by the information-dependent acquisition method. Thirty-one phenol compounds, including hydrolysable tannins such as ellagitannins and gallotannins, and phenol acids from the peel extract of pomegranate, were identified based on previous studies [39–43]. Compounds were observed by their $[M - H]^-$ ions in negative ESI mode and their $[M]^+$ ions in positive mode, which was helpful to detect their precise mass measurement. To organize the description of peak assignment and further characterization of individual compounds, all the phenol compounds were divided according to their compound classes in hydrolysable tannins, hydroxybenzoic acids, hydroxycinnamic acids, and dihydroflavonol for the portion of the negative ion. While for positive ions, four anthocyanins were identified. To obtain the most valuable chemical information and achieve better attribution of every peak of specific substances, G1 peel extract was selected out of all the extract samples for the investigation.

A comprehensive description of the method adopted for identifying all the substances detected by UPLC-MS/MS can be found in the Supplementary Materials (Tables S1 and S2). Accurate mass measurements, retention time, and main MS/MS product ions for all phenol compounds were reported in the Supplementary Materials.

### 3.2. Quantitative Analysis Using UPLC-ESI-MS/MS

The methodology established in this research may be used for the quantitative measurement of secondary metabolites contained in the peel and male flower extracts of four different pomegranate varieties. Quantitative data were expressed as mg/kg dry weight matter for all extract samples. Selected polyphenol compounds were quantified and compared among pomegranate peel and male flower extracts. The contents of the nine chosen compounds (gallic acid, punicalin, punicalagin A and B, ellagic acid, cyanidin 3,5-diglucoside, cyanidin 3-glucoside, pelargonidin 3,5 diglucoside, pelargonidin 3-glucoside) in the peel and male flower extracts of pomegranate from four different varieties of pomegranate are shown in Table 2 and Figure 1.

Among phenol compounds, ellagitannins represent the predominant class of phenol compounds present in pomegranate peel and flowers [15,40]. The ellagitannins, punicalagin A and B, punicalin, and ellagic acid were quantified in our work. High levels of punicalagin A and B (7206.4 mg/kg and 5812.9 mg/kg, respectively) were detected in the Mollar de Elche 2020 peel extract. The content of punicalagin A was also high in the Parfianka (3767.3 mg/kg) and G1 (3622.3 mg/kg) 2020 peel extracts. The Mollar de Elche 2019 peel extract (2176.7 mg/kg) and the Wonderful 2020 peel extract (2754.8 mg/kg) revealed a content of punicalagin A higher than the other pomegranate varieties taken into account in this study. The Parfianka peel extract reported a high content in punicalagin B (5367.8 mg/kg), followed by the peel extracts of Mollar de Elche 2019 and Wonderful 2020. Our results are consistent with previous research that found extremely high levels of punicalagin (10,543.4 mg/g) in pomegranate peels. [13,15,44]. The punicalagin in pomegranate peel extract represented the most dominant component in the extract compared with ellagic acid, gallic acid, catechin, and epicatechin [45].

A high concentration of punicalin was detected in the Mollar de Elche 2020 peel extract, followed by the Wonderful and G1 2020 peel extracts. The Wonderful peel extract demonstrated a higher ellagic acid content than the other pomegranate varieties. The ellagic acid in the Wonderful 2020 peel extract was 428.9 mg/kg, while in the Mollar de

Elche and G1 2020 peel extracts it was 289.7 and 337.3 mg/kg, respectively. The peel extract of Mollar de Elche 2019 resulted in an ellagic acid concentration of 231.2 mg/kg, while in the Wonderful and G1 peel extracts, the concentration of ellagic acid was lower.

Among the male flower extract, Wonderful 2020 showed a high level of punicalin followed by punicalagin B and A. In Mollar de Elche, the content of punicalin was higher compared to the concentration of punicalagin A and B. In the male flower extract, the ellagic acid content was low for Wonderful (42.4 mg/kg) and Mollar de Elche (87.1 mg/kg). Our work investigated the content of gallic acid, which belongs to the hydroxybenzoic acid class. Among the peel extracts, the samples with a high gallic acid content were the G1 and Parfianka 2020 peel extracts, followed by the Mollar de Elche 2020 and 2019 peel extracts. The concentration of gallic acid in male flower extract of Wonderful showed a high concentration of 925.2 mg/kg compared to the concentration obtained for the Mollar de Elche male flower extract (789.5 mg/kg). The concentrations of phenols detected in our study matched those observed in previous studies. The range of punicalagin in the peel extract was higher than the range of ellagitannins and hydroxybenzoic acids [46,47]. We also investigated the content of anthocyanins which revealed differences in their contents in different cultivars. In detail, the amount of cyanidin 3,5-diglucoside was 25.2 mg/kg in the Wonderful 2019 peel extract, while in the 2020 variety, the concentration was much lower (4.7 mg/kg). This compound was also detected in the Wonderful male flower extract, where the concentration was 6.1 mg/kg, while in Mollar de Elche peel and male flower extract, the concentration of cyanidin 3,5-diglucoside and cyanidin 3-glucoside was under the limit of quantification. Pelargonidin 3,5-diglucoside was detected in high concentration in G1 2019 peel extract (20.6 mg/kg), although, in the G1 2020 peel extract the concentration was under the detection limit. In the Wonderful and Mollar de Elche male flower extract, the pelargonidin 3,5-diglucoside was 2.4 mg/kg and 8.1 mg/kg, respectively. The mono glycoside pelargonidin was 13.1 mg/kg in the Wonderful 2019 peel extract, followed by Parfianka 2020 peel extract and Mollar de Elche 2020 and 2019 peel extracts. In the male flower extract, the content of pelargonidin 3-glucoside was 1.1 mg/kg for the Wonderful variety, while it was under the limit of detection in Mollar de Elche.

Our results confirmed that the phenol composition of pomegranate is strongly influenced by the fruit part (such as peel, mesocarp, and arils), cultivar, environmental conditions, solvent, and methods used for the extraction, as also reported in other studies [21,48,49]. Additionally, our work used a green extraction technique based on US technology, one of the most advantageous greenways for bioactive component extraction. This technology employs cavitation to collapse membrane cells, enabling the extraction of many molecules. The US technology provides various advantages, including low prices and the capacity to reduce extraction time, solvent use, and energy consumption [16,21,50–52]. Many studies highlight how US technology is used mainly on natural products, including pomegranate phenol extraction [12,16,21,48,51,53].

**Table 2.** Quantification of polyphenols by using UPLC-ESI-MS/MS. Each sample was isolated in triplicate and analysed separately. Three repetitions were used to determine the standard deviation.

| Analyte | Wonderful Peel | | Mollar de Elche Peel | | G1 Peel | | Wonderful Male Flowers | | Mollar de Elche Male Flowers | | Wonderful Peel | | Mollar de Elche Peel | | G1 Peel | | Parfianka Peel | |
|---|---|---|---|---|---|---|---|---|---|---|---|---|---|---|---|---|---|---|
| | 2019 | | 2019 | | 2019 | | 2020 | | 2020 | | 2020 | | 2020 | | 2020 | | 2020 | |
| | Conc. [1] | DS% | Conc. [1] | DS% | Conc. [1] | DS% | Conc. [1] | DS% | Conc. [1] | DS% | Conc. [1] | DS% | Conc. [1] | DS% | Conc. [1] | DS% | Conc. [1] | DS% |
| Gallic acid | 9.7 | 12.6 | 33.6 | 2.2 | 17.2 | 1.7 | 925.2 | 5.2 | 789.5 | 1.4 | 28.5 | 1.9 | 47.8 | 13.1 | 53.3 | 16.8 | 58.5 | 0.4 |
| Punicalin | 7.7 | 0.5 | 34.1 | 18.8 | <LOQ | <LOQ | 5948.2 | 1.0 | 2143.8 | 3.3 | 638.7 | 1.2 | 946.4 | 3.7 | 670.1 | 8.1 | 67.6 | 3.1 |
| Punicalagin A | 478.9 | 6.5 | 2176.7 | 4.9 | 325.3 | 9.5 | 3562.2 | 0.4 | 430.4 | 1.32 | 2754.8 | 7.9 | 7206.4 | 4.6 | 3622.3 | 4.2 | 3767.3 | 6.7 |
| Punicalagin B | 947.8 | 1.9 | 3343.6 | 0.6 | 540.7 | 3.8 | 4757.8 | 0.5 | 667.5 | 3.3 | 3320.1 | 1.3 | 5812.9 | 1.3 | 2805.7 | 3.1 | 5367.8 | 0.4 |
| Ellagic acid | 48.9 | 12.8 | 231.2 | 0.4 | 19.7 | 6.6 | 42.4 | 2.4 | 87.1 | 3.5 | 418.9 | 0.2 | 289.7 | 2.7 | 337.3 | 1.5 | 123.2 | 3.0 |
| Cyanidin 3,5-diglucoside | 25.2 | 3.6 | <LOQ | <LOQ | 3.4 | 14.1 | 6.1 | 14.4 | <LOQ | <LOQ | 4.7 | 9.7 | <LOQ | <LOQ | <LOQ | <LOQ | 5.7 | 4.2 |
| Cyanidin 3-glucoside | 23.9 | 13.2 | 8.3 | 32.5 | 0.5 | 9.2 | <LOQ | <LOQ | <LOQ | <LOQ | 7.1 | 14.1 | <LOQ | <LOQ | <LOQ | <LOQ | 8.3 | 13.7 |
| Pelargonidin 3,5-diglucoside | 8.4 | 9.8 | 9.4 | 9.4 | 20.6 | 13.4 | 2.4 | 14.5 | 8.1 | 14.9 | 3.5 | 15.7 | <LOQ | <LOQ | <LOQ | <LOQ | 3.9 | 12.1 |
| Pelargonidin 3-glucoside | 13.1 | 18.6 | 7.2 | 18.6 | 1.8 | 0.2 | 1.1 | 14.4 | <LOQ | <LOQ | 7.0 | 10.1 | <LOQ | <LOQ | <LOQ | <LOQ | 8.1 | 13.2 |

[1] The mean value is expressed as mg/Kg of DM (dry matter).

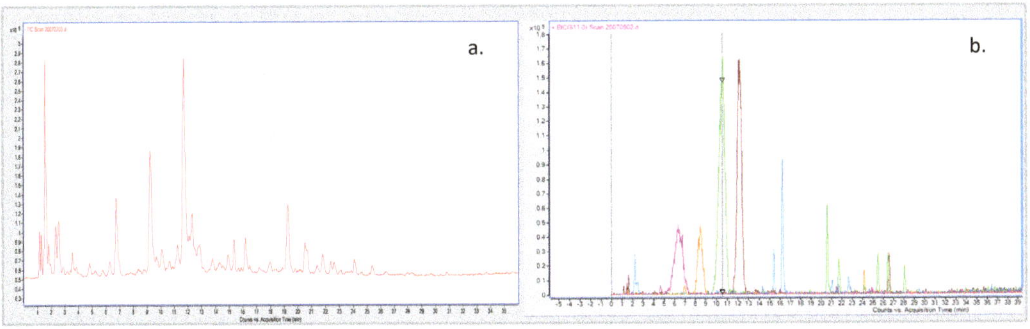


**Figure 1.** HPLC chromatographic profile of the phenol compounds (**a**), and quantified anthocyanins (**b**) (A4 in brown, A2 in green, A3 in orange, A1 in fuchsia), present in pomegranate peel extracts (variety G1). For peaks identification see Tables S1 and S2.

### *3.3. The Total Phenol Content (TPC) and Antioxidant Capacity (AC) of Pomegranate Extracts*

Table 3 reports the TPC and AC results obtained for the four pomegranate cultivars and their peel and male flower extracts. The Folin–Ciocalteu colorimetric method determined total phenol content and demonstrated impacts that ranged between 0.50 to 12.34 μmol GAE/g for peel extracts, and 0.46 and 0.778 μmol GAE/g for male flower extracts. Among the analysed samples, the peel extract from Mollar de Elche 2020 reported content of 12.341 μmol GAE/g, which represented the highest TPC, while Wonderful 2019 showed the lowest TPC. For the male flower extract samples, the TPC results were 0.778 and 0.746 μmol GAE/g for Wonderful and Mollar de Eche, respectively. The results show that pomegranate peel extracts have a phenol content 20 times higher than the corresponding extracts obtained from the male flowers of the same cultivars of Wonderful and Mollar de Elche. Our results are in line with the TPC of pomegranate peel extract reported in other works [54,55]. Based on the obtained results, we might highlight the impact of the type of cultivars and seasonal variation genotype of TPC in the biosynthetic pathway, as reported in previous studies [26,56–58]. The antioxidant capacity was obtained by two methods based on the evaluation of the free-radical scavenging capacity of the peel and male flower extract (DPPH and ABTS) and one method based on the reducing power of the extract samples (FRAP).

To determine whether the pomegranate extracts produced had bio-active antioxidant properties due to their high content of phenol compound, which is widely known as being responsible for antioxidant activity, the in vitro antioxidant activity of all extracts was evaluated using the DPPH, ABTS, and FRAP assays [59]. Our results show that pomegranate peel and male flower extracts display variability in inhibitory activity against DPPH radicals ranging between 0.242–34.361 μmolTEA/g. Among the tested pomegranate peel extract, the highest radical scavenging activity was detected for the Wonderful 2020 peel extract, followed by the G1 2020 peel extract and the Parfianka 2020 peel extract. Meanwhile, male flower extracts show their highest inhibitory activity against DPPH radicals in the following order: Wonderful > Mollar de Elche.

Our work also evaluated the antioxidant activity by ABTS radical cationic decolorization assay, showing that the highest radical activity was detected for the Wonderful 2020 peel extract (29.301 μmol TEA/g) the lowest one in the Wonderful 2019 peel extract. Among male flowers, it was found that the antioxidant activity was 6.808 and 3.168 μmolTEA/g for the Wonderful and Mollar de Elche male flower extracts, respectively.

The ferric reducing and antioxidant power assay was employed. The FRAP results of pomegranate peel and male flower extract were determined to compare the four varieties of pomegranate in our investigation. The Mollar de Elche 2020 and G1 2020 peel extract reported the highest FRAP value among all peel extracts tested, followed by Wonderful 2020, Parfianka 2020, G1 2019, Mollar de Elche 2019, and Wonderful 2019.

We found a positive and high Pearson correlation with a significant $p$ value (* $p < 0.05$; **** $p < 0.0001$) between antioxidant activity measured with FRAP and ABTS and total phenolic contents, suggesting that the antioxidant activity is dependent on the number of phenolic compounds present in the extracts.

Results show that the peel extracts displayed higher AC than the male flower extracts. The AC found in pomegranate peel extracts was connected with the punicalagin concentration determined by our quantitative study, and it was also consistent with other studies [15,55].

**Table 3.** Total phenol content (TPC) and antioxidant capacity (AC) of pomegranate extracts.

| Samples | Folin–Ciocalteu | ABTS | | FRAP | DPPH | |
|---|---|---|---|---|---|---|
| | (µmol GAE/g) | (µmol TEA/g) | $IC_{50(mg/mL)}$ | (µmol TEA/g) | (µmol TEA/g) | $IC_{50(mg/mL)}$ |
| Wonderful 2019 Peel | 0.500 ± 0.004 | 0.076 ± 0.002 | 0.016 ± 0.001 | 2.170 ± 0.003 | 0.242 ± 0.056 | 0.065 ± 0.056 |
| Mollar de Elche 2019 Peel | 2.304 ± 0.006 | 3.290 ± 0.001 | 0.001 ± 0.001 | 3.299 ± 0.028 | 0.455 ± 0.007 | 0.035 ± 0.007 |
| G1 2019 Peel | 1.872 ± 0.002 | 2.121 ± 0.001 | 0.001 ± 0.001 | 3.730 ± 0.001 | 1.524 ± 0.012 | 0.011 ± 0.012 |
| Wonderful 2020 male flowers | 0.778 ± 0.003 | 6.808 ± 0.002 | 0.002 ± 0.001 | 0.615 ± 0.022 | 1.149 ± 0.014 | 0.014 ± 0.013 |
| Mollar de Elche 2020 male flowers | 0.746 ± 0.003 | 3.168 ± 0.002 | 0.001 ± 0.001 | 0.458 ± 0.013 | 0.444 ± 0.020 | 0.036 ± 0.023 |
| Wonderful 2020 Peel | 6.346 ± 0.001 | 29.301 ± 0.001 | 0.001 ± 0.002 | 7.015 ± 0.024 | 34.361 ± 0.001 | 0.001 ± 0.001 |
| Mollar de Elche 2020 Peel | 12.341 ± 0.002 | 18.862 ± 0.004 | 0.001 ± 0.001 | 12.435 ± 0.801 | 3.230 ± 0.003 | 0.003 ± 0.001 |
| G1 2020 Peel | 9.283 ± 0.015 | 21.754 ± 0.001 | 0.002 ± 0.001 | 12.407 ± 0.739 | 5.029 ± 0.010 | 0.002 ± 0.003 |
| Parfianka 2020 Peel | 6.098 ± 0.001 | 15.875 ± 0.001 | 0.001 ± 0.001 | 4.860 ± 0.237 | 4.393 ± 0.002 | 0.003 ± 0.001 |

### 3.4. Antimicrobial Activity Evaluation

To explore the possibilities of using pomegranate by-products as a natural preservative in cosmetic formulations alone or as a booster to reintroduce agri-food waste into the cosmetics formulation as a natural bioactive component, we investigated the antimicrobial activities of peel and male flower extracts derived from Wonderful, Mollar de Elche, Parfianka, and G1 cultivars. The evaluation of our extracts' activity against the Gram-positive and Gram-negative was obtained by the disk-diffusion testing [60] (Table 4). The extracts activities were tested against *Staphylococcus aureus, Escherichia coli, Pseudomonas aeruginosa,* and *Candida albicans* which are microbial species referenced in the challenge test for cosmetic products. The Wonderful 2020 peel extract was the most efficient against *Staphylococcus aureus,* with a 70% growth inhibition in microdilution tests and an inhibition zone diameter of 3 mm by disk diffusion. The highest antimicrobial activity against the *Escherichia coli* was demonstrated by the Wonderful 2020 and G1 2020 peel extracts, while all cultivars were active against *Candida albicans.* In particular, a good activity was detected for the G1 2019 and 2020 peel extracts, the Wonderful 2020 peel extract, and the Mollar de Elche 2020 peel extract. The antimicrobial activity obtained for peel extracts was in line with a previous work by Khan et al. [61] reporting high antimicrobial activity for pomegranate peel extract against Gram-negative bacteria. Besides, Kupnik et al. [62] found that *E. coli* and *P. aeruginosa* were more susceptible to pomegranate peel extract than to other parts of the fruit. In accordance with other studies, peel and male flower extract were effective also against Gram-positives [5,26,62–64].

The peel G1 extract was chosen as the most promising among the tested extracts for further antibacterial investigation using a turbidimetric test. Table 5 shows the $IC_{50}$ obtained against *S. aureus* and *P. aeruginosa* in G1 2020 peel extract. It was detected that a 1:2 dilution of the extract was effective in reducing 97% of the bacteria.

Our pomegranate extract demonstrated a variable antimicrobial activity against the four microbial species considered. This might be attributed to the pH values ranging from $3.9 \pm 0.2$ to $4.6 \pm 0.3$ that were found in all of the extracts and the high concentration of polyphenols that can inhibit microbial growth. Data herein indicated that peel extract, mainly from the G1 and Wonderful varieties, had an effective antimicrobial activity, due to its inhibitory effect on bacterial growth. At a concentration of 10 mg/disc, the G1 extract induced the formation of a clear inhibition zone ranging from 1 to 4 mm against all microorganisms tested. Another extract with a measurable antimicrobial activity was the Mollar de Elche 2020 peel extract, with an ability to inhibit the growth of the *Aspergillus brasiliensis* (data not shown). Furthermore, this conclusion is supported by the TVC of the yeasts and moulds, which were found lower than 10 UFC/g in raw materials from the Mollar de Elche 2019 and 2020 peel extract.

Our results concur with several studies on the antimicrobial activity of pomegranate extracts [5,26,64].

Accordingly, we can correlate the antimicrobial activity of pomegranate peel extract with the high concentration of polyphenol compounds such as punicalagin and punicalin as evidenced by the UPLC-ESI-MS/MS analysis. The pomegranate peel extract obtained from the G1 cultivars demonstrated interesting inhibitory activity and had among the highest concentrations of punicalagin (3622.3 and 2805.7 mg/kg for two isomers A and B, respectively). Punicalagins use their hydrophilic part to interact with the polar regions of the cell membrane compromising its ability to transport substances inside the cell [5,63]. Additionally, phenols can render substrates unavailable to microorganisms or interfere with protein secretion.

The results against the Gram-negative for the G1 and Wonderful pomegranate extracts pave the way for their further application as booster antimicrobial agents to be used in combination with common synthetic antimicrobials, which might be added in lower amounts thereby increasing skin tolerance and product safety. Furthermore, since these extracts are obtained from the peel and male flowers of pomegranate, they provide a concrete example of natural products made from agri-food waste that have been reintroduced into the cosmetics' circular economy concept.

**Table 4.** Evaluation of the antimicrobial activity by using *S. aureus*, *E. coli*, and *C. albicans*.

| **Samples** | **Diameter of Inhibition (mm)** | **Evaluation** [a] |
|---|---|---|
| | *S. aureus* | |
| Wonderful 2019 Peel | 8 | ++ |
| Wonderful 2020 Peel | 12 | +++ |
| Mollar de Elche 2019 Peel | <8 | ++ |
| Mollar de Elche 2020 Peel | 8 | ++ |
| G1 2019 Peel | 8 | ++ |
| G1 2020 Peel | 8 | + |
| Parfianka 2020 Peel | 8 | + |
| Mollar de Elche Male flower | - | + |
| Wonderful Male flower | - | + |

**Table 4.** *Cont.*

| Samples | Diameter of Inhibition (mm) | Evaluation [a] |
|---|---|---|
| | *E. coli* | |
| Wonderful 2019 Peel | 10 | |
| Wonderful 2020 Peel | 10 | + |
| Mollar de Elche 2019 Peel | 10 | |
| Mollar de Elche 2020 Peel | 10 | + |
| G1 2019 Peel | 8 | ++ |
| G1 2020 Peel | 10 | + |
| | *C. albicans* | |
| Wonderful 2019 Peel | 10 | ++ |
| Wonderful 2020 Peel | 10 | ++ |
| Mollar de Elche 2019 Peel | 10 | ++ |
| Mollar de Elche 2020 Peel | 12 | ++ |
| G1 2019 Peel | 12 | +++ |
| G1 2020 Peel | 14 | +++ |

[a] The evaluation considers two parameters: the area of inhibition and the growth of inhibition, with + ranging from 0% to 40% with 1 mm, ++ ranging from 50% to 60% with 1 to 2 mm, and +++ ranging from 60% to 70% with >2 mm.

**Table 5.** Determination of $IC_{50}$ against *S. aureus* and *P. aeruginosa* in G1 2020 peel extract.

| Sample % [a] | OD/mL | Rid. % |
|---|---|---|
| | *S. aureus* | |
| 0 | 0.4467 | - |
| 5 | 0.2733 | 23 |
| 10 | 0.2232 | 34 |
| 18 | 0.2932 | 20 |
| 20 | 0.2398 | 30 |
| 25 | 0.2318 | 32 |
| 30 | 0.1520 | 52 |
| 50 | 0.0052 | 97 |
| | *P. aeruginosa* | |
| 0 | 0.7013 | - |
| 35 | 0.4600 | 34 |
| 40 | 0.3837 | 45 |
| 45 | 0.4630 | 34 |
| 50 | 0.4500 | 35 |

[a] Tested sample dilution expressed in (%).

### 3.5. Cytocompatibility of Pomegranate Extract in Human Keratinocytes

The cytocompatibility was evaluated in the pomegranate peel and male flower extracts 2019 and 2020 in keratinocytes HaCaT cells to find the nontoxic concentration of pomegranate extracts and further exploitation of these extracts as antimicrobial agents in cosmetics formulations. The use of keratinocytes was based on the concept that these cells play a significant role in creating skin structure and the sustenance of homeostasis, including the production of the skin barrier and the extracellular matrix (ECM). The cells were treated with extract concentrations ranging from 0.15 to 5.00 mg/mL for 24 h. The vitality of the cells was then determined by the MTT test. The results showed no significant difference in cell viability between the different pomegranate extracts ($p > 0.05$) varieties at any extract concentration. Figure 2 shows the cytotoxicity results obtained for the extracts selected with the most interesting antimicrobial properties, such as the G1 2020 and 2019 peel extracts and the Wonderful 2020 and 2019 peel extracts. Additionally, no change in the morphology was observed after treatment of keratinocytes with the extracts (data not shown).

Our results agreed with a previous study that reported null cytotoxicity of pomegranate extract against HaCaT cells [65]. Liu et al. [66] also demonstrated that the pomegranate extract resulted in no cytotoxicity in a concentration range from 6.25 to 100.00 μg/mL. These promising results obtained in the cytotoxicity tests confirm the suitability of the extracts as a natural booster ingredient for cosmetic formulations.

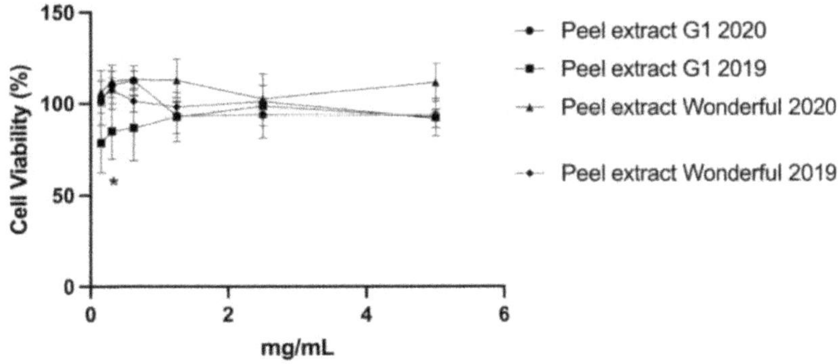


**Figure 2.** Cytotoxicity of the pomegranate peel extracts G1 and Wonderful 2019 and 2020 in HaCaT cells evaluated by MTT assay. For 24 h, cells were treated with an extract at different concentrations (0.15–5.00 mg/mL). The data are shown as a percentage of control cells and as the mean ± SEM of four separate experiments. (* $p < 0.01$ vs. untreated cells; one-way ANOVA with Dunnett post hoc test).

## 4. Conclusions

The main phenol components profile, antioxidant capacity, antibacterial activity, and cell viability of four pomegranate types cultivated in Marche were investigated. The UPLC-ESI-MS/MS study revealed that Mollar de Elche had a high concentration of punicalagin A and B, whereas G1 and Parfianka had high concentrations of gallic acid. Moreover, antibacterial activity testing revealed that all extracts were promising. Finally, this work shows the value of studying bioactive chemicals in agri-food waste products to generate innovative natural antibacterial and antioxidant components for cosmetic formulation. Further research will focus on the antibacterial activity of suitable pomegranate extract mixes and pomegranate extracts combined with conventional preservatives to minimise their concentration in cosmetic products.

**Supplementary Materials:** The following are available online at https://www.mdpi.com/article/10.3390/antiox11040768/s1, Table S1: Identification of polyphenols in pomegranate by-product extracts by UPLC-ESI-MS/MS analysis, Table S2: Identification of anthocyanins in pomegranate by-products extracts by UPLC-ESI-MS/MS analysis.

**Author Contributions:** Conceptualization, P.D.M. and M.R.G.; methodology, P.D.M., M.R.G., M.C., G.L. and E.B.; software, M.R.G., M.C. and E.B.; validation, M.C., E.B., P.D.M. and G.L.; formal analysis M.C., M.R.G., L.D.N. and S.N.; investigation, M.R.G., M.C., L.D.N. and S.N.; resources, P.D.M. and R.C.; data curation, M.R.G., P.D.M., M.C., L.D.N. and S.N.; writing—original draft preparation, M.R.G. and P.D.M.; writing—review and editing, P.D.M., M.R.G., L.A.V., E.B. and S.N.; visualization, P.D.M., M.R.G., D.V.P. and L.A.V.; supervision, R.C., M.R.G., L.D.N. and L.A.V.; project administration, P.D.M. and R.C.; funding acquisition, P.D.M. All authors have read and agreed to the published version of the manuscript.

**Funding:** This project has received funding from the European Union's Horizon 2020 research and innovation program under the Marie Sklodowska Curie grant agreement No. 777682 (CANCER), 734684 (CHARMED), 872391 (CONCRETE), 872860 (PRISAR2), 807281 (ACORN), 852985 (SIMICA), 952520 (BIOSAFETY), 101029908 (GELNANODEP).

**Institutional Review Board Statement:** Not applicable.

**Informed Consent Statement:** Not applicable.

**Data Availability Statement:** Data is contained within the article and Supplementary Materials.

**Acknowledgments:** The authors acknowledge the Onori Maria Rosaria Sant'Elpidio a Mare (Fermo, Marche) farm for providing the pomegranate waste used in this study.

**Conflicts of Interest:** The authors declare no conflict of interest.

## References

1. Commission, Legislation. Available online: http://ec.europa.eu/growth/sectors/cosmetics/legislation/index_en.htm (accessed on 10 March 2022).
2. Fournière, M.; Latire, T.; Souak, D.; Feuilloley, M.G.J.; Bedoux, G. Staphylococcus epidermidis and Cutibacterium acnes: Two Major Sentinels of Skin Microbiota and the Influence of Cosmetics. *Microorganisms* **2020**, *8*, 1752. [CrossRef]
3. Pinto, D.; Ciardiello, T.; Franzoni, M.; Pasini, F.; Giuliani, G.; Rinaldi, F. Effect of commonly used cosmetic preservatives on skin resident microflora dynamics. *Sci. Rep.* **2021**, *11*, 8695. [CrossRef]
4. Halla, N.; Fernandes, I.P.; Heleno, S.A.; Costa, P.; Boucherit-Otmani, Z.; Boucherit, K.; Rodrigues, A.E.; Ferreira, I.C.F.R.; Barreiro, M.F. Cosmetics Preservation: A Review on Present Strategies. *Molecules* **2018**, *23*, 1571. [CrossRef]
5. Rosas-Burgos, E.C.; Burgos-Hernández, A.; Noguera-Artiaga, L.; Kačániová, M.; Hernández-García, F.; Cárdenas-López, J.L.; Carbonell-Barrachina, Á.A. Antimicrobial activity of pomegranate peel extracts as affected by cultivar. *J. Sci. Food Agric.* **2017**, *97*, 802–810. [CrossRef]
6. Morganti, P.; Morganti, G.; Gagliardini, A.; Lohani, A. From Cosmetics to Innovative Cosmeceuticals—Non-Woven Tissues as New Biodegradable Carriers. *Cosmetics* **2021**, *8*, 65. [CrossRef]
7. Barbulova, A.; Colucci, G.; Apone, F. New Trends in Cosmetics: By-Products of Plant Origin and Their Potential Use as Cosmetic Active Ingredients. *Cosmetics* **2015**, *2*, 82–92. [CrossRef]
8. Bolaji, I.; Nejad, B.; Billham, M.; Mehta, N.; Smyth, B.; Cunningham, E. Multi-criteria decision analysis of agri-food waste as a feedstock for biopolymer production. *Resour. Conserv. Recycl.* **2021**, *172*, 105671. [CrossRef]
9. Herman, A.; Herman, A.P.; Domagalska, B.W.; Młynarczyk, A. Essential Oils and Herbal Extracts as Antimicrobial Agents in Cosmetic Emulsion. *Indian J. Microbiol.* **2013**, *53*, 232–237. [CrossRef]
10. Kalouta, K.; Eleni, P.; Boukouvalas, C.; Vassilatou, K.; Krokida, M. Dynamic mechanical analysis of novel cosmeceutical facial creams containing nano-encapsulated natural plant and fruit extracts. *J. Cosmet. Dermatol.* **2020**, *19*, 1146–1154. [CrossRef]
11. Fayeulle, A.; Trudel, E.; Damiens, A.; Josse, A.; Ben Hadj Youssef, N.; Vigneron, P.; Vayssade, M.; Rossi, C.; Ceballos, C. Antimicrobial and antioxidant activities of amines derived from vanillin as potential preservatives: Impact of the substituent chain length and polarity. *Sustain. Chem. Pharm.* **2021**, *22*, 100471. [CrossRef]
12. Akhtar, S.; Ismail, T.; Fraternale, D.; Sestili, P. Pomegranate peel and peel extracts: Chemistry and food features. *Food Chem.* **2015**, *174*, 417–425. [CrossRef]
13. Ding, W.; Wang, H.; Zhou, Q.; Wu, C.; Gao, X.; Cheng, X.; Tian, L.; Wang, C. Simultaneous determination of polyphenols and triterpenes in pomegranate peel based on high-performance liquid chromatography fingerprint by solvent extraction and ratio blending method in tandem with wavelength switching. *Biomed. Chromatogr.* **2019**, *33*, e4690. [CrossRef]
14. Viuda-Martos, M.; Fernández-López, J.; Pérez-Álvarez, J.A. Pomegranate and its Many Functional Components as Related to Human Health: A Review. *Compr. Rev. Food Sci. Food Saf.* **2010**, *9*, 635–654. [CrossRef]
15. Young, J.E.; Pan, Z.; Teh, H.E.; Menon, V.; Modereger, B.; Pesek, J.J.; Matyska, M.T.; Dao, L.; Takeoka, G. Phenolic composition of pomegranate peel extracts using an liquid chromatography-mass spectrometry approach with silica hydride columns. *J. Sep. Sci.* **2017**, *40*, 1449–1456. [CrossRef]
16. Boggia, R.; Turrini, F.; Villa, C.; Lacapra, C.; Zunin, P.; Parodi, B. Green Extraction from Pomegranate Marcs for the Production of Functional Foods and Cosmetics. *Pharmaceuticals* **2016**, *9*, 63. [CrossRef]
17. Tehranifar, A.; Zarei, M.; Nemati, Z.; Esfandiyari, B.; Vazifeshenas, M.R. Investigation of physico-chemical properties and antioxidant activity of twenty Iranian pomegranate (*Punica granatum* L.) cultivars. *Sci. Hortic.* **2010**, *126*, 180–185. [CrossRef]
18. El-Hadary, A.E.; Ramadan, M.F. Phenolic profiles, antihyperglycemic, antihyperlipidemic, and antioxidant properties of pomegranate (Punica granatum) peel extract. *J. Food Biochem.* **2019**, *43*, e12803. [CrossRef]
19. Kowalska, H.; Czajkowska, K.; Cichowska, J.; Lenart, A. What's new in biopotential of fruit and vegetable by-products applied in the food processing industry. *Trends Food Sci. Technol.* **2017**, *67*, 150–159. [CrossRef]
20. Das, A.K.; Nanda, P.K.; Chowdhury, N.R.; Dandapat, P.; Gagaoua, M.; Chauhan, P.; Pateiro, M.; Lorenzo, J.M. Application of Pomegranate by-Products in Muscle Foods: Oxidative Indices, Colour Stability, Shelf Life and Health Benefits. *Molecules* **2021**, *26*, 467. [CrossRef]
21. Tabaraki, R.; Heidarizadi, E.; Benvidi, A. Optimization of ultrasonic-assisted extraction of pomegranate (*Punica granatum* L.) peel antioxidants by response surface methodology. *Sep. Purif. Technol.* **2012**, *98*, 16–23. [CrossRef]

22. Kaderides, K.; Papaoikonomou, L.; Serafim, M.; Goula, A.M. Microwave-assisted extraction of phenolics from pomegranate peels: Optimization, kinetics, and comparison with ultrasounds extraction. *Chem. Eng. Processing-Process Intensif.* **2019**, *137*, 1–11. [CrossRef]
23. Zarfeshany, A.; Asgary, S.; Javanmard, S.H. Potent health effects of pomegranate. *Adv. Biomed. Res.* **2014**, 3, 100. [PubMed]
24. Hayrapetyan, H.; Hazeleger, W.C.; Beumer, R.R. Inhibition of Listeria monocytogenes by pomegranate (*Punica granatum*) peel extract in meat paté at different temperatures. *Food Control* **2012**, *23*, 66–72. [CrossRef]
25. Martínez, L.; Castillo, J.; Ros, G.; Nieto, G. Antioxidant and Antimicrobial Activity of Rosemary, Pomegranate and Olive Extracts in Fish Patties. *Antioxidants* **2019**, *8*, 86. [CrossRef]
26. Pagliarulo, C.; De Vito, V.; Picariello, G.; Colicchio, R.; Pastore, G.; Salvatore, P.; Volpe, M.G. Inhibitory effect of pomegranate (*Punica granatum* L.) polyphenol extracts on the bacterial growth and survival of clinical isolates of pathogenic Staphylococcus aureus and Escherichia coli. *Food Chem.* **2016**, *190*, 824–831. [CrossRef]
27. Kaur, G.; Jabbar, Z.; Athar, M.; Alam, M.S. *Punica granatum* (pomegranate) flower extract possesses potent antioxidant activity and abrogates Fe-NTA induced hepatotoxicity in mice. *Food Chem. Toxicol.* **2006**, *44*, 984–993. [CrossRef]
28. Jun, X.; Lang, H.; Liang-gong, Y. Continuous extraction of phenolic compounds from pomegranate peel using high voltage electrical discharge. *Food Chem.* **2017**, *230*, 354–361.
29. Mitsagga, C.; Petrotos, K.; Giavasis, I. Antimicrobial Properties of Lyophilized Extracts of Olive Fruit, Pomegranate and Orange Peel Extracts against Foodborne Pathogenic and Spoilage Bacteria and Fungi In Vitro and in Food Matrices. *Molecules* **2021**, *26*, 7038. [CrossRef]
30. Gigliobianco, M.R.; Cortese, M.; Peregrina, D.V.; Villa, C.; Lupidi, G.; Pruccoli, L.; Angeloni, C.; Tarozzi, A.; Censi, R.; Di Martino, P. Development of New Extracts of Crocus sativus L. By-Product from Two Different Italian Regions as New Potential Active Ingredient in Cosmetic Formulations. *Cosmetics* **2021**, *8*, 51. [CrossRef]
31. Cortese, M.; Gigliobianco, M.R.; Peregrina, D.V.; Sagratini, G.; Censi, R.; Di Martino, P. Quantification of phenolic compounds in different types of crafts beers, worts, starting and spent ingredients by liquid chromatography-tandem mass spectrometry. *J. Chromatogr. A* **2020**, *1612*, 460622. [CrossRef]
32. Zorzetto, C.; Sánchez-Mateo, C.C.; Rabanal, R.M.; Lupidi, G.; Petrelli, D.; Vitali, L.A.; Bramucci, M.; Quassinti, L.; Caprioli, G.; Papa, F.; et al. Phytochemical analysis and in vitro biological activity of three *Hypericum* species from the Canary Islands (*Hypericum reflexum*, *Hypericum canariense* and *Hypericum grandifolium*). *Fitoterapia* **2015**, *100*, 95–109. [CrossRef] [PubMed]
33. Singleton, V.L.; Rossi Joseph, A. Colorimetry of Total Phenolics with Phosphomolybdic-Phosphotungstic Acid Reagents. *Am. J. Enol. Vitic.* **1965**, *16*, 144.
34. Brand-Williams, W.; Cuvelier, M.E.; Berset, C. Use of a free radical method to evaluate antioxidant activity. *LWT—Food Sci. Technol.* **1995**, *28*, 25–30. [CrossRef]
35. Venditti, A.; Bianco, A.; Quassinti, L.; Bramucci, M.; Lupidi, G.; Damiano, S.; Papa, F.; Vittori, S.; Maleci Bini, L.; Giuliani, C.; et al. Phytochemical Analysis, Biological Activity, and Secretory Structures of *Stachys annua* (L.) L. subsp *annua* (Lamiaceae) from Central Italy. *Chem. Biodivers.* **2015**, *12*, 1172–1183. [CrossRef] [PubMed]
36. Re, R.; Pellegrini, N.; Proteggente, A.; Pannala, A.; Yang, M.; Rice-Evans, C. Antioxidant activity applying an improved ABTS radical cation decolorization assay. *Free Radic. Biol. Med.* **1999**, *26*, 1231–1237. [CrossRef]
37. Benzie, I.F.; Strain, J.J. The ferric reducing ability of plasma (FRAP) as a measure of "antioxidant power": The FRAP assay. *Anal. Biochem.* **1996**, *239*, 70–76. [CrossRef]
38. Ornano, L.; Venditti, A.; Ballero, M.; Sanna, C.; Quassinti, L.; Bramucci, M.; Lupidi, G.; Papa, F.; Vittori, S.; Maggi, F.; et al. Chemopreventive and antioxidant activity of the chamazulene-rich essential oil obtained from Artemisia arborescens L. growing on the Isle of La Maddalena, Sardinia, Italy. *Chem. Biodivers.* **2013**, *10*, 1464–1474. [CrossRef]
39. Rahima, A.; Sanawar, M.; Haizhong, L.; Ablikim, U.; Guangying, S.; Guozheng, H.; Akber, A.H. Qualitative Analysis of Polyphenols in Macroporous Resin Pretreated Pomegranate Husk Extract by HPLC-QTOF-MS. *Phytochem. Anal.* **2017**, *28*, 465–473.
40. Fischer Ulrike, A.; Reinhold, C.; Kammerer Dietmar, R. Identification and quantification of phenolic compounds from pomegranate (*Punica granatum* L.) peel, mesocarp, aril and differently produced juices by HPLC-DAD–ESI/MSn. *Food Chem.* **2011**, *127*, 807–821. [CrossRef]
41. García-Villalba, R.; Espín, J.C.; Aaby, K.; Alasalvar, C.; Heinonen, M.; Jacobs, G.; Voorspoels, S.; Koivumäki, T.; Kroon, P.A.; Pelvan, E.; et al. Validated Method for the Characterization and Quantification of Extractable and Nonextractable Ellagitannins after Acid Hydrolysis in Pomegranate Fruits, Juices, and Extracts. *J. Agric. Food Chem.* **2015**, *63*, 6555–6566. [CrossRef]
42. Clarisse, G.; Minjie, Z.; Sonia, L.; Saïd, E. Identification of punicalagin as the bioactive compound behind the antimicrobial activity of pomegranate (*Punica granatum* L.) peels. *Food Chem.* **2021**, *352*, 129396.
43. Hernández-Corroto, E.; Marina, M.L.; García, M.C. Extraction and identification by high resolution mass spectrometry of bioactive substances in different extracts obtained from pomegranate peel. *J. Chromatogr. A* **2019**, *1594*, 82–92. [CrossRef] [PubMed]
44. Valeria, S.; Lucia, R.C.; Cinzia, C.; Gabriele, B.; Valeria, R.F.; Simona, F.; Nicolina, T.; Marco, R.; Luca, V. Beneficial Effects of Pomegranate Peel Extract and Probiotics on Pre-adipocyte Differentiation. *Front. Microbiol.* **2019**, *10*, 660.
45. Li, J.; He, X.; Li, M.; Zhao, W.; Liu, L.; Kong, X. Chemical fingerprint and quantitative analysis for quality control of polyphenols extracted from pomegranate peel by HPLC. *Food Chem.* **2015**, *176*, 7–11. [CrossRef]

46. Talal, S.; Taleb, K.; Boubker, N.; Rabiaa, E.; Abderrahman, M.; Maryam, B.; Abdelkhalid, E. Determination of Punicalagins Content, Metal Chelating, and Antioxidant Properties of Edible Pomegranate (*Punica granatum* L.) Peels and Seeds Grown in Morocco. *Int. J. Food Sci.* **2020**, *2020*, 8885889.
47. Kazemi, M.; Karim, R.; Mirhosseini, H.; Abdul Hamid, A. Optimization of pulsed ultrasound-assisted technique for extraction of phenolics from pomegranate peel of Malas variety: Punicalagin and hydroxybenzoic acids. *Food Chem.* **2016**, *206*, 156–166. [CrossRef] [PubMed]
48. Faizan, A.; Amardeep, K. Optimization of The Ultrasonic Assisted Extraction Process to Obtain Phenolic Compounds from Pomegranate (*Punica granatum*) Peels Using Response Surface Methodology. *Int. J. Agric. Sci.* **2018**, *10*, 7581–7585.
49. Jelena, Ž.; Katarina, Š.; Teodora, J.; Nada, Ć.; Nebojša, M. Optimization of ultrasound-assisted extraction of polyphenolic compounds from pomegranate peel using response surface methodology. *Sep. Purif. Technol.* **2018**, *194*, 40–47.
50. Farid, C.; Natacha, R.; Alice, M.; Mohammad, T.; Sandrine, P.; Anne-Sylvie, F.; Maryline, A. Review of Green Food Processing techniques. Preservation, transformation, and extraction. *Innov. Food Sci. Emerg. Technol.* **2017**, *41*, 357–377.
51. Parvin, S.; Elham, A.; Shahin, Z.; Ramaswamy Hosahalli, S. Ultrasound assisted extraction of bioactive compounds from pomegranate (*Punica granatum* L.) peel. *LWT* **2019**, *101*, 342–350.
52. Rajha, H.N.; Mhanna, T.; Kantar, S.; Khoury, A.; Louka, N.; Maroun, R.G. Innovative process of polyphenol recovery from pomegranate peels by combining green deep eutectic solvents and a new infrared technology. *LWT* **2019**, *111*, 138–146. [CrossRef]
53. Santos Mariana, P.; Souza Mariana, C.; Sumere Beatriz, R.; da Silva Laise, C.; Cunha Diogo, T.; Bezerra Rosangela Maria, N.; Rostagno Mauricio, A. Extraction of bioactive compounds from pomegranate peel (*Punica granatum* L.) with pressurized liquids assisted by ultrasound combined with an expansion gas. *Ultrason. Sonochemistry* **2019**, *54*, 11–17. [CrossRef] [PubMed]
54. Hulya, O.H.; Hulya, Y.; Selen, I.S. Comparison of antioxidant activities of juice, peel, and the seed of pomegranate (*Punica granatum* L.) and inter-relationships with total phenolic, Tannin, anthocyanin, and flavonoid contents. *Food Sci. Biotechnol.* **2012**, *21*, 373–387.
55. Zahra, D.; Margherita, F.; Marzieh, T.; Farnoosh, A.; Ali, H.; Sadat, H.M.; Oliveri, C.G.; Khalili, S.E. Antioxidant activity and total phenolic content of ethanolic extract of pomegranate peels, juice and seeds. *Food Chem. Toxicol.* **2018**, *114*, 108–111.
56. Sadiye, G.; Onur, S.; Ebru, O.; Mustafa, O. Total phenolic distribution of juice, peel, and seed extracts of four pomegranate cultivars. *Pharmacogn. Mag.* **2011**, *7*, 161–164.
57. Mouna, A.; Héla, Y.; Salma, C.; Ibtihel, K.; Mohamed, B.; Hamadi, A.; Ayadi, M.A. Antioxidant properties and phenolic profile characterization by LC-MS/MS of selected Tunisian pomegranate peels. *J. Food Sci. Technol.* **2017**, *54*, 2890–2901.
58. Shirin, S.; Ali, K.; Brecht Jeffrey, K.; Ali, S. Chemical and physical attributes of fruit juice and peel of pomegranate genotypes grown in Florida, USA. *Food Chem.* **2021**, *342*, 128302.
59. Dong, H.; Zheng, L.; Yu, P.; Jiang, Q.; Wu, Y.; Huang, C.; Yin, B. Characterization and Application of Lignin−Carbohydrate Complexes from Lignocellulosic Materials as Antioxidants for Scavenging In Vitro and In Vivo Reactive Oxygen Species. *ACS Sustain. Chem. Eng.* **2020**, *8*, 256–266. [CrossRef]
60. Mounyr, B.; Moulay, S.; Koraichi, I.S. Methods for in vitro evaluating antimicrobial activity: A review. *J. Pharm. Anal.* **2016**, *6*, 71–79.
61. Khan, J.A.; Hanee, S. Antibacterial properties of *Punica granatum* peels. *Int. J. Appl. Biol. Pharm. Technol.* **2011**, *2*, 23–27.
62. Kupnik, K.; Primožič, M.; Vasić, K.; Knez, Ž.; Leitgeb, M. A Comprehensive Study of the Antibacterial Activity of Bioactive Juice and Extracts from Pomegranate (*Punica granatum* L.) Peels and Seeds. *Plants* **2021**, *10*, 1554. [CrossRef] [PubMed]
63. Jing, C.; Chunling, L.; Xiaolu, O.; Ibrahim, K.; Yudi, G.; Mingxi, L. Antimicrobial Activity of Pomegranate Peel and Its Applications on Food Preservation. *J. Food Qual.* **2020**, *2020*, 8850339.
64. Subramaniam, P.; Dwivedi, S.; Uma, E.; Girish Babu, K. Effect of pomegranate and aloe vera extract on streptococcus mutans: An in vitro study. *Dent. Hypotheses* **2012**, *3*, 99–105. [CrossRef]
65. Yepes-Molina, L.; Hernández, J.A.; Carvajal, M. Nanoencapsulation of Pomegranate Extract to Increase Stability and Potential Dermatological Protection. *Pharmaceutics* **2021**, *13*, 271. [CrossRef]
66. Liu, C.; Guo, H.; DaSilva, N.A.; Li, D.; Zhang, K.; Wan, Y.; Gao, X.-H.; Chen, H.-D.; Seeram, N.P.; Ma, H. Pomegranate (*Punica granatum*) Phenolics Ameliorate Hydrogen Peroxide-Induced Oxidative Stress and Cytotoxicity in Human Keratinocytes. *J. Funct. Foods* **2019**, *54*, 559–567. [CrossRef]

antioxidants

MDPI

*Article*

# Anti-Inflammation and Anti-Melanogenic Effects of Maca Root Extracts Fermented Using *Lactobacillus* Strains

**Jisun Yang [1], Hyeijin Cho [2], Minchan Gil [2] and Kyung Eun Kim [1,2,*]**

1 Department of Cosmetic Sciences, Sookmyung Women's University, Seoul 04310, Republic of Korea
2 Department of Health Industry, Sookmyung Women's University, Seoul 04310, Republic of Korea
* Correspondence: kyungeun@sookmyung.ac.kr; Tel.: +82-2-710-9211

**Abstract:** Maca is a well-known biennial herb with various physiological properties, such as antioxidant activity and immune response regulation. In this study, the antioxidant, anti-inflammatory, and anti-melanogenic effects of fermented maca root extracts were investigated. The fermentation was carried out using *Lactobacillus* strains, such as *Lactiplantibacillus plantarum subsp. plantarum*, *Lacticaseibacillus rhamnosus*, *Lacticaseibacillus casei*, and *Lactobacillus gasseri*. In RAW 264.7 cells, the non-fermented maca root extracts increased the secretion of nitric oxide (NO), an inflammatory mediator, in a dose-dependent manner. In contrast, the fermented extracts showed considerably lower NO secretion than the non-fermented extracts at concentrations of 5% and 10%. This indicates the effective anti-inflammatory effects of fermented maca. The fermented maca root extracts also inhibited tyrosinase activity, melanin synthesis, and melanogenesis by suppressing MITF-related mechanisms. These results show that fermented maca root extracts exhibit higher anti-inflammatory and anti-melanogenesis effects than non-fermented maca root extracts. Thus, maca root extracts fermented using *Lactobacillus* strains have the potential to be used as an effective cosmeceutical raw material.

**Keywords:** maca root; *Lepidium meyenii*; postbiotics; anti-melanogenesis; anti-inflammation

**Citation:** Yang, J.; Cho, H.; Gil, M.; Kim, K.E. Anti-Inflammation and Anti-Melanogenic Effects of Maca Root Extracts Fermented Using *Lactobacillus* Strains. *Antioxidants* **2023**, *12*, 798. https://doi.org/10.3390/antiox12040798

Academic Editors: Irene Dini and Sonia Laneri

Received: 3 February 2023
Revised: 22 March 2023
Accepted: 23 March 2023
Published: 24 March 2023



## 1. Introduction

Postbiotics, also known as 'non-viable probiotics' or 'inactivated probiotics,' are functional bioactive compounds released after bacteria are lysed or produced through the fermentation of living bacteria, which have an indirect or direct beneficial effect on host cells [1]. In contrast, probiotics are live microorganisms that affect the health of the host when administered in appropriate amounts and prebiotics are food ingredients that have beneficial effects on the gut microbiome, and include human milk oligosaccharides, lactulose, and inulin derivatives [2,3]. The impacts of probiotics on intestinal immunity and improving gastrointestinal health are well known [4–6]. Compared to probiotics, research on postbiotics is in its infancy. However, studies have shown that postbiotics may improve human health by strengthening the intestinal barrier, promoting antibacterial activity against intestinal pathogens, and reducing inflammation [7,8]. Evidence that these postbiotics have the potential to modulate human health points to the importance of postbiotic research [6].

Postbiotics do not contain live microorganisms, reducing the risks associated with ingestion, and confer health benefits to the host through mechanisms similar to those of probiotics [9]. Therefore, postbiotics are considered effective because compared to probiotics, they retain similar benefits but without the side effects that can occur due to the presence of living microbes [2]. In particular, in the case of cosmetics, since live microorganisms cannot be used due to safety and stability concerns, postbiotics are used as cosmetic ingredients. Several studies have shown that postbiotics have a positive effect on skin function. Indeed, the topical application of postbiotics, such as *Lactobacillus*, *Bifidobacterium*, and *Vitrescilla filiformis* microbial lysates, may provide benefits to the skin [10–12]. The application of *Bifidobacterium longum sp.* extract, a probiotic lysate, improved skin sensitivity and increased skin resistance to chemical and physical aggression, whereas *V. filiformis* lysate promoted the restoration of a healthy skin microbiome by reducing *Staphylococcus aureus*'s colonization of the skin, and improved skin barrier function, thus improving transepidermal water loss (TEWL) [13,14]. Additionally, a probiotic-derived ingredient (CLS02021)—a postbiotic mixture of metabolites including enzymes, organic acids, and peptides resulting from the co-fermentation of three probiotic strains—has contributed to increasing the elasticity, moisture, pore-size reduction, and wrinkle depth of the skin. Similarly, heat-killed *Lactococcus chungagnensis* CAU 1447 has shown beneficial effects on wound healing by inducing the expression of wound-healing-promoting cytokines, chemokines, and growth factors [15,16]. This suggests that the topical application of postbiotics has beneficial effects on the skin; consequently, there is a growing market trend for the inclusion of postbiotics in cosmetic ingredients and/or products [17]. In fact, several postbiotics are already widely used in cosmetics [18].

The most commonly used probiotic strains for the development of postbiotics belong to the *Lactobacillus* and *Bifidobacterium* genera [16]. Postbiotics developed using *Lactobacilli* or *Bifidobacterium* have shown the potential for immunoregulation [19–21]. Postbiotics using *Lactobacilli* also have shown anti-senescence potential, and antioxidant, anti-inflammation, and anti-biofilm activity [22–25]. Most of the probiotics used in the development of postbiotic products in the cosmetic market also include *Lactobacillus*, *Bifidobacterium*, and *Saccharomyces*, and these products provide functions such as skin conditioning, skin smoothing, skin regeneration, anti-inflammation, anti-wrinkling, etc. [16]. Among these, strains belonging to *Lactobacilli* are the most commonly used [26]. *Lactobacilli* include lactic acid bacteria (LAB), which are non-spore-forming, Gam-positive, and microaerobic bacteria that produce lactic acid, the main product of carbohydrate fermentation. The use of LAB improves the regulation of the skin's immune system and maintains skin homeostasis [27–29]. *Lactobacilli* contribute to wound healing, resistance to infection by interfering with pathogens, and defense against inflammatory processes affecting the skin [30–33].

Maca (*Lepidium meyenii*) is a biennial herbaceous plant belonging to the Brassicaceae family that includes cabbage, cauliflower, and garden cress. Maca root is native to the Andes, has been cultivated for at least 2000 years, and grows at various altitudes ranging from 2800 to 5000 m above sea level [34]. It adapts well to harsh high-altitude conditions with low temperatures, strong ultraviolet (UV) rays, low oxygen levels, and varied climates [35]. Moreover, polysaccharides from maca exhibit high antioxidant activity [36,37]. In particular, macamide and macaene fractions isolated from maca have shown antioxidant activity when evaluated through DPPH (2,2-diphenyl-1-picrylhydrazyl) and ABST ((2,2′-azino-bis(3-ethylbenzothiazoline-6-sulfonic acid)) radical scavenging [38]. Maca effectively controls sexual dysfunction, memory enhancement, skin protection, and also has neuroprotective, antidepressant, and anticancer properties [34,39–41]. In addition, polysaccharides, or peptides, from maca regulate immunomodulatory mechanisms by enhancing the secretion of nitric oxide (NO), tumor necrosis factor-alpha (TNF-$\alpha$), and interleukin-6 (IL-6) in macrophages [42,43]. There are few studies on the effects of maca on the skin. It has been reported to prevent and improve skin damage caused by UV rays and promote high-altitude wound healing by regulating the immune system [44,45].

This study aimed to determine the possibility of developing cosmetic materials with skin-whitening and anti-inflammatory functions using fermented maca root extract, a postbiotic created by fermenting maca root with *Lactobacillus* bacterial strains.

## 2. Materials and Methods

### 2.1. Maca Root Fermented by Lactobacillus

Maca root powder (GUAYAPI, Paris, France) was dissolved in tertiary distilled water at concentrations of 2.5, 5, and 10% (*w*/*w*). The mixture was sterilized and extracted under high pressure and temperature in an autoclave (Universal Scientific, Madison, OH, USA) at 121 °C for 15 min. *Lactiplantibacillus plantarum subsp. plantarum* (previously *Lactobacillus plantarum*) KCTC 3108, *Lacticaseibacillus rhamnosus* (previously *Lactobacillus rhamnosus*) KCTC 5033, *Lacticaseibacillus casei* (previously *Lactobacillus casei*) KCTC 3109, and *Lactobacillus gasseri* KCTC 3143 cultured in MRS (deMan Rogosa Sharpe) broth (Difco Laboratories, New York, NY, USA) were washed with tertiary distilled water and inoculated in the mixture at a concentration of $1 \times 10^9$ CFU (Colony forming unit)/mL (5% *w/v*). These *Lactobacilli* were selected based on studies that suggested they would have a positive effect on skin condition. *Lactiplantibacillus plantarum subsp. plantarum* has a regulating effect on human skin health by acting against skin aging and improving skin microbiome, and *Lacticaseibacillus casei* improves skin barrier and reduces skin flakiness [46–48]. *Lacticaseibacillus rhamnosus* improves skin wound healing and reduces scar formation in mice, and lysates of *Lacticaseibacillus rhamnosus* improve skin barrier function in reconstructed human epidermis model [12,49]. Moreover, biosurfactants produced by *Lactobacillus gasseri* show antimicrobial properties [50]. Fermentation was performed in a shaking incubator (VISION SCIENTIFIC, Daejeon, Republic of Korea) for 72 h at 37 °C. After fermentation, the supernatant was filtered using Whatman® Grade 2 qualitative filter paper (Sigma-Aldrich, St. Louis, MO, USA) and a 0.2 µm-pore-size syringe filter (HYUNDAI MICRO, Seoul, Republic of Korea) and stored at −80 °C until use (Figure 1).

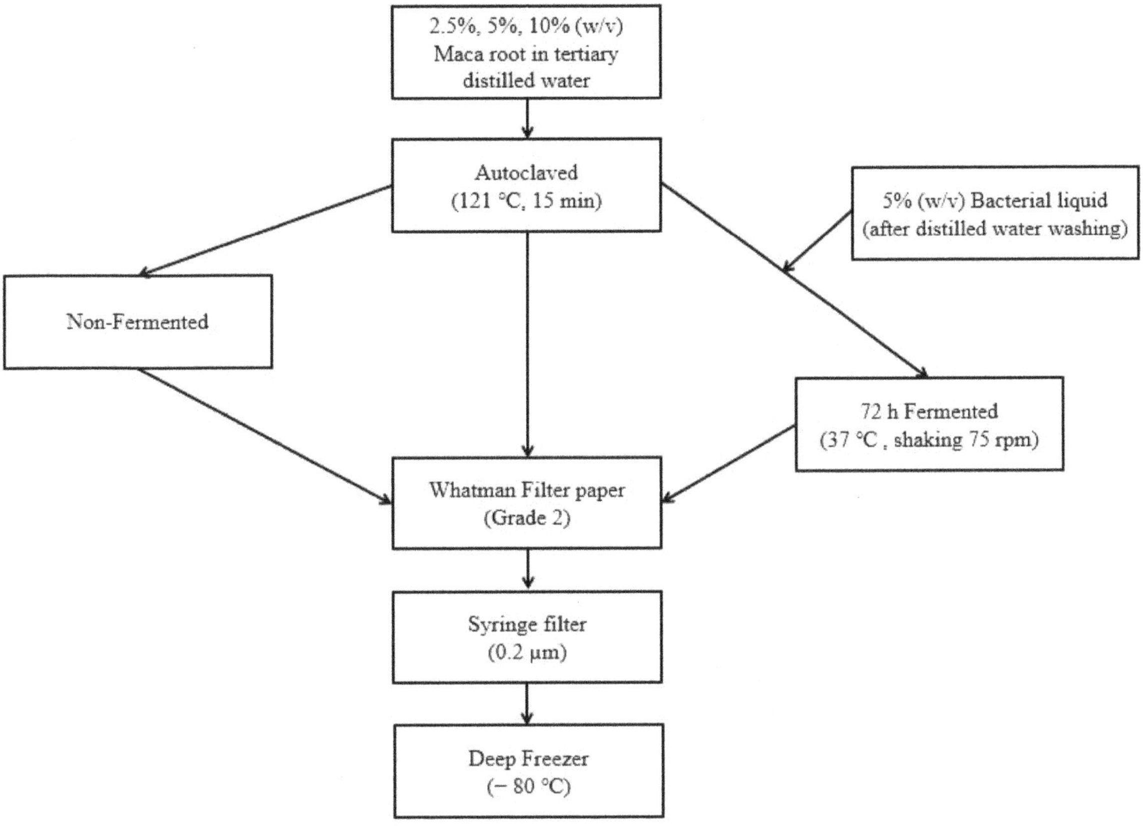


**Figure 1.** The preparation of non-fermented and fermented maca root extracts using lactic acid strains.

### 2.2. Cell Culture

B16F10 melanoma cells were purchased from American Type Culture Collection (ATCC, Manassas, VA, USA) and RAW 264.7 cell lines were purchased from the Korean Cell Line Bank (KCBL; Seoul, Republic of Korea). Cells were maintained in high-glucose D MEM (Dulbecco Modified Eagle Medium; Cytiva Life Sciences, Logan, MA, USA) containing 1% Anti-Anti (Antibiotic-Antimycotic, Gibco, NY, USA) and 10% fetal bovine serum (FBS). The cell lines were grown in a 37 °C $CO_2$ incubator (VISIONBIONEX, Bucheon, Republic of Korea) under a 5% $CO_2$ atmosphere.

The lactic acid strains used for fermentation, namely, *Lactiplantibacillus plantarum subsp. plantarum* KCTC 3108, *Lacticaseibacillus rhamnosus* KCTC 5033, *Lacticaseibacillus casei* KCTC 3109, and *Lactobacillus gasseri* KCTC 3143, were purchased from Korean Collection for Type Cultures (KCTC; Seoul, Republic of Korea). These strains were cultivated in Lactobacilli MRS (deMan Rogosa Sharpe) broth (Difco Laboratories, NY, USA) at 37 °C and 75 rpm, and then sub-cultured for 48 h.

### 2.3. Cell Viability

The cytotoxicity of fermented maca root extracts on B16F10 and RAW 264.7 cells was determined using CCK-8 (Cell Counting Kit-8) solution (Donginbio, Seoul, Republic of Korea). B16F10 cells were seeded at a concentration of $2 \times 10^3/100$ µL and RAW 264.7 cell lines were seeded at a concentration of $4 \times 10^3/100$ µL in a culture medium in 96-well plates (SPL, New York, NY, USA). After 24 h, non-fermented and fermented maca root extracts were diluted in the culture medium by 1/100 of the total volume and treated with 100 µL in each well. After 24 h and 48 h of treatment, 20 µL of CCK-8 solution was added and incubated for 2 h at 37 °C. The absorbance of the samples was measured at 450 nm using a microplate reader (Thermo Fisher, Waltham, MA, USA). The results were the averages of triplicate samples.

$$\text{Cell viability } (\%) = \frac{\text{Absorbance 450 nm of Sample}}{\text{Absorbance 450 nm of Control}} \times 100$$

### 2.4. NO Assay

RAW 264.7 cells were seeded at a concentration of $3 \times 10^5$ in a 6-well plate. After 24 h of seeding, 1 µg/mL lipopolysaccharide (LPS) from *Escherichia coli* (Sigma-Aldrich, St. Louis, MO, USA), or 1/100 of the total volume of non-fermented and fermented maca root extracts, was administered for treatment for 24 h. After 24 h, 100 µL of the supernatant was transferred to a 96-well plate, and 50 µL of N1 buffer from the NO Plus Detection Kit (iNtRON, Seongnam, Republic of Korea) was added and reacted at room temperature for 10 min. Then, N2 buffer was added, incubation was performed at room temperature for 10 min, and the absorbance was measured at 520 nm using a microplate reader (Thermo Fisher, Waltham, MA, USA). The amount of nitrate produced was calculated using a nitrate standard curve. The results were the averages of triplicate samples.

### 2.5. Mushroom Tyrosinase Inhibition Assay

The tyrosinase inhibition effect of the mushroom tyrosinase assay was determined based on a modified previous study [51]. Tyrosinase from mushrooms (Sigma-Aldrich, St. Louis, MO, USA) was dissolved in phosphate buffer (pH 6.6; Sigma-Aldrich, St. Louis, MO, USA) at a concentration of 7500 U/mL. L-tyrosine (Sigma-Aldrich, St. Louis, MO, USA) was used as a substrate and subsequently dissolved in 1 M HCl (hydrochloric acid) at a concentration of 15 mM. Next, 660 µL of 0.1 M sodium phosphate buffer (SPB; pH 6.8) was dispensed in a 1.5 mL Eppendorf tube and 60 µL samples were treated. Then, 90 µL of 7500 U/mL mushroom tyrosinase and 15 mM L-tyrosine were dispensed sequentially and incubated at 37 °C for 12 min. After incubation, the Eppendorf tubes were placed on ice for 1 min and 150 µL of each sample was placed in each well of a 96-well plate. Finally, absorbance was measured at 490 nm using a microplate reader (Thermo Fisher,

Waltham, MA, USA). In the control, tertiary diluted water was used instead of the sample, and arbutin (500 μg/mL; Sigma-Aldrich, St. Louis, MO, USA) was used as a positive control. The results were the averages of triplicate samples.

$$\text{Tyrosinase inhibition rate (\%)} = 100 \times (1 - \frac{B - B'}{A - A'})$$

A: Tertiary diluted water instead of sample;
A′: Without tyrosinase in condition A;
B: Treated sample;
B′: Without tyrosinase in condition B.

### 2.6. *Intracellular Melanin Content Assay*

Analysis of melanin content was carried out based on modified previous reports [52]. B16F10 cells were seeded at a density of $2 \times 10^5$ in a 60 mm culture dish (SPL, New York, NY, USA). After 24 h, 200 nM α-melanocyte stimulating hormone (α-MSH; Sigma-Aldrich, St. Louis, MO, USA) and 1/100 of the total volume of maca root extracts and fermented maca root extracts or arbutin (500 μg/mL) were administered for treatment for 48 h. The culture medium was removed. Next, the cells were removed using 1 mL PBS (phosphate buffered saline) and transferred to a 1.5 mL Eppendorf tube. The collected cells were centrifuged at 13,000 rpm for 10 min. After centrifugation, the supernatant was removed, and 650 μL of 1 N NaOH (Sigma-Aldrich, St. Louis, MO, USA) solution, in which 10% DMSO (Dimethyl sulfoxide; Sigma-Aldrich, St. Louis, MO, USA) was dissolved, was added to the pellet. The mixture was dissolved at 80 °C for 1 h. Then, the tubes were vortexed and 200 μL samples were transferred to a 96-well plate and the absorbance was measured at 405 nm using a microplate reader (Thermo Fisher, Waltham, MA, USA). Moreover, 500 μg/mL arbutin (Sigma-Aldrich, St. Louis, MO, USA) was used as a positive control. The results were the averages of triplicate samples.

### 2.7. *Extracellular Melanin Content Assay*

Extracellular melanin content was determined with reference to modified previous reports [53]. B16F10 cells were seeded at a density of $2 \times 10^5$ in a 60 mm culture dish (SPL, New York, NY, USA). After 24 h, the medium was removed. Then, 200 nM α-MSH and 1/100 of the total volume of maca root extracts and fermented maca root extracts or arbutin (500 μg/mL) were administered for treatment for 48 h using DMEM without phenol red (Cytiva Life Sciences, Logan, MA, USA). After 48 h of treatment, 200 μL of the medium was transferred to a 96-well plate. Absorbance was subsequently measured at 405 nm using a microplate reader (Thermo Fisher, Waltham, MA, USA). The results were the averages of triplicate samples.

### 2.8. *Intracellular Tyrosinase Activity*

Analysis of intracellular tyrosinase activity was carried out with reference to modified previous research [54]. B16F10 cells were seeded in 60 mm culture dishes at a concentration of $2 \times 10^5$. After 24 h, 200 nM α-MSH and 1/100 of the total volume of non-fermented and fermented maca root extracts were administered for treatment for 48 h. Cells were collected and centrifuged at 13,000 rpm for 10 min. The supernatant was removed. Then, pellets were suspended in 100 μL phosphate buffer (pH 6.6) containing 1% (*v*/*v*) Triton X-100 (Sigma-Aldrich, St. Louis, MO, USA) and shaken on ice for 40 min. Next, the cells were centrifuged at 13,000 rpm for 20 min and the supernatant was collected for quantification. Sodium phosphate buffer (0.1 M; pH 7) was used as a solvent, and 40 μg of protein was added to a total volume of 40 μL. A total of 40 μg of protein was incubated with 200 μL of 10 mM L-DOPA (L-3,4-dihydroxyphenylalanine) in the dark at 37 °C for 30 min. After the reaction, absorbance was measured at 475 nm using a microplate reader (Thermo Fisher,

Waltham, MA, USA). Arbutin (500 μg/mL) was used as a positive control. The results were the averages of triplicate samples.

$$\text{Intracellular tyrosinase activity } (\%) = \frac{\text{Absorbance of Sample at 475 nm}}{\text{Absorbance of Control at 475 nm}} \times 100$$

### 2.9. Quantitative Real-Time PCR

There are three essential enzymes for melanin synthesis: *tyrosinase (TYR), tyrosinase-related protein 1 (TRP-1),* and *tyrosinase related protein 2 (TRP-2)* [55]. These enzymes are transcriptionally regulated by *microphthalmia-associated transcription factor (MITF)*, a master gene for melanocyte differentiation [56]. Therefore, in this study, *MITF, tyrosinase* and *TRP-2* mRNA expression was determined. B16F10 RNA was extracted using TRIzol reagent (Thermo Fisher, Waltham, MA, USA). The extracted RNA was reverse-transcribed to synthesize cDNA using Superscript VILO (Invitrogen, Waltham, MA, USA). Next, PCR was performed using a 2× qPCRBIO SyGreen Mixture (PCRBIOSYSTEMS, Wayne, PA, USA) as well as forward and reverse primers. PCR conditions were as follows: after pre-incubation at 95 °C for 2 min, denaturation at 95 °C for 5 s and annealing at 60 °C for 30 s were repeated for 40 cycles, and extension was performed at 72 °C for 5 min. qRT-PCR was performed using a The LightCycler® 96 instrument (Roche Life Science, Pleasanton, CA, USA). GAPDH was used as the housekeeping gene. The results were the averages of duplicate samples. The primers used in the experiments are listed in Table 1.

**Table 1.** Primer sequences for RT-PCR.

| Gene Name | Sequence | Base Pairs |
|---|---|---|
| mGAPDH [57] | F: 5′—ACATCAAGAAGGTGGTGAAG—3′<br>R: 5′—ATTCAAGAGAGTAGGGAGGG—3′ | 392 bp |
| mMITF [58] | F: 5′—AGCGTGTATTTTCCCCACAG—3′<br>R: 5′—TAGCTCCTTAATGCGGTCGT—3′ | 124 bp |
| mTyrosinase | F: 5′—CAGGCTCCCATCTTCAGCAGAT—3′<br>R: 5′—ATCCCTGTGAGTGGACTGGCAA—3′ | 132 bp |
| mTRP-2 | F: 5′—GCAAGATTGCCTGTCTCTCCAG—3′<br>R: 5′—CTTGAGAGTCCAGTGTTCCGTC—3′ | 119 bp |

### 2.10. DPPH Antioxidant Assay

The antioxidant effects of non-fermented and fermented maca root extracts were determined using the OxiTec™ DPPH antioxidant assay kit (BIOMAX, Guri, Republic of Korea) according to the manufacturer's protocol. Non-fermented and fermented maca root extracts (2.5% and 5%, respectively) were prepared, and the assay buffer and DPPH working solution were individually added to a 96-well plate. The plate was incubated for 30 min at room temperature in the dark. Finally, absorbance was measured at 517 nm using a microplate reader. Ascorbic acid (500 μg/mL) was used as a positive control. The results were the averages of triplicate samples.

$$\text{Inhibition ratio of sample } (\%): \left(\frac{\text{Acs} - \text{As}}{\text{Acs}}\right) \times 100$$

$A_{cs}$: Blank 1—Blank 2;
$A_S$: Abs of Sample—Blank 2.

### 2.11. Determination of Total Phenolic Content

The total phenolic content (TPC) was determined according to the method described previously with slight modifications [59]. After mixing 2N Folin-Ciocalteu's phenol reagent (Sigma-Aldrich, St. Louis, MO, USA) and distilled water at a ratio of 1:2, 10 μL of the mixture and 10 μL of sample were mixed and left in the dark for 3 min. Then, 150 μL of 20% $Na_2CO_3$ (Sigma-Aldrich, St. Louis, MO, USA) was added and reacted in the dark for

1 h, and absorbance was measured at 750 nm using a microplate reader (Thermo Fisher, Waltham, MA, USA). The concentration of TPC in the samples was obtained by substituting the measured absorbance into a standard calibration curve prepared using gallic acid (Sigma-Aldrich, St. Louis, MO, USA) as a standard material. The results were the averages of triplicate samples.

### *2.12. Statistical Analysis*

All data were analyzed using the GraphPad Prism software (version 8.0). Statistical significance was assessed using unpaired two-tailed Student's *t*-tests. $p < 0.05$ was considered statistically significant (* $p < 0.05$, ** $p < 0.005$, *** $p < 0.0005$).

## 3. Results

### *3.1. Effects of Antioxidants of Non-Fermented Maca Root Extracts Maintained through Fermentation by Lactobacilli*

The antioxidant effects of 2.5% and 5% of the non-fermented and fermented maca root extracts are shown in Figure 2. Here, 2.5% and 5% concentrations of the non-fermented maca root extracts showed approximately 32% and 39% DPPH inhibition ratios, respectively. At a concentration of 2.5%, the maca root extracts fermented using *Lactiplantibacillus plantarum subsp. plantarum* KCTC 3108, *Lacticaseibacillus rhamnosus* KCTC 5033, *Lacticaseibacillus casei* KCTC 3109, and *Lactobacillus gasseri* KCTC 3143 showed DPPH inhibition ratios of approximately 36.9%, 40.6%, 36.1%, and 37.9%, respectively. Similarly, the DPPH inhibition ratios were approximately 43.9%, 44.6%, 40.2%, and 40.9%, respectively, at a concentration of 5%. This suggests that the fermentation of maca root extracts maintains the antioxidative effect of this plant.

### *3.2. Total Phenolic Content of Non-Fermented Maca Root Extracts Maintained through Fermentation by Lactobacilli*

It is well known that plant phenolic compounds form one of the major groups of compounds that act as major antioxidants or free radical scavengers, and whole plant polyphenols are effective as singlet-reducing agents, oxygen scavengers, and hydrogen atom donators [60]. Therefore, the total phenolic content (TPC) of the non-fermented and fermented maca root extracts were measured for each concentration. The content of phenolic compounds was calculated using calibration curve, expressed in gallic acid equivalents (GAE) as milligrams per 100 µL of the extracts (mg GAE/100 µL extracts). The TPC of the non-fermented maca root extracts was approximately 3.06, 4.91, and 9.19 mg GAE/100 µL extracts at concentration of 2.5, 5, and 10%, respectively (Table 2). The fermented maca root extracts using *Lactiplantibacillus plantarum subsp. plantarum* KCTC 3108, *Lacticaseibacillus rhamnosus* KCTC 5033, *Lacticaseibacillus casei* KCTC 3109, and *Lactobacillus gasseri* KCTC 3143 showed no significant difference in TPC compared to the non-fermented group at concentrations of 2.5, 5, and 10%.

**Table 2.** Total phenolic contents of the non-fermented and fermented maca root extracts.

| Treatment | Total Phenolic Content (mg GAE/100 µL Extracts) | | |
|---|---|---|---|
| | **2.5% of Maca Root** | **5% of Maca Root** | **10% of Maca Root** |
| Non-fermented | 3.06 ± 0.02 | 4.91 ± 0.04 | 9.19 ± 0.01 |
| *L. plantarum* KCTC 3108 | 3.01 ± 0.06 | 5.06 ± 0.06 | 8.93 ± 0.16 |
| *L. rhamnosus* KCTC 5033 | 3.14 ± 0.10 | 5.05 ± 0.09 | 8.95 ± 0.02 |
| *L. casei* KCTC 3109 | 2.94 ± 0.05 | 4.95 ± 0.08 | 8.97 ± 0.05 |
| *L. gasseri* KCTC 3143 | 3.1 ± 00.09 | 5.18 ± 0.06 | 8.91 ± 0.16 |

Data are the results of three independent experiments.

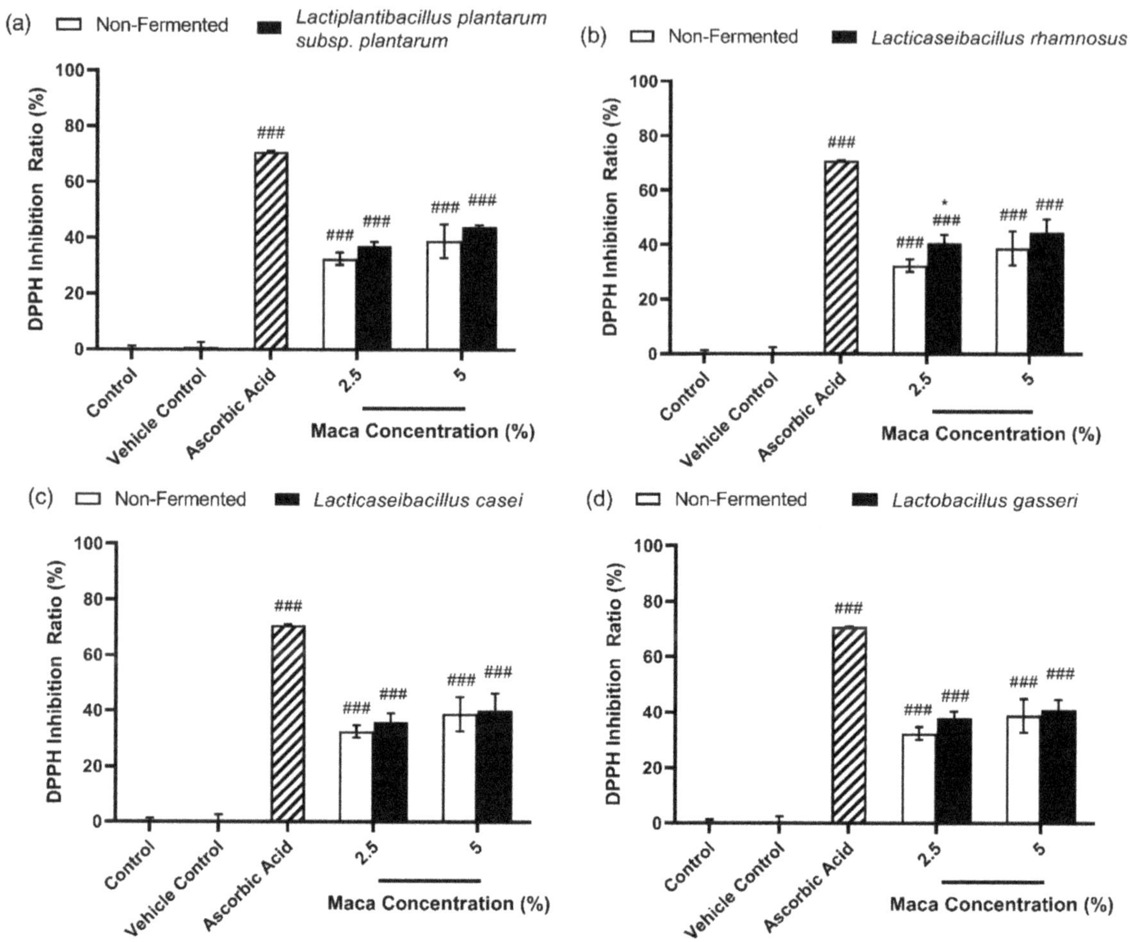


**Figure 2.** Effects of non-fermented and fermented maca root extracts on antioxidant activity evaluated using the DPPH assay. Treated with maca root extracts fermented using (**a**) *Lactiplantibacillus plantarum subsp. plantarum* KCTC 3108, (**b**) *Lacticaseibacillus rhamnosus* KCTC 5033, (**c**) *Lacticaseibacillus casei* KCTC 3109, and (**d**) *Lactobacillus gasseri* KCTC 3143. Error bars, mean ± SD. ### $p < 0.0005$ compared with the vehicle control group and * $p < 0.05$ compared to the same concentration of non-fermented maca root extracts.

### *3.3. Fermented Maca Root Extracts Suppresses NO Production Compared to Non-Fermented Maca Root Extracts*

Some maca root components may play an immunoregulatory role by releasing pro-inflammatory cytokines and producing NO [42,43]. However, these pro-inflammatory cytokines contribute to the exacerbation of allergic asthma [61]. In addition, NO is a highly reactive molecule/free radical that can have oxidative properties [62]. Therefore, the NO production must be suppressed to be developed as a cosmetic ingredient. In this study, the NO concentration was measured to determine whether the inflammatory response was suppressed during the maca root fermentation.

To confirm the concentration without cytotoxicity in RAW 264.7, cell viability was first measured for each concentration of the non-fermented and fermented maca root extracts using CCK-8 reagent. The non-fermented and fermented maca root extracts were treated with 1/100 of the total medium volume for 24 h and 48 h and compared with the control group, which was treated with tertiary diluted water. Figure 3 shows the viability percentages of RAW 264.7 cells after (a) 24 h and (b) 48 h of treatment. The non-fermented and fermented maca root extracts showed over 100% cell viability compared to the control. As a result, the treatment of the non-fermented and fermented maca root extracts did not show any cytotoxicity on RAW 264.7 cells for 24 h and 48 h. Moreover, those concentrations of non-fermented and fermented maca root extracts were used to measure the NO production in RAW 264.7 cells.

(a)

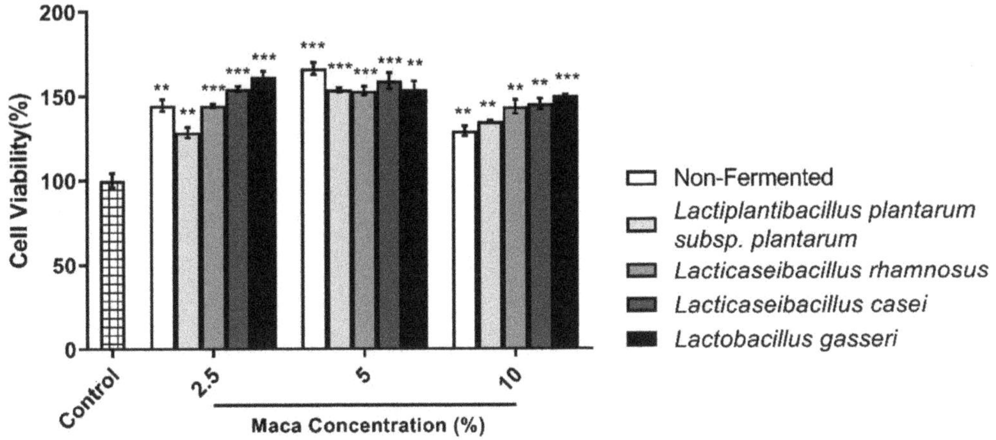


(b)

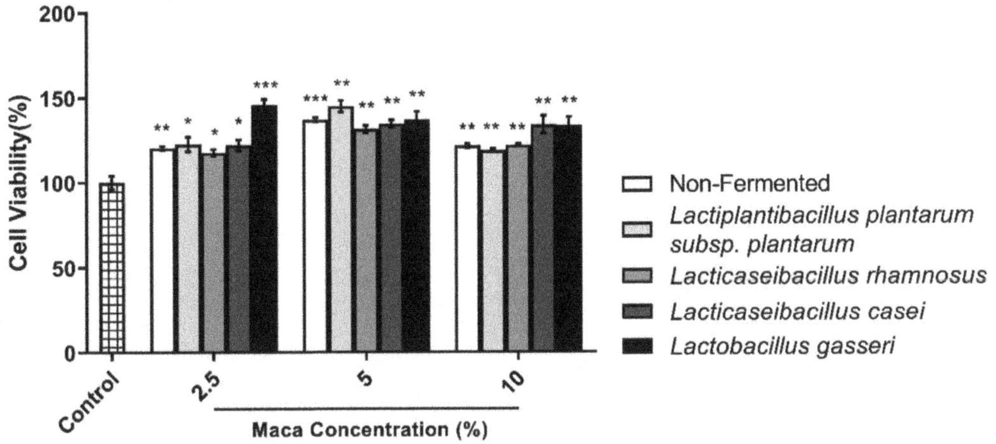


**Figure 3.** Cell viability of non-fermented and fermented maca root extracts on the RAW 264.7 cell line. Non-fermented and fermented maca root extracts' effects on the cell viability of RAW 264.7 cells were examined for (**a**) 24 h and (**b**) 48 h using CCK-8 reagent. The results shown are the averages of triplicate samples. Error bars, mean ± SD. * $p < 0.05$, ** $p < 0.005$, *** $p < 0.0005$ compared to the control group.

Figure 4 compares the NO production of RAW 264.7 cells when treated with the non-fermented and fermented maca root extracts. LPS (1 μg/mL) was used as a positive control. In this study, the non-fermented maca root extracts increased the NO production in a dose-dependent manner. At 5% and 10% concentrations, the fermented maca root extracts showed an extremely low NO production compared to the non-fermented maca root extracts. This indicates that the fermented maca root extract inhibits inflammation, whereas the non-fermented maca root extract does not.

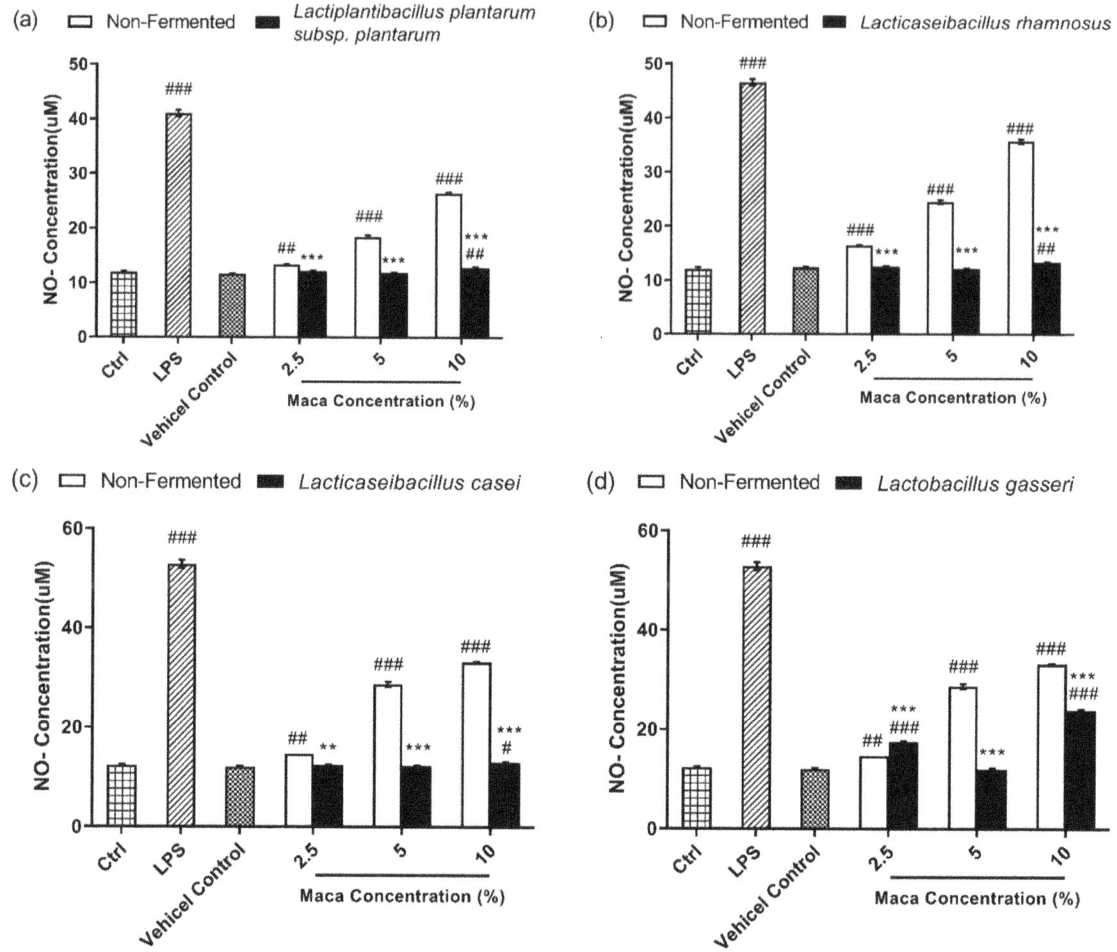


**Figure 4.** Effects of non-fermented and fermented maca root extracts on NO production in RAW 264.7 cells. NO concentrations in RAW 264.7 cells treated with maca root extracts fermented using (**a**) *Lactiplantibacillus plantarum subsp. plantarum* KCTC 3108, (**b**) *Lacticaseibacillus rhamnosus* KCTC 5033, (**c**) *Lacticaseibacillus casei* KCTC 3109, and (**d**) *Lactobacillus gasseri* KCTC 3143. Error bars, mean ± SD. # $p < 0.05$, ## $p < 0.005$, and ### $p < 0.0005$ compared to the control group, ** $p < 0.005$, *** $p < 0.0005$ compared to the same concentration of non-fermented maca root extracts.

### 3.4. Fermented Maca Root Extracts Inhibit Mushroom Tyrosinase Activity

To confirm the effect of non-fermented and fermented maca roots on mushroom tyrosinase, the inhibition rate of mushroom tyrosinase activity was investigated using L-tyrosine as a substrate. Here, 500 μg/mL arbutin was used as a positive control and tertiary diluted water was added to the control group instead of the sample. As shown in Figure 5, the 2.5%, 5%, and 10% non-fermented maca root extracts showed approximately 25%, 34%, and 47% tyrosinase inhibition, respectively. These results indicate that the non-fermented maca root increased the tyrosinase inhibition rate in a dose-dependent manner. Moreover, 500 μg/mL arbutin showed approximately 35% tyrosinase inhibition. *Lactiplantibacillus plantarum subsp. plantarum* KCTC 3108 showed 69.44%, 81.72%, and 86.59% tyrosinase inhibition, respectively, in a dose-dependent manner (Figure 5a). *Lacticaseibacillus rhamnosus* KCTC 5033 showed 39.18%, 54.12%, and 82.17% tyrosinase inhibition, respectively, in a dose-dependent manner (Figure 5b). As shown in Figure 5c, *Lacticaseibacillus casei* KCTC 3109 showed tyrosinase inhibition rates of 38.5%, 50.58%, and 69.08%, respectively. Moreover, *Lactobacillus gasseri* KCTC 3143 showed 54.83%, 63.92%, and 73.17% tyrosinase inhibition rates, respectively, in a dose-dependent manner (Figure 5d). In this study, the maca root extracts exhibited tyrosinase inhibition in a dose-dependent manner, with the fermented maca root extracts showing a higher tyrosinase inhibition rate than the non-fermented maca root extracts.

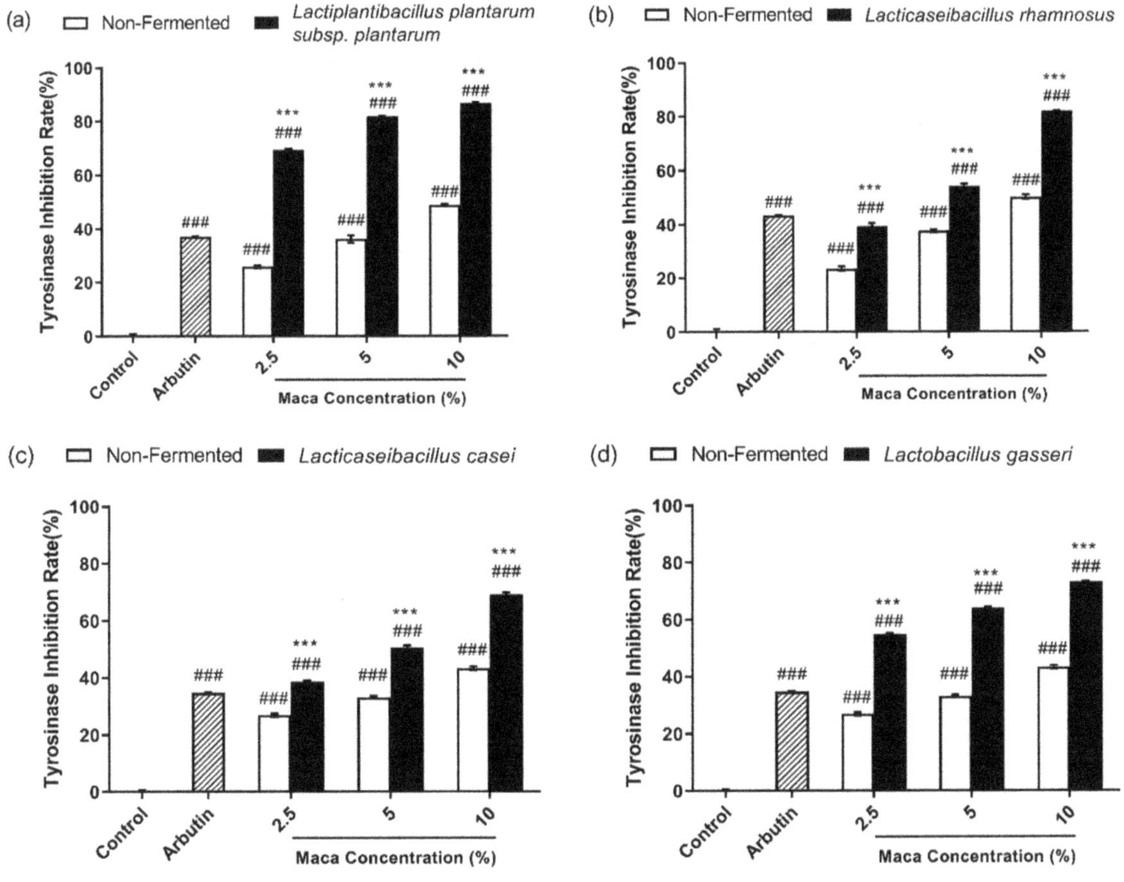


**Figure 5.** Effects of non-fermented and fermented maca root extracts on the inhibition rate (%) of mushroom tyrosinase. Treated with maca root extracts fermented using (**a**) *Lactiplantibacillus plantarum subsp. plantarum* KCTC 3108, (**b**) *Lacticaseibacillus rhamnosus* KCTC 5033, (**c**) *Lacticaseibacillus casei* KCTC 3109, and (**d**) *Lactobacillus gasseri* KCTC 3143. Error bars, mean ± SD. ### $p < 0.0005$ compared to the control group, and *** $p < 0.0005$ compared to the same concentration of non-fermented maca root extracts.

### *3.5. Fermented Maca Root Extracts Suppress Melanin Synthesis*

To determine whether the non-fermented and fermented maca root extracts is cytotoxic to B16F10 melanoma cells, cell viability was confirmed using the CCK-8 assay. The non-fermented and fermented maca root extracts were treated with 1/100 of the total volume of the media for 24 h and 48 h. They were subsequently compared to the control group that was treated with tertiary diluted water instead of the sample. Figure 6 shows the viability (%) of B16F10 cells after (a) 24 h and (b) 48 h of treatment, and all samples showed more than 95% viability compared with the control. The CCK-8 assay confirmed the safety of the non-fermented and fermented maca concentrations in B16F10 cells after 24 h and 48 h of treatment.

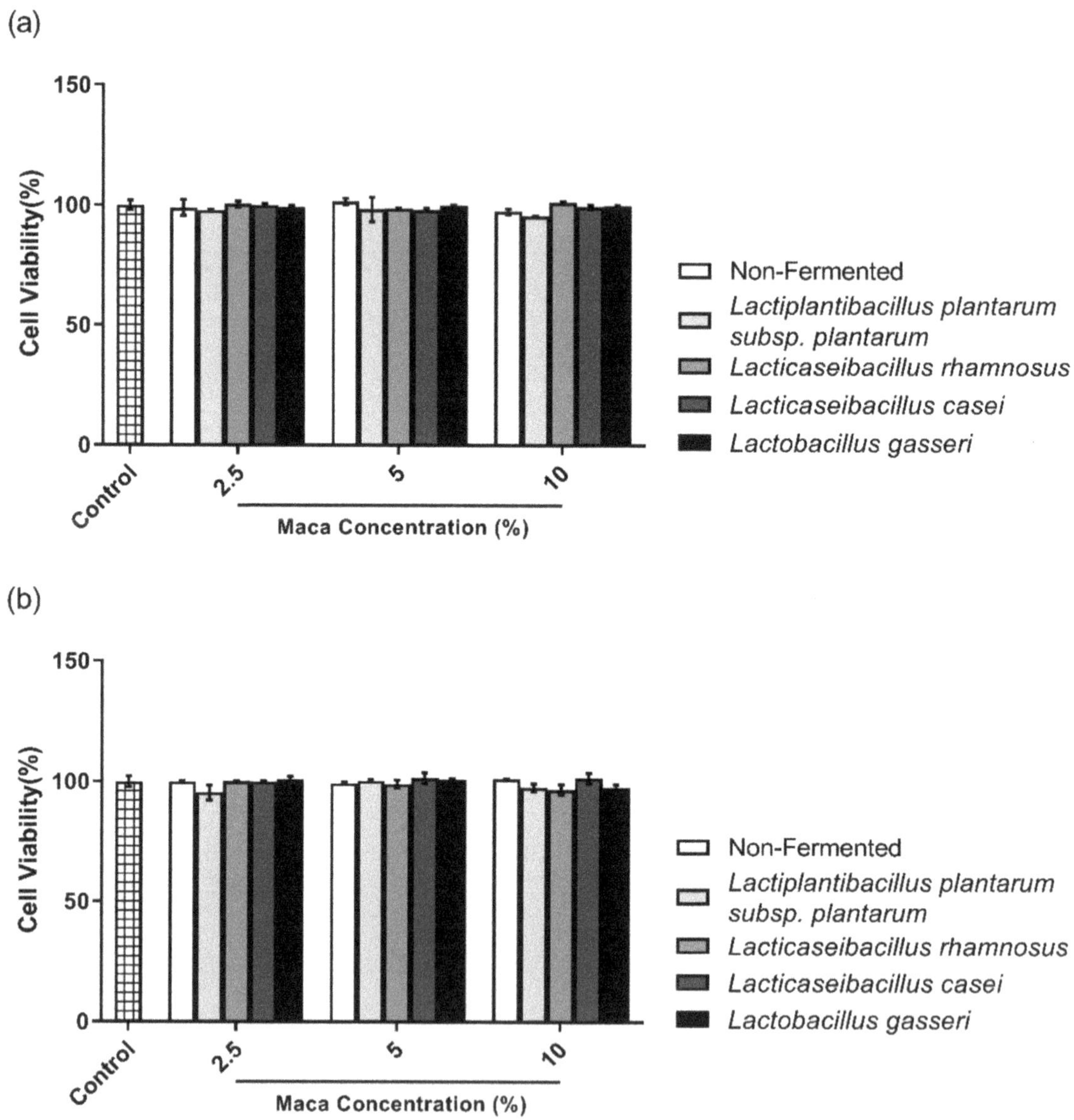


**Figure 6.** Cell viability of non-fermented and fermented maca root extracts on the B16F10 cell line. CCK-8 was used to confirm the non-fermented and fermented maca root extracts' effects on the cell viability of B16F10 when treated for (**a**) 24 h and (**b**) 48 h.

As the fermented maca root extracts showed a higher tyrosinase inhibition rate in an extracellular experiment (Figure 5), we examined whether these maca root extracts influenced melanin pigment production in B16F10 cells (Figure 7). In addition, the melanin content in the medium was confirmed using DMEM without phenol red (Figure 8). The intracellular melanin content, which was treated with the non-fermented maca root extracts in a dose-dependent manner, was reduced (Figure 7). The fermented maca root extract postbiotic treatment showed a lower intracellular melanin content than the non-fermented maca root extract treatment. Figure 8 shows the relative melanin content in DMEM without phenol red after 48 h of treatment. Both the non-fermented and the fermented maca root extracts caused low extracellular melanin content in a dose-dependent manner. The fermented maca root extracts caused a lower extracellular melanin content than the non-fermented maca root extracts. These results suggest that non-fermented and fermented maca reduce both the intracellular and extracellular melanin content. Furthermore, the fermented conditions show lower melanin content than the non-fermented conditions. This indicates that the non-fermented and fermented maca root extracts reduce melanogenesis, and that the fermented maca root extracts are more effective in their anti-melanogenesis effects than the non-fermented maca root extracts.

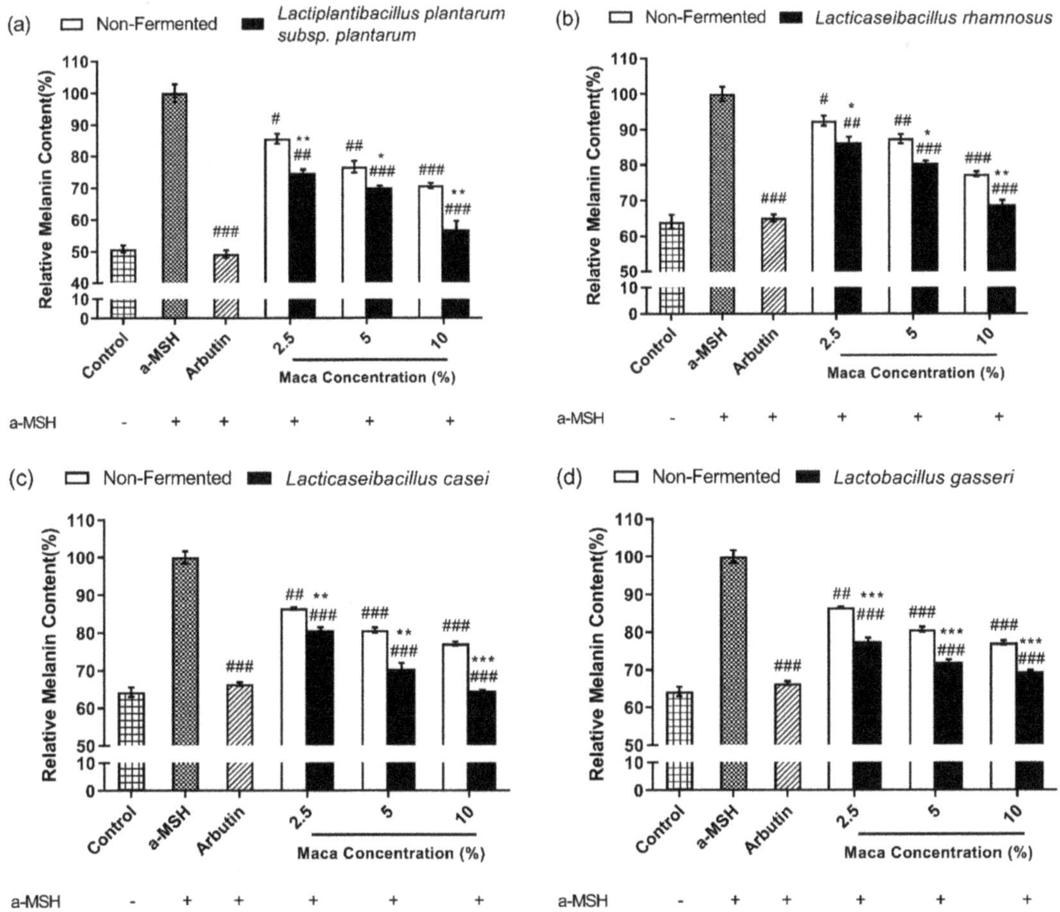


**Figure 7.** Effect of non-fermented and fermented maca root extracts on intracellular melanin content. Intracellular melanin content of B16F10 treated with maca root extracts fermented using (**a**) *Lactiplantibacillus plantarum subsp. plantarum* KCTC 3108, (**b**) *Lacticaseibacillus rhamnosus* KCTC 5033, (**c**) *Lacticaseibacillus casei* KCTC 3109, and (**d**) *Lactobacillus gasseri* KCTC 3143. Error bars, mean ± SD. # $p < 0.05$, ## $p < 0.005$ and ### $p < 0.0005$ compared to the α-MSH group, * $p < 0.05$, ** $p < 0.005$, and *** $p < 0.0005$ compared to the same concentration of non-fermented maca root extracts.

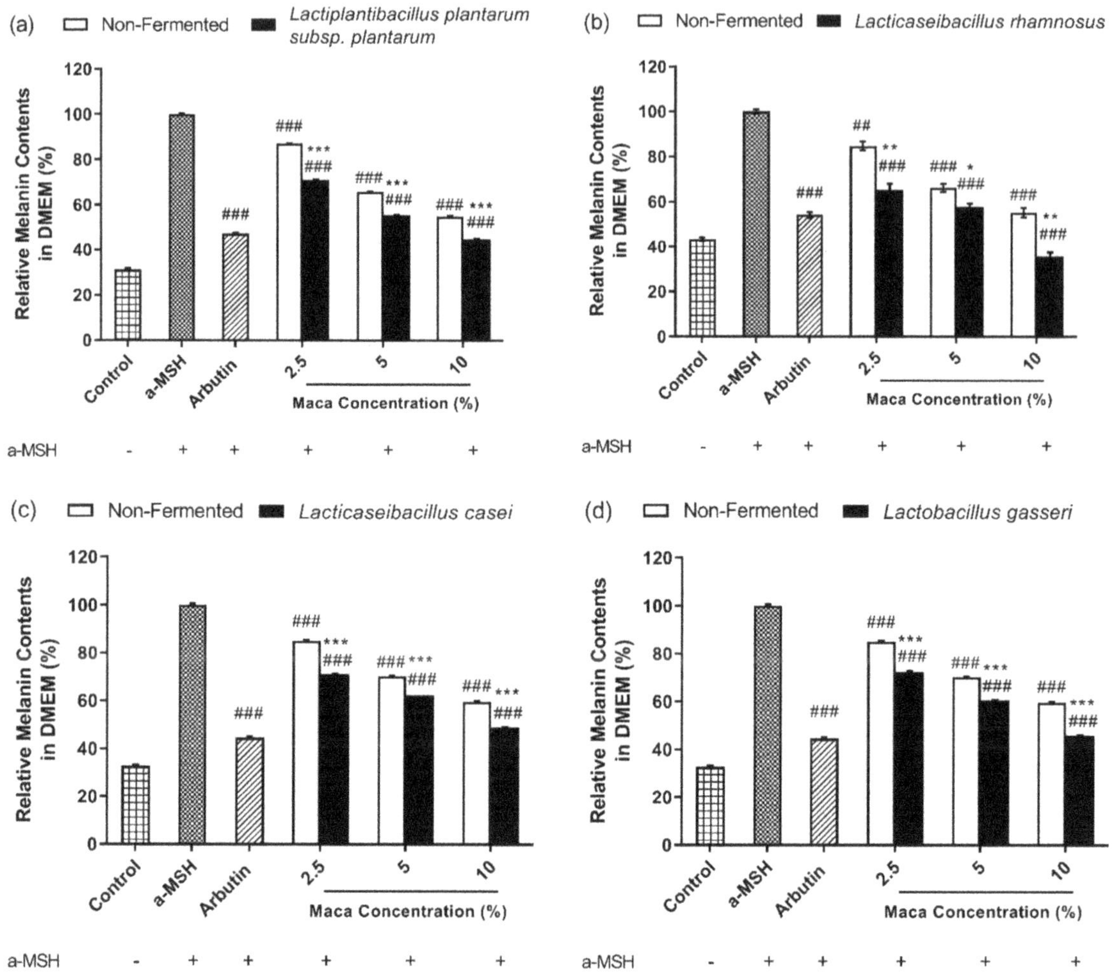


**Figure 8.** Effects of non-fermented and fermented maca root extracts on extracellular melanin content. Extracellular melanin content of B16F10 cells treated with maca root extracts fermented using (**a**) *Lactiplantibacillus plantarum subsp. plantarum* KCTC 3108, (**b**) *Lacticaseibacillus rhamnosus* KCTC 5033, (**c**) *Lacticaseibacillus casei* KCTC 3109, and (**d**) *Lactobacillus gasseri* KCTC 3143. Error bars, mean ± SD. ## $p < 0.005$ and ### $p < 0.0005$ compared to α-MSH group, * $p < 0.05$, ** $p < 0.005$, and *** $p < 0.0005$ compared to the same concentration of non-fermented maca root extracts.

### *3.6. Fermented Maca Root Extracts Inhibit Intracellular Tyrosinase Activity*

Melanin pigments are derived from L-tyrosine, and melanin-producing enzymes, such as *tyrosinase (TYR)*, *tyrosinase-related protein 1(TRP1)* and *tyrosinase-relate d protein 2 (TRP2)*, are involved in melanogenesis. Tyrosinase is a copper-dependent enzyme that catalyzes the conversion of L-tyrosine to L-DOPA, which is a rate-limiting step in melanin synthesis [63]. As shown in Figures 5, 7 and 8, the non-fermented and fermented maca root extracts inhibited melanogenesis. Therefore, the inhibitory effect of the non-fermented and fermented maca root extracts on tyrosinase enzyme activity was investigated under intracellular conditions. Arbutin (500 μg/mL) was used as the positive control. The non-fermented and fermented maca root extracts inhibited tyrosinase activity under intracellular conditions in a dose-dependent manner (Figure 9). Furthermore, the fermented conditions were more effective at inhibiting intracellular tyrosinase than the non-fermented conditions.

This shows that non-fermented and fermented maca root extracts affect tyrosinase activity and reduce melanin production, indicating that both types of maca root extract are possible anti-melanogenesis agents.

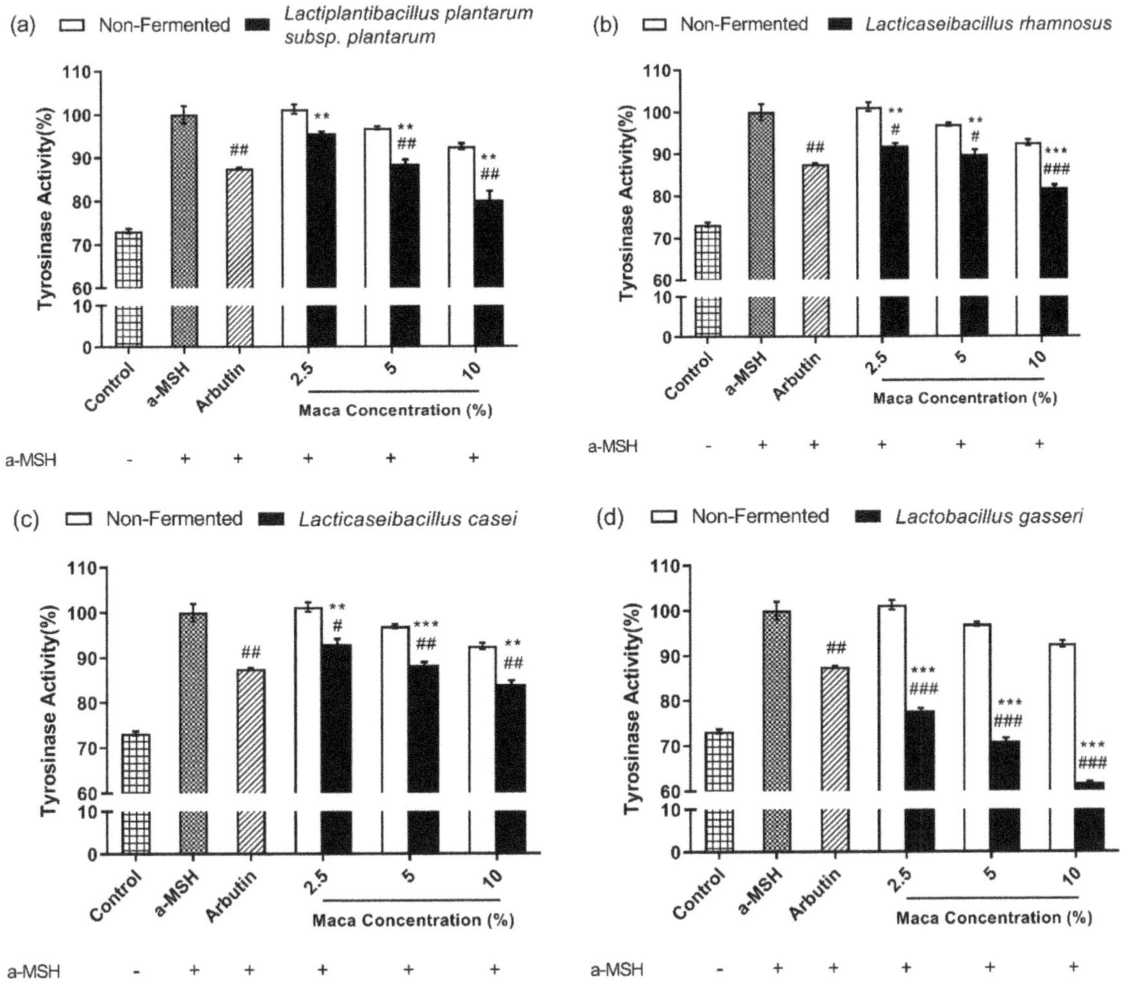


**Figure 9.** Effects of non-fermented and fermented maca root extracts on intracellular tyrosinase activity. Intracellular tyrosinase activity of B16F10 cells treated with maca root extracts fermented using (**a**) *Lactiplantibacillus plantarum subsp. plantarum* KCTC 3108, (**b**) *Lacticaseibacillus rhamnosus* KCTC 5033, (**c**) *Lacticaseibacillus casei* KCTC 3109, and (**d**) *Lactobacillus gasseri* KCTC 3143. Error bars, mean ± SD. # $p < 0.05$, ## $p < 0.005$, and ### $p < 0.0005$ compared to the α-MSH group. ** $p < 0.005$ and *** $p < 0.0005$ compared to the same concentration of the non-fermented maca root extract group.

### 3.7. Fermented Maca Root Extracts Suppressed mRNA expression of MITF, Tyrosinase and TRP-2 in B16F10

The expression of the MITF, tyrosinase, and TRP-2 regulates melanogenesis by regulating the enzymatic cascade [64]. In this study, the fermented maca root extracts showed higher anti-melanogenic effects than the non-fermented maca root extracts (Figures 5 and 7–9). To determine whether the fermented maca root extracts showed anti-melanogenic effects by regulating the expression of genes involved in melanogenesis, the mRNA expression of the *MITF*, *tyrosinase*, and *TRP-2* in B16F10 cells treated with the fermented maca root extracts was confirmed. As shown in Figure 10, a 10% concentration of fermented maca root extracts considerably reduced the *MITF, tyrosinase*, and *TRP-2* mRNA expression compared to the same concentration of non-fermented maca root extracts. These results indicate that the fermented maca root extracts suppress melanin synthesis by inhibiting the expression of these three genes, potentially resulting in whitened skin.

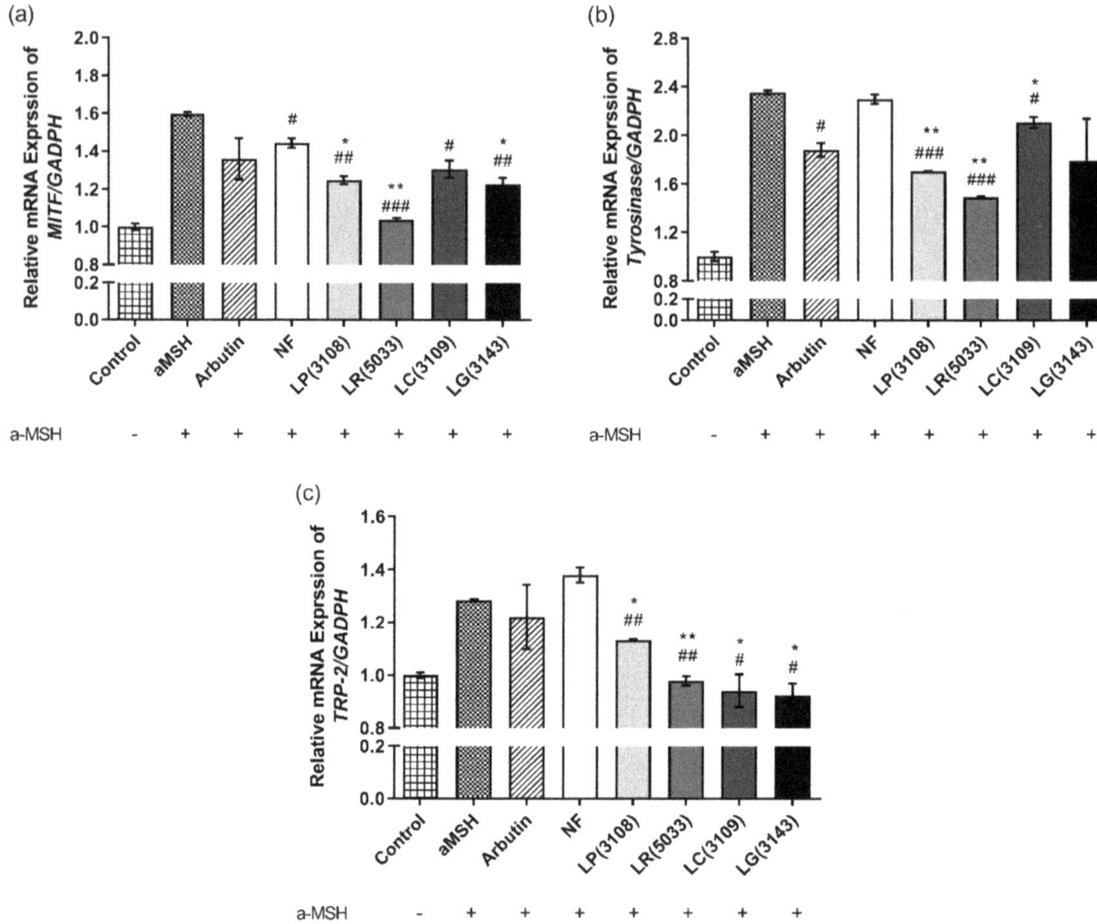


**Figure 10.** Effects of non-fermented and fermented maca root extracts on the relative mRNA expression of (**a**) MITF, (**b**) tyrosinase, and (**c**) TRP-2 in B16F10. Error bars, mean $\pm$ SD. # $p < 0.05$, ## $p < 0.005$, and ### $p < 0.0005$ compared to the $\alpha$-MSH group. * $p < 0.05$ and ** $p < 0.005$ compared to the non-fermented maca root extract group. NF, non-fermented; LP(3108), *Lactiplantibacillus plantarum subsp. plantarum* KCTC 3108; LR(5033), *Lacticaseibacillus rhamnosus* KCTC 5033; LC(3109), *Lacticaseibacillus casei* KCTC 3109; LG(3143), *Lactobacillus gasseri* KCTC 3143.

## 4. Discussion

Maca root (*Lepidium meyenii*) is well known to have several physiologically active properties, including antioxidant, immunomodulatory, and anticancer effects, but there have been fewer studies on the effect of maca root on the skin [40]. However, maca root contains various polysaccharides, amino acids, phenolics, and minerals, implying that it is suitable as a raw material for cosmetic products [65]. Therefore, in this study, we tried to verify the skin-whitening effect of maca root. Especially, fermented plant extracts have better antioxidant, anti-melanogenic, anti-inflammatory, and anti-wrinkle effects compared with non-fermented plant extracts [66–69]. Fermented red ginseng showed whitening and anti-wrinkle effects [70], whereas fermented *Fructus acrtii* extracts suppressed the expression of MMP-1 in human fibroblasts (HDFs), and fermented hydroponic ginseng showed improved antioxidant, anti-inflammatory, and anti-melanin production [67]. Furthermore, fermented unpolished black rice (*Oryza sativa* L.) and fermented Viola mandshurica inhibited melanogenesis in B16F10 [71,72], while fermented soybean extracts showed higher antioxidant and skin-whitening effects than those in a non-fermented condition [73]. Furthermore, fermented black ginseng showed anti-wrinkle effects on HDF [74]. This suggests that fermented plant extracts are more effective cosmetic ingredients than their non-fermented counterparts. Based on these studies, we fermented maca root using *Lactobacilli* to investigate its effects on inflammation and melanogenesis.

*Lactobacilli* cause changes in components by breaking down macromolecules or amino acids during fermentation to produce several metabolites, such as bacteriocins, vitamins, and exopolysaccharides [75]. In particular, among the component changes caused by plant fermentation, changes in phenolic compounds mainly cause changes in antioxidant, anti-inflammatory, or anti-melanogenic activity [76]. Phenolic compounds are found in various fruit and vegetables and may provide health benefits by reducing the risk of developing metabolic disorders [77]. The biological properties of phenolic compounds are diverse and include antioxidant, anti-inflammatory, or properties that inhibit enzymes involved in the treatment of human diseases such as hypertension, type 2 diabetes, skin hyperpigmentation, or Alzheimer's disease [78]. Fermentation can cause either an increase or a decrease in phenolic compounds. For example, the fermentation of Cactus Cladodes with lactic acid produced flavonoid derivatives (isorhamnetin and kaempferol), resulting in the production of antioxidant and anti-inflammatory properties [79]. Additionally, compared to non-fermented hydroponic ginseng (HPG), the total phenolic and flavonoid content of HPG fermented by Leuconostoc mesenteroides increased, showing antioxidant activity and anti-inflammatory effects by inhibiting the NO production and the expression of inflammatory cytokines, such as tumor necrosis factor-$\alpha$, interleukin (IL)-1$\beta$, and IL-6 [80]. In addition, the increase in polyphenols, flavonoids, and polysaccharides in fermented vine tea using saccharomyces cerevisiae showed a strong scavenging ability for reactive oxygen species (ROS). Furthermore, it inhibited the intracellular tyrosinase activity and melanin synthesis in B16F10 and reduced inflammatory cell recruitment in a zebrafish [81]. Additionally, a further study has shown that cocoa seeds fermented with both yeast and LAB decreased in total polyphenol content (TPC) and antioxidant power [82]. These studies suggest that the increase or decrease in phenolic compounds during fermentation affects their antioxidant, anti-inflammatory, and anti-melanogenic effects. It is well known that maca root also contains phenolic compounds such as flavonoid, epigallocatechin gallate (EGCG), and epicatechin [65], and based on the above studies, we investigated the TPC of maca root extracts before and after fermentation and there was no significant difference in total polyphenol contents between the non-fermented and fermented maca root extracts at concentrations of 2.5, 5, and 10% (Table 2). Since these results were based on the total phenolic compound, further studies are needed to determine which phenolic-component changes during fermentation affect the anti-inflammatory and anti-melanin effects of fermented maca root extracts.

In addition to phenolic compounds, maca root is composed of several components such as amino acids, polysaccharides, and secondary metabolites [65]. Leucine, arginine, and glutamic acid account for a high proportion of the amino acids in maca root, and MC-1 and MC-2 are well known polysaccharides isolated from maca root [83,84]. In addition, macaenes and macamides are well-known secondary metabolites of maca root that exhibit antioxidant activity [38]. Numerous amino acids participate directly or indirectly in melanogenesis, and it is known that leucine and glutamic acid affect anti-melanogenic effects [85]. Leucine is important for tyrosinase inhibitory activity and glutamic acid is one of the constituent amino acids of glutathione which inhibits tyrosinase enzyme and converts eumelanin to pheomelanin production, which has a skin-whitening effect [85,86]. Additionally, arginine is known as a precursor to the synthesis of NO, an inflammatory mediator [87], and MC-1 and MC-2 exhibited immunomodulatory activity increasing the levels of NO, TNF-a and IL-6 in RAW 264.7 cells [42,88]. Figures 7 and 9 show that the non-fermented and fermented maca root extracts dose dependently inhibited melanin production and intracellular tyrosinase activity in B16F10 cells. In addition, as shown in Figure 4, the non-fermented maca root extracts increased NO secretion in macrophages in a dose-dependent manner, and the fermented maca root extracts inhibited NO secretion compared to the non-fermented group at the same concentration. These results suggested that the fermentation using Lactobacilli increased their anti-melanogenic and anti-inflammatory effects. Based on the studies showing that Lactobacilli can use and produce amino acids [75,89,90] and polysaccharides [91,92], and the fact that maca root consists of various components, our results imply that during fermentation, Lactobacilli may change the components contained in maca root, causing a reduction in NO in macrophages and melanin in melanocytes. Therefore, further studies are needed to study which components of maca root extract change through fermentation.

Postbiotics can be obtained specifically through a fermentation process, meaning they contain many different types of metabolites. These metabolites confer various bioactive properties to postbiotics. Most metabolites with a physiological activity are produced by LAB and show various health benefit activities, such as antioxidant, anti-inflammatory, immunomodulatory, and anti-microbial activities [19–21,23,25]. Additionally, compared to probiotics, postbiotics show similar beneficial effects and higher stability and safety, meaning postbiotics are highly sustainable ingredients in cosmetic products [17]. In particular, previous studies have clinically proven that postbiotics can improve skin health. For instance, the oral intake of heat-killed cells of *Lactococcus lactis strain H61* improves women's skin health [93]. In clinical trials, postbiotics of TAC/collagen have been shown to improve the skin's moisture score and inflammatory index in vivo and accelerate the wound healing of acne symptoms in acne-prone volunteers. Additionally, this TAC/collagen gel reduced the number of brown spots and porphyrins on the facial skin [94]. Topical application of an anti-acne lotion containing a fermented lysate produced by *Lactiplantibacillus plantarum* has been shown to improve moisturization and regulate skin pH, with positive results in the treatment of acne [95]. Additionally, various products using postbiotics as ingredients exist in the cosmetic market [17]. Therefore, fermented maca root extract is highly likely to be applied as a cosmetic ingredient, and additional safety and clinical-efficacy tests are required.

The skin protects itself from solar irradiation through melanogenesis. When the skin is exposed to UV light, keratinocytes increase the transcription of the pro-opiomelanocortin gene in response to cellular damage. Pro-opiomelanocortin is a precursor protein that is cleaved to form α-MSH [96]. α-MSH binds to MC1R located on the surface of melanocytes as a factor that regulates melanocytes and skin pigmentation. The binding activates several pathways. The PKA-CREB and mitogen-activated protein kinase (MAPK) pathways are primarily involved in melanin production [63,97]. In melanogenesis associated with PKA-CREB signaling, α-MSH binding elevates intracellular adenylate cyclase, which increases cAMP concentration and PKA cell-signaling. Moreover, PKA phosphorylates the cAMP response element (CREB), which acts as a transcription factor for several genes, including the *MITF* [98,99]. The MAPK family comprises three types of protein kinases: extracellular reactive kinases (ERKs), c-Jun N-terminal kinases (JNKs), and p38 MAPKs [97]. The JUN MAPK and p38 MAPK pathways can stimulate melanogenesis by upregulating the *MITF* expression, and the ERK MAPK-dependent *MITF* expression pathway is involved in melanogenesis [100,101]. *Tyrosinase, TRP-1*, and *TRP-2* are the three main enzymes required for melanogenesis. The minor-eye-associated transcription factor (MITF), the master gene for melanocyte differentiation, regulates these three enzymes [102]. B16F10 cells treated with both fermented maca root extracts and α-MSH showed a suppressed *MITF, tyrosinase*, and *TRP-2* mRNA expression compared to those treated with only α-MSH, or both α-MSH and the non-fermented maca root extracts. These results show that the fermented maca root extracts improved anti-melanogenesis by inhibiting the expression of the *MITF*, *tyrosinase*, and *TRP-2*. Therefore, maca root extracts fermented using *Lactobacillus* have the potential to be used as a raw material in the production of cosmetic products. However, further research is needed to determine which ingredients and pathways exhibit the beneficial effects of these fermented plant extracts.

## 5. Conclusions

This study identified the antioxidative, anti-inflammatory, and anti-melanogenic effects of fermented maca root extract using *Lactobacilli*. Fermented maca root extracts maintained the antioxidant ability of the non-fermented maca root extracts and considerably reduced the production of NO, thereby demonstrating anti-inflammatory effects. In addition, fermented maca root extracts inhibited melanogenesis by inhibiting the expression and tyrosinase activity of the *MITF*, *tyrosinase*, and *TRP-2* mRNA. This shows that fermentation of maca root using *Lactobacillus* maintained its antioxidant effects and improved its anti-inflammatory and anti-melanogenic effects compared to the non-fermented maca root extracts. Postbiotics are currently in the limelight in the cosmetics industry for their safety, stability, and functional properties, and clinical trial research results on postbiotics as cosmetic ingredients are accumulating. Based on the results for fermented maca root extracts used in this study, clinical trials and cosmetic-product development using these postbiotics can be expected. Furthermore, the application of this postbiotic in minimally invasive procedures can be expected.

**Author Contributions:** Conceptualization, K.E.K.; methodology, J.Y. and H.C.; validation, J.Y.; formal analysis, K.E.K., M.G., J.Y. and H.C.; investigation, J.Y. and H.C.; data curation, K.E.K. and M.G.; writing—original draft preparation, J.Y.; writing—review and editing, K.E.K. and M.G.; visualization, J.Y.; supervision, K.E.K.; project administration, K.E.K.; funding acquisition, K.E.K. All authors have read and agreed to the published version of the manuscript.

**Funding:** This research was funded by the National Research Foundation of Korea (NRF), grant-funded by the Korea government (Ministry of Science and ICT); grant number NRF-2020R1A2C1101129.

**Institutional Review Board Statement:** Not applicable.

**Informed Consent Statement:** Not applicable.

**Data Availability Statement:** Data are contained within the article.

**Conflicts of Interest:** The authors declare no conflict of interest.

## References

1. Wegh, C.A.M.; Geerlings, S.Y.; Knol, J.; Roeselers, G.; Belzer, C. Postbiotics and Their Potential Applications in Early Life Nutrition and Beyond. *Int. J. Mol. Sci.* **2019**, *20*, 4673. [CrossRef] [PubMed]
2. Zolkiewicz, J.; Marzec, A.; Ruszczynski, M.; Feleszko, W. Postbiotics-A Step Beyond Pre- and Probiotics. *Nutrients* **2020**, *12*, 2189. [CrossRef] [PubMed]
3. Scarpellini, E.; Rinninella, E.; Basilico, M.; Colomier, E.; Rasetti, C.; Larussa, T.; Santori, P.; Abenavoli, L. From Pre- and Probiotics to Post-Biotics: A Narrative Review. *Int. J. Environ. Res. Public Health* **2021**, *19*, 37. [CrossRef] [PubMed]
4. Khan, S.; Moore, R.J.; Stanley, D.; Chousalkar, K.K. The Gut Microbiota of Laying Hens and Its Manipulation with Prebiotics and Probiotics To Enhance Gut Health and Food Safety. *Appl. Environ. Microbiol.* **2020**, *86*, e00600-20. [CrossRef]
5. Kim, S.K.; Guevarra, R.B.; Kim, Y.T.; Kwon, J.; Kim, H.; Cho, J.H.; Kim, H.B.; Lee, J.H. Role of Probiotics in Human Gut Microbiome-Associated Diseases. *J. Microbiol. Biotechnol.* **2019**, *29*, 1335–1340. [CrossRef]
6. Li, H.Y.; Zhou, D.D.; Gan, R.Y.; Huang, S.Y.; Zhao, C.N.; Shang, A.; Xu, X.Y.; Li, H.B. Effects and Mechanisms of Probiotics, Prebiotics, Synbiotics, and Postbiotics on Metabolic Diseases Targeting Gut Microbiota: A Narrative Review. *Nutrients* **2021**, *13*, 3211. [CrossRef]
7. Scott, E.; De Paepe, K.; Van de Wiele, T. Postbiotics and Their Health Modulatory Biomolecules. *Biomolecules* **2022**, *12*, 1640. [CrossRef]
8. Hernandez-Granados, M.J.; Franco-Robles, E. Postbiotics in human health: Possible new functional ingredients? *Food Res. Int.* **2020**, *137*, 109660. [CrossRef]
9. Vinderola, G.; Sanders, M.E.; Salminen, S.; Szajewska, H. Postbiotics: The concept and their use in healthy populations. *Front. Nutr.* **2022**, *9*, 1002213. [CrossRef]

10. Puebla-Barragan, S.; Reid, G. Probiotics in Cosmetic and Personal Care Products: Trends and Challenges. *Molecules* **2021**, *26*, 1249. [CrossRef]
11. Gueniche, A.; Liboutet, M.; Cheilian, S.; Fagot, D.; Juchaux, F.; Breton, L. *Vitreoscilla filiformis* Extract for Topical Skin Care: A Review. *Front. Cell. Infect. Microbiol.* **2021**, *11*, 747663. [CrossRef]
12. Jung, Y.O.; Jeong, H.; Cho, Y.; Lee, E.O.; Jang, H.W.; Kim, J.; Nam, K.; Lim, K.M. Lysates of a Probiotic, *Lactobacillus rhamnosus*, Can Improve Skin Barrier Function in a Reconstructed Human Epidermis Model. *Int. J. Mol. Sci.* **2019**, *20*, 4289. [CrossRef]
13. Gueniche, A.; Bastien, P.; Ovigne, J.M.; Kermici, M.; Courchay, G.; Chevalier, V.; Breton, L.; Castiel-Higounenc, I. Bifidobacterium longum lysate, a new ingredient for reactive skin. *Exp. Dermatol.* **2010**, *19*, e1–e8. [CrossRef]
14. Gueniche, A.; Knaudt, B.; Schuck, E.; Volz, T.; Bastien, P.; Martin, R.; Rocken, M.; Breton, L.; Biedermann, T. Effects of nonpathogenic gram-negative bacterium *Vitreoscilla filiformis* lysate on atopic dermatitis: A prospective, randomized, double-blind, placebo-controlled clinical study. *Br. J. Dermatol.* **2008**, *159*, 1357–1363. [CrossRef] [PubMed]
15. Catic, T.; Pehlivanovic, B.; Pljakic, N.; Balicevac, A. The Moisturizing Efficacy of a Proprietary Dermo-Cosmetic Product (CLS02021) Versus Placebo in a 4-week Application Period. *Med. Arch.* **2022**, *76*, 108–114. [CrossRef]
16. Nam, Y.; Kim, J.; Baek, J.; Kim, W. Improvement of Cutaneous Wound Healing via Topical Application of Heat-Killed Lactococcus chungangensis CAU 1447 on Diabetic Mice. *Nutrients* **2021**, *13*, 2666. [CrossRef]
17. Duarte, M.; Oliveira, A.L.; Oliveira, C.; Pintado, M.; Amaro, A.; Madureira, A.R. Current postbiotics in the cosmetic market-an update and development opportunities. *Appl. Microbiol. Biotechnol.* **2022**, *106*, 5879–5891. [CrossRef]
18. Gueniche, A.; Perin, O.; Bouslimani, A.; Landemaine, L.; Misra, N.; Cupferman, S.; Aguilar, L.; Clavaud, C.; Chopra, T.; Khodr, A. Advances in Microbiome-Derived Solutions and Methodologies Are Founding a New Era in Skin Health and Care. *Pathogens* **2022**, *11*, 121. [CrossRef]
19. Tejada-Simon, M.V.; Pestka, J.J. Proinflammatory cytokine and nitric oxide induction in murine macrophages by cell wall and cytoplasmic extracts of lactic acid bacteria. *J. Food Prot.* **1999**, *62*, 1435–1444. [CrossRef]
20. Matsuguchi, T.; Takagi, A.; Matsuzaki, T.; Nagaoka, M.; Ishikawa, K.; Yokokura, T.; Yoshikai, Y. Lipoteichoic acids from *Lactobacillus* strains elicit strong tumor necrosis factor alpha-inducing activities in macrophages through Toll-like receptor 2. *Clin. Diagn. Lab. Immunol.* **2003**, *10*, 259–266. [CrossRef]
21. Kim, H.G.; Lee, S.Y.; Kim, N.R.; Lee, H.Y.; Ko, M.Y.; Jung, B.J.; Kim, C.M.; Lee, J.M.; Park, J.H.; Han, S.H.; et al. *Lactobacillus plantarum* lipoteichoic acid down-regulated Shigella flexneri peptidoglycan-induced inflammation. *Mol. Immunol.* **2011**, *48*, 382–391. [CrossRef]
22. Kumar, R.; Sharma, A.; Gupta, M.; Padwad, Y.; Sharma, R. Cell-Free Culture Supernatant of Probiotic *Lactobacillus fermentum* Protects Against H(2)O(2)-Induced Premature Senescence by Suppressing ROS-Akt-mTOR Axis in Murine Preadipocytes. *Probiotics Antimicrob. Proteins* **2020**, *12*, 563–576. [CrossRef] [PubMed]
23. Wang, K.; Niu, M.; Song, D.; Song, X.; Zhao, J.; Wu, Y.; Lu, B.; Niu, G. Preparation, partial characterization and biological activity of exopolysaccharides produced from *Lactobacillus fermentum* S1. *J. Biosci. Bioeng.* **2020**, *129*, 206–214. [CrossRef] [PubMed]
24. Tsilingiri, K.; Barbosa, T.; Penna, G.; Caprioli, F.; Sonzogni, A.; Viale, G.; Rescigno, M. Probiotic and postbiotic activity in health and disease: Comparison on a novel polarised ex-vivo organ culture model. *Gut* **2012**, *61*, 1007–1015. [CrossRef] [PubMed]
25. Merghni, A.; Dallel, I.; Noumi, E.; Kadmi, Y.; Hentati, H.; Tobji, S.; Ben Amor, A.; Mastouri, M. Antioxidant and antiproliferative potential of biosurfactants isolated from *Lactobacillus casei* and their anti-biofilm effect in oral *Staphylococcus aureus* strains. *Microb. Pathog.* **2017**, *104*, 84–89. [CrossRef]
26. Teame, T.; Wang, A.; Xie, M.; Zhang, Z.; Yang, Y.; Ding, Q.; Gao, C.; Olsen, R.E.; Ran, C.; Zhou, Z. Paraprobiotics and Postbiotics of Probiotic Lactobacilli, Their Positive Effects on the Host and Action Mechanisms: A Review. *Front. Nutr.* **2020**, *7*, 570344. [CrossRef]
27. Ong, J.S.; Taylor, T.D.; Yong, C.C.; Khoo, B.Y.; Sasidharan, S.; Choi, S.B.; Ohno, H.; Liong, M.T. *Lactobacillus plantarum* USM8613 Aids in Wound Healing and Suppresses *Staphylococcus aureus* Infection at Wound Sites. *Probiotics Antimicrob. Proteins* **2020**, *12*, 125–137. [CrossRef]
28. Alves, E.; Gregorio, J.; Rijo, P.; Rosado, C.; Rodrigues, L.M. The Impact of Kefir on Epidermal Water Homeostasis in Healthy Human Skin. *Life* **2022**, *12*, 1075. [CrossRef]
29. Kimoto-Nira, H. New lactic acid bacteria for skin health via oral intake of heat-killed or live cells. *Anim. Sci. J.* **2018**, *89*, 835–842. [CrossRef]
30. Lukic, J.; Chen, V.; Strahinic, I.; Begovic, J.; Lev-Tov, H.; Davis, S.C.; Tomic-Canic, M.; Pastar, I. Probiotics or pro-healers: The role of beneficial bacteria in tissue repair. *Wound Repair Regen.* **2017**, *25*, 912–922. [CrossRef]
31. Zettam, A.; Taleb, A.; Sauvage, S.; Boithias, L.; Belaidi, N.; Sanchez-Perez, J.M. Applications of a SWAT model to evaluate the contribution of the Tafna catchment (north-west Africa) to the nitrate load entering the Mediterranean Sea. *Environ. Monit. Assess.* **2020**, *192*, 510. [CrossRef]
32. Baquerizo Nole, K.L.; Yim, E.; Keri, J.E. Probiotics and prebiotics in dermatology. *J. Am. Acad. Dermatol.* **2014**, *71*, 814–821. [CrossRef]
33. Al-Ghazzewi, F.H.; Tester, R.F. Impact of prebiotics and probiotics on skin health. *Benef. Microbes* **2014**, *5*, 99–107. [CrossRef]
34. Wang, S.; Zhu, F. Chemical composition and health effects of maca (*Lepidium meyenii*). *Food Chem.* **2019**, *288*, 422–443. [CrossRef]
35. Huarancca Reyes, T.; Esparza, E.; Crestani, G.; Limonchi, F.; Cruz, R.; Salinas, N.; Scartazza, A.; Guglielminetti, L.; Cosio, E. Physiological responses of maca (*Lepidium meyenii* Walp.) plants to UV radiation in its high-altitude mountain ecosystem. *Sci. Rep.* **2020**, *10*, 2654. [CrossRef] [PubMed]

36. Zha, S.; Zhao, Q.; Chen, J.; Wang, L.; Zhang, G.; Zhang, H.; Zhao, B. Extraction, purification and antioxidant activities of the polysaccharides from maca (*Lepidium meyenii*). *Carbohydr. Polym.* **2014**, *111*, 584–587. [CrossRef] [PubMed]
37. Caicai, K.; Limin, H.; Liming, Z.; Zhiqiang, Z.; Yongwu, Y. Isolation, purification and antioxidant activity of polysaccharides from the leaves of maca (*Lepidium Meyenii*). *Int. J. Biol. Macromol.* **2018**, *107*, 2611–2619. [CrossRef] [PubMed]
38. Fu, L.; Wei, J.; Gao, Y.; Chen, R. Antioxidant and antitumoral activities of isolated macamide and macaene fractions from *Lepidium meyenii* (Maca). *Talanta* **2021**, *221*, 121635. [CrossRef]
39. Zhang, Y.; Zhou, F.; Ge, F. Effects of combined extracts of *Lepidium meyenii* and *Allium tuberosum* Rottl. on erectile dysfunction. *BMC Complement. Altern. Med.* **2019**, *19*, 135. [CrossRef]
40. da Silva Leitao Peres, N.; Cabrera Parra Bortoluzzi, L.; Medeiros Marques, L.L.; Formigoni, M.; Fuchs, R.H.B.; Droval, A.A.; Reitz Cardoso, F.A. Medicinal effects of Peruvian maca (*Lepidium meyenii*): A review. *Food Funct.* **2020**, *11*, 83–92. [CrossRef]
41. Zhou, Y.; Zhu, L.; Li, H.; Xie, W.; Liu, J.; Zhang, Y.; Li, Y.; Wang, C. In vivo and in vitro neuroprotective effects of maca polysaccharide. *Front. Biosci. Landmark Ed.* **2022**, *27*, 8. [CrossRef]
42. Chen, W.; Zhu, X.; Wang, L.; Xin, X.; Zhang, M. Effects of two polysaccharides from *Lepidium meyenii* (maca) on intestinal immunity and inflammation in vitro. *Food Funct.* **2022**, *13*, 3441–3452. [CrossRef]
43. He, P.; Pan, L.; Wu, H.; Zhang, L.; Zhang, Y.; Zhang, Y.; Yang, J.; Lin, Z.; Zhang, M. Isolation, Identification, and Immunomodulatory Mechanism of Peptides from *Lepidium meyenii* (Maca) Protein Hydrolysate. *J. Agric. Food Chem.* **2022**, *70*, 4328–4341. [CrossRef]
44. Castaneda-Alarcon, M.; Bell-Cortez, C.; Hidalgo-Ascensios, J.; Moreno-Exebio, L. Photoprotective activity of a cream containing lyophilized aqueous extract of *Lepidium meyenii* (MACA) against ultraviolet irradiation on mouse skin. *Rev. Peru Med. Exp. Salud Publica* **2021**, *38*, 434–441. [CrossRef]
45. Nunez, D.; Olavegoya, P.; Gonzales, G.F.; Gonzales-Castaneda, C. Red Maca (*Lepidium meyenii*), a Plant from the Peruvian Highlands, Promotes Skin Wound Healing at Sea Level and at High Altitude in Adult Male Mice. *High Alt. Med. Biol.* **2017**, *18*, 372–383. [CrossRef]
46. Tsai, W.H.; Chou, C.H.; Chiang, Y.J.; Lin, C.G.; Lee, C.H. Regulatory effects of *Lactobacillus plantarum*-GMNL6 on human skin health by improving skin microbiome. *Int. J. Med. Sci.* **2021**, *18*, 1114–1120. [CrossRef]
47. Saito, Y.; Mihara, T.; Maruyama, K.; Saito, J.; Ikeda, M.; Tomonaga, A.; Kumagai, T. Effects of intake of *Lactobacillus casei* subsp. casei 327 on skin conditions: A randomized, double-blind, placebo-controlled, parallel-group study in women. *Biosci. Microbiota Food Health* **2017**, *36*, 111–120. [CrossRef]
48. Lee, K.; Kim, H.J.; Kim, S.A.; Park, S.D.; Shim, J.J.; Lee, J.L. Exopolysaccharide from *Lactobacillus plantarum* HY7714 Protects against Skin Aging through Skin-Gut Axis Communication. *Molecules* **2021**, *26*, 1651. [CrossRef]
49. Moreira, C.F.; Cassini-Vieira, P.; Canesso, M.C.C.; Felipetto, M.; Ranfley, H.; Teixeira, M.M.; Nicoli, J.R.; Martins, F.S.; Barcelos, L.S. *Lactobacillus rhamnosus* CGMCC 1.3724 (LPR) Improves Skin Wound Healing and Reduces Scar Formation in Mice. *Probiotics Antimicrob. Proteins* **2021**, *13*, 709–719. [CrossRef]
50. Morais, I.M.C.; Cordeiro, A.L.; Teixeira, G.S.; Domingues, V.S.; Nardi, R.M.D.; Monteiro, A.S.; Alves, R.J.; Siqueira, E.P.; Santos, V.L. Biological and physicochemical properties of biosurfactants produced by *Lactobacillus jensenii* P(6A) and *Lactobacillus gasseri* P(65). *Microb. Cell Fact.* **2017**, *16*, 155. [CrossRef]
51. No, J.K.; Soung, D.Y.; Kim, Y.J.; Shim, K.H.; Jun, Y.S.; Rhee, S.H.; Yokozawa, T.; Chung, H.Y. Inhibition of tyrosinase by green tea components. *Life Sci* **1999**, *65*, PL241–PL246. [CrossRef] [PubMed]
52. Kang, M.H.; Jang, G.Y.; Ji, Y.J.; Lee, J.H.; Choi, S.J.; Hyun, T.K.; Kim, H.D. Antioxidant and Anti-Melanogenic Activities of Heat-Treated Licorice (Wongam, *Glycyrrhiza glabra* × *G. uralensis*) Extract. *Curr. Issues Mol. Biol.* **2021**, *43*, 1171–1187. [CrossRef] [PubMed]
53. Oh, T.I.; Yun, J.M.; Park, E.J.; Kim, Y.S.; Lee, Y.M.; Lim, J.H. Plumbagin Suppresses alpha-MSH-Induced Melanogenesis in B16F10 Mouse Melanoma Cells by Inhibiting Tyrosinase Activity. *Int. J. Mol. Sci.* **2017**, *18*, 320. [CrossRef] [PubMed]
54. Eghbali-Feriz, S.; Taleghani, A.; Al-Najjar, H.; Emami, S.A.; Rahimi, H.; Asili, J.; Hasanzadeh, S.; Tayarani-Najaran, Z. Anti-melanogenesis and anti-tyrosinase properties of Pistacia atlantica subsp. mutica extracts on B16F10 murine melanoma cells. *Res. Pharm. Sci.* **2018**, *13*, 533–545. [CrossRef]
55. Ye, Y.; Wang, H.; Chu, J.H.; Chou, G.X.; Yu, Z.L. Activation of p38 MAPK pathway contributes to the melanogenic property of apigenin in B16 cells. *Exp. Dermatol.* **2011**, *20*, 755–757. [CrossRef]
56. Tachibana, M. Cochlear melanocytes and MITF signaling. *J. Investig. Dermatol. Symp. Proc.* **2001**, *6*, 95–98. [CrossRef]
57. Jung, H.J.; Park, S.H.; Cho, K.M.; Jung, K.I.; Cho, D.; Kim, T.S. Threonyl-tRNA Synthetase Promotes T Helper Type 1 Cell Responses by Inducing Dendritic Cell Maturation and IL-12 Production via an NF-kappaB Pathway. *Front. Immunol.* **2020**, *11*, 571959. [CrossRef]
58. Kang, W.; Choi, D.; Park, S.; Park, T. Carvone Decreases Melanin Content by Inhibiting Melanoma Cell Proliferation via the Cyclic Adenosine Monophosphate (cAMP) Pathway. *Molecules* **2020**, *25*, 5191. [CrossRef]
59. Lee, J.; Renita, M.; Fioritto, R.J.; St Martin, S.K.; Schwartz, S.J.; Vodovotz, Y. Isoflavone characterization and antioxidant activity of ohio soybeans. *J. Agric. Food Chem.* **2004**, *52*, 2647–2651. [CrossRef]
60. Hatami, T.; Emami, S.A.; Miraghaee, S.S.; Mojarrab, M. Total Phenolic Contents and Antioxidant Activities of Different Extracts and Fractions from the Aerial Parts of Artemisia biennis Willd. *Iran. J. Pharm. Res.* **2014**, *13*, 551–559.
61. Yang, T.; Li, Y.; Lyu, Z.; Huang, K.; Corrigan, C.J.; Ying, S.; Wang, W.; Wang, C. Characteristics of Proinflammatory Cytokines and Chemokines in Airways of Asthmatics: Relationships with Disease Severity and Infiltration of Inflammatory Cells. *Chin. Med. J.* **2017**, *130*, 2033–2040. [CrossRef]

62. Dweik, R.A.; Boggs, P.B.; Erzurum, S.C.; Irvin, C.G.; Leigh, M.W.; Lundberg, J.O.; Olin, A.C.; Plummer, A.L.; Taylor, D.R.; American Thoracic Society Committee on Interpretation of Exhaled Nitric Oxide Levels for Clinical Applications. An official ATS clinical practice guideline: Interpretation of exhaled nitric oxide levels (FENO) for clinical applications. *Am. J. Respir. Crit. Care Med.* **2011**, *184*, 602–615. [CrossRef]
63. D'Mello, S.A.; Finlay, G.J.; Baguley, B.C.; Askarian-Amiri, M.E. Signaling Pathways in Melanogenesis. *Int. J. Mol. Sci.* **2016**, *17*, 1144. [CrossRef]
64. Schiaffino, M.V. Signaling pathways in melanosome biogenesis and pathology. *Int. J. Biochem. Cell Biol.* **2010**, *42*, 1094–1104. [CrossRef]
65. Li, Y.; Wang, S.; Xin, Y.; Zheng, M.; Xu, F.; Xi, X.; Cao, H.; Cui, X.; Guo, H.; Han, C. Maca Cosmetics: A Review on Constituents, Therapeutics and Advantages. *J. Oleo Sci.* **2018**, *67*, 789–800. [CrossRef]
66. Kim, J.H.; Bae, J.T.; Song, M.H.; Lee, G.S.; Choe, S.Y.; Pyo, H.B. Biological activities of Fructus arctii fermented with the basidiomycete Grifola frondosa. *Arch. Pharm. Res.* **2010**, *33*, 1943–1951. [CrossRef]
67. Park, J.Y.; Song, M.W.; Kim, K.T.; Paik, H.D. Improved Antioxidative, Anti-Inflammatory, and Antimelanogenic Effects of Fermented Hydroponic Ginseng with Bacillus Strains. *Antioxidants* **2022**, *11*, 1848. [CrossRef]
68. Bae, J.T.; Ko, H.J.; Kim, G.B.; Pyo, H.B.; Lee, G.S. Protective effects of fermented Citrus unshiu peel extract against ultraviolet-A-induced photoageing in human dermal fibrobolasts. *Phytother. Res.* **2012**, *26*, 1851–1856. [CrossRef]
69. Hsu, M.F.; Chiang, B.H. Stimulating effects of Bacillus subtilis natto-fermented Radix astragali on hyaluronic acid production in human skin cells. *J. Ethnopharmacol.* **2009**, *125*, 474–481. [CrossRef]
70. Lee, H.S.; Kim, M.R.; Park, Y.; Park, H.J.; Chang, U.J.; Kim, S.Y.; Suh, H.J. Fermenting red ginseng enhances its safety and efficacy as a novel skin care anti-aging ingredient: In vitro and animal study. *J. Med. Food* **2012**, *15*, 1015–1023. [CrossRef]
71. Sangkaew, O.; Yompakdee, C. Fermented Unpolished Black Rice (*Oryza sativa* L.) Inhibits Melanogenesis via ERK, p38, and AKT Phosphorylation in B16F10 Melanoma Cells. *J. Microbiol. Biotechnol.* **2020**, *30*, 1184–1194. [CrossRef] [PubMed]
72. Kwak, Y.J.; Kim, K.S.; Kim, K.M.; Yu, H.Y.; Chung, E.; Kim, S.J.; Cha, J.Y.; Lee, Y.C.; Lee, J.H. Fermented Viola mandshurica inhibits melanogenesis in B16 melanoma cells. *Biosci. Biotechnol. Biochem.* **2011**, *75*, 841–847. [CrossRef] [PubMed]
73. Chae, G.Y.; Ha, B.J. The Comparative Evaluation of Fermented and Non-fermented Soybean Extract on Antioxidation and Whitening. *Toxicol. Res.* **2011**, *27*, 205–209. [CrossRef] [PubMed]
74. Pham, Q.L.; Jang, H.J.; Kim, K.B. Anti-wrinkle effect of fermented black ginseng on human fibroblasts. *Int. J. Mol. Med.* **2017**, *39*, 681–686. [CrossRef]
75. Wang, Y.; Wu, J.; Lv, M.; Shao, Z.; Hungwe, M.; Wang, J.; Bai, X.; Xie, J.; Wang, Y.; Geng, W. Metabolism Characteristics of Lactic Acid Bacteria and the Expanding Applications in Food Industry. *Front. Bioeng. Biotechnol.* **2021**, *9*, 612285. [CrossRef]
76. Leonard, W.; Zhang, P.; Ying, D.; Adhikari, B.; Fang, Z. Fermentation transforms the phenolic profiles and bioactivities of plant-based foods. *Biotechnol. Adv.* **2021**, *49*, 107763. [CrossRef]
77. Sales, P.M.; Souza, P.M.; Simeoni, L.A.; Silveira, D. alpha-Amylase inhibitors: A review of raw material and isolated compounds from plant source. *J. Pharm. Pharm. Sci.* **2012**, *15*, 141–183. [CrossRef]
78. Rahman, M.M.; Rahaman, M.S.; Islam, M.R.; Rahman, F.; Mithi, F.M.; Alqahtani, T.; Almikhlafi, M.A.; Alghamdi, S.Q.; Alruwaili, A.S.; Hossain, M.S.; et al. Role of Phenolic Compounds in Human Disease: Current Knowledge and Future Prospects. *Molecules* **2021**, *27*, 233. [CrossRef]
79. Filannino, P.; Cavoski, I.; Thlien, N.; Vincentini, O.; De Angelis, M.; Silano, M.; Gobbetti, M.; Di Cagno, R. Lactic Acid Fermentation of Cactus Cladodes (*Opuntia ficus*-indica L.) Generates Flavonoid Derivatives with Antioxidant and Anti-Inflammatory Properties. *PLoS ONE* **2016**, *11*, e0152575. [CrossRef]
80. Hwang, J.E.; Kim, K.T.; Paik, H.D. Improved Antioxidant, Anti-inflammatory, and Anti-adipogenic Properties of Hydroponic Ginseng Fermented by Leuconostoc mesenteroides KCCM 12010P. *Molecules* **2019**, *24*, 3359. [CrossRef]
81. Xu, J.; Hussain, M.; Su, W.; Yao, Q.; Yang, G.; Zhong, Y.; Zhou, L.; Huang, X.; Wang, Z.; Gu, Q.; et al. Effects of novel cellulase (Cel 906) and probiotic yeast fermentation on antioxidant and anti-inflammatory activities of vine tea (*Ampelopsis grossedentata*). *Front. Bioeng. Biotechnol.* **2022**, *10*, 1006316. [CrossRef]
82. Albertini, B.; Schoubben, A.; Guarnaccia, D.; Pinelli, F.; Della Vecchia, M.; Ricci, M.; Di Renzo, G.C.; Blasi, P. Effect of fermentation and drying on cocoa polyphenols. *J. Agric. Food. Chem.* **2015**, *63*, 9948–9953. [CrossRef]
83. Gonzales, G.F. Ethnobiology and Ethnopharmacology of *Lepidium meyenii* (Maca), a Plant from the Peruvian Highlands. *Evid. Based Complement. Alternat. Med.* **2012**, *2012*, 193496. [CrossRef]
84. Li, Y.; Xu, F.; Zheng, M.; Xi, X.; Cui, X.; Han, C. Maca polysaccharides: A review of compositions, isolation, therapeutics and prospects. *Int. J. Biol. Macromol.* **2018**, *111*, 894–902. [CrossRef]
85. Boo, Y.C. Up- or Downregulation of Melanin Synthesis Using Amino Acids, Peptides, and Their Analogs. *Biomedicines* **2020**, *8*, 322. [CrossRef]
86. Sonthalia, S.; Daulatabad, D.; Sarkar, R. Glutathione as a skin whitening agent: Facts, myths, evidence and controversies. *Indian J. Dermatol. Venereol. Leprol.* **2016**, *82*, 262–272. [CrossRef]
87. Lorin, J.; Zeller, M.; Guilland, J.C.; Cottin, Y.; Vergely, C.; Rochette, L. Arginine and nitric oxide synthase: Regulatory mechanisms and cardiovascular aspects. *Mol. Nutr. Food Res.* **2014**, *58*, 101–116. [CrossRef]
88. Zhang, M.; Wang, G.; Lai, F.; Wu, H. Structural Characterization and Immunomodulatory Activity of a Novel Polysaccharide from *Lepidium meyenii*. *J. Agric. Food Chem.* **2016**, *64*, 1921–1931. [CrossRef]

89. Sugahara, H.; Nagayama, K.; Ikeda, S.; Hirota, T.; Nakamura, Y. D- and l-amino acid concentrations in culture broth of *Lactobacillus* are highly dependent on the phylogenetic group of *Lactobacillus*. *Biochem. Biophys. Rep.* **2021**, *27*, 101073. [CrossRef]
90. Biz, C.; Fantoni, I.; Crepaldi, N.; Zonta, F.; Buffon, L.; Corradin, M.; Lissandron, A.; Ruggieri, P. Clinical practice and nursing management of pre-operative skin or skeletal traction for hip fractures in elderly patients: A cross-sectional three-institution study. *Int. J. Orthop. Trauma Nurs.* **2019**, *32*, 32–40. [CrossRef]
91. Lee, I.C.; Caggianiello, G.; van Swam, I.I.; Taverne, N.; Meijerink, M.; Bron, P.A.; Spano, G.; Kleerebezem, M. Strain-Specific Features of Extracellular Polysaccharides and Their Impact on *Lactobacillus plantarum*-Host Interactions. *Appl. Environ. Microbiol.* **2016**, *82*, 3959–3970. [CrossRef]
92. Ganzle, M.G.; Follador, R. Metabolism of oligosaccharides and starch in lactobacilli: A review. *Front. Microbiol.* **2012**, *3*, 340. [CrossRef] [PubMed]
93. Kimoto-Nira, H.; Aoki, R.; Sasaki, K.; Suzuki, C.; Mizumachi, K. Oral intake of heat-killed cells of Lactococcus lactis strain H61 promotes skin health in women. *J. Nutr. Sci.* **2012**, *1*, e18. [CrossRef] [PubMed]
94. Ho, H.H.; Chen, C.W.; Yi, T.H.; Huang, Y.F.; Kuo, Y.W.; Lin, J.H.; Chen, J.F.; Tsai, S.Y.; Chan, L.P.; Liang, C.H. Novel application of a Co-Fermented postbiotics of TYCA06/AP-32/CP-9/collagen in the improvement of acne vulgaris-A randomized clinical study of efficacy evaluation. *J. Cosmet. Dermatol.* **2022**, *21*, 6249–6260. [CrossRef] [PubMed]
95. Cui, H.; Guo, C.; Wang, Q.; Feng, C.; Duan, Z. A pilot study on the efficacy of topical lotion containing anti-acne postbiotic in subjects with mild-to-moderate acne. *Front. Med.* **2022**, *9*, 1064460. [CrossRef]
96. Millington, G.W. Proopiomelanocortin (POMC): The cutaneous roles of its melanocortin products and receptors. *Clin. Exp. Dermatol.* **2006**, *31*, 407–412. [CrossRef]
97. Huang, H.C.; Chou, Y.C.; Wu, C.Y.; Chang, T.M. [8]-Gingerol inhibits melanogenesis in murine melanoma cells through down-regulation of the MAPK and PKA signal pathways. *Biochem. Biophys. Res. Commun.* **2013**, *438*, 375–381. [CrossRef]
98. Hacker, E.; Boyce, Z.; Kimlin, M.G.; Wockner, L.; Pollak, T.; Vaartjes, S.A.; Hayward, N.K.; Whiteman, D.C. The effect of MC1R variants and sunscreen on the response of human melanocytes in vivo to ultraviolet radiation and implications for melanoma. *Pigment. Cell Melanoma Res.* **2013**, *26*, 835–844. [CrossRef]
99. Rouzaud, F.; Kadekaro, A.L.; Abdel-Malek, Z.A.; Hearing, V.J. MC1R and the response of melanocytes to ultraviolet radiation. *Mutat. Res.* **2005**, *571*, 133–152. [CrossRef]
100. Feng, Z.C.; Riopel, M.; Popell, A.; Wang, R. A survival Kit for pancreatic beta cells: Stem cell factor and c-Kit receptor tyrosine kinase. *Diabetologia* **2015**, *58*, 654–665. [CrossRef]
101. Li, P.H.; Liu, L.H.; Chang, C.C.; Gao, R.; Leung, C.H.; Ma, D.L.; David Wang, H.M. Silencing Stem Cell Factor Gene in Fibroblasts to Regulate Paracrine Factor Productions and Enhance c-Kit Expression in Melanocytes on Melanogenesis. *Int. J. Mol. Sci.* **2018**, *19*, 1475. [CrossRef]
102. Liu, L.; Fu, M.; Pei, S.; Zhou, L.; Shang, J. R-Fluoxetine Increases Melanin Synthesis Through a 5-HT1A/2A Receptor and p38 MAPK Signaling Pathways. *Int. J. Mol. Sci.* **2018**, *20*, 80. [CrossRef]

antioxidants

Article

# The Effect of Carnosine on UVA-Induced Changes in Intracellular Signaling of Human Skin Fibroblast Spheroids

Gilda Aiello [1,2], Francesca Rescigno [3], Marisa Meloni [3], Beatrice Zoanni [2], Giancarlo Aldini [2], Marina Carini [2] and Alfonsina D'Amato [2,*]

1 Department of Human Science and Quality of Life Promotion, Telematic University San Raffaele, 00166 Rome, Italy
2 Department of Pharmaceutical Sciences, University of Milan, Via L. Mangiagalli 25, 20133 Milan, Italy
3 VitroScreen, In Vitro Innovation Center, 20149 Milan, Italy
* Correspondence: alfonsina.damato@unimi.it

**Abstract:** Dermis fibroblasts are very sensitive to penetrating UVA radiation and induce photodamage. To protect skin cells against this environmental damage, there is an urgent need for effective compounds, specifically targeting UVA-induced mitochondrial injury. This study aimed to analyze the effect of carnosine on the proteome of UVA-irradiated human skin fibroblast, cultured in a three-dimensional (3D) biological system recapitulating dermal compartment as a test system to investigate the altered cellular pathways after 48 h and 7 days of culture with or without carnosine treatment. The obtained results indicate that UVA dysregulates Oxidative Phosphorylation, the Fibrosis Signaling Pathway, Glycolysis I and Nrf2-mediated Oxidative Stress Response. Carnosine exercises provide a protective function against the harmful effects of UVA radiation by activating the Nrf2 pathway with the upregulations of some ROS-detoxifying enzymes such as the glutathione S-transferase (GST) protein family. Additionally, carnosine regulates the activation of the Epithelial Adherens Junction and Wound Healing Signaling Pathway by mediating the activation of structural proteins such as vinculin and zyxin as well as fibronectin 1 and collagen type XVIII alpha 1 chain against UVA-induced changes.

**Keywords:** 3D dermis spheroids; proteomics; carnosine; label free quantification; UVA

**Citation:** Aiello, G.; Rescigno, F.; Meloni, M.; Zoanni, B.; Aldini, G.; Carini, M.; D'Amato, A. The Effect of Carnosine on UVA-Induced Changes in Intracellular Signaling of Human Skin Fibroblast Spheroids. *Antioxidants* **2023**, *12*, 300. https://doi.org/10.3390/antiox12020300

Academic Editors: Irene Dini, Sonia Laneri and Alexandros Georgakilas

Received: 23 November 2022
Revised: 18 January 2023
Accepted: 23 January 2023
Published: 28 January 2023



## 1. Introduction

UV radiation has been described as one of the main physical factors that affects skin cells daily, resulting in inflammation, premature aging and cancer [1]. The UV light which naturally reaches the Earth's surface contains UVB (280–320 nm) and UVA (320–400 nm) radiation. Each has unique properties, including energy, depth of penetration and biological effects. Among them, UVA is able to deeply penetrate into the skin dermis and it is considered the main cause of skin photoaging. UV-induced cellular damages are the main trigger of cellular senescence in skin, mainly in dermal fibroblasts, modulating extracellular matrix (ECM) deposition and remodeling.

Exposure of skin cells to UV radiation promotes the generation of reactive oxygen species (ROS) and subsequent lipid peroxidation, protein modifications and DNA damage, leading to the alteration of the antioxidant cellular system [2]. Contrary to the direct effect of ionizing radiations which corresponds to a direct ionization of DNA due to one electron ejection, UVA (and visible) light, by interaction with endogenous or exogenous photoreceptors, induces the formation of DNA damage through photomediated toxic reactions, characterized by low energy of the photons in respect to the ionizing radiations. As a result, proteins, lipids, and nucleic acids are altered by oxidative modifications which could induce the aggregation and formation of complex adducts. For example, reactive products of lipid peroxidation often bind to proteins causing a loss of antioxidant, proinflammatory and proapoptotic properties [3].

Skin mitochondria are highly involved in UVA skin damage. UVA induces the formation of mitochondrial ROS (mtROS) and mtDNA mutations and the perturbation of the electron transport chain, leading to ATP depletion and membrane depolarization [4]. Functional modules of mitochondria such as energy metabolism, biogenesis, fission and fusion are strongly affected by UV radiation. In this context, there is an urgent need for effective compounds, specifically targeting UVA-induced mitochondrial injury, to protect skin cells against this environmental damage. To date, several natural compounds endowed with antioxidants and carbonyl scavenger properties have been used to protect the skin against UV radiation [5,6]. Among them, carnosine, an endogenous β-alanyl-L-histidine dipeptide, has shown promising results in the prevention of Oxidative Stress modulation on the scaffold-free human dermis spheroids [7] and on UVA-irradiated nude mice skin [8]. Carnosine pretreatment in 3D-aged fibroblast resulted in modulation of important pathways such as free radical scavenging, apoptosis, mitochondrial function and ECM reorganization. In UVA-exposed nude mice, skin carnosine was able to neutralize the effects of reactive oxygen species by modifying proteins and impairing their functions. To further investigate the mechanism of action of carnosine in the prevention of UVA damage on human skin, the proteome of dermis spheroids of human fibroblast pretreated by UVA and carnosine was analyzed by quantitative mass spectrometry and network analyses. Three-dimensional cultured scaffold-free dermis spheroid represents the most suitable and micro-physiological model to mirror the complex homeostatic equilibrium and dynamical communicability between the cellular and extracellular space mimicking the physiological and pathological conditions of native tissue.

The application of VitroScreen ORA® 3D scaffold-free dermis spheroids as a micro-physiological testing platform coupled with quantitative label-free proteomics enables a description of an in depth, in human cell-based physiological tissue system, the dynamic functional and structural response to photodamage induced by UVA exposure in the dermal stroma mirroring a native human dermis behavior. In particular, VitroScreen ORA® spheroids are advanced tissue systems based on a bottom-up approach, able to better preserve the natural cell donor phenotype and consequently in the dermis deposition and architectural organization. The use of fibroblasts as target cells to study photodamage has been shown to be relevant by using targeted methods such as ELISA, qPCR or microscopy [9]. By exploiting the potential of a quantitative proteomics approach applied to the 3D spheroid model, the present work lays the foundations for its potential application to personalized care and medicine approaches as well as in dermatological research on aging and photoaging processes. Advanced analytical approaches based on genomics and proteomics allows a comprehensive study of the molecular pathways describing the alteration of protein phenotype by UV exposure and restored by carnosine treatment. By combining the advantages of the label-free quantitative proteomics with network analyses, the present study provides an in-depth investigation of the human dermis proteome signature after exposure to UVA radiation and the subsequent protective and antioxidant effect of carnosine.

## 2. Materials and Methods

### 2.1. Cell Source and Culture

Primary dermal fibroblasts isolated from a 59-year-old donor purchased from Innoprot (Derio, Spain), were cultured and amplified at early passages in CnT-PR F serum-free medium (CELLnTEC, Bern, Switzerland) in order to obtain $12 \times 10^6$ cells.

### 2.2. VitroScreen ORA® Dermis Spheroid Production

VitroScreen ORA® dermis models were developed according to the VitroScreen internal procedure. In brief, primary human dermal fibroblasts were detached once 90% of confluence was reached. Cells were washed twice with DPBS 1X (Merck Life Science S.r.l., Milan, Italy) and incubated with T/E solution (Primary Detach Kit, Innoprot, Derio, Spain) for 5 min. Cells were counted, centrifuged and seeded in Akura® Plate (InSphero AG, Schlieren, Switzerland) by ViaFlo Assist Plus (Integra Bioscience AG, Zizers, Switzerland) in order to obtain 10,000 cells/spheroids. Thanks to hanging drop technology, specific geometrical guidance allows cells to self-assemble forming round-shaped tridimensional scaffold-free spheroids. After 3 days in the hanging drop culture, spheroids were well formed and ready to be transferred into Akura® V2 plate (InSphero AG, Schlieren, Switzerland).

### 2.3. Nucleic Acid Extractions for Nanostring Analysis and RNA QC

The 3D scaffold-free spheroids were lysed in proper buffer and the extraction of the nucleic acids was performed following the manufacturer's instructions (RNeasy MicroKit by Qiagen, cat. 74004). RNA was eluted in RNAse-free water and stored at −80 °C until use. A total of 2 µL of RNA was used to perform QC by a 2100 Bioanalyzer (Agilent) on Agilent RNA 6000 Nano Chip with Agilent RNA 6000 Lader and Reagent. The method is based on capillary gel electrophoresis. All RNA samples were verified against the RNA acceptance criteria: a total RNA concentration ≥20 ng/µL, an RNA integrity number value ≥8.

### 2.4. Gene Expression Analysis Using Nanostring nCounter System

The hybridized sample was prepared on chip using nCounter MAX Prep Station and GPS transcripts were quantified using a NanoString Digital Analyzer (NanoString Technologies, Seattle, WA, USA). A total of 100 ng of RNA was used as sample input in a hybridization assay (NanoString Technologies, Seattle, WA, USA). Each hybridized sample was prepared on a cartridge using nCounter MAX Prep Station and an individual transcript of a biomarker signature was quantified using nCounter MAX Digital Analyzer (NanoString Technologies, Seattle, WA, USA). The control criteria for the NanoString analysis were imaging quality: >0.75; linearity: >0.95; limit of detection (LOD): (POS_E/LOD) >1 and binding density of 0.05 < BD < 2.25.

### 2.5. L-Carnosine Treatment and UVA Exposure

Following 72 h from transfer, the treated spheroid series were pre-incubated with 150 µM L-carnosine, previously observed as an efficacy dose to protect the dermis induced by aging [7], and added to the culture medium for 4 h. After pre-treatment, UVA series were prepared for UV exposure: the culture media with/or without treatment were discarded and DPBS1X was added to all tissues in order to avoid interference with UV exposure. ORA® dermis spheroids were irradiated with UVA at a working distance of 12 cm, at 30 J/cm$^2$ (Xenon Lamp Oriel Solar Simulator, Filter WG 335 3 mm) aiming to induce a perturbation of protein assessment in scaffold-free dermis spheroids; the dose was defined considering the complex physiology of the model and according to internal validated data [10]. Untreated and unexposed samples (NC) were rinsed with DPBS 1X and placed at RT in the dark in the same conditions of the exposed series. After UVA exposure, fresh medium with or without treatment was added to all series for 48 h and 7 days (D7) of recovery. All treated series were compared to NC and to untreated exposed samples (UVA). Fresh treatment was added every other day until 48 h and 7 days of culture. At each time point, three biological replicates of pooled 20 spheroids were collected, washed twice with cold DPBS 1X and allowed to settle to the bottom of conical tubes. After sedimentation, DPBS was discarded, and tissues were stored as pellets at −80 °C until the analysis.

### 2.6. Protein Extraction and In-Solution Trypsin Digestion

The cell pellets were resuspended in the solubilization buffer (8 M urea in 50 mM Tris–HCl, 30 mM NaCl, pH 8.5 and 1% protease inhibitor) and incubated on ice for

5–10 min. Tissue lysates were further homogenized by sonication in an ice bath, three times each for 15 s with 1 min intervals, using an ultra sonicator. Samples were centrifuged at 14,000× *g* for 20 min at 4 °C. The protein supernatant was collected into the new Eppendorf tube and the pelleted cell debris were discarded. Samples were stored at −80 °C until we used it for further experiments. The protein estimation was carried out by using the BCA assay. A total of 10 μg of proteins was diluted in 50 mM $NH_4HCO_3$ and then reduced with 5 mM DL-dithiothreitol (DTT, Sigma-Aldrich) for 30 min at 52 °C, shaken at 500 rpm and alkylated with 15 mM iodoacetamide (Sigma-Aldrich) for 20 min in the dark at room temperature. The trypsin digestion was performed in a 1:20 enzyme/protein ratio (*w*/*w*) (Trypsin Sequencing Grade; Roche, Monza, Italy) overnight at 37 °C. The obtained peptides were desalted using zip-tip C18, then dried and stored at −20 °C before the analysis.

### 2.7. High-Resolution LC-MS/MS Analysis and Data Elaboration

Tryptic peptides were analyzed using a Dionex Ultimate 3000 nano-LC system (Sunnyvale CA, USA) connected to an Orbitrap Fusion Tribrid Mass Spectrometer (Termo Scientifc, Bremen, Germany) equipped with a nano-electrospray ion source. Peptide mixtures were pre-concentrated onto an Acclaim PepMap 100–100 μm × 2 cm C18 and separated on an EASY-Spray column, 15 cm × 75 μm ID packed with Thermo Scientific Acclaim PepMap RSLC C18, 3 μm, 100 Å. The chromatographic separation was performed at 35 °C and the flow rate was 300 nL/min. Mobile phases were the following: 0.1% formic acid (FA) in water (solvent A); 0.1% FA in water/acetonitrile (solvent B) with 2/8 ratio. Peptides were pre-concentrated at 96% of solvent A and eluted from the column with the following gradient: 4% to 28% of B for 90 min and then 28% to 40% of B in 10 min, and to 95% within the following 6 min to rinse the column. The column was re-equilibrated for 20 min. Total run time was 130 min. One blank was run between triplicates to prevent sample carryover. MS spectra were collected over on an m/z range of 375–1500 at 120,000 resolutions, operating in the data-dependent mode, cycle time of 3 s. Higher-energy collisional dissociation (HCD) was performed with collision energy set at 35%. Each sample was analyzed in three technical triplicates. The resulting MS raw data from all the technical and biological replicates were analyzed by using MaxQuant software (version 1.6.2.3). The Andromeda search engine was used to identify MS/MS-based peptides and proteins in MaxQuant comprising a target-decoy approach with less than a 1% False Discovery Rate (FDR). In the present study we used the Uniprot_Homosapiens database. Trypsin was selected as the cutting enzyme, two missed cleavages and maximum number of five modifications per peptide were allowed. Methionine oxidation and acetylation (N terminus) was used as a variable modification. Carbamidomethylation was used as a fixed modification. The proteins were selected with a minimum of two peptides. For the label-free quantification of proteins, we applied the MaxLFQ algorithm. The match between the runs option was enabled and the remaining default parameters were permitted. The data are available on request from the authors. An open-source Perseus software (version 1.6.1.3; Max Planck Institute of Biochemistry, Germany) was used for the identification of statistically significantly, differentially regulated proteins. The interpretation and visualization of the results obtained from the MaxQuant software were performed by a two-sample *t*-test using Perseus (v1.6.1.3, Max Planck Institute of Biochemistry, Germany). Statistical parameters ($p < 0.05$; $q < 0.05$, q = FDR adjusted *p*-value) were set to identify the differentially expressed proteins between samples. Variabilities of biological replicates were measured with Pearson correlation coefficient values of the LFQ intensities. The differentially regulated proteins contained a minimum of two peptides and FDR adjusted *p*-value.

### 2.8. Protein Network Analysis

The protein network analysis related to significantly modulated proteins was carried out by Ingenuity Pathways Analysis (IPA) (QIAGEN Inc., June 2022, https://www.qiagenbioinformatics.com/products/ingenuitypathway-analysis). The statistical enrichment of involved pathways was performed by the right-tailed Fisher's exact test, in correlation with QIAGEN Knowledge Base, assigning a $p$-value. The core analyses performed by IPA, using the differentially expressed proteins in the uploaded dataset, assess signaling pathways, molecular interaction networks and biological functions that can be likely perturbed. The overall activation/inhibition states of canonical pathways are predicted through a z-score algorithm. This z-score is used to statistically compare the uploaded dataset with the pathway patterns. The pathways are colored to indicate their activation z-scores: orange predicts a gain of function, while blue predicts a loss of function. The pathway is activated when molecules' causal relationships with each other (i.e., activation edge and the inhibition edge between the molecules based on literature findings) generate an activity pattern for the molecules and the end-point functions in the pathway [11].

## 3. Results and Discussion

Dermis fibroblasts are one of the most representative cellular types of the human skin, are able to secrete structural and functional proteins, and are key molecules for the assembly of the deep stromal compartment. Due to their localization, dermis stromal fibroblasts are more protected from external factors than epidermal cells, forming the upper layer of the skin. As a result, dermis fibroblasts are very sensitive to penetrating UVA radiation. In this study, we evaluated the effect on a photo-damaged VitroScreen ORA® dermis model by an omics approach for proteome remodeling after 48 h and 7 days of culture with or without carnosine treatment.

### *3.1. VitroScreen ORA® Dermis Model to Study the UVA Dermis Damage and Protection Effect of Carnosine*

VitroScreen ORA® dermis models are miniaturized human biomimetic systems developed to mirror the physiology and the dynamic features of the native human dermis. The tissue engineered bottom-up approach allowed ORA® models to mimic the natural dermis architecture in a 3D microscale: In this defined configuration, primary dermis fibroblasts at early passages aggregate according to their natural phenotype. During spheroids' development, cells start to secrete endogenous key structural and functional molecules building a compact stromal ECM, without the addition of an exogenous support.

In order to characterize the cellular model during the time culture, gene expression analysis was employed. The selected NanoString CodeSet included 59 genes, 49 of which were endogenous and related to ECM remodeling and to Oxidative Stress induced by photo-exposure. The remaining 10 genes were termed housekeeping genes and can potentially be used for purposes such as normalization and quality assessment. In addition to the probes targeting transcripts in the RNA samples, positive and negative control probes were also included in the CodeSet for assessing nonspecific background binding and to assess the effectiveness of hybridization. The hybridization reactions were successful as verified by NanoString quality control. The within-sample stability of the housekeeping genes was also assessed and a smaller subset of housekeeping genes was selected for the calculation of normalization scaling factors. These genes included NUBP1, EDC3, HDAC3 and PRPF38A. The observed expression of the CodeSet genes is shown in Figure 1. The ECM genes resulting in being differentially regulated during the defined experimental window such as HMOX1, COL4A1, CD44 and MME were upregulated while COL1A1, ELN and COL3A1 were downregulated, attesting to the induction of UVA-induced physiological aging of the extracellular matrix (Figure 1). The activation of metalloproteases such as MME suggests a disruption of ECM accompanied by a decrease in collagen isoform COL1A1 and COL3A1 during the time (48 h versus 7 days). However, the modulation of BSG and DCN, the latter binding collagen in the same region as the metalloprotease's cleavage site, affords collagen protection against protease degradation indicating an ECM remodeling.

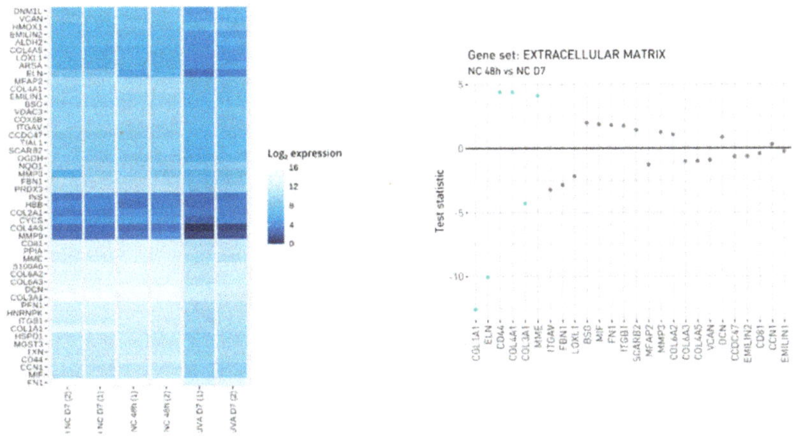


**Figure 1.** Visualization of observed expression levels. The left figure is a heatmap that displays the log2-transformed expression values and the right figure displays gene level statistics for genes within the manually curated gene sets. Statistics correspond to those from the differential expression analysis comparing NC_48 h vs. NC_D7 (analysis performed in triplicate, each replicate is a pool of 20 spheroids on 150–200 μm in size diameter).

### *3.2. Protein Profiling of Human Dermis Spheroids Model after Exposure to UVA and Carnosine Treatment*

Thanks to a high biological relevance and their physiological profile, ORA® dermis models are suitable as a tissue platform for in-depth investigation on UVA photo-damage induction and for evaluation of positive antioxidant role of carnosine in preserving tissue structural and functional integrity [7]. Spheroids showed a proteome capable of being induced by UVA external stimuli, described by 2912 proteins of which several were specifically differentially regulated. The exposure to 30 J/cm$^2$ UVA radiations severely affected the basal proteomic profile of dermis spheroids inducing a modulation of key molecules related to Oxidative Stress Signaling, adhesion molecules and apoptotic events, showing a dynamic evolution during the time. The UVA treatment at 48 h resulted in the damage of spheroids characterized by 420 upregulated and 536 downregulated proteins, while the same treatment after D7 induced a variation of proteome described by 353 over expressed and 612 down expressed proteins (Figure 2A,C and Supplementary Table S1).

The UVA damage of spheroids was prevented by 150 μM carnosine added every other day, inducing the upregulation of 165 proteins and the downregulation of 303 proteins (48 h) and the upregulation of 320 proteins and the downregulation of 202 proteins (D7) (Figure 2B,D and Supplementary Table S1). Carnosine added as pre-treatment and during the recovery after UVA exposure, allowed a chemical protection against UVA photo-damage by preserving tissue viability, metabolism and ECM integrity, according to the duration of treatments. Specifically, network analyses showed UVA damages and their carnosine prevention, by specific functional modules, such as "Epithelial Adherens Junction Signaling", "Oxidative Phosphorylation", "Wound Healing Signaling Pathway" and "Nrf2-mediated Oxidative Stress Response" (Table 1) which are herein described.

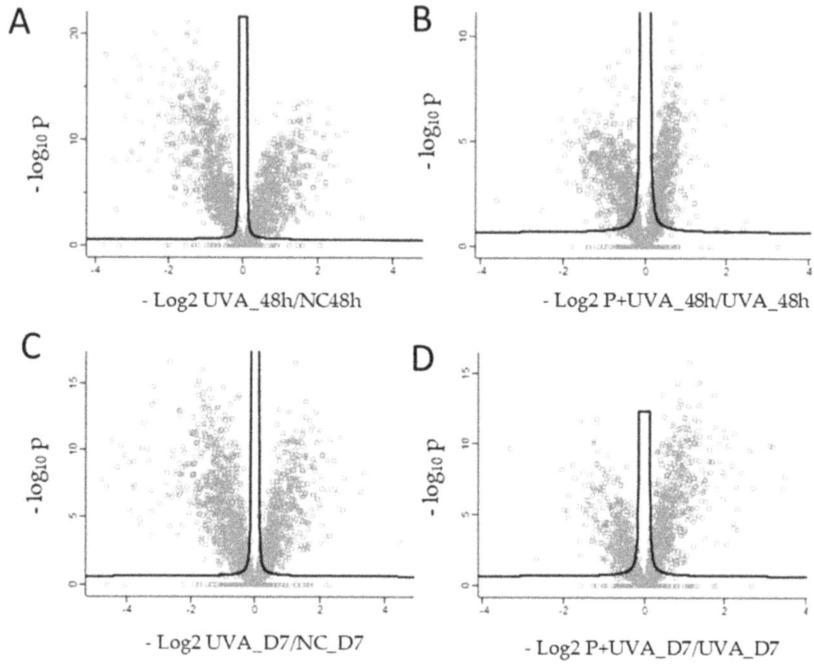


**Figure 2.** Scatter plots of log2 ratio on x-axis against -log10 *p*-value on y-axis of significantly quantified proteins resulted from two sided *t*-test. (**A**) Quantified proteins in human dermis spheroids after 48 h of UVA exposure (UVA_48hvsNC_48 h). (**B**) Quantified proteins in carnosine pre-treated dermis versus control at 48 h (P+UVA_48hvsUVA_48 h). (**C**) Quantified proteins in human dermis spheroids after D7 of UVA exposure versus control (UVA_D7vsNC_D7). (**D**) Quantified proteins in carnosine pre-treated dermis versus control at D7 (P+UVA_D7 vs. UVA_D7). Statistical parameters ($p < 0.05$; $q < 0.05$, q = FDR adjusted *p*-value) were set to identify the differentially expressed proteins between samples. Biological and technical triplicates were analyzed.

**Table 1.** Functional modules evoked by UVA exposure. Positive z-score indicates the activation pathway; negative z-score indicates the downregulation of the pathway (the analysis was performed by IPA (Qiagen) using the quantitative LFQ dataset).

| | Z-Score | | | |
|---|---|---|---|---|
| **Canonical Pathways** | **UVA_48h** | **P_UVA_48h** | **UVA_D7** | **P_UVA_D7** |
| Oxidative Phosphorylation | 5.39 | −1.16 | 4.71 | −1.67 |
| Hepatic Fibrosis Signaling Pathway | 2.41 | −1.60 | 1.57 | −2.31 |
| Wound Healing Signaling Pathway | 1.53 | −2.53 | 1.61 | −3.16 |
| Estrogen Receptor Signaling | 3.14 | −2.11 | 1.89 | −2.89 |
| Macropinocytosis Signaling | 2.45 | −2.45 | 2.24 | −2.45 |
| GP6 Signaling Pathway | 2.14 | −1.63 | 0.91 | −2.12 |
| Pulmonary Fibrosis Idiopathic Signaling Pathway | 1.23 | −2.31 | 1.00 | −2.53 |
| G Beta Gamma Signaling | 2.11 | −2.65 | 0.91 | −2.65 |
| Gαi Signaling | 1.90 | −2.45 | 1.51 | −2.45 |
| Regulation of eIF4 and p70S6K Signaling | 1.67 | −2.45 | 1.73 | −2.45 |
| p70S6K Signaling | 1.90 | −2.45 | 1.39 | −2.24 |
| Thrombin Signaling | 1.51 | −2.83 | 0.91 | −2.65 |

**Table 1.** *Cont.*

| Canonical Pathways | Z-Score | | | |
|---|---|---|---|---|
| | UVA_48h | P_UVA_48h | UVA_D7 | P_UVA_D7 |
| NF-κB Activation by Viruses | 1.89 | −2.24 | 1.41 | −2.24 |
| Regulation of Cellular Mechanics by Calpain Protease | 2.12 | −1.89 | 0.91 | −2.45 |
| Glioma invasiveness Signaling | 1.89 | −2.45 | 0.63 | −2.24 |
| HGF Signaling | 1.16 | −2.45 | 0.28 | −2.45 |
| UVA-induced MAPK signaling | 1.13 | −2.00 | 1.13 | −2.00 |
| CDK5 Signaling | 1.51 | −2.24 | 0.26 | −2.24 |
| PEDF Signaling | 1.14 | −1.63 | 0.71 | −2.24 |
| Fc Epsilon RI signaling | −1.13 | −2.-24 | 0.38 | −2.-24 |
| Glycolysis I | −3.46 | 1.41 | −3.61 | 2.83 |
| PTEN Signaling | −1.73 | 2.65 | −1.16 | 2.45 |
| Epithelial Adherens Junction Signaling | −2.20 | 1.27 | −2.35 | 2.00 |
| CLEAR Signaling Pathway | −1.67 | 1.70 | −1.41 | 2.45 |
| Gluconeogenesis I | −2.11 | 2.45 | −2.11 | 2.83 |
| Sirtuin Signaling Pathway | −0.85 | −2.32 | −1.26 | −0.83 |
| Apotosis | 1.25 | −0.94 | 2.19 | −2.82 |

### 3.3. UVA Induces the Modulation of Epithelial Adherens Junction Signaling

Network analyses, based on differentially regulated proteins, showed a negative modulation of epithelial Adherens Junction Signaling induced by UVA radiation after 48 h (z-score = −2.20) and 7 days (D7) (z-score = −2.35) of recovery (Table 1, Figure 3). The Adherens Junctions in epithelial cells are specialized structures belonging to the machinery of the cell–cell adhesion and consist of nectin and cadherin proteins, bound to actin cytoskeleton. In physiological and health conditions, actinin, zyxin, tubulins and myosins modulate cytoskeletal reorganization and actin polymerization, leading to the formation of Adherens Junctions. UVA radiation induces a downregulation of epithelial adhesion formation. Specifically, different actin isoforms such as actin-related proteins ARPC4 (log2 ratio = −0.93), ARPC2 (log2 ratio = −1.0) and ARPC1A (log2 ratio = −0.65), actin alpha 1, ACTN1 (log2 ratio = −0.88), actin alpha 4, ACTN4 (log2 ratio = −0.68), zyxin ZYX (log2 ratio = −2.13) and vinculin VCL (log2 ratio = −0.86), were downregulated by UVA at 48 h (Table 2). A similar trend was observed upon 7 days of recovery after UVA exposure, mostly for zyxin and vinculin whose log2 ratio dropped down to −2.42 and −1.11, respectively. The main function of zyxin is to form a bridge between the adhesion components of the cell membrane and the cytoskeleton, and its absence characterizes the early adhesions [12]. The UVA treatment resulted in a decrease of zyxin expression at 48 and 7 days after photo-damage (log2 = −2.13 and −2.42), while prevention of carnosine was efficient to restore its expression at both time points (ZYX, log2 ratio = 0.38 and 1.52), demonstrating the time-depending effects of carnosine on photo-damaged system. On the contrary, RAS-like proto-oncogene A, (RALA, log 2ratio = 1.53 at 48 h, log2 ratio =1.59 at D7) and the proteins belonging to the RAS oncogene family such as RAB7A (log 2ratio = 0.79 at 48 h, log 2ratio = 0,67 at D7) and RAB5C (log 2ratio = 0.72 at 48 h, log 2ratio = 0.82 at D7) were found overexpressed after both time of recovery after UVA exposure. RAS-mediated proliferative overdrive may induce replicative stress and activation of DNA damage responses [13]. This upregulation was observed to be modulated by carnosine reducing the overexpression of RALA (log2 ratio = −0.66 at 48 h, log2 ratio = −0.78 at D7) and RAB7 (log2 ratio = −0.61 at 48 h, log2 ratio = −0.44 at D7). The impairment of cell adhesion and cellular motility was also confirmed by the activation of calpain protease at a long time point (D7) (z score = 0.91). Calpains are intracellular $Ca^{2+}$-dependent cysteine proteases which regulate cell adhesion by modulating the spreading and motility of many cell types. It has been observed that calpain-deficient fibroblasts have enhanced membrane protrusion and filopodia formation [14]. In the meantime, the

Integrin Signaling which shows a pivotal role in cell adhesion and adhesive signaling mostly for tissue repairing, was compromised by UVA exposure (z score = −1.71 at D7). Since the best protective effects of carnosine were observed after long-term treatment (7 days); from this point on, the discussion focuses on the description of the effects of carnosine treatment at D7.

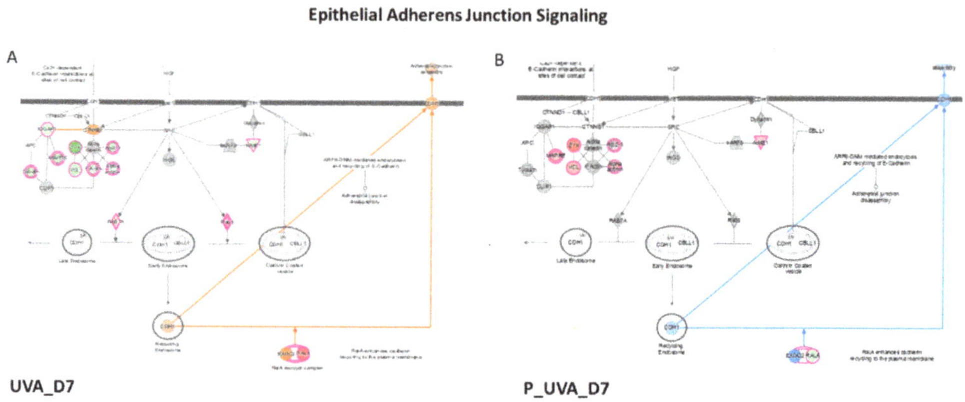


**Figure 3.** Epithelial Adherens Junction Signaling enhanced in UVA vs. control in (**A**) and decreased in (**B**) P_UVA vs. UVA (IPA). Red shows the increased features covered by our input database and green shows those decreased features. The orange color of the central hub indicates an upregulation of the module. For the node shapes: Knowledge: IPA Legend (salesforce-sites.com) (Qiagen, USA).

**Table 2.** Proteins involved in the Epithelial Adherens Junction Signaling. Log2 ratio ≥ 0.6 indicates upregulation, log2 ratio ≤ − 0.6 represents downregulation.

| **Epithelial Adherens Junction Signaling** | | | **UVA_48 h** | **P_UVA_48 h** | **UVA_D7** | **P_UVA_D7** |
|---|---|---|---|---|---|---|
| **Gene Name** | **Protein Name** | **Accession Number** | **Log Ratio** | **Log Ratio** | **Log Ratio** | **Log Ratio** |
| RALA | RAS-like proto-oncogene A | P11233 | 1.53 | −0.66 | 1.59 | −0.78 |
| RAB7A | RAB7A, member RAS oncogene family | A0A158RFU6 | 0.79 | −0.61 | 0.67 | −0.44 |
| RAB5C | RAB5C, member RAS oncogene family | A0A024R1U4 | 0.72 | −0.38 | 0.82 | −0.30 |
| ARPC4 | Actin-related protein 2/3 complex subunit 4 | P59998 | −0.93 | 0.46 | −0.53 | 0.26 |
| ARPC1A | Actin-related protein 2/3 complex subunit 1A | Q75MY0 | −0.65 | 0.25 | −0.76 | 0.31 |
| ARPC2 | Actin-related protein 2/3 complex subunit 2 | Q53R19 | −1.00 | 0.36 | −0.40 | 0.36 |
| NME1 | NME/NM23 nucleoside diphosphate kinase 1 | P15531 | −1.01 | 0.62 | −0.77 | 0.64 |
| MAPRE1 | Microtubule-associated protein RP/EB family member 1 | Q15691 | −1.29 | 0.62 | −1.18 | 0.70 |
| ACTN4 | Actinin alpha 4 | A0A0S2Z3G9 | −0.68 | 0.54 | −0.69 | 0.79 |
| ACTN1 | Actinin alpha 1 | A0A024R694 | −0.88 | 0.57 | −0.83 | 0.87 |
| VCL | Vinculin | A0A024QZN4 | −0.86 | 0.35 | −1.11 | 0.87 |
| ZYX | Zyxin | Q15942 | −2.13 | 0.38 | −2.42 | 1.52 |

### 3.4. UVA Induces the Modulation of Wound Healing Signaling Pathway

Chronic UV exposure causes photoaging with profound alterations to the dermal connective tissue [15]. The morphological changes in skin fibroblasts exposed to UVA have already been demonstrated by scanning electron microscopy and immunofluorescence approaches. As fibroblasts are the main source of ECM, the functional changes in fibroblasts in photoaged skin are related to a reduction in collagen due to the inhibition of the polymerization of actin filaments accelerating the reduction in collagen synthesis [15]. In our study, both the UVA exposure reveals the dysregulation of collagen metabolism with the upregulation of proteins involved in the collagen degradation process, verified by gene expression too (Figure 4A). The upregulation of MMP-3 to the detriment of structural proteins as ELN, FN1, DCN, EMLIN1 and different isoform of collagen confirms the establishment of the well-known UVA-induced photoaging process negatively affecting ECM proteins. Wounds have chronic inflammation, impaired re-epithelialization to close the wound and abnormal dermal–epidermal connectivity that results in poor dermal–epidermal interaction, damaged microvasculature and abnormal collagen matrix deposition in the wound tissue [16]. Specifically, different isoforms of collagen, such as collagen type I, (COL1A1 log2 ratio = 0.96; COL1A2, log2 ratio = 0.49) III (COL3A1 log2 ratio = 0.45), IV (COL4A2, log2 ratio =1.00) V (COL5A1, log2 ratio = 0.31) and VI (COL6A1, log2 ratio = 1.71; COL6A2 log2 ratio = 2.02, COL6A3 log2 ratio = 4.48) resulted in being overexpressed in the carnosine treatment at D7 (Table 3). Among those, COL6A3 has the capability to direct matrix assembly and influence dermal cell behavior by interacting with other ECM molecules. Within ECM connective tissue, type VI collagen forms a highly branched filamentous meshwork that encircles the fibers of principal fibrillar collagens type I, II, and III [17].

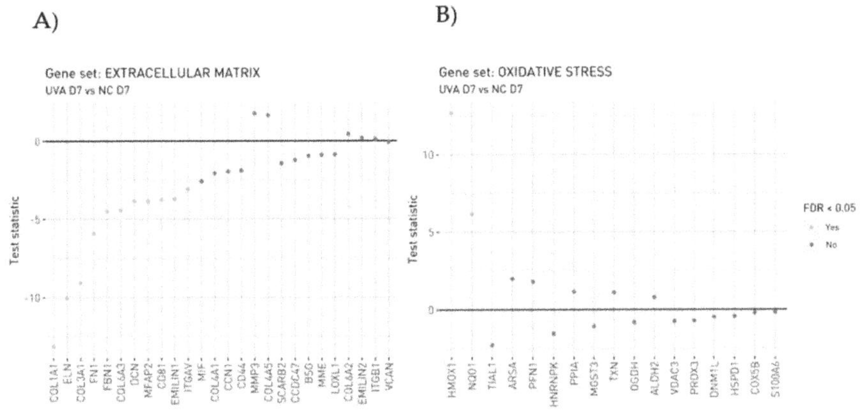


**Figure 4.** Gene level statistics for genes within the manually curated gene sets. (**A**) ECM genes altered by UVA_D7 vs. NC_D7. (**B**) Oxidative Stress genes altered by UVA_D7 vs. NC_D7. Statistics correspond to those from the differential expression analysis comparing UVA_D7 vs. NC_D7. Analysis was performed in triplicate; each replicate is a pool of 20 spheroids on 150–200 μm in size diameter.

**Table 3.** Proteins involved in the Wound Healing Signaling Pathway.

| **Wound Healing Signaling Pathway** | | | **UVA_48 h** | **P_UVA_48 h** | **UVA_D7** | **P_UVA_D7** |
|---|---|---|---|---|---|---|
| **Gene Name** | **Protein Name** | **Accession Number** | **Log Ratio** | **Log Ratio** | **Log Ratio** | **Log Ratio** |
| COL1A1 | Collagen type I alpha 1 chain | D3DTX7 | 1.99 | −0.99 | 0.96 | −1.27 |
| COL3A1 | Collagen type III alpha 1 chain | P02461 | 0.79 | 0.53 | 0.45 | 0.00 |
| COL1A2 | Collagen type I alpha 2 chain | A0A087WTA8 | 1.20 | 0.00 | 0.49 | 0.00 |
| COL5A1 | Collagen type V alpha 1 chain | A0A024R8E5 | 0.99 | 0.24 | 0.31 | 0.00 |
| COL15A1 | Collagen type XV alpha 1 chain | P39059 | −1.74 | 0.51 | −2.35 | 0.48 |
| LAMA2 | Laminin subunit alpha 2 | A0A087WYF1 | 0.60 | 0.00 | 0.00 | 0.00 |
| COL6A2 | Collagen type VI alpha 2 chain | P12110 | 2.30 | −1.04 | 2.02 | −1.26 |
| COL6A1 | Collagen type VI alpha 1 chain | A0A087X0S5 | 1.77 | −0.50 | 1.71 | −0.93 |
| RRAS | RAS related | A0A024QZF2 | 0.93 | −0.41 | 1.05 | −0.85 |
| RALA | RAS-like proto-oncogene A | P11233 | 1.53 | −0.66 | 1.59 | −0.78 |
| RAP1B | Rap1b Member of RAS oncogene family | P61224 | 1.28 | −0.84 | 1.04 | −0.75 |

**Table 3.** *Cont.*

| Wound Healing Signaling Pathway | | | UVA_48 h | P_UVA_48 h | UVA_D7 | P_UVA_D7 |
|---|---|---|---|---|---|---|
| **Gene Name** | **Protein Name** | **Accession Number** | **Log Ratio** | **Log Ratio** | **Log Ratio** | **Log Ratio** |
| ITGB1 | Integrin subunit beta 1 | P05556 | 1.36 | −0.74 | 1.30 | −0.71 |
| RAC1 | Rac family small gtpase 1 | A4D2P1 | 0.81 | −0.88 | 0.52 | −0.70 |
| COL4A2 | Collagen type IV alpha 2 chain | P08572 | 0.75 | 0.00 | 1.00 | −0.28 |
| MAP2K2 | Mitogen-activated protein kinase kinase 2 | P36507 | −1.49 | 0.00 | −2.00 | 0.00 |
| MAPK1 | Mitogen-activated protein kinase 1 | Q1HBJ4 | −0.73 | 0.00 | −0.94 | 0.00 |
| STAT1 | Signal transducer and activator of transcription 1 | P42224 | −0.95 | 0.00 | −0.60 | 0.00 |
| RRAS2 | RAS related 2 | P62070 | 0.77 | −1.04 | 0.95 | 0.00 |
| FN1 | Fibronectin 1 | A0A024R462 | −1.08 | 0.00 | −1.28 | 0.28 |
| COL18A1 | Collagen type XVIII alpha 1 chain | D3DSM4 | −1.36 | 0.58 | −1.38 | 0.55 |
| COL6A3 | Collagen type VI alpha 3 chain | B7ZW00 | 3.19 | −3.60 | 4.48 | −2.66 |

### *3.5. UVA Induces Nrf2-Mediated Oxidative Stress Response*

Considered as a "master regulator of the antioxidant response", Nrf2 controls the transcription of genes encoding ROS-detoxifying enzymes and various other antioxidant proteins [18,19]. Nrf2 has a crucial role in the maintenance of cellular redox homeostasis by regulating the biosynthesis, utilization and regeneration of glutathione, thioredoxin and NADPH and by controlling the production of reactive oxygen species by mitochondria and NADPH oxidase. Under homeostatic conditions, Nrf2 affects the mitochondrial membrane potential, fatty acid oxidation, availability of substrates (NADH and FADH2/succinate) for respiration and ATP synthesis [20]. In fibroblasts, the activation of Nrf2 induces cellular senescence and its activation is qualified as a marker of the cancer-associated fibroblast phenotype. Previously published data showed that the Nrf2 activation in fibroblasts alters ECM protein expression and secretion by decreasing collagens type I and type III the most and several major proteoglycans (e.g., asporin, biglycan and versican) [21]. In UVB-treated keratinocytes, ROS-mediated apoptosis was reduced by the activation of Nrf2 [22]. In UVB-treated fibroblasts, instead, the defensive response of cellular death was induced by phosphorylated Akt by MAPK through the EGF receptor, independently to Oxidative Stress [23]. Indeed, in our work, UVA-exposed 3D spheroids have shown cell death (Table 1) due to the increasing concentration of ROS and downregulation of the Nrf2 pathway. Despite several publications reporting that PI3K/Akt pathway mediates the Nrf2 activation, in our dataset, the dephosphorylated AKT protein was found to be not regulated (Supplementary Table S1), whereas MAPK1 and MAP2K2 were down expressed (Table 4).

The UVA exposure at D7 modules the activation of Nrf2, responsible to the inducible expression of a group of detoxication enzymes, such as glutathione S-transferase including Glutathione S-transferase mu1 (GSTM1, log2 ratio = −2.43), Glutathione S-transferase mu3 (GSTM3, log2 ratio = −1.84), Glutathione S-transferase omega1 (GSTO1, log2 ratio = −1.67), Glutathione S-transferase pi1 (GSTP1, log2 ratio = −1.77) and Glutathione disulphide reductase (GSR, log2 ratio = −0.72) (Figure 5 and Table 4). Our finding underlines the role of Nrf2 as a truly pleiotropic transcription factor by playing an important role in the downregulation of several glutathione S-transferases, which are the enzymes mediating the elimination of reactive oxygen species (ROS). Reduced glutathione (GSH) interacts directly with ROS to form oxidized glutathione (GSSG).

**Table 4.** Proteins involved in the Nrf2-mediated Oxidative Stress Response.

| **Nrf2-Mediated Oxidative Stress Response** | | | **UVA_48h** | **P_UVA_48h** | **UVA_D7** | **P_UVA_D7** |
|---|---|---|---|---|---|---|
| **Gene Name** | **Protein Name** | **Accession Number** | **Log Ratio** | **Log Ratio** | **Log Ratio** | **Log Ratio** |
| AKR1A1 | Aldo-keto reductase family 1 member A1 | V9HWI0 | −0.94 | 0.40 | −1.32 | 0.83 |
| CLPP | Caseinolytic mitochondrial matrix peptidase proteolytic subunit | Q16740 | 0.73 | −0.33 | 0.58 | 0.00 |
| EPHX1 | Epoxide hydrolase 1 | R4SBI6 | 1.30 | −1.18 | 0.97 | −0.89 |
| GSTM1 | Glutathione S-transferase mu 1 | X5DR03 | −1.29 | −0.61 | −2.43 | 1.00 |
| GSTM3 | Glutathione S-transferase mu 3 | Q6FGJ9 | −1.20 | 0.00 | −1.84 | 0.71 |
| GSTO1 | Glutathione S-transferase omega 1 | V9HWG9 | −1.69 | 0.73 | −1.67 | 1.24 |
| GSTP1 | Glutathione S-transferase pi 1 | V9HWE9 | −1.44 | 0.51 | −1.77 | 1.10 |
| GSR | Glutathione-disulphide reductase | V9HW90 | −0.92 | 0.64 | −0.72 | 0.77 |
| RAP1B | RAP1B, member of RAS oncogene family | P61224 | 1.28 | −0.84 | 1.04 | −0.75 |
| RALA | RAS-like proto-oncogene A | P11233 | 1.53 | −0.66 | 1.59 | −0.78 |
| RRAS | RAS related | A0A024QZF2 | 0.93 | −0.41 | 1.05 | −0.85 |
| RRAS2 | RAS related 2 | P62070 | 0.77 | −1.04 | 0.95 | 0.00 |
| SOD1 | Superoxide dismutase 1 | V9HWC9 | −1.91 | 0.62 | −2.19 | 1.30 |
| TXN | Thioredoxin | H9ZYJ2 | −1.22 | 0.00 | −1.87 | 1.19 |
| TXNRD1 | Thioredoxin reductase 1 | Q16881 | −1.15 | 0.74 | −0.93 | 1.09 |
| SOD2 | Superoxide dismutase 2 | P04179 | 1.44 | 0.00 | 1.65 | 0.00 |
| MAPK1 | Mitogen-activated protein kinase 1 | Q1HBJ4 | −0.728 | 0.00 | −0.939 | 0.00 |
| MAP2K2 | Mitogen-activated protein kinase kinase 2 | P36507 | −1.488 | 0.00 | −1.996 | 0.00 |

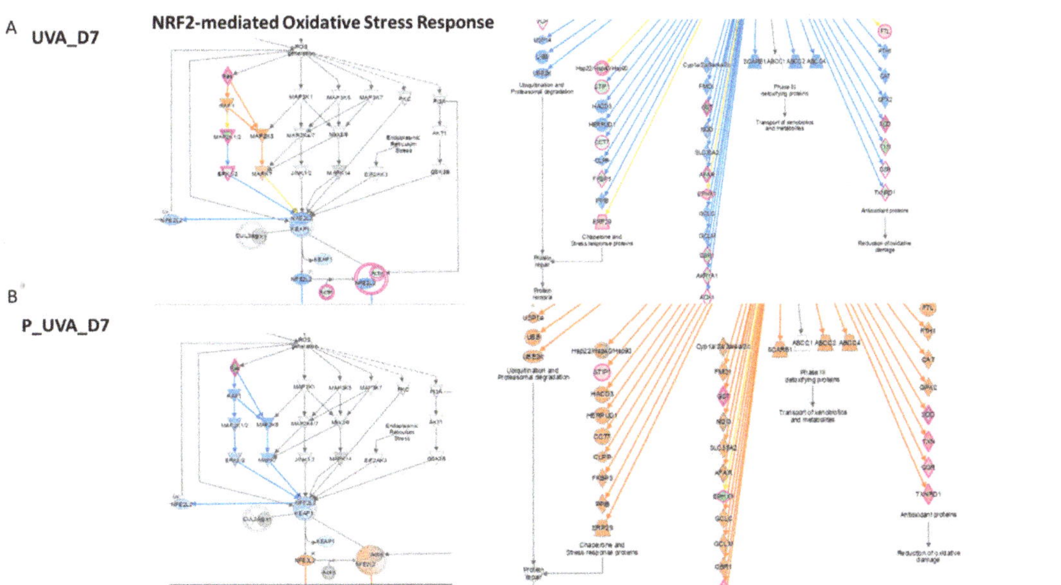


**Figure 5.** Genes involved in Nrf2 modulation after UVA exposure vs. control (**A**) after carnosine treatment (**B**) P_UVA vs. UVA (IPA). Red shows the increased genes, blue shows the decreased genes, grey shows the unaltered genes.

GSH is also a co-substrate for several detoxication enzymes such as glutathione peroxidase (GPx) and glutathione S-transferase (GST). Regeneration of oxidized GSH (GSSG) is mediated by an enzyme, glutathione reductase (GSR), that in this way participates in maintaining redox homeostasis [24]. Decreased GSH concentration, due to UVA-stimulated ROS production, results in the alteration of redox homeostasis in cells.

Moreover, the evidence of strong UVA-induced Oxidative Stress is found in the downregulation of Superoxide dismutase1 (SOD1, log2 ratio = −2.19), thioredoxin (TXN, log2 ratio = −1.87) and thioredoxin reductase1 (TXNRD1 log2 ratio = −0.93), proteins involved in redox cellular homeostasis and verified by gene expression too Specifically, UVA induced the increment of HMOX1, considered as critical ferroptosis-related gene (Figure 4B).

However, this trend has been observed to be reversed by carnosine after 7 days of treatment which inhibits UV-induced damage by restoring the proteins' level as the following: Glutathione S-transferase mu1 (GSTM1, log2 ratio = 1.00), Glutathione S-transferase mu3 (GSTM3, log2 ratio = 0.71), Glutathione S-transferase omega1 (GSTO1, log2 ratio = 1.24), Glutathione S-transferase pi1 (GSTP1, log2 ratio = 1.10) and Glutathione disulphide reductase (GSR, log2 ratio = 0.77). As well as a strong upregulation, the irradiated and treated series for 7 days was observed for Superoxide dismutase1 (SOD1, log2 ratio = 1.30), thioredoxin (TXN, log2 ratio = 1.19) and thioredoxin reductase1 (TXNRD1 log2 ratio = 1.09). Specifically, the variation of SOD1 and TXT is involved in the downregulation of the quantity of reactive oxygen species pathway (z score = −2.75) induced by carnosine. The activation of Nrf2 in fibroblasts also induces cellular senescence due to the activation of plasminogen activator inhibitor-1 being a key inducer of the senescence program which induces a deposit of a senescence-promoting matrix [21]. It has been also observed that Nrf2 mediated the fibroblast senescence with a faster re-epithelialization and accelerated wound closure. Additionally, UVA induces the activation of the Mitogen Activated Protein Kinase (MAPK) pathway which was restored by carnosine treatment. In addition, carnosine induced a reduction of apoptosis (Table 1) at both incubation time and the upregulation of the Nrf2 pathway to reduce Oxidative Stress and senescence too.

### 3.6. UVA Regulates the Glycolysis and the Oxidative Phosphorylation

In excess, oxidation can provoke metabolic failure, compromising cell viability by inactivating enzymes of Glycolysis and the Krebs cycle. ROS can inhibit Glycolysis allowing the cells to divert flux into the pentose phosphate pathway (ox-PPP) to promote NADPH synthesis and protect against Oxidative Stress [25].

The indirect evidence of the ROS effect evoked by UVA radiation after 7 days of carnosine treatment concerns the downregulation of multiple glycolytic enzymes, including glyceraldehyde 3-phosphate dehydrogenase (GAPDH, log2 ratio = −1.24), pyruvate kinase M2 (PKM, log2 ratio = −1.64), phosphofructokinase (PFKL, log2 ratio = −1.88) and pyruvate kinase M1/2 (PKM, log2 ratio = −1.64). It is well known that Glycolysis can divert to combat the Oxidative Stress. ROS inhibits multiple glycolytic enzymes, including glyceraldehyde 3-phosphate dehydrogenase, pyruvate kinase M2 and phosphofructokinase-1, directly to target cysteine residues [25]. While the nicotinamide adenine dinucleotide (NADH/NAD+) drives ATP production in the cytosol by Glycolysis and in the mitochondria by Oxidative Phosphorylation, nicotinamide adenine dinucleotide phosphate (NADPH/ NADP+) is mainly involved in the defense against reactive oxygen species [26,27]. Consistently, glycolytic inhibition could promote flux into the oxidative arm of the pentose phosphate pathway to generate NADPH. NADPH provides the reducing power that fuels the protein-based antioxidant system and its concentration is regulated by the pentose phosphate pathway. In addition, NADPH is the donor of the reductive potential to thioredoxins and glutathione, which in turn neutralize ROS. NADPH is, in fact, consumed by glutathione reductase (GSR) to recycle oxidized glutathione (GSSG). A metabolic switch from Glycolysis to Oxidative Phosphorylation or Oxidative Stress possibly correlated to ROS generation has been already observed [28]. NAD(P)H biomarker showed an early change at 30 min after exposure in dermal fibroblasts, while a significant change in epidermal keratinocytes was only observed at 2 h after exposure to UVA [28]. The results here highlighted the activation, mediated by UVA, of Oxidative Phosphorylation (OXPHOS) with the upregulation of several core subunits of NADH: ubiquinone oxidoreductase, complex I. Considering its heterogeneity, due to its 13 subunits, complex I is very susceptible to oxidative damage. Here, 10 subunits were detected as upregulated upon 7 days of treatment after UVA radiation. Specifically, NDUFS2 (core subunit 2 log2 ratio = 0.253), NDUFS3 (core subunit 3 log2 ratio = 0.658), NDUFS8 (core subunit 8 log2ratio = 0.93), NDUFV2 (core subunit V2 log2 ratio = 0.607), NDUFA13 (subunit A13 log2 ratio = 2.653), NDUFA2 (subunit A2 log2 ratio = 0.663), NDUFA5 (subunit A5 log2 ratio = 1.018), NDUFA8 (subunit A8 log2 ratio = 1.562), NDUFA9 (subunit A9 log2 ratio = 0.429) and NDUFB10 (subunit B10, log2 ratio = 1.413) (Table 5). Many therapeutic properties of carnosine as a pharmacological and cosmetic substance have been described either on 3D fibroblast cells or animals [29]. By penetrating at multilayer/spheroid fibroblast, carnosine is able to exert it effect on the mitochondrial metabolism, inducing an enhancement of glycolytic enzymes characterized by an increase in log 2 ratio as follows: (GAPDH, log2 ratio = 0.63), glucose-6-phosphate isomerase (GPI, log2 ratio = 1.28) and phosphoglycerate kinase1 (PGK1, log2 ratio = 0.78).

**Table 5.** Proteins involved in the Glycolysis pathway.

| Glycolysis | | | UVA_48 h | P_UVA_48 h | UVA_D7 | P_UVA_D7 |
|---|---|---|---|---|---|---|
| **Gene Name** | **Protein Name** | **Accession Number** | **Log Ratio** | **Log Ratio** | **Log Ratio** | **Log Ratio** |
| ALDOA | Aldolase, fructose-bisphosphate A | P04075 | −1.33 | 0.63 | −1.61 | 1.02 |
| ALDOC | Aldolase, fructose-bisphosphate C | A0A024QZ64 | −1.30 | 0.76 | −1.70 | 0.96 |
| ENO1 | Enolase 1 | A0A024R4F1 | −1.24 | 0.66 | −1.59 | 0.92 |
| ENO2 | Enolase 2 | Q6FHV6 | −1.26 | 0.74 | −2.14 | 1.58 |
| GAPDH | Glyceraldehyde-3-phosphate dehydrogenase | V9HVZ4 | −1.14 | 0.34 | −1.24 | 0.63 |
| GPI | Glucose-6-phosphate isomerase | K7EIL4 | −1.66 | 0.71 | −1.85 | 1.28 |
| PFKL | Phosphofructokinase, liver type | P17858 | −1.22 | −1.07 | −1.88 | 0.00 |
| PFKM | Phosphofructokinase, muscle | P08237 | −1.00 | 0.00 | −0.93 | 0.00 |
| PGAM1 | Phosphoglycerate mutase 1 | Q6FHU2 | −0.88 | 0.46 | −1.14 | 0.52 |
| PGK1 | Phosphoglycerate kinase 1 | V9HWF4 | −1.51 | 0.54 | −1.82 | 0.78 |
| PKM | Pyruvate kinase M1/2 | A0A024R5Z9 | −1.00 | −0.23 | −1.64 | 0.00 |
| TPI1 | Triosephosphate isomerase 1 | V9HWK1 | −1.35 | 0.85 | −1.61 | 1.35 |

### 3.7. *UVA Radiations Regulates the Sirtuin Signaling Pathway*

The network analysis reveals the modulation of 44 genes involved in the Sirtuin pathway which overall is downregulated (z score = −1.25) after 7 days related to the control. Sirtuins (SIRTs) comprise one of four classes of histone deacetylases (HDACs; I–IV) that play important roles in a variety of cellular functions. Here, three isoforms of linker histone, namely H1.0 (log 2 ratio = 1.91), H1.4 (log2 ratio = 1.50), H1.5 (log2 ratio = 1.94), were found to be strongly upregulated upon the UVA exposure. Several studies have demonstrated that SIRT1 levels decrease with age in dermal fibroblasts isolated from female donors who ranged from 20–67 years old [30]. Specifically, the case for SIRT1 involvement in UVB-mediated DNA damage has been observed in several in vitro experiments using human fibroblasts. The UVB radiation has been shown to decrease SIRT1 protein levels [31]. However, although SIRT1 has been observed to decrease in in vitro studies, Lang et al. showed that SIRT4 levels increase in fibroblasts exposed to UVB radiation and this correlates with an increase in cellular senescence [32]. On the basis of these observations, the role of SIRT1 as a tumor promoter or suppressor in UVB-induced cancer initiation is unclear and might vary with cell/tissue type or protein levels. However, the finding of this study on 3D spheroids agrees with those observed in nude mice skin upon UVA exposure [8].

## 4. Conclusions

UVA exposure is a powerful attack on the skin, in particular, the dermal compartment determining a non-reversible dermal damage affecting the skin elasticity and inducing photoaging and photo-carcinogenesis. Within the very early events of the photoaging process, the ROS generation, by promoting mitochondrial electron transport chain damage, can be considered the most relevant. The present study, based on a combination of network analyses and high-resolution mass spectrometry applied to advanced micro-physiological scaffold-free 3D human spheroids underlines the protective role of carnosine against UVA-induced damages and reveals wide mechanisms of action on several pathways and molecular signaling. Indeed, the response to carnosine treatment highlighted the modulation of several pathways, in particular, related to mitochondrial protection. The modulation of multiple pathways such as Oxidative Phosphorylation, Glycolysis I and the Nrf2-mediated Oxidative Stress Response, highlight both the role of mitochondrial dysfunctions induced by UVA and the protective role of carnosine against them; as well, it was associated with a positive effect on fibrillin, which as early marker of dermal matrix remodeling, confirms the response to UVA stimulus and the protective role of carnosine.

In conclusion, due to the advanced micro-physiological dermal spheroids system conjugated by genomics and proteomics, it has been possible in a relatively short experimental window to mirror the cascade of complex events that follow UVA exposure and to observe the recovery from the damage as it physiologically occurs in vivo thanks to an antioxidant reference molecule, carnosine, that was confirmed and able to offer the protection of dermis against early and delayed UVA damages at the dermal level.

**Supplementary Materials:** The following supporting information can be downloaded at: https://www.mdpi.com/article/10.3390/antiox12020300/s1, Table S1. List of quantified proteins obtained by MaxQuant software.

**Author Contributions:** Conceptualization, M.C., A.D. and M.M.; methodology, F.R., G.A. (Gilda Aiello) and A.D.; formal analysis, F.R., G.A. (Gilda Aiello), A.D. and B.Z. resources, M.C., G.A (Giancarlo Aldini) and M.M.; data curation, F.R., G.A. (Gilda Aiello) and A.D.; writing—original draft preparation and review, all authors. All authors have read and agreed to the published version of the manuscript.

**Funding:** This research received no external funding.

**Data Availability Statement:** Data is contained within the article and supplementary material.

**Acknowledgments:** We thank UNITECH OMICs, mass spectrometry platform of Università degli Studi di Milano, for running mass spectrometry analyses and Robin Gradin, Data Scientist at SenzaGen AB, Sweden for NanoString™ data biostatistical analysis.

**Conflicts of Interest:** The authors declare no conflict of interest.

## References


1. D'Orazio, J.; Jarrett, S.; Amaro-Ortiz, A.; Scott, T. Uv radiation and the skin. *Int. J. Mol. Sci.* **2013**, *14*, 12222–12248. [CrossRef] [PubMed]
2. de Jager, T.L.; Cockrell, A.E.; Du Plessis, S.S. Ultraviolet light induced generation of reactive oxygen species. In *Ultraviolet Light in Human Health, Diseases and Environment*; Springer: Cham, Switzerland, 2017; Volume 996, pp. 15–23. [CrossRef]
3. Spickett, C.M.; Pitt, A.R. Modification of proteins by reactive lipid oxidation products and biochemical effects of lipoxidation. *Protein Oxid.* **2020**, *64*, 19–31. [CrossRef] [PubMed]
4. He, H.L.; Xiong, L.D.; Jian, L.E.; Li, L.M.; Wu, Y.L.; Qiao, S.Y. Role of mitochondria on uv-induced skin damage and molecular mechanisms of active chemical compounds targeting mitochondria. *J. Photochem. Photobiol. B-Biol.* **2022**, *232*, 112464. [CrossRef] [PubMed]
5. Liu, Y.H.; Xu, G.Z.; Sayre, L.M. Carnosine inhibits (e)-4-hydroxy-2-nonenal-induced protein cross-linking: Structural characterization of carnosine-hne adducts. *Chem. Res. Toxicol.* **2003**, *16*, 1589–1597. [CrossRef]
6. Davies, S.S.; Brantley, E.J.; Voziyan, P.A.; Amarnath, V.; Zagol-Ikapitte, I.; Boutaud, O.; Hudson, B.G.; Oates, J.A.; Roberts, L.J. Pyridoxamine analogues scavenge lipid-derived gamma-ketoaldehydes and protect against $H_2O_2$-mediated cytotoxicity. *Biochemistry* **2006**, *45*, 15756–15767. [CrossRef]
7. Aiello, G.; Rescigno, F.; Meloni, M.; Baron, G.; Aldini, G.; Carini, M.; D'Amato, A. Oxidative stress modulation by carnosine in scaffold free human dermis spheroids model: A proteomic study. *Int. J. Mol. Sci.* **2022**, *23*, 1468. [CrossRef]
8. Silvia, R.; Marina, C.; Baron, G.; Giancarlo, A.; Anne, N.S.; Alfonsina, D. Study of carnosine's effect on nude mice skin to prevent uv-a damage. *Free. Radic. Biol. Med.* **2021**, *173*, 97–103. [CrossRef]

9. Nakyai, W.; Tissot, M.; Humbert, P.; Grandmottet, F.; Viyoch, J.; Viennet, C. Effects of repeated uva irradiation on human skin fibroblasts embedded in 3d tense collagen matrix. *Photochem. Photobiol.* **2018**, *94*, 715–724. [CrossRef]
10. Meloni, M.; Farina, A.; de Servi, B. Molecular modifications of dermal and epidermal biomarkers following uva exposures on reconstructed full-thickness human skin. *Photochem. Photobiol. Sci.* **2010**, *9*, 439–447. [CrossRef]
11. Fabregat, A.; Jupe, S.; Matthews, L.; Sidiropoulos, K.; Gillespie, M.; Garapati, P.; Haw, R.; Jassal, B.; Korninger, F.; May, B.; et al. The reactome pathway knowledgebase. *Nucleic Acids Res.* **2018**, *46*, D649–D655. [CrossRef]
12. Beckerle, M.C. Zyxin: Zinc fingers at sites of cell adhesion. *Bioessays* **1997**, *19*, 949–957. [CrossRef] [PubMed]
13. Pylayeva-Gupta, Y.; Grabocka, E.; Bar-Sagi, D. Ras oncogenes: Weaving a tumorigenic web. *Nat. Rev. Cancer* **2011**, *11*, 761–774. [CrossRef] [PubMed]
14. Perrin, B.J.; Amann, K.J.; Huttenlocher, A. Proteolysis of cortactin by calpain regulates membrane protrusion during cell migration. *Mol. Biol. Cell* **2006**, *17*, 239–250. [CrossRef]
15. Battie, C.; Jitsukawa, S.; Bernerd, F.; Del Bino, S.; Marionnet, C.; Verschoore, M. New insights in photoaging, uva induced damage and skin types. *Exp. Dermatol.* **2014**, *23*, 7–12. [CrossRef]
16. Basu, P.; Martins-Green, M. Signaling pathways associated with chronic wound progression: A systems biology approach. *Antioxidants* **2022**, *11*, 1506. [CrossRef]
17. Keene, D.R.; Engvall, E.; Glanville, R.W. Ultrastructure of type-vi collagen in human-skin and cartilage suggests an anchoring function for this filamentous network. *J. Cell Biol.* **1988**, *107*, 1995–2006. [CrossRef]
18. Kansanen, E.; Kuosmanen, S.M.; Leinonen, H.; Levonen, A.-L. The keap1-nrf2 pathway: Mechanisms of activation and dysregulation in cancer. *Redox Biol.* **2013**, *1*, 45–49. [CrossRef] [PubMed]
19. Sykiotis, G.P.; Bohmann, D. Stress-activated cap'n'collar transcription factors in aging and human disease. *Sci. Signal.* **2010**, *3*, re3. [CrossRef]
20. Ludtmann, M.H.R.; Angelova, P.R.; Zhang, Y.; Abramov, A.Y.; Dinkova-Kostova, A.T. Nrf2 affects the efficiency of mitochondrial fatty acid oxidation. *Biochem. J.* **2014**, *457*, 415–424. [CrossRef]
21. Hiebert, P.; Wietecha, M.S.; Cangkrama, M.; Haertel, E.; Mavrogonatou, E.; Stumpe, M.; Steenbock, H.; Grossi, S.; Beer, H.D.; Angel, P.; et al. Nrf2-mediated fibroblast reprogramming drives cellular senescence by targeting the matrisome. *Dev. Cell* **2018**, *46*, 145–161.e10. [CrossRef]
22. Ryu, H.C.; Kim, C.; Kim, J.Y.; Chung, J.H.; Kim, J.H. UVB Radiation Induces Apoptosis in Keratinocytes by Activating a Pathway Linked to "BLT2-Reactive Oxygen Species". *J. Investig. Dermatol.* **2010**, *130*, 1095–1106. [CrossRef] [PubMed]
23. Mavrogonatou, E.; Angelopoulou, M.; Rizou, S.V.; Pratsinis, H.; Gorgoulis, V.G.; Kletsas, D. Activation of the JNKs/ATM-p53 axis is indispensable for the cytoprotection of dermal fibroblasts exposed to UVB radiation. *Cell Death Dis.* **2022**, *13*, 647. [CrossRef] [PubMed]
24. Panich, U.; Sittithumcharee, G.; Rathviboon, N.; Jirawatnotai, S. Ultraviolet radiation-induced skin aging: The role of dna damage and oxidative stress in epidermal stem cell damage mediated skin aging. *Stem Cells Int.* **2016**, *2016*, 7370642. [CrossRef]
25. Mullarky, E.; Cantley, L.C.; Nakao, K.; Minato, N.; Uemoto, S. Diverting glycolysis to combat oxidative stress. In *Innovative Medicine: Basic Research and Development*; Springer: Tokyo, Japan, 2015.
26. Ying, W.H. $NAD^+$/NADH and $NADP^+$/NADPH in cellular functions and cell death: Regulation and biological consequences. *Antioxid. Redox Signal.* **2008**, *10*, 179–206. [CrossRef]
27. Blacker, T.S.; Mann, Z.F.; Gale, J.E.; Ziegler, M.; Bain, A.J.; Szabadkai, G.; Duchen, M.R. Separating nadh and nadph fluorescence in live cells and tissues using flim. *Nat. Commun.* **2014**, *5*, 3936. [CrossRef]
28. Ung, T.P.L.; Lim, S.; Solinas, X.; Mahou, P.; Chessel, A.; Marionnet, C.; Bornschlögl, T.; Beaurepaire, E.; Bernerd, F.; Pena, A.M.; et al. Simultaneous NAD(P)H and FAD fluorescence lifetime microscopy of long uva-induced metabolic stress in reconstructed human skin. *Sci. Rep.* **2021**, *11*, 22171. [CrossRef] [PubMed]
29. Boldyrev, A.A.; Aldini, G.; Derave, W. Physiology and pathophysiology of carnosine. *Physiol. Rev.* **2013**, *93*, 1803–1845. [CrossRef] [PubMed]
30. Kalfalah, F.; Sobek, S.; Bornholz, B.; Götz-Rösch, C.; Tigges, J.; Fritsche, E.; Krutmann, J.; Köhrer, K.; Deenen, R.; Ohse, S.; et al. Inadequate mito-biogenesis in primary dermal fibroblasts from old humans is associated with impairment of pgc1a-independent stimulation. *Exp. Gerontol.* **2014**, *56*, 59–68. [CrossRef]
31. Wahedi, H.M.; Lee, T.H.; Moon, E.-Y.; Kim, S.Y. Juglone up-regulates sirt1 in skin cells under normal and uvb irradiated conditions. *J. Dermatol. Sci.* **2016**, *81*, 210–212. [CrossRef]
32. Lang, A.; Grether-Beck, S.; Singh, M.; Kuck, F.; Jakob, S.; Kefalas, A.; Altinoluk-Hambüchen, S.; Graffmann, N.; Schneider, M.; Lindecke, A.; et al. Microrna-15b regulates mitochondrial ros production and the senescence-associated secretory phenotype through sirtuin 4/sirt4. *Aging-Us* **2016**, *8*, 484–505. [CrossRef]

*antioxidants*

MDPI

*Article*

# Cosmeceutical Effects of *Ishige okamurae* Celluclast Extract

**Fengqi Yang [1,2], Jimin Hyun [1], D. P. Nagahawatta [1], Young Min Kim [3], Moon-Soo Heo [1,*] and You-Jin Jeon [1,*]**

1 Department of Marine Life Sciences, Jeju National University, Jeju 63243, Republic of Korea
2 Marine Science Institute, Jeju National University, Jeju 63333, Republic of Korea
3 Aqua Green Technology Co., Ltd., Smart Bldg., Jeju Science Park, Cheomdan-ro, Jeju 63309, Republic of Korea
* Correspondence: msheo@jejunu.ac.kr (M.-S.H.); youjinj@jejunu.ac.kr (Y.-J.J.); Tel.: +82-64-754-3475 (Y.-J.J.)

**Abstract:** Sulfated polysaccharides extracted from brown algae are unique algal polysaccharides and potential ingredients in the cosmeceutical, functional food, and pharmaceutical industries. Therefore, the present study evaluated the cosmeceutical effects, including antioxidant, anti-wrinkle, anti-inflammation, and photoprotective activities, of *Ishige okamurae* Celluclast extract (IOC). The IOC was abundant in sulfated polysaccharides (48.47%), polysaccharides (44.33%), and fucose (43.50%). Moreover, the IOC effectively scavenged free radicals, and its anti-inflammatory properties were confirmed in lipopolysaccharide-induced RAW 264.7 macrophages; therefore, the IOC may produce auxiliary effects by inhibiting reactive oxygen species (ROS). In vitro (Vero cells) and in vivo (zebrafish) studies further confirmed that the IOC produced a protective effect against hydrogen-peroxide-induced oxidative stress in a dose-dependent manner. In addition, the IOC suppressed intracellular ROS and apoptosis and enhanced HO-1 and SOD-1 expression through transcriptional activation of Nrf2 and downregulation of Keap1 in HaCaT cells. Furthermore, the IOC exhibited a potent protective effect against ultraviolet-B-induced skin damage and photoaging. In conclusion, the IOC possesses antioxidant, anti-inflammatory, and photoprotective activities, and can, therefore, be utilized in the cosmeceutical and functional food industries.

**Keywords:** *Ishige okamurae*; Celluclast; sulfated polysaccharides; cosmeceutical effect; antioxidant activity; anti-inflammatory activity; photoprotective effect

**Citation:** Yang, F.; Hyun, J.; Nagahawatta, D.P.; Kim, Y.M.; Heo, M.-S.; Jeon, Y.-J. Cosmeceutical Effects of *Ishige okamurae* Celluclast Extract. *Antioxidants* **2022**, *11*, 2442. https://doi.org/10.3390/antiox11122442

Academic Editors: Irene Dini and Sonia Laneri

Received: 16 November 2022
Accepted: 8 December 2022
Published: 10 December 2022



## 1. Introduction

Reactive oxygen species (ROS), which are present in various forms, are physiological metabolites generated through oxygen metabolism under normal physiological conditions. Various factors, including ultraviolet-B (UVB) radiation, heavy metal ions, drugs, pollutants, and fine dust particles, promote ROS production [1–3]. However, ROS overproduction can interfere with the mitochondrial membrane potential and disrupt the cellular redox balance, resulting in the induction of oxidative stress and development of numerous diseases, including neurological disorders, cardiovascular disease, digestive diseases, and certain cancers [4,5]. As the skin is inevitably exposed to UVB rays in daily life, this can lead to oxidative damage of the skin tissues and cells. Excessive ROS accumulation leads to photoaging and inflammation of the human skin, eventually resulting in dermal malignancies [6,7]. Moreover, some cosmetics containing specific chemicals, such as heavy metals, steroids, hydroquinones, and nitrosamines, albeit effective in the short-term, can harm human health in the long-term [8]. Therefore, eco-friendly cosmetics containing natural ingredients must be developed. In recent years, cosmetic products with non-toxic and highly bioactive natural ingredients have attracted increasing research attention.

In this context, seaweeds have found wide pharmaceutical and industrial applications as natural sources of compounds because of their non-toxicity, easy cultivation, low cost, and presence of an array of active compounds, such as polysaccharides, proteins, minerals, vitamins, and pigments; therefore, seaweeds are the best candidates to replace synthetic compounds [9–11]. The majority of the algal species contain polysaccharides, polyphenols, and peptides, and most of these compounds exhibit biological functions, such as antioxidant, antimicrobial, antitumoral, anti-inflammatory, and UVB protective effects [12,13].

*Ishige okamurae* is one of the most common edible brown algae. It is widely distributed in the coastal areas of East Asia and contains abundant polyphenols, fucoxanthin, diphlorethohydroxycarmalol, and ishigoside bioactive compounds [7,14,15]. Many activities of *I. okamurae* have been investigated. For instance, the ethanolic extract of *I. okamurae* was effective against particulate matter-induced skin damage and produced antibacterial effects. In addition, the protective effects of the methanolic extract of *I. okamurae* against inflammatory myopathy as well as its anti-obesity, antioxidant, and anticancer activities have been reported [11,15–17]. Although there are many studies on *I. okamurae*, there is little information available on the cosmeceutical effects of *I. okamurae*, especially the Celluclast extract. A variety of extraction technologies are widely used, such as enzyme-assisted extraction, ultra-high-pressure extraction, microwave-assisted extraction technology, and supercritical fluid extraction. Compared with organic and water extraction, enzymatic extraction has a high catalytic efficiency and is more conducive to destroying the cell walls, which can not only improve the extraction yield but also show better biological activity, which is suitable for algae extraction [18,19]. In addition, enzyme-assisted extraction technology has mild reaction conditions and is safe. Compared with microwave and ultra-high-pressure extraction technologies, it has low equipment requirements, simple operation, is solvent-free, and has a moderate cost, so it is more suitable for food and cosmetic industrial production.

To this end, in the present study, Celluclast, a food-grade carbohydrase, was used to hydrolyze *I. okamurae*, and cosmeceutical effects of the *I. okamurae* Celluclast extract (IOC) were evaluated. Specifically, combining chemical technology, zebrafish, and cellular-level research, the antioxidant, anti-inflammatory, anti-wrinkle, and photoprotective activities of the IOC were evaluated to expand the applications of this brown alga in the cosmetic industry.

## 2. Materials and Methods

### *2.1. Materials*

The collagenase from clostridium histolyticum, elastase from porcine pancreas, hyaluronidase, 2′,7′-dichlorodihydroflurescin diacetate (DCFH-DA), acridine orange, 3-(4-5-dimethyl-2yl)-2-5-diphynyltetrasolium bromide (MTT), dimethyl sulfoxide (DMSO), phosphate-buffered saline (PBS), standard gallic acid, 2,2′-azino-bis(3-ethylbenzothiazoline-6-sulphonic acid) (ABTS), $H_2O_2$, and the Human matrix metalloproteinases (MMP)-1, 2, 9 Enzyme-Linked Immunosorbent Assay (ELISA) kit were purchased from Sigma Co. (St. Louis, MO, USA). Penicillium/streptomycin (P/S), Dulbecco's modified Eagle's medium (DMEM), Roswell Park Memorial Institute-1640 (RPMI-1640) medium, trypsin-EDTA, and fetal bovine serum (FBS) were purchased from Gibco BRL (Life Technologies, Burlington, ON, Canada). The PGE2 kit used in the experiments was purchased from R&D Systems (Minneapolis, MN, USA). Anti-heme oxygenase 1 (HO-1), anti-superoxide dismutase 1 (SOD1), and anti-nuclear factor-erythroid 2-related factor-2 (Nrf2) antibodies were purchased from Cell Signaling Technology (Beverly, MA, USA). Kelch-like ECH-associated protein 1 (keap1) and β-actin antibody were purchased from Santa Cruz Biotechnology (Santa Cruz, CA, USA). All other chemicals used in this study were of analytical grade.

### *2.2. Sample Extraction*

*I. okamurae* was harvested in April 2020 from Seongsan, Jeju Island, Republic of Korea. Following immersion in water for 24 h, the algae were washed three times with tap water to remove salt, sand, and debris. The algae were then freeze-dried and crushed. Lyophilized *I. okamurae* powder (100 g) was mixed with distilled water (1 L) at pH 4.5. Next, the mixture was extracted using Celluclast (1 g) in a continuous shaking incubator for 6 h at 50 °C. Subsequently, the enzyme was inactivated (100 °C) for 15 min. The pH of the extract was adjusted to 7.0, and the supernatant was separated using centrifugation (12,000× *g*, 15 min, 4 °C). The supernatant was then concentrated by evaporation and stored in a −80 °C deep freezer before freeze-drying. IOC powder was obtained by freeze-drying the mentioned supernatant.

### 2.3. Chemical Composition Analysis

Protein content was quantified using a commercial BCA protein assay kit (Thermo Scientific, Rockford, IL, USA). Sulfate content was analyzed using the $BaCl_2$ gelatin method. The samples were hydrolyzed using 4 M trifluoroacetic acid for 5 h at 100 °C [20]. Gallic acid and glucose were used as the standards to evaluate the total phenolic and polysaccharide content, respectively [21,22]. The monosaccharide content of samples was determined using the Bio-LC (HPAEC-PAD system; Dionex, Sunnyvale, CA, USA) with the CarboPacTM PA1 column following pretreatment with 2 M trifluoroacetic acid. The samples were detected with the ED50 Dionex electrochemical detector using a standard containing fucose, arabinose, galactose, rhamnose, glucose, xylose, and fructose.

### 2.4. Radical Scavenging Assay

The radical scavenging activity was investigated using a previously published method [23]. In brief, ABTS (7 mM) and potassium persulfate (2.45 mM) were reacted in the dark for 12 h at 25 °C to generate the radicals. The ABTS and sample solutions were mixed and provided 10 min of incubation in the dark. The radical scavenging activity of the samples was evaluated at 734 nm.

### 2.5. Measurement of Enzyme Inhibitory Effects of IOC

#### 2.5.1. Collagenase Inhibitory Effect Assay

The collagenase inhibitory activity was estimated using 1 mg of Azocoll with 800 μL of 0.1 M Tris–HCl buffer (pH 7.0). An amount of 100 μL of 200 units·$mL^{-1}$ of collagenase and 100 μL of the sample were added with 1 h incubation at 43 °C. Following incubation, the mixture was centrifuged (3000 rpm, 10 min) and the supernatant was measured at 540 nm [24].

#### 2.5.2. Elastase Inhibitory Effect Assay

The elastase inhibitory activity was estimated according to the previous method [25]. In brief, elastase was treated with a mixture of the sample and 1.015 mM N-succinyl-ala-ala-ala-ala-p-nitroanilide, and 10 min of incubation was provided at 25 °C. The microplate reader (BioTek Synergy HT, BioTek Instrument, Winooski, UT, USA) was utilized to measure the absorbance at 410 nm.

#### 2.5.3. Hyaluronidase (HAase) Inhibitory Assay

The HAase inhibitory activity was estimated following the previous method with slight modifications [26]. Briefly, the sample was mixed with HAase (10 mg·$mL^{-1}$) dissolved in acetate buffer (pH 3.5) at a 1:1 ratio and provided 20 min of incubation at 37 °C. A further 20 min of incubation was provided with $CaCl_2$ (12.5 mM). Then, hyaluronate was treated as a substrate and provided another 40 min of incubation. Finally, the mixture solution was incubated at 100 °C for 3 min after treating NaOH (0.4 N) and potassium tetraborate (0.4 M), and 180 μL of p-dimethylaminobenzaldehyde was added at 25 °C. The microplate reader (BioTek Synergy HT) was utilized to measure the absorbance at 540 nm.

### 2.6. Cell Viability Analysis

RAW 264.7, Monkey kidney fibroblasts (Vero), and human keratinocyte (HaCaT) cell lines were purchased from the Korean Cell Line Bank (KCLB, Seoul, Republic of Korea). Vero cells were maintained in RPMI-1640, and RAW 264.7 and HaCaT cells were cultured in DMEM under a controlled environment (37 °C, 5% $CO_2$). All media contained 1% antibiotics and 10% fetal bovine serum. All cells were seeded at the density of $1 \times 10^5$ cells·$mL^{-1}$

in 96-well plates and cultured for 24 h. The cells were treated with varying concentrations of IOC (3.125, 6.25, 12.5, 25, 50, 100, and 200 $\mu g \cdot mL^{-1}$) and cell viability was evaluated using the MTT assay.

### *2.7. Effect of the IOC on $H_2O_2$-Induced Intracellular ROS Production and Cell Viability*

After 24 h of Vero cells seeding, different concentrations of the IOC (3.125–200 $\mu g \cdot mL^{-1}$) were treated, incubated for 24 h, and then co-treated with hydrogen peroxide (600 μM). The cytoprotective effect of the IOC against $H_2O_2$-induced cell damage was assessed using cell viability. Cells were incubated with $H_2O_2$ for 1 h, and then 10 μL of DCFH-DA (25 $\mu g \cdot mL^{-1}$) was added and incubated for another 10 min. Excitation at 485 nm and emission at 530 nm were measured using a microplate reader (BioTek Synergy HT) [27].

### *2.8. Effect of the IOC NO and PGE2 Generation on Lipopolysaccharide (LPS)-Induced RAW 264.7 Macrophages*

To evaluate the anti-inflammatory effect of the IOC, RAW 264.7 cells were treated with IOC for 1 h, following co-treatment with 1 $\mu g \cdot mL^{-1}$ of LPS for 23 h. Cell viability was assessed using the MTT assay. The supernatant was collected. NO and PGE2 production were analyzed using the Griess reagent and commercial ELISA kit, respectively [25].

### *2.9. Cytoprotective Activity of the IOC on UVB-Induced HaCaT Cells*

A UVB meter containing a fluorescent bulb (wavelength 315–280 nm, peak 313 nm) was used to irradiate (30 $mJ \cdot cm^{-2}$) HaCaT cells to induce photodamage. Serum-free DMEM medium was used for the incubation of cells until analysis. The effect of the IOC (3.125–200 $\mu g \cdot mL^{-1}$) on UVB-induced photodamage was determined based on the measurement of intracellular ROS levels, apoptotic body formation, and UVB-irradiated HaCaT cell viability using the DCFH-DA assay, Hoechst 33342 staining, and MTT assay, respectively [24,28,29].

### *2.10. Western Blotting*

The effects of the IOC on the expression of SOD-1, Nrf2, keap1, and HO-1 were assessed using Western blotting. IOC-treated and UVB-irradiated HaCaT cells were harvested and lysed. The BCA protein assay was used to quantify protein content in the lysate and it was subjected to 7.5% sodium dodecyl sulfate (SDS)–polyacrylamide gel electrophoresis (PAGE). The proteins were transferred onto nitrocellulose membranes blocked with 5% skim milk at 27 °C. Overnight incubation of membranes with primary antibodies at 4 °C was performed following incubation with secondary antibodies (Santa Cruz Biotechnology, Paso Robles, CA, USA) for 2 h at room temperature. Bands were detected using an ECL Western blotting detection kit and photographed using the FUSION SOLO Vilber Lourmat system (Paris, France).

### *2.11. Determination of Metalloproteinase (MMPs) Expression Levels Using ELISA*

HaCaT cells were incubated with the IOC (50, 100, and 200 $\mu g \cdot mL^{-1}$) for 2 h and washed with PBS before UVB irradiation (30 $mJ \cdot cm^{-2}$). Following UVB irradiation, the cells were incubated with serum-free DMEM for 24 h. Culture media were collected, and MMP expression levels were quantified using a commercial ELISA kit according to the manufacturer's instructions.

### *2.12. Origin and Maintenance of Zebrafish*

Adult zebrafish were purchased from the commercial market (Jeju Aquarium, Republic of Korea). The zebrafish were separately maintained in 3 L tanks under a 14 h light/10 h dark cycle at 28.5 °C and fed twice daily (Tetra GmgH D-49304, Melle, Germany). In the morning, embryos were obtained from natural spawning through the breeding of two males and one female.

### 2.13. Measurement of Heart Beating Rate, Survival Rate, ROS Generation, and Cell Death in Zebrafish Embryos

The protective effect of IOC against $H_2O_2$ was investigated using approximately 7–9 h post-fertilization (hpf) embryos by treating IOC (50, 100, and 200 $\mu g \cdot mL^{-1}$); controls were untreated. Following treatment for 1 h, the embryos were stimulated with $H_2O_2$ (5 mM), and the plate was incubated for 3 dpf. The survival rate and heart beating rate were counted according to the previously optimized method [30]. Acridine orange and DCFH-DA were utilized to evaluate the cell death and intracellular ROS generation, respectively [31]. Zebrafish larvae were photographed using a fluorescence microscope (CoolSNAP-Pro Color Digital Camera; Olympus, Japan) and the fluorescence intensity was quantified using ImageJ software (version 1.50i, NIH, USA).

### 2.14. Statistical Analysis

The data were analyzed using one-way ANOVA and Dunnett's multiple comparisons test by GraphPad Prism 8 (GraphPad Software, Inc., San Diego, CA, USA). All the experiments were repeated in triplicate. All data are expressed as the mean values with the standard error of the mean (SEM). $p < 0.05$ was considered significantly different.

## 3. Results

### 3.1. Chemical Composition of IOC

Chemical composition was investigated after *I. okamurae* was hydrolyzed using Cellucalst. As shown in Table 1, the IOC contained 5.32 ± 0.43% protein, 3.82 ± 0.31% phenolics, 4.14 ± 0.12% sulfates, and 44.33 ± 0.65% polysaccharides. Overall, the IOC contained 48.47% sulfated polysaccharides. Furthermore, the monosaccharide content of the IOC was determined. The IOC contained 43.50% fucose, 36.39% glucose, 7.96% galactose, 11.74% xylose, 0.17% rhamnose, and 0.25% arabinose.

**Table 1.** Chemical composition of IOC obtained from *I. okamurae*.

| Sample | | IOC |
|---|---|---|
| Protein content% | | 5.32 ± 0.43 |
| Phenolic content% | | 3.82 ± 0.31 |
| Sulfate content% | | 4.14 ± 0.12 |
| Polysaccharide content% | | 44.33 ± 0.65 |
| Sulfated polysaccharide% | | 48.47 |
| Monosaccharide % | Fucose | 43.50 |
| | Glucose | 36.39 |
| | Galactose | 7.96 |
| | Xylose | 11.74 |
| | Rhamnose | 0.17 |
| | Arabinose | 0.25 |

### 3.2. Enzyme Inhibitory Effects of the IOC

The inhibitory effects of the IOC on commercial collagenase, elastase, and HAase were examined, and the results are presented in Figure 1. The IOC significantly and dose-dependently increased collagenase and HAase inhibitory rates. The elastase inhibition rate did not differ among the concentration groups but was significantly increased relative to that in the control sample. At the IOC concentrations of 25, 50, and 100 $\mu g \cdot mL^{-1}$, the inhibitory rates were 18.01%, 34.11%, and 55.62% against collagenase; 18.85%, 23.20%, and 23.43% against elastase; 1.83%, 13.56%, and 40.23% against HAase, respectively. Therefore, the IOC may produce anti-wrinkle effects through the inhibition of collagenase, elastase, and HAase.

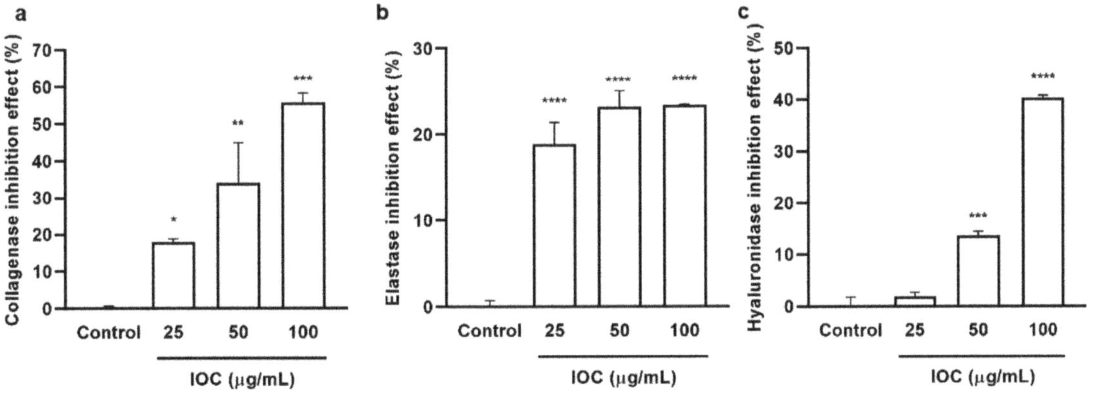


**Figure 1.** IOC inhibits commercial collagenase, elastase, and hyaluronidase: (**a**) collagenase inhibitory activity of IOC; (**b**) elastase inhibitory activity of IOC; (**c**) hyaluronidase inhibitory activity of IOC. All experiments were triplicated, and data are expressed as the mean ± SEM. * $p < 0.05$, ** $p < 0.01$, *** $p < 0.001$, and **** $p < 0.0001$ as compared to control group.

### 3.3. *Antioxidant Effect of the IOC*

#### 3.3.1. Free Radical Scavenging Activities of IOC

The free radical scavenging activity of the IOC was assessed using the ABTS assay. The activity of the IOC against ABTS radical cations was measured using concentrations ranging between 25 and 100 $\mu g \cdot mL^{-1}$. The percent radical scavenging activity at different concentrations of the extract is shown in Figure 2a. Significant scavenging activity was noted at 50 and 100 $\mu g \cdot mL^{-1}$ concentrations compared with the control value. In particular, the ABTS radical scavenging activity exceeded 96% at the concentration of 100 $\mu g \cdot mL^{-1}$.

#### 3.3.2. Effect of IOC on $H_2O_2$-Induced Intracellular ROS Generation and Cytotoxicity in Vero Cells

Chemical methods revealed that the IOC possesses antioxidant and anti-wrinkle activities. To further detect IOC activities, in vitro experiments were performed, and the sample concentration range was expanded. Specifically, seven concentrations (3.125–200 $\mu g \cdot mL^{-1}$) were tested to determine cytotoxicity. Cell viability did not differ significantly among the different concentrations. Hence, these concentrations (3.125–200 $\mu g \cdot mL^{-1}$) were used in further experiments. The intracellular ROS levels and viability of Vero cells following $H_2O_2$ stimulation are shown in Figure 2. There were no significant differences in intracellular ROS levels and cell viability between the 3.125 and 6.25 $\mu g \cdot mL^{-1}$ concentration groups; however, 12.5 to 200 $\mu g \cdot mL^{-1}$ concentration groups significantly differed from the control group and the significance of differences was dose-dependent. ROS generation tended to decrease (Figure 2b), whereas cell viability tended to increase in 12.5–200 $\mu g \cdot mL^{-1}$ concentration groups compared with values in the control group (Figure 2c).

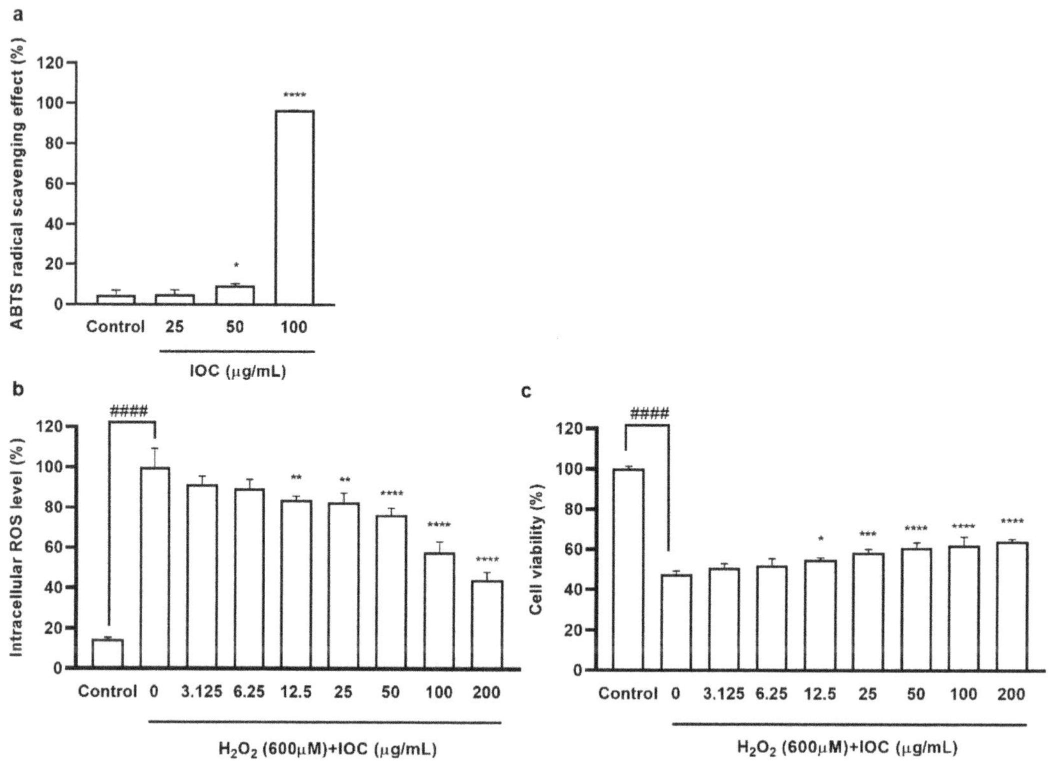


**Figure 2.** (**a**) ABTS free radical scavenging activity of IOC; (**b**) the intracellular ROS scavenging effect of IOC during $H_2O_2$ stimulated oxidative stress in Vero cells; (**c**) the protective effect of IOC against $H_2O_2$ stimulated cell death in Vero cells. All experiments were triplicated, and data are expressed as the mean ± SEM. In ABTS radical scavenging activity results: * $p < 0.05$ and **** $p < 0.0001$ as compared to control group. In intracellular ROS level and cell viability results: * $p < 0.05$, ** $p < 0.01$, *** $p < 0.001$, and **** $p < 0.0001$ as compared to the $H_2O_2$-treated group, and #### $p < 0.0001$ as compared to the control group.

### 3.4. Effect of the IOC on NO and PGE2 Production in LPS-Induced RAW 264.7 Cells

To evaluate the anti-inflammatory effects of the IOC, we assessed the suppression of NO and PGE2 production in LPS-induced RAW 264.7 macrophages. We selected the concentration of 3.125–200 μg·mL$^{-1}$ based on its non-cytotoxicity to macrophages (data not shown). As shown in Figure 3, when cells were exposed to LPS, NO and PGE2 production was significantly increased but cell viability was reduced compared with the control values. NO secretion was downregulated following IOC pretreatment in a dose-dependent manner, and cell viability increased significantly following treatment with 50, 100, and 200 μg·mL$^{-1}$ of the IOC. In addition, the IOC (25, 50, 100, and 200 μg·mL$^{-1}$) inhibited LPS-induced PGE2 production in RAW 264.7 cells.

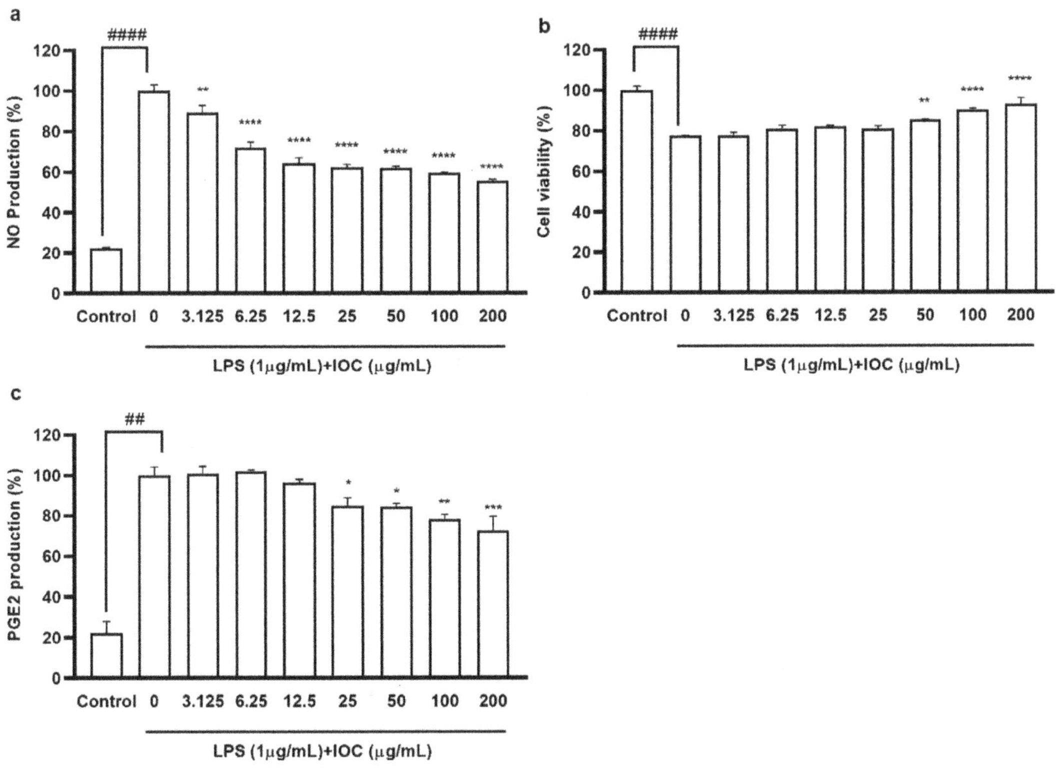


**Figure 3.** Effect of IOC on LPS-induced NO and PGE2 release by RAW 264.7 cells. (**a**) NO production inhibitory effect of IOC; (**b**) cytoprotective; (**c**) the level of PGE2. All experiments were triplicated, and data are expressed as the mean ± SEM. * $p < 0.05$, ** $p < 0.01$, *** $p < 0.001$, and **** $p < 0.0001$ as compared to the LPS-treated group; ## $p < 0.01$ and #### $p < 0.0001$ as compared to the control group.

### *3.5. Effects of the IOC on Intracellular ROS Levels and Cell Viability in UVB-Induced HaCaT Cells*

According to the cytotoxicity results, there were no significant differences in cell viability among 3.125 to 200 μg·mL$^{-1}$ concentration groups, confirming that the IOC was non-toxic to HaCaT cells in this range (data not shown); thus, concentration ranges were selected for subsequent experiments. As shown in Figure 4a, UVB exposure increased intracellular ROS levels. However, treatment with the IOC reversed this effect in a dose-dependent manner. The cell viability of UVB-induced HaCaT cells was drastically reduced. However, this effect was restored following treatment with 100 and 200 μg·mL$^{-1}$ of the IOC (Figure 4b).

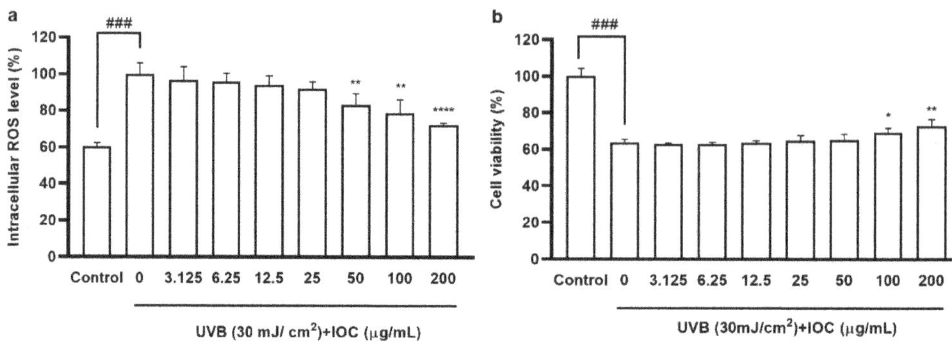


**Figure 4.** IOC inhibits intracellular ROS levels in UVB-irradiated HaCaT cells. (**a**) Intracellular ROS scavenging effect of IOC in UVB-induced HaCaT cells; (**b**) protective effects of IOC against UVB-irradiated HaCaT cells damage. All experiments were triplicated, and data are expressed as the mean ± SEM. * $p < 0.05$, ** $p < 0.01$, and **** $p < 0.0001$ as compared to the UVB-irradiated group, and ### $p < 0.001$ as compared to the control group.

### 3.6. Effect of the IOC against UVB-Induced Apoptosis

Based on the above experimental results, higher concentrations of the IOC were found to present stronger activity, especially the concentrations from 50 to 200 μg·mL$^{-1}$, which were significantly different from the other groups. Therefore, a concentration range of 50–200 μg·mL$^{-1}$ was selected for subsequent experiments to verify the activity of the IOC. Figure 5 shows the protective effects of the IOC against UVB-induced apoptosis. Cells in the control group, which were untreated and not exposed to UVB, contained intact nuclei, whereas cells in the negative control group, which were untreated but exposed to UVB, showed significantly fragmented nuclei. However, nuclear fragmentation was reduced in the IOC treatment groups, especially in the 100 μg·mL$^{-1}$ and 200 μg·mL$^{-1}$ treatment groups. Therefore, the IOC suppresses apoptosis following UVB irradiation, protecting the HaCaT cells from UVB-induced photodamage.

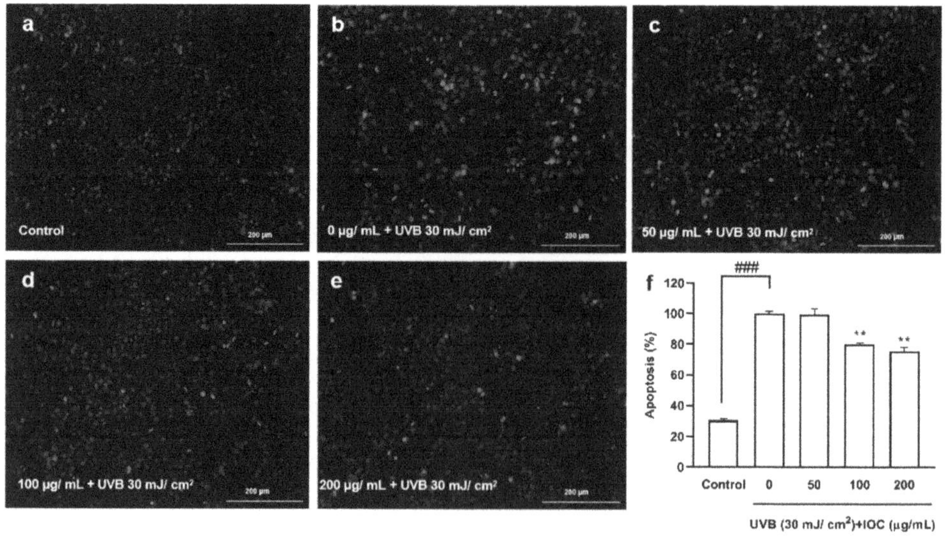


**Figure 5.** The protective effects of IOC against UVB-induced apoptosis in HaCaT cells. (**a**) Non-UVB-irradiated cells; (**b**) 30 mJ·cm$^{-2}$ UVB-irradiated cells; (**c**) cells treated with 50 μg·mL$^{-1}$ of IOC and

UVB; (**d**) cells treated with 100 μg·mL$^{-1}$ of IOC and UVB; (**e**) cells treated with 200 μg·mL$^{-1}$ of IOC and UVB; (**f**) quantification of apoptotic cells. ** $p < 0.01$ as compared to the UVB-irradiated group and ### $p < 0.001$ as compared to the control group.

### *3.7. Effect of the IOC on UVB-Induced Antioxidant Enzyme Expression through Nrf2 Activation in HaCaT Cells*

The Nrf2–Keap1 pathway plays pivotal roles in regulating the induction of antioxidant enzymes. To evaluate the effects of the IOC on UVB-irradiation-induced antioxidant enzyme expression, Western blotting was performed to analyze protein levels. Nrf2, SOD-1, and HO-1 expression was significantly inhibited following UVB irradiation. However, co-treatment with the IOC significantly restored the expression of these antioxidant enzymes in a dose-dependent manner (Figure 6a,b). Treatment with the IOC activated the Nrf2–Keap1 pathway, enhanced Nrf2 expression, and suppressed Keap1 expression. While UVB irradiation upregulated Keap1 expression, high concentrations of IOC downregulated it (Figure 6a). Therefore, the IOC inhibited intracellular ROS generation by promoting the activation of Nrf2 and inducing the expression of antioxidant enzymes, such as SOD-1 and HO-1.

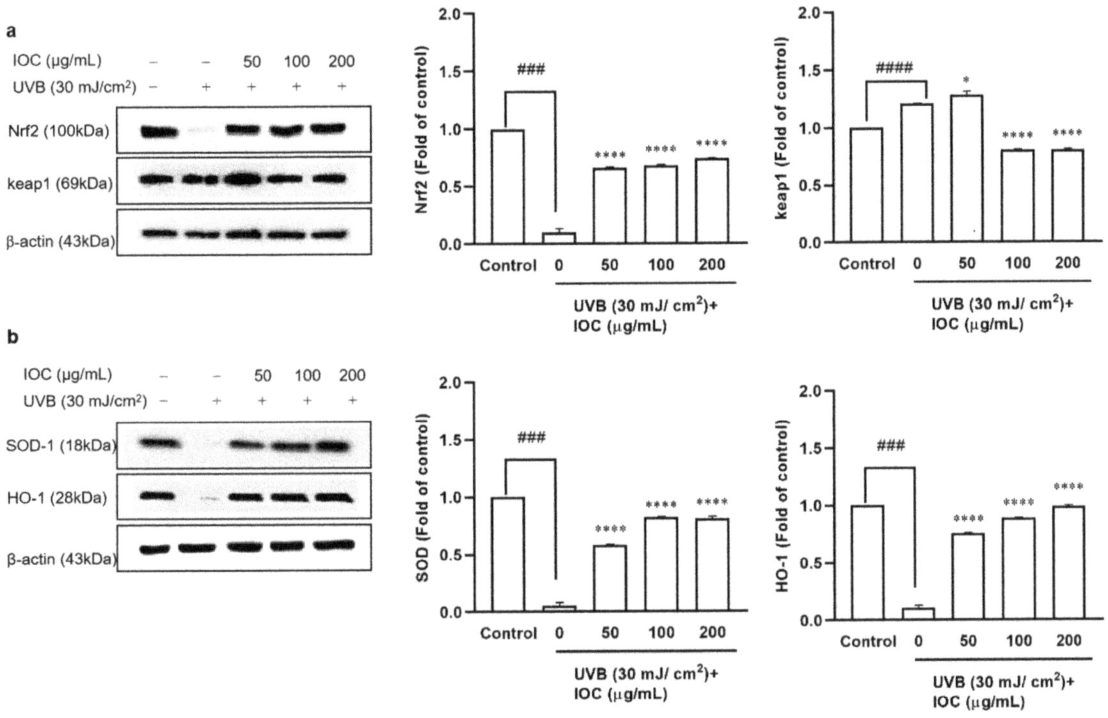


**Figure 6.** Effect of IOC on antioxidant-related protein in UVB-irradiated HaCaT cells. (**a**) Protein levels of Nrf2 and Keap1; (**b**) protein levels of SOD-1 and HO-1. β-actin was used as internal control. Quantification was assisted with the ImageJ software. Results are represented as mean ± SEM. * $p < 0.05$ and **** $p < 0.0001$ as compared to the UVB-irradiated group; ### $p < 0.001$ and #### $p < 0.0001$ as compared to the control group.

### 3.8. Effects of IOC on MMPs Expression in UVB-Induced HaCaT Cells

Collagen acts as an important structural support system providing strength and elasticity to the skin. UVB exposure activates MMPs secretion, leading to collagen degradation, which is a hallmark of skin aging [32]. As shown in Figure 7, UVB irradiation significantly increased MMP1, 2, and 9 secretion, However, MMPs' expression decreased in a dose-dependent manner in IOC-pretreated HaCaT cells. Therefore, the IOC inhibited MMPs expression, thereby playing an anti-wrinkle role, which is important to prevent photoaging.

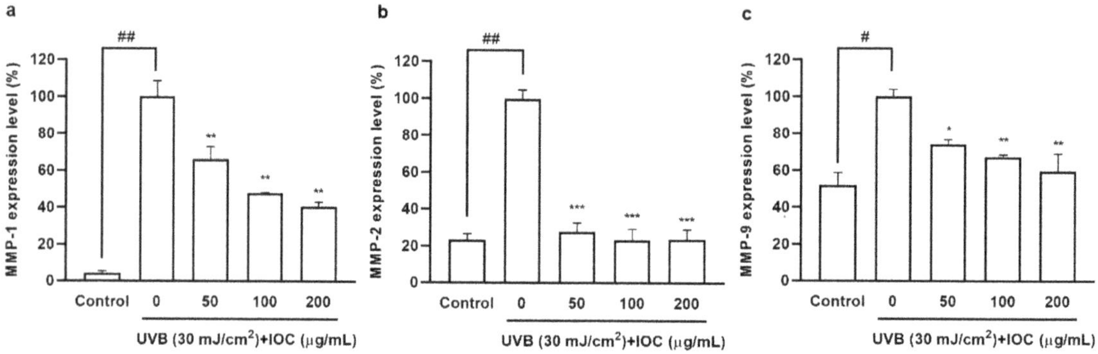


**Figure 7.** IOC reduces MMPs expression in UVB-irradiated HaCaT cells. (**a**) MMP-1 expression level in UVB-irradiated HaCaT cells; (**b**) MMP-2 expression level in UVB-irradiated HaCaT cells; (**c**) MMP-9 expression level in UVB-irradiated HaCaT cells. All experiments were triplicated, and data are expressed as the mean ± SEM. * $p < 0.05$, ** $p < 0.01$, and *** $p < 0.001$ as compared to the UVB-irradiated group; # $p < 0.05$ and ## $p < 0.01$ as compared to the control group.

### 3.9. Effects of the IOC on $H_2O_2$-Induced Oxidative Stress in Zebrafish

A zebrafish embryonic model was used to determine the effect of the IOC on $H_2O_2$-induced oxidative stress in vivo. The survival rate, heart beating rate, ROS generation, and cell death of zebrafish embryos were investigated. Figure 8 illustrated that the intracellular ROS and cell death decreased in a dose-dependent manner following treatment with the IOC. Additionally, compared with the control group, it was found that the survival rate was significantly decreased when stimulated with $H_2O_2$. However, this survival rate increased after treatment with 100 and 200 $\mu g \cdot mL^{-1}$ of the IOC. Furthermore, the heart beating rate of zebrafish was decreased in a dose-dependent manner in IOC-treated groups (50–200 $\mu g \cdot mL^{-1}$) compared with that in the $H_2O_2$-induced group. These observations confirmed that the IOC may produce protective effects against $H_2O_2$-induced oxidative stress.

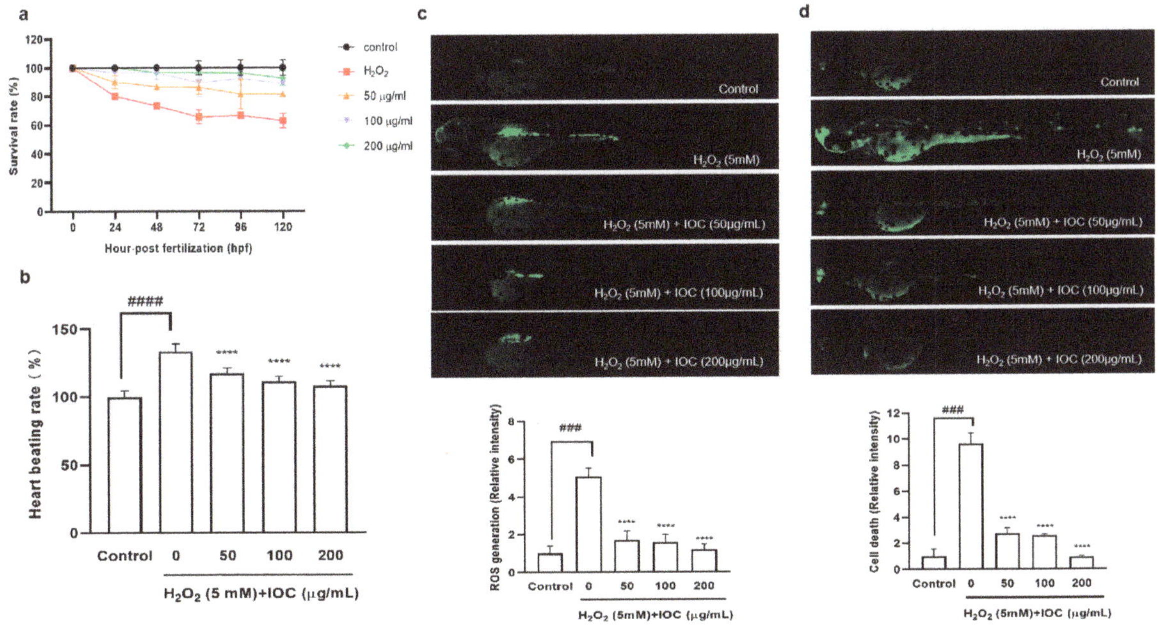


**Figure 8.** IOC depresses $H_2O_2$-induced oxidative damage in vivo in zebrafish. (**a**) Survival rate; (**b**) heart beating rate of both atrium and ventricle; (**c**) intracellular ROS generation; (**d**) cell death. All experiments were triplicated, and data are expressed as the mean ± SEM. **** $p < 0.0001$ as compared to the $H_2O_2$-induced group, ### $p < 0.001$ and #### $p < 0.0001$ as compared to the control group.

## 4. Discussion

Among many biological processes, inflammation, oxidation, and photodamage are linked to one another. UVB radiation from the sun can activate complex biochemical reactions that induce skin damage and aging. Simultaneously, excessive UVB irradiation decreases cellular antioxidant levels, resulting in ROS accumulation [33]. When the ROS concentration exceeds the basal level, cellular defenses against oxidative stress are weakened, stimulating the expression of pro-inflammatory factors and MMPs, and leading to an inflammatory response and accelerating skin aging [34,35].

*I. okamurae* is an edible brown alga belonging to the Ishige genus of the Ishigeaceae family. *I. okamurae* contains a number of bioactive compounds and shows various bioactivities. For instance, Kang et al. [11] reported that isophloroglucin A derived from *I. okamurae* shows anti-obesity activity. Moreover, fucoxanthin has been reported to inhibit LPS-induced inflammatory responses in RAW 264.7 cells through the suppression of NF-kB activation and MAPK phosphorylation [14]. However, the cosmeceutical effects of the sulfated polysaccharide extract from *I. okamurae* remained to be assessed. Therefore, we used different cell lines as well as in vitro and in vivo methods to explore the cosmeceutical effects of IOC from multiple perspectives, such as anti-inflammatory, antioxidant, and photoprotective effects.

The IOC contains abundant sulfated polysaccharides (48.47%), fucose (43.50%), and glucose (36.39%). The extraction of bioactive compounds from seaweeds is restricted by the complexity of their cell wall polysaccharides. As Celluclast can degrade the seaweed cell wall, enzyme-assisted extraction helps the release of biologically active substances and contributes to increasing the content of polysaccharides and fucose in the extract [19]. As shown in Table 1, carbohydrates were concentrated during the precipitation process, and high amounts of sulfated polysaccharides were confirmed, which contributed to the cosmeceutical effects. Previous in vitro and in vivo studies have shown that sulfated polysaccharide extracts from Hizikia fusiforme possess antioxidant, anti-inflammatory, and photoprotective activities [36]. Sanjeewa et al. [37] reported that fucoidan—a sulfated polysaccharide isolated from Sargassum horneri—produced anti-inflammatory effects in LPS-induced cells. Moreover, Kim et al. [38] demonstrated that polysaccharides isolated from Psidium guajava exhibited antioxidant effects; simultaneously, fucose, galactose, and rhamnose levels were increased in the sulfate polysaccharide fraction. Fucose, galactose, rhamnose, and arabinose are associated with antioxidant activity. Organic free radicals, such as DPPH, ABTS, and ORAC, are commonly used to assess antioxidant activity. Therefore, in the present study, the antioxidant activity of the IOC was evaluated using the ABTS assay. As expected, the IOC possessed potent free radical scavenging activity, particularly at the concentration of 100 $\mu g \cdot mL^{-1}$. The balance between enzymatic antioxidants and free radicals is important for effective intracellular oxidative stress relief. In the present study, a significant and dose-dependent decline in intracellular ROS levels was noted following IOC treatment. Oxidative stress affects cell viability. Abundant fucose in the IOC implies greater antioxidant potential; accordingly, increasing the IOC concentration produced a protective effect against $H_2O_2$-stimulated cell death in Vero cells.

Inflammation is a necessary component of physiological defense processes and is a response to cellular damage caused by oxidative stress, radiation (e.g., UV radiation), and endotoxins (e.g., LPSs) [6]. LPS, a crucial cell wall component of Gram-negative bacteria, induces an inflammatory response in macrophages, promoting the production of inflammatory mediators, such as NO, TNF-$\alpha$, PGE2, IL-6, and IL-1$\beta$ [29]. Simultaneously, increased NO production promotes ROS generation, which, in turn, induces apoptosis. Oxidative stress and inflammatory responses are interrelated. Macrophage production of PGE2 and proinflammatory cytokines plays key roles in the inflammatory process. To confirm the cytotoxicity and anti-inflammatory activity of the IOC, the viability of LPS-induced RAW 264.7 macrophages and inhibition of the production of NO and PGE2, which are inflammatory response indicators, were confirmed. At all concentrations, the IOC suppressed NO secretion in LPS-stimulated RAW 264.7 cells, and PGE2 levels decreased significantly in a dose-dependent manner with increasing IOC concentrations. Similarly, Sanjeewa et al. [37] showed that fucoidan—a complex sulfated polysaccharide isolated from Sargassum horneri—exhibits potent anti-inflammatory activity by blocking the MAPK and NF-kB signaling pathways. In addition, Jayawardena et al. [39] demonstrated that the anti-inflammatory potential of fucoidan extracted from the brown alga Turbinaria ornata was assisted by enzymatic hydrolysis in both in vivo and in vitro models.

UV radiation can cause photodamage, wrinkles, and skin diseases by inducing excessive production of intracellular ROS and stimulating the expression of matrix MMPs and pro-inflammatory cytokines [34,40]. Natural aging and photoaging are the major causes of skin aging, manifesting as dryness, hyperpigmentation, laxity, and wrinkles [41]. Additionally, UV-radiation-mediated ROS production has been shown to promote the expression of specific genes involved in signaling pathways, resulting in various physiological effects, such as inflammatory responses. Increased ROS and $H_2O_2$ concentrations activate the NF-$\kappa$B pathway, thereby increasing PGE2 and cytokine levels [6,35]. Therefore, we examined the inhibitory effects of the IOC on wrinkling-related enzymes. The increased activity of skin enzymes, such as collagenase, elastase, and HAase, leads to the proteolysis of the extracellular matrix (ECM). Furthermore, upregulated expression of MMPs, particularly MMP-2 and MMP-9, can lead to ECM degradation. Another important collagenase is MMP-1, which mainly degrades type I and type III collagen [36]. Chronic UV exposure denatures collagen and enzymes in the dermis, and elastin and collagen degradation results in the loss of skin elasticity and reduction in skin thickness, respectively. These are the major reasons for skin aging and wrinkling [24,42,43]. Our results showed that the IOC produced inhibitory effects on collagenase, elastase, and HAase and dose-dependently downregulated the expression of MMP-1, MMP-2, and MMP-9. Therefore, the IOC may act as a potential anti-wrinkle agent by interrupting the degradation of these skin enzymes under UV exposure. HaCaT cells, which are immortalized human keratinocytes, are widely used to study epidermal homeostasis. Therefore, we investigated the photoprotective activity of the IOC following UVB stimulation of HaCaT cells. The protective effects of the IOC against UVB irradiation were observed through cell viability assays. Our experiments revealed strong photoprotective activity of the IOC. As such, IOC treatment delayed skin aging by downregulating ROS production and protecting against nuclear fragmentation in UVB-stimulated HaCaT cells.

Nearly 80% of ROS production is induced by UV radiation. Therefore, antioxidant substances are critical for inhibiting oxidative damage and protecting the skin from photodamage. Activation of the Nrf2–Keap1 signaling pathway maintains high levels of antioxidant enzymes, such as SOD-1 and HO-1, which play pivotal roles in protection against photoaging [44]. The IOC significantly enhanced the levels of SOD-1 and HO-1 antioxidants through upregulating Nrf2 expression and significantly downregulating Keap1 expression. Therefore, the IOC produced protective effects against UVB-induced oxidative damage and apoptosis in HaCaT cells, primarily through the activation of the Nrf2 signaling pathway. In a similar study, Oh et al. [45] showed that 3,5-dicaffeoyl-epi-quinic acid reduced oxidative stress and prevented photoaging through upregulation of the antioxidant enzyme transcription factor Nrf2. These results imply that the IOC can serve as an effective skin protective ingredient in cosmetics.

ROS plays a significant role in oxidative stress, causing the breakdown of DNA, proteins, cell membranes, and other constituents. The accumulation of molecular damage leads to apoptosis, necrosis, and death. $H_2O_2$ is associated with the formation of hydroxyl and singlet oxygen radicals, which can stimulate intracellular ROS production and cause cell damage and senescence [46,47]. As a result, antioxidative substances are critical for preventing oxidative stress. Antioxidants in cosmeceuticals play critical roles in suppressing and inhibiting oxidative damage reactions. Therefore, the in vivo antioxidant effects of the IOC were investigated in a zebrafish embryo model. Because of their suitability for studying human disease processes and development, zebrafish are considered a painless in vitro alternative, becoming one of the most widely used vertebrate models [48]. Following $H_2O_2$ stimulation, excess intracellular ROS are generated, leading to cell death. A dose-dependent reduction in ROS levels and remarkably reduced cell death rate were observed in zebrafish following treatment with the IOC. The survival rate of $H_2O_2$-stimulated zebrafish was significantly decreased; however, the rate increased following co-treatment with IOC. In addition, $H_2O_2$ stimulation caused heartbeat disorder in zebrafish; however, co-treatment with IOC effectively downregulated the heart rate of zebrafish. In summary, our in vitro and in vivo results indicate that the IOC possesses potent anti-inflammatory and antioxidant activities and presents photoprotective potential.

## 5. Conclusions

We investigated the effect of sulfated polysaccharides from Celluclast-assisted extract of the brown seaweed *I. okamurae* as a source of natural cosmetic ingredients. IOC produced protective effects against $H_2O_2$-induced oxidative stress both in vitro (Vero cells) and in vivo (zebrafish). Furthermore, IOC produced anti-inflammatory effects in RAW 264.7 macrophages, and IOC exhibited antioxidant, anti-wrinkle, and photoprotective effects by suppressing UVB-induced oxidative stress, activating the Nrf2–Keap1 signaling pathway, and reducing MMPs expression in HaCaT cells. Therefore, the IOC may be used as an effective ingredient in the functional food and cosmetic industries.

**Author Contributions:** Conceptualization, F.Y. and Y.-J.J.; methodology, F.Y., J.H., D.P.N. and Y.M.K.; software, F.Y. and D.P.N.; validation, F.Y. and J.H.; investigation, F.Y., M.-S.H. and Y.M.K.; resources, Y.-J.J.; data curation, F.Y. and J.H.; supervision, Y.-J.J. and M.-S.H.; writing—original draft preparation, F.Y.; writing—review and editing, F.Y. and Y.-J.J.; funding acquisition, Y.-J.J. All authors have read and agreed to the published version of the manuscript.

**Funding:** This work was supported by the 2022 education, research, and student guidance grant funded by Jeju National University.

**Institutional Review Board Statement:** The zebrafish experiment was approved by the Animal Care and Use Committee of Jeju National University (Approval No. 2020-0049).

**Informed Consent Statement:** Not applicable.

**Data Availability Statement:** The data are contained within the article.

**Conflicts of Interest:** The authors declare no conflict of interest.

## References

1. Sahoo, B.M.; Banik, B.K.; Borah, P.; Jain, A. Reactive oxygen species (ROS): Key components in cancer therapies. *Anti-Cancer Agents Med. Chem.* **2022**, *22*, 215–222. [CrossRef]
2. Aslam, A.; Bahadar, A.; Liaquat, R.; Saleem, M.; Waqas, A.; Zwawi, M. Algae as an attractive source for cosmetics to counter environmental stress. *Sci. Total Env.* **2021**, *772*, 144905. [CrossRef]
3. Juan, C.A.; Pérez de la Lastra, J.M.; Plou, F.J.; Pérez-Lebeña, E. The Chemistry of Reactive Oxygen Species (ROS) Revisited: Outlining Their Role in Biological Macromolecules (DNA, Lipids and Proteins) and Induced Pathologies. *Int. J. Mol. Sci.* **2021**, *22*, 4642. [CrossRef] [PubMed]
4. Liu, T.; Sun, L.; Zhang, Y.; Wang, Y.; Zheng, J. Imbalanced GSH/ROS and sequential cell death. *J. Biochem. Mol. Toxicol.* **2022**, *36*, e22942. [CrossRef]
5. Liu, Z.; Ren, Z.; Zhang, J.; Chuang, C.-C.; Kandaswamy, E.; Zhou, T.; Zuo, L. Role of ROS and nutritional antioxidants in human diseases. *Front. Physiol.* **2018**, *9*, 477. [CrossRef] [PubMed]
6. Kageyama, H.; Waditee-Sirisattha, R. Antioxidative, Anti-Inflammatory, and Anti-Aging Properties of Mycosporine-Like Amino Acids: Molecular and Cellular Mechanisms in the Protection of Skin-Aging. *Mar. Drugs* **2019**, *17*, 222. [CrossRef]
7. Wang, L.; Kim, H.S.; Oh, J.Y.; Je, J.G.; Jeon, Y.-J.; Ryu, B. Protective effect of diphlorethohydroxycarmalol isolated from Ishige okamurae against UVB-induced damage in vitro in human dermal fibroblasts and in vivo in zebrafish. *Food Chem. Toxicol.* **2020**, *136*, 110963. [CrossRef] [PubMed]
8. Nduka, J.K.; Kelle, H.I.; Odiba, I.O. Review of health hazards and toxicological effects of constituents of cosmetics. In *Poisoning in the Modern World-New Tricks for an Old Dog?* IntechOpen: London, UK, 2019; pp. 59–76.
9. Lomartire, S.; Gonçalves, A.M.M. An Overview of Potential Seaweed-Derived Bioactive Compounds for Pharmaceutical Applications. *Mar. Drugs* **2022**, *20*, 141. [CrossRef]
10. Pradhan, B.; Bhuyan, P.P.; Patra, S.; Nayak, R.; Behera, P.K.; Behera, C.; Behera, A.K.; Ki, J.-S.; Jena, M. Beneficial effects of seaweeds and seaweed-derived bioactive compounds: Current evidence and future prospective. *Biocatal. Agric. Biotechnol.* **2022**, *39*, 102242. [CrossRef]
11. Kang, N.; Oh, S.; Kim, H.-S.; Ahn, H.; Choi, J.; Heo, S.-J.; Byun, K.; Jeon, Y.-J. Ishophloroglucin A, derived from Ishige okamurae, regulates high-fat-diet-induced fat accumulation via the leptin signaling pathway, associated with peripheral metabolism. *Algal. Res.* **2020**, *50*, 101974. [CrossRef]
12. Jimenez-Lopez, C.; Pereira, A.G.; Lourenço-Lopes, C.; Garcia-Oliveira, P.; Cassani, L.; Fraga-Corral, M.; Prieto, M.A.; Simal-Gandara, J. Main bioactive phenolic compounds in marine algae and their mechanisms of action supporting potential health benefits. *Food Chem.* **2021**, *341*, 128262. [CrossRef] [PubMed]
13. Julião, D.R.; Afonso, C.; Gomes-Bispo, A.; Bandarra, N.M.; Cardoso, C. The effect of drying on undervalued brown and red seaweed species: Bioactivity alterations. *Phycol. Res.* **2021**, *69*, 246–257. [CrossRef]
14. Kim, K.-N.; Heo, S.-J.; Yoon, W.-J.; Kang, S.-M.; Ahn, G.; Yi, T.-H.; Jeon, Y.-J. Fucoxanthin inhibits the inflammatory response by suppressing the activation of NF-κB and MAPKs in lipopolysaccharide-induced RAW 264.7 macrophages. *Eur. J. Pharm.* **2010**, *649*, 369–375. [CrossRef] [PubMed]
15. Wang, L.; Kim, H.S.; Je, J.-G.; Oh, J.Y.; Kim, Y.-S.; Cha, S.-H.; Jeon, Y.-J. Protective Effect of Diphlorethohydroxycarmalol Isolated from Ishige okamurae Against Particulate Matter-Induced Skin Damage by Regulation of NF-κB, AP-1, and MAPKs Signaling Pathways In Vitro in Human Dermal Fibroblasts. *Molecules* **2020**, *25*, 1055. [CrossRef]
16. Kim, M.-S.; Oh, G.-W.; Jang, Y.-M.; Ko, S.-C.; Park, W.-S.; Choi, I.-W.; Kim, Y.-M.; Jung, W.-K. Antimicrobial hydrogels based on PVA and diphlorethohydroxycarmalol (DPHC) derived from brown alga Ishige okamurae: An in vitro and in vivo study for wound dressing application. *Mater. Sci. Eng.* **2020**, *107*, 110352. [CrossRef]
17. Kim, S.-Y.; Ahn, G.; Kim, H.-S.; Je, J.-G.; Kim, K.-N.; Jeon, Y.-J. Diphlorethohydroxycarmalol (DPHC) Isolated from the Brown Alga Ishige okamurae Acts on Inflammatory Myopathy as an Inhibitory Agent of TNF-α. *Mar. Drugs* **2020**, *18*, 529. [CrossRef]
18. Li, B.B.; Smith, B.; Hossain, M.M. Extraction of phenolics from citrus peels: II. Enzyme-assisted extraction method. *Sep. Purif. Technol.* **2006**, *48*, 189–196. [CrossRef]
19. Quitério, E.; Grosso, C.; Ferraz, R.; Delerue-Matos, C.; Soares, C. A Critical Comparison of the Advanced Extraction Techniques Applied to Obtain Health-Promoting Compounds from Seaweeds. *Mar. Drugs* **2022**, *20*, 677. [CrossRef]
20. Saito, H.; Yamagata, T.; Suzuki, S. Enzymatic Methods for the Determination of Small Quantities of Isomeric Chondroitin Sulfates. *J. Biol. Chem.* **1968**, *243*, 1536–1542. [CrossRef]
21. Chandler, S.F.; Dodds, J.H. The effect of phosphate, nitrogen and sucrose on the production of phenolics and solasodine in callus cultures of solanum laciniatum. *Plant Cell Rep.* **1983**, *2*, 205–208. [CrossRef]

22. DuBois, M.; Gilles, K.A.; Hamilton, J.K.; Rebers, P.A.; Smith, F. Colorimetric Method for Determination of Sugars and Related Substances. *Anal. Chem.* **1956**, *28*, 350–356. [CrossRef]
23. Re, R.; Pellegrini, N.; Proteggente, A.; Pannala, A.; Yang, M.; Rice-Evans, C. Antioxidant activity applying an improved ABTS radical cation decolorization assay. *Free Radic. Biol. Med.* **1999**, *26*, 1231–1237. [CrossRef] [PubMed]
24. Wang, L.; Oh, J.Y.; Yang, H.-W.; Kim, H.S.; Jeon, Y.-J. Protective effect of sulfated polysaccharides from a Celluclast-assisted extract of Hizikia fusiforme against ultraviolet B-induced photoaging in vitro in human keratinocytes and in vivo in zebrafish. *Mar. Life Sci. Technol.* **2019**, *1*, 104–111. [CrossRef]
25. Kraunsoe, J.A.E.; Claridge, T.D.W.; Lowe, G. Inhibition of Human Leukocyte and Porcine Pancreatic Elastase by Homologues of Bovine Pancreatic Trypsin Inhibitor. *Biochemistry* **1996**, *35*, 9090–9096. [CrossRef] [PubMed]
26. Elson, L.A.; Morgan, W.T.J. A colorimetric method for the determination of glucosamine and chondrosamine. *Biochem. J.* **1933**, *27*, 1824. [CrossRef] [PubMed]
27. Yang, X.; Kang, M.-C.; Lee, K.-W.; Kang, S.-M.; Lee, W.-W.; Jeon, Y.-J. Antioxidant activity and cell protective effect of loliolide isolated from *Sargassum ringgoldianum* subsp. *coreanum*. *Algae* **2011**, *26*, 201–208. [CrossRef]
28. Kim, H.-S.; Shin, B.-O.; Kim, S.-Y.; Wang, L.; Lee, W.; Kim, Y.T.; Rho, S.; Cho, M.; Jeon, Y.-J. Antioxidant activity of pepsin hydrolysate derived from edible Hippocampus abdominalis in vitro and in zebrafish models. *Korean J. Fish. Aquat. Sci.* **2016**, *49*, 445–453. [CrossRef]
29. Wang, L.; Oh, J.Y.; Fernando, S.; Sanjeewa, K.A.; Kim, E.-A.; Lee, W.; Jeon, Y.-J. Soft corals collected from Jeju Island; a potential source of anti-inflammatory phytochemicals. *J. Chitin Chitosan* **2016**, *21*, 247–254. [CrossRef]
30. Ko, J.-Y.; Kim, E.-A.; Lee, J.-H.; Kang, M.-C.; Lee, J.-S.; Kim, J.-S.; Jung, W.-K.; Jeon, Y.-J. Protective effect of aquacultured flounder fish-derived peptide against oxidative stress in zebrafish. *Fish Shellfish. Immunol.* **2014**, *36*, 320–323. [CrossRef]
31. Fernando, I.P.S.; Sanjeewa, K.K.A.; Kim, H.-S.; Kim, S.-Y.; Lee, S.-H.; Lee, W.W.; Jeon, Y.-J. Identification of sterols from the soft coral Dendronephthya gigantea and their anti-inflammatory potential. *Environ. Toxicol. Pharm.* **2017**, *55*, 37–43. [CrossRef]
32. Fisher, G.J.; Kang, S.; Varani, J.; Bata-Csorgo, Z.; Wan, Y.; Datta, S.; Voorhees, J.J. Mechanisms of Photoaging and Chronological Skin Aging. *Arch. Dermatol.* **2002**, *138*, 1462–1470. [CrossRef] [PubMed]
33. Pourzand, C.; Albieri-Borges, A.; Raczek, N.N. Shedding a New Light on Skin Aging, Iron- and Redox-Homeostasis and Emerging Natural Antioxidants. *Antioxidants* **2022**, *11*, 471. [CrossRef] [PubMed]
34. Zhang, D.; Lu, C.; Yu, Z.; Wang, X.; Yan, L.; Zhang, J.; Li, H.; Wang, J.; Wen, A. Echinacoside alleviates UVB irradiation-mediated skin damage via inhibition of oxidative stress, DNA damage, and apoptosis. *Oxid Med. Cell. Longev.* **2017**, *2017*, 6851464. [CrossRef] [PubMed]
35. Jordan, C.T.; Hilt, J.Z.; Dziubla, T.D. Polyphenol conjugated poly (beta-amino ester) polymers with hydrogen peroxide triggered degradation and active antioxidant release. *J. Appl. Polym. Sci.* **2020**, *137*, 48647. [CrossRef]
36. Wang, L.; Lee, W.; Oh, J.Y.; Cui, Y.R.; Ryu, B.; Jeon, Y.-J. Protective Effect of Sulfated Polysaccharides from Celluclast-Assisted Extract of Hizikia fusiforme Against Ultraviolet B-Induced Skin Damage by Regulating NF-κB, AP-1, and MAPKs Signaling Pathways In Vitro in Human Dermal Fibroblasts. *Mar. Drugs* **2018**, *16*, 239. [CrossRef]
37. Sanjeewa, K.K.A.; Jayawardena, T.U.; Kim, S.-Y.; Kim, H.-S.; Ahn, G.; Kim, J.; Jeon, Y.-J. Fucoidan isolated from invasive Sargassum horneri inhibit LPS-induced inflammation via blocking NF-κB and MAPK pathways. *Algal. Res.* **2019**, *41*, 101561. [CrossRef]
38. Kim, S.-Y.; Kim, E.-A.; Kim, Y.-S.; Yu, S.-K.; Choi, C.; Lee, J.-S.; Kim, Y.-T.; Nah, J.-W.; Jeon, Y.-J. Protective effects of polysaccharides from Psidium guajava leaves against oxidative stresses. *Int. J. Biol. Macromol.* **2016**, *91*, 804–811. [CrossRef]
39. Jayawardena, T.U.; Fernando, I.P.S.; Lee, W.W.; Sanjeewa, K.K.A.; Kim, H.-S.; Lee, D.-S.; Jeon, Y.-J. Isolation and purification of fucoidan fraction in Turbinaria ornata from the Maldives; Inflammation inhibitory potential under LPS stimulated conditions in in-vitro and in-vivo models. *Int. J. Biol. Macromol.* **2019**, *131*, 614–623. [CrossRef]
40. Ries, C.; Egea, V.; Karow, M.; Kolb, H.; Jochum, M.; Neth, P. MMP-2, MT1-MMP, and TIMP-2 are essential for the invasive capacity of human mesenchymal stem cells: Differential regulation by inflammatory cytokines. *Blood* **2007**, *109*, 4055–4063. [CrossRef]
41. Gu, Y.; Han, J.; Jiang, C.; Zhang, Y. Biomarkers, oxidative stress and autophagy in skin aging. *Ageing Res. Rev.* **2020**, *59*, 101036. [CrossRef]
42. Maity, N.; Nema, N.K.; Abedy, M.K.; Sarkar, B.K.; Mukherjee, P.K. Exploring Tagetes erecta Linn flower for the elastase, hyaluronidase and MMP-1 inhibitory activity. *J. Ethnopharmacol.* **2011**, *137*, 1300–1305. [CrossRef] [PubMed]
43. Azmi, N.; Hashim, P.; Hashim, D.M.; Halimoon, N.; Majid, N.M.N. Anti–elastase, anti–tyrosinase and matrix metalloproteinase–1 inhibitory activity of earthworm extracts as potential new anti–aging agent. *Asian Pac. J. Trop. Biomed.* **2014**, *4*, S348–S352. [CrossRef] [PubMed]
44. Park, C.; Park, J.; Kim, W.-J.; Kim, W.; Cheong, H.; Kim, S.-J. Malonic Acid Isolated from Pinus densiflora Inhibits UVB-Induced Oxidative Stress and Inflammation in HaCaT Keratinocytes. *Polymers* **2021**, *13*, 816. [CrossRef] [PubMed]
45. Oh, J.H.; Karadeniz, F.; Lee, J.I.; Seo, Y.; Kong, C.S. Protective effect of 3,5-dicaffeoyl-epi-quinic acid against UVB-induced photoaging in human HaCaT keratinocytes. *Mol. Med. Rep.* **2019**, *20*, 763–770. [CrossRef]
46. Karapetsas, A.; Voulgaridou, G.-P.; Konialis, M.; Tsochantaridis, I.; Kynigopoulos, S.; Lambropoulou, M.; Stavropoulou, M.-I.; Stathopoulou, K.; Aligiannis, N.; Bozidis, P.; et al. Propolis Extracts Inhibit UV-Induced Photodamage in Human Experimental In Vitro Skin Models. *Antioxidants* **2019**, *8*, 125. [CrossRef]

47. Jiratchayamaethasakul, C.; Ding, Y.; Hwang, O.; Im, S.-T.; Jang, Y.; Myung, S.-W.; Lee, J.M.; Kim, H.-S.; Ko, S.-C.; Lee, S.-H. In vitro screening of elastase, collagenase, hyaluronidase, and tyrosinase inhibitory and antioxidant activities of 22 halophyte plant extracts for novel cosmeceuticals. *Fish. Aquat. Sci.* **2020**, *23*, 6. [CrossRef]
48. Watzke, J.; Schirmer, K.; Scholz, S. Bacterial lipopolysaccharides induce genes involved in the innate immune response in embryos of the zebrafish (Danio rerio). *Fish Shellfish Immunol.* **2007**, *23*, 901–905. [CrossRef]

*antioxidants*

MDPI

*Article*

# Cosmeceutical Potential of Extracts Derived from Fishery Industry Residues: Sardine Wastes and Codfish Frames

**Martim Cardeira [1,2], Ana Bernardo [1], Inês C. Leonardo [1,2], Frédéric B. Gaspar [1,2], Marta Marques [3], Rodrigo Melgosa [3], Alexandre Paiva [3], Pedro Simões [3], Naiara Fernández [1] and Ana Teresa Serra [1,2,*]**

1 IBET—Instituto de Biologia Experimental e Tecnológica, Apartado 12, 2781-901 Oeiras, Portugal
2 Instituto de Tecnologia Química e Biológica António Xavier, Universidade Nova de Lisboa (ITQB NOVA), Av. da República, 2780-157 Oeiras, Portugal
3 LAQV-REQUIMTE—Associated Laboratory for Green Chemistry (LAQV) of the Network of Chemistry and Technology (REQUIMTE), Chemistry Department, NOVA School of Science and Technology, Universidade Nova de Lisboa, 2829-516 Caparica, Portugal
* Correspondence: tserra@ibet.pt

**Abstract:** The fishery industry generates large amounts of waste (20–75% (*w*/*w*) of the total caught fish weight). The recovery of bioactive compounds from residues and their incorporation in cosmetics represents a promising market opportunity and may contribute to a sustainable valorisation of the sector. In this work, protein-rich extracts obtained by high-pressure technologies (supercritical $CO_2$ and subcritical water) from sardine (*Sardina pilchardus*) waste and codfish (*Gadus morhua*) frames were characterized regarding their cosmeceutical potential. Antioxidant, anti-inflammatory and antibacterial activities were evaluated through chemical (ORAC assay), enzymatic (inhibition of elastase and tyrosinase), antimicrobial susceptibility (*Klebsiella pneumoniae*, *Staphylococcus aureus* and *Cutibacterium acnes*) and cell-based (in keratinocytes-HaCaT) assays. Sardine extracts presented the highest antibacterial activity, and the extract obtained using higher extraction temperatures (250 °C) and without the defatting step demonstrated the lowest minimum inhibitory concentration (MIC) values (1.17; 4.6; 0.59 mg/mL for *K. pneumoniae*, *S. aureus* and *C. acnes*, respectively). Codfish samples extracted at lower temperatures (90 °C) were the most effective anti-inflammatory agents (a concentration of 0.75 mg/mL reduced IL-8 and IL-6 levels by 58% and 47%, respectively, relative to the positive control). Threonine, valine, leucine, arginine and total protein content in the extracts were highlighted to present a high correlation with the reported bioactivities ($R^2 \geq 0.7$). These results support the potential application of extracts obtained from fishery industry wastes in cosmeceutical products with bioactive activities.

**Keywords:** fish waste streams valorisation; antioxidant activity; anti-inflammatory activity; antimicrobial activity; anti-ageing; anti-hyperpigmentation; cosmeceuticals

**Citation:** Cardeira, M.; Bernardo, A.; Leonardo, I.C.; Gaspar, F.B.; Marques, M.; Melgosa, R.; Paiva, A.; Simões, P.; Fernández, N.; Serra, A.T. Cosmeceutical Potential of Extracts Derived from Fishery Industry Residues: Sardine Wastes and Codfish Frames. *Antioxidants* **2022**, *11*, 1925. https://doi.org/10.3390/antiox11101925

Academic Editors: Irene Dini and Sonia Laneri

Received: 31 July 2022
Accepted: 23 September 2022
Published: 28 September 2022



## 1. Introduction

With the constant search for innovation, especially for active ingredients, the cosmetic industry is growing and has demonstrated the intention to replace petroleum-derived components moving forward toward natural compounds [1]. The antioxidant properties of natural active ingredients can help in the prevention of several skin issues caused by oxidative stress and ageing [2,3]. Skin ageing can be induced by both intrinsic (such as inflammation or telomere shortening) and extrinsic (environmental) factors [4]. Skin ageing leads to the loss of mature collagen and alterations at the extracellular matrix (ECM) which compromises the barrier function, resulting in a dry appearance and susceptibility to external aggressors, increasing the risk for skin disorders [5]. This process can be accelerated by several enzymes, such as elastases, matrix metalloproteinases (MMPs) and hyaluronidases that can induce ECM degradation [6], or even by the accumulation of excessive reactive oxygen species (ROS) that can compromise the normal cell function [7]. Environmental factors, such as exposure to UV radiation, leads to the generation of high quantities of ROS that induces the same molecular and cellular responses as intrinsic ageing, but with amplified effects. Importantly, ROS can intensify the activity of enzymes related to skin ageing or skin pigmentation processes [8,9], and thus the presence of antioxidants can play an important role in the cosmetic field.

In 2018, world fish consumption was estimated by FAO to stand at 20.5 kg per capita [10], which leads to large quantities of by-products, mostly skin and bones. The generated residues correspond to 20–75% (*w*/*w*) of the total caught fish weight, potentially leading to environmental problems [11,12]. However, these residues still contain a significant amount of lipids, proteins, and minerals and should be adequately valorised. In recent years, extracts derived from waste generated by the fish industry have shown bioactive properties such as antihypertensive, antioxidative, antimicrobial, neuroprotective, antihyperglycemic, anti-ageing, and anti-inflammatory [13–19]. Atlantic codfish (*Gadus morhua*) and sardine (*Sardina pilchardus*) are among the most consumed fish in Portugal and extracts derived from its residues have shown promising nutraceutical potential, such as antioxidant, antiproliferative or anti-inflammatory activities [11,20–22]. However, since the exploitation and valorisation of fish industry wastes is still in an early stage, there is plenty of room to explore opportunities for the industry to convert this waste into high-value market bioproducts, including cosmetic ingredients.

In a previous work, we explored the use of high-pressure technologies (supercritical $CO_2$ and subcritical water), to isolate bioactive fractions from sardine waste and codfish frames with promising health benefits [11,22]. For sardine wastes, we demonstrated that by applying a first step with supercritical carbon ($ScCO_2$) (to remove lipid fraction) followed by an extraction process with subcritical water (SW) it was possible to obtain protein hydrolysates with high antioxidant potential and antiproliferative effect in colorectal cancer cells [22]. Subcritical water extraction/hydrolysis were also applied to obtain proteins-, peptides- and amino acid-enriched extracts from codfish frames and we showed that lower processing temperatures (90 °C) favour the extraction of compounds with anti-inflammatory potential in a human intestinal epithelial cell model [11]. Most of the proteins present in codfish frames extracts were collagen and collagen fragments. Other compounds include minor quantities of lipids, ash and some sugars. Sardine extracts were rich in peptides and amino acids, and lipids, ash and sugars were also present. Since fish-derived proteins and peptides may become an important resource for cosmetic industries, the present study aims to further evaluate the bioactive potential of these extracts derived from fish-processing wastes and by-products focused on the assessment of their cosmeceutical potential [23]. For this purpose, a range of chemical, enzymatic, and cell-based assays were applied to explore the antioxidant, anti-ageing, anti-hyperpigmentation, anti-inflammatory, and antimicrobial effects of the extracted samples. Correlation studies were also performed to identify the main bioactive constituents with cosmeceutical potential.

## 2. Materials and Methods

### 2.1. Reagents

3,4-dihydroxy-l-phenylalanine (L-DOPA), mushroom tyrosinase, porcine pancreatic elastase (PPE) type III, N-succinyl-Ala-Ala-Ala-p-nitroanilide (AAAPVN), Tris (2-amino-2-hydroxymethyl-propane-1,3-diol), 2,2′-azobis (2-methylpropionamidine)dihydrochloride (AAPH), and 2′,7′-dichlorofluorescein diacetate (DCFH-DA) were purchased from Sigma-Aldrich (St. Louis, MO, USA). Calcium-adjusted Mueller Hinton broth (CAMHB) was purchased from BD (Sparks, MD, USA). Brain-heart infusion (BHI) was purchased from Avantor (Radnor, PA, USA). AnaeroGen™ Compact sachets were purchased from Oxoid (Hampshire, UK). PrestoBlue™, Dulbecco's Modified Eagle Medium (DMEM), heat-inactivated Fetal Bovine Serum (FBS) and Penicillin-Streptomycin were obtained from Invitrogen (San Diego, CA, USA). Human immortalized non-tumorigenic keratinocyte cell line HaCaT was obtained from Cell Line Service (Eppelheim, Germany). Human IL-8 and IL-6 Mini TMB ELISA Development Kits were obtained from Peprotech (London, UK).

All other reagents and solvents used in the present study were of analytical grade and purchased from available suppliers.

## 2.2. Samples

The extracts used in this work were the ones developed in our previous studies focused on process optimization [11,22]. Briefly, codfish frames were supplied by Pascoal and Filhos S.A. (Gafanha da Nazaré, Portugal) and consisted of fish backbone and adhered muscle. Sardine waste, made of heads, spines and viscera, was supplied by Conservas A Poveira S.A. (Póvoa de Varzim, Portugal). The proximate composition of the raw materials used have been presented in our earlier works [11,22]. Protein (47 wt %) and ash (39 wt %) were the major components of codfish frames, with small quantity of lipids and carbohydrates (2 wt % and 0.3 wt %, respectively). Collagen is found to be the major protein in codfish frames, and in this case, it accounts for ca. 65% of the total protein content of original waste. In contrast, sardine is an oily fish, thus its waste is much richer in lipids than codfish frames. Sardine wastes showed a lipid content of 26 wt % and a protein content of 52 wt %, the rest being ash (17 wt %) and carbohydrates (3 wt %).

The extracts from codfish frames (Cf1, Cf2, Cf3 and Cf4) and sardine wastes (S1, S2 and S3) selected for this work were obtained by high pressure technology in a lab-scale apparatus as previously described [11,22,24] using the conditions summarized in Table 1. Briefly, 60 g of ground codfish frames or sardine waste (defatted or non-defatted) were loaded into a high-pressure reactor that was put inside an oven. The water pump was switch on at desired flowrate (ca. 10 mL/min) and pressure was set to 100 bar. As soon as pressure reached that value, the electrical oven was switch on, and the experiment started. The different extracts were collected during 30 min at different temperatures (90–250 °C). S1 and S3 extracts were obtained after a defatting process of the sardine waste by $ScCO_2$ before SW extraction. Subcritical water extraction experiments were duplicated. For each extract sample, 25 mL were taken in triplicate, lyophilized, and weighed to calculate the corresponding extraction yield. Analytical data—protein content—are expressed as mean ± standard deviation (SD) of triplicates. The information regarding the characterization of these extracts in terms of protein content, amino acid profile, major mineral compounds or toxic and heavy metals is described in our previous works [11,22].

**Table 1.** Extraction process techniques and parameters used for each sample. Extraction yield and protein content are expressed as mean ± standard deviation (SD).

| Sample | Defatting Conditions | SW Extraction Conditions | Extraction Yield (g/100 g Feed) [11,22] | Protein Content (wt %) [11,22] |
|---|---|---|---|---|
| Cf1 | - | 90 °C, 100 bar | 13.2 ± 0.5 | 81.6 ± 0.3 |
| Cf2 | - | 140 °C, 100 bar | 27.7 ± 0.5 | 93.6 ± 0.3 |
| Cf3 | - | 190 °C, 100 bar | 41.4 ± 0.5 | 95 ± 0.3 |
| Cf4 | - | 250 °C, 100 bar | 53.9 ± 0.5 | 84.4 ± 0.3 |
| S1 | $ScCO_2$ (40 °C, 250 bar) | 190 °C, 100 bar | 45.7 ± 2.8 | 87.5 ± 2.7 |
| S2 | - | 250 °C, 100 bar | 58.5 ± 0.4 | 57.5 ± 1.8 |
| S3 | $ScCO_2$ (40 °C, 250 bar) | 250 °C, 100 bar | 61.7 ± 2.0 | 85.2 ± 0.6 |

Cf1, Cf2, Cf3, Cf4—extracts from codfish frames; S1, S2, S3—extracts from sardine wastes.

Stock solutions of Cf1, Cf2, Cf3, Cf4, and S2 were prepared in Milli-Q $H_2O$ at a concentration of 100 mg/mL. The other samples, namely S1 and S3, were dissolved in DMSO (300 and 550 mg/mL, respectively) due to their lower solubility in water. Samples were frozen and kept at −20 °C until further use. For cellular assays, the samples were previously sterilized by heat (121 °C, 15 min) in an autoclave (Tuttnauer 3870 el, Breda, Netherlands).

## 2.3. Oxygen Radical Absorbance Capacity (ORAC) Assay

ORAC assay was performed to evaluate the antioxidant capacity of the samples towards peroxyl radicals ($ROO^\bullet$), following the method developed by Huang et al. [25], with some adjustments as reported previously [26]. Briefly, in a black 96-well microplate, 150 μL disodium fluorescein (0.3 μM) was added to 25 μL of sample dilutions and incubated for 10 min at 37 °C. Afterwards, the reaction was initiated by the addition of 25 μL of 2,2′-Azobis (2-amidinopropane) dihydrochloride (AAPH, 153 mM) and fluorescence (Ex/Em 485 ± 20/528 ± 20 nm) was measured for 40 min at 37 °C in a FLx800 fluorescence microplate reader (FL800 Bio-Tek Instruments, Winooski, VT, USA). A standard curve was prepared using 5, 10, 20, 30 and 40 μM of (6-hydroxy-2,5,7,8-tetramethylchroman-2-carboxylic acid (Trolox)). All solutions were prepared in phosphate-buffered saline (PBS), 75 mM, pH 7.4. The results are expressed as micromoles of Trolox equivalent antioxidant capacity per gram of extract (μ mol TEAC/g extract).

### 2.4. *Enzymatic Assays*

#### 2.4.1. Elastase Inhibition Assay

This assay was based on the work of Wittenauer et al. [27] with some modifications as described previously [6]. Elastase inhibitory activity is determined by a spectrophotometric method using porcine pancreatic elastase (PPE) and N-succinyl-Ala-Ala-Ala-p-nitroanilide (AAAPVN) as the enzyme-substrate, by monitoring the release of p-nitroaniline at 410 nm. PPE was dissolved in 100 mM Tris (2-amino-2-hydroxymethyl-propane-1,3-diol)-HCl buffer (pH = 8.0) to a concentration of 1 mg/mL and stored at 20 °C in aliquots. On the day of the assay, an aliquot was taken and diluted in buffer to a concentration of 0.03 U/mL, 10 μL was loaded in the wells of the microtiter plates together with 100 μL of the Tris-HCl buffer and 30 μL of each sample. After 20 min of pre-incubation at 25 °C, the reaction was initiated by the addition of 40 μL of the substrate AAAPVN (0.55 mM). Absorbance was measured for 20 min after the addition of AAAPVN at a BioTek Instruments EPOCH 2 spectrophotometer microplate reader. The calculations were made as described in Equation (1), where $A_{control}$ and $A_{sample}$ represent the absorbance at 410 nm in the absence or presence of the sample, respectively. Since DMSO was used to dissolve samples S1 and S3, this solvent was also tested and used as control for these samples. The potential of the extracts to inhibit elastase was evaluated with increasing concentrations, to determine dose-dependent relations and establish the half maximal inhibitory concentrations ($IC_{50}$) values, indicating the capacity of each sample in enzymatic activity inhibition to an extent of 50%.

$$\%_{inhibition} = \frac{\left(A_{control} - A_{sample}\right)}{A_{control}} * 100 \quad (1)$$

All results are expressed as $IC_{50}$ mean value with the lower and upper limits of a 95% confidence interval, obtained from at least three independent experiments.

#### 2.4.2. Tyrosinase Inhibition Assay

The tyrosinase inhibitory potential of the extracts was evaluated spectrophotometrically using mushroom tyrosinase and L-DOPA as the substrate [28]. Tyrosinase converts L-DOPA to dopaquinone, which will sequentially cyclize to form dopachrome. The dopachrome formation can be observed by measurement of the absorbance at 475 nm. The substrate was added to the enzyme in the presence of the sample dilutions, to a final concentration of 30 U/mL tyrosinase and 2.5 mM L-DOPA. Since DMSO was used to dissolve samples S1 and S3, this solvent was also tested and used as control for these samples. After incubation at 37 °C for 30 min, absorbance was measured at 475 nm on a BioTek Instruments EPOCH 2 microplate spectrophotometer. All the reagents were prepared in sodium phosphate buffer (SPB; 0.1 M, pH 6.8), prepared by mixing sodium phosphate dibasic dihydrate and sodium phosphate monobasic monohydrate, and the calculations were made as described in Equation (1). All results are expressed as $IC_{50}$ mean value with the lower and upper limits of a 95% confidence interval, obtained from at least three independent experiments.

### *2.5. Antimicrobial Susceptibility Testing*

The target microorganisms selected for the antibacterial activity assays were the gram-negative bacteria *Klebsiella pneumoniae* CECT 8453 and the gram-positive bacteria *Staphylococcus aureus* ATCC 6538 and *Cutibacterium acnes* ATCC 6919$^T$. For *K. pneumoniae* CECT 8453 and *S. aureus* ATCC 6538, assays were performed according to the broth microdilution method of CLSI M07-A10 guidelines as previously described by Rodrigues et al. [29]. In short, extract stock solutions were distributed in a round bottom microtiter 96-well plate and 2-fold serially diluted in calcium-adjusted Mueller Hinton broth (CAMHB; BD, Sparks, MD, USA) to obtain a concentration range of solutions. The inoculum was prepared using the growth method to achieve a homogenous suspension in saline solution. The adjusted inoculum was additionally diluted in CAMHB to guarantee that, following inoculation, each well contained around $5 \times 10^4$ CFU. The plates were incubated under aerobic conditions at 37 °C for 16 to 20 h. For *C. acnes*, the assays were performed as previously described with the use of brain-heart infusion (BHI) (Avantor, Radnor, PA, USA) broth instead of CAMHB and by incubating the microtiter plates for 70–74 h at 37 °C in anaerobic jars containing the atmosphere generation system AnaeroGen™ Compact (Oxoid, Hampshire, UK).

For each stock solution analysed, a positive control (CAMHB or BHI and diluted inoculum), a medium sterility control (uninoculated CAMHB or BHI), and an extract sterility control (uninoculated 2-fold extract stock solution in CAMHB or BHI) were performed accordingly. Minimum inhibitory concentration (MIC) values were the lowest concentration of a sample that visibly inhibited microbial growth after incubation. When needed, MIC values were confirmed using the cell viability reagent PrestoBlue™ (Invitrogen, San Diego, CA USA) following the manufacturer's guidelines. An additional MIC value, designated MIC*, was also established and defined as the lowest concentration of a sample at which bacterial growth was visually and differentially affected in comparison to the positive control. Minimum bactericidal concentration (MBC) values were reported as the lowest concentration of a sample leading to at least 99.9% reduction in viable bacterial counts in comparison to the initial inoculum and for equal incubation time. Results were expressed as a median of the values obtained after three biological replicates. Since DMSO was used to dissolve samples S1 and S3, this solvent was also tested to ensure that the final concentration used did not interfere with the target microorganism, hence the assay.

### *2.6. Cell-Based Assays*

#### 2.6.1. Cell Culture

Human keratinocyte cell line HaCaT (CLS, Germany) was cultured in a standard Dulbecco's Modified Eagle Medium (DMEM) supplemented with 10% ($v/v$) fetal bovine serum (FBS) and 1% ($v/v$) penicillin-streptomycin. The cells were routinely maintained as monolayers in 75 cm$^2$ culture flasks and incubated at 37 °C with 5% $CO_2$ in a humidified atmosphere.

#### 2.6.2. In Vitro Cytotoxicity

Cytotoxicity assays were performed according to previous works [6]. Briefly, HaCaT cells were seeded at a density of $1.4 \times 10^5$ cells/cm$^2$ in 96 well plates. After 3 days, cells were incubated with different concentrations of each sample (Cf1; Cf2; Cf3; Cf4; S2—50, 25, 12.5, 6.25, 3.13, 1.56, 0.78, 0.39; S1—3, 1.5, 0.75, 0.38, 0.19, 0.09, 0.05, 0.02; S3—5.5, 2.75, 1.38, 0.69, 0.34, 0.17, 0.09, 0.04 mg/mL) diluted in culture medium (DMEM medium containing 0.5% FBS). Wells containing cells incubated only with culture medium supplemented with 0.5% ($v/v$) of FBS were used as control. Solvent controls with 50% water or 1% DMSO in culture medium were also performed to exclude solvent toxicity. After 24 h of incubation, the cell viability was evaluated using PrestoBlue® (5% $v/v$ in culture medium) for 2 h at 37 °C, 5% $CO_2$, according to the manufacturer's instructions. After this, the fluorescence of each well was measured (Ex./Em. 560 ± 20/590 ± 20 nm) in an FLx800 fluorescence microplate reader (BioTek Instruments, Winooski, VT, USA). Cell viability was expressed as the percentage of viable cells relative to the control. Three independent experiments were performed in triplicate.

#### 2.6.3. Cellular Antioxidant Activity

Cellular antioxidant activity was evaluated following previously described methods [6,30], with some modifications. Briefly, HaCaT cells were seeded at a density of $1.4 \times 10^5$ cells/cm$^2$ in 96 well plates and the formation of intracellular ROS was monitored using 2′,7′-dichlorofluorescein diacetate (DCFH-DA) as a fluorescent probe. 72 h after seeding, cells were washed with PBS and incubated with non-toxic concentrations of the samples (0.1875 mg/mL; 0.375 mg/mL; 0.75 mg/mL) plus 25 µM DCFH-DA in PBS for 1 h. Subsequently, cells were washed again with PBS and incubated with the stress inducer (600 µM AAPH in PBS) for 1 h. After that, fluorescence was measured in an FL800 microplate fluorescence reader (Bio-Tek Instruments, Winooski, VT, USA) (Ex/Em 485 $\pm$ 20/528 $\pm$ 20 nm). The results are expressed as ROS percentage relative to the untreated control (cells treated with DCFH-DA and AAPH). Three independent experiments were performed in triplicate.

#### 2.6.4. Evaluation of IL-6 and IL-8 Secretion

Experiments were performed as previously described [31], with several modifications. Briefly, HaCaT cells were seeded at a density of $1 \times 10^5$ cells/cm$^2$ in 12 well plates. After 3 days, cells were stimulated with 15 µg/mL of lipopolysaccharides (LPS) from *Escherichia coli* and co-incubated with three different concentrations of each extract (0.1875 mg/mL; 0.375 mg/mL; 0.75 mg/mL) diluted in culture medium (DMEM medium containing 0.5% FBS). Cells incubated only with LPS and cells incubated with only culture media were used as positive and negative controls, respectively. After 24 h, supernatants were collected, centrifuged for 10 min at 2000 g and stored at −80 °C until further analysis. IL-6 and IL-8 levels were assessed by enzyme-linked immunosorbent assay (ELISA), using commercially available kits (PeproTech; London, UK), according to the manufacturer's instructions, with absorbance measured at 450 nm with wavelength correction set at 620 nm in a microplate spectrophotometer (EPOCH 2, BioTek Instruments, Winooski, VT, USA). The results are expressed as IL-6 or IL-8 percentage relative to the positive control (cells stimulated with LPS). Three independent experiments were performed in triplicate.

### *2.7. Statistical Analysis*

ORAC and cell-based assays results are expressed as the mean value $\pm$ SD, obtained from at least three independent experiments. For the enzymatic and cytotoxicity assays, the $IC_{50}$ values were determined from dose-response curves through log10 plots using GraphPad Prism 8.4.3. software (GraphPad Software, Inc., La Jolla, CA, USA). Results are presented as $IC_{50}$ with a 95% confidence interval. Statistical analysis of the results was performed using the former software. When homogeneous variance and a normal distribution of the data were verified, the results were analysed by one-way analysis of variance (ANOVA), followed by the Tukey test for multiple comparisons. In the case of heterogeneous variances or if the data were not normally distributed, an appropriate unpaired Student's *t*-test was performed to determine whether the means were significantly different. A $p$-value $\leq 0.05$ was accepted as statistically significant in all cases. Antimicrobial susceptibility testing results are expressed as the median value, obtained from at least three independent experiments.

## 3. Results and Discussion

This study aims to investigate the cosmeceutical potential of protein-rich extracts that were produced by high-pressure technologies from fishery industry wastes, namely sardine wastes and codfish frames [11,22]. The selection of extracts was based on previous results regarding their characterization and process conditions. For codfish, all extracts were selected aiming at evaluating the impact of the extraction temperature on the recovery of compounds with promising bioactive effects on the skin. For sardine extracts, only three extracts derived by both defatted and non-defatted raw materials, processed at higher temperatures (190 and 250 °C) and with the highest extraction yield (>45.7 g/100 g) were chosen. The results regarding the total protein content of each extract are presented in Table 1.

### *3.1. Antioxidant, Anti-Ageing and Anti-Hyperpigmentation Activities*

The potential cosmeceutical effect of the extracts was initially screened using chemical and enzymatic assays to evaluate their antioxidant, anti-ageing, and anti-hyperpigmentation effect. For the antioxidant capacity, the ORAC assay was selected as it measures the ability of samples to scavenge biologically relevant ROS, namely peroxyl radicals, which are considered one of the main inducers of skin ageing [2,32]. The anti-ageing effect was also evaluated through elastase inhibition since this enzyme is reported to be responsible for the degradation of elastin and other ECM proteins [6]. For the anti-hyperpigmentation effect, the tyrosinase assay was used, to evaluate the capacity of samples to inhibit melanin production. Table 2 summarizes the ORAC and $IC_{50}$ values of all extracts.

**Table 2.** Antioxidant, anti-ageing, and anti-hyperpigmentation activities of codfish and sardine residues extracts. ORAC values are expressed as the mean value ± SD. Enzymatic assays $IC_{50}$ values are expressed as mean with a 95% confidence interval.

| Sample | Antioxidant Activity | Anti-Ageing Activity | Anti-Hyperpigmentation Activity |
|---|---|---|---|
| | ORAC (µmol TEAC/mg extract) | Elastase Inhibition ($IC_{50}$, mg extract/mL) | Tyrosinase Inhibition ($IC_{50}$, mg extract/mL) |
| Cf1 | 0.64 ± 0.18 | 42.58 (37.38, 49.88) | >100 ** |
| Cf2 | 0.54 ± 0.22 | 60.35 (50.71, 73.43) | 78.59 (61.41, 100.70) |
| Cf3 | 0.59 ± 0.26 | 34.74 (24.57, 49.16) | >100 ** |
| Cf4 | 1.29 ± 0.26 | 38.11 (30.42, 49.52) | 40.89 (30.26, 54.43) |
| S1 | 1.94 ± 0.08 | 44.29 (39.10, 53.44) | 82.51 (57.12, 108.39) |
| S2 | 1.19 ± 0.10 | >300 * | 3.70 (3.26, 4.42) |
| S3 | 1.24 ± 0.06 | 17.96 (12.36, 25.83) | 10.40 (5.02, 18.15) |

* The maximum concentration tested was 300 mg/mL; ** The maximum concentration tested was 100 mg/mL.

Our results show that sardine and codfish extracts presented antioxidant activity and inhibition effects on elastase and tyrosinase enzymes' activity. Among sardine samples, S1 showed the highest ORAC value (1.94 ± 0.08 µmol TEAC/mg extract) followed by S3 and S2. These results are in accordance with a previous antioxidant evaluation through an alternative method (2,2-diphenyl-1-picrylhydrazyl-DPPH assay) where the extract S1 presented the lowest $IC_{50}$ values [22]. The higher scavenging capacity towards peroxyl radicals of the S1 sample could be derived from peptides with different amino acid sequences present in the extracts since interactions among them can influence the radical scavenging ability [33]. Additionally, compounds generated in Maillard or other thermo-oxidation reactions might influence the antioxidant activity of the samples [34]. Despite the lowest ORAC value, samples S2 and S3 were the ones with the highest capacity in inhibiting tyrosinase and elastase activities, respectively.

Among codfish extracts, Cf4 was shown to have the highest antioxidant and anti-hyperpigmentation activities. SEC-GPC analysis of the extracts has shown that peptides of decreasing molecular weight were obtained when increasing the extraction temperature up to 250 °C [11]. The elastase inhibition capacity of this sample is within the range of values found for other Cf extracts, namely Cf1 and Cf3. However, for the concentrations tested, these two extracts showed no inhibition activity towards tyrosinase.

In the literature there are some reports showing that extracts from marine by-products presents relevant antioxidant activity and inhibition of tyrosinse. For instance, extracts derived from marine (*Scophthalmus maximus*) by-products by alkaline hydrolysis, have demonstrated antioxidant activity accessed by different chemical assays: 1,1-diphenyl-2-picrylhydrazyl (DPPH) radical-scavenging ability (36.12% in relation to control), ABTS (2,2′-azinobis-(3-ethyl-benzothiazoline-6-sulphonic acid) method (12.81 µg BHT/mL) and crocin bleaching assay (8.03 µg Trolox/mL) [35]. In another study, enzymatic extraction of alum-salted jellyfish (*Lobonema smithii*) showed to produce hydrolysates with high antioxidant activity ($IC_{50}$ = 0.9 mg/mL for ABTS e DPPH assays) and tyrosinase inhibitory potential, with $IC_{50}$ values ranging between 14.1 and 24.5 mg/mL [36], which are in the same order of magnitude as the ones obtained in this work. Additionally, extracts with similar antioxidant values (ORAC values – 0.4 to 3.5 µmol TEAC/mg extract) were obtained from another type of food industry residues, namely winemaking waste streams [6]. However, these winery residues extracts presented higher tyrosinase ($IC_{50}$ from 4.0 to 0.14 mg extract/mL) and elastase ($IC_{50}$ from 3.4 to 0.1 mg extract/mL) inhibitory capacities than those obtained in this work, probably due to the presence of phenolic compounds that are recognized to have several bioactivities. It is important to mention that in our study we used mushroom tyrosinase to screen the anti-hyperpigmentation effect of extracts, as this enzyme has been widely used in high throughput assays [37]. Nevertheless, since there are some controversies regarding the similarity and homology of this enzyme with mammalian/human tyrosinase [38–40], future studies involving mammalian cell lines should be considered to evaluate the potential anti-hyperpigmentation effect of sardine and codfish extracts.

To identify which compounds could be responsible for the bioactive response of extracts, correlation studies between bioactivity data and the extracts' amino acid composition reported previously [11,22] were performed (Table S1). For antioxidant activity, the highest correlations ($R^2 \geq 0.7$) were obtained between ORAC value and total threonine, free valine, as well as free and total leucine for all extracts. In the case of sardine samples, a high correlation ($R^2 \geq 0.94$) was also obtained for free and total tryptophan content. Accordingly, all these amino acids have been previously reported to have antioxidant properties in several model systems [41–43]. For elastase and tyrosinase inhibition activities, the highest correlation coefficients ($R^2 \geq 0.8$) were obtained for total protein content and free arginine, respectively, suggesting that these compounds could have an important role in the potential anti-ageing and anti-hyperpigmentation effect of fish industry waste streams extracts.

### 3.2. Antibacterial Activity

*S. aureus* and *K. pneumoniae* were chosen as representative gram-positive and -negative bacteria to evaluate the antimicrobial capacity of the extracts. The antibacterial activity assays performed with codfish frame extracts revealed that all of them were able to affect both bacterial strains' growth behaviour (MIC* results in Tables 3 and 4). However, true growth inhibition (MIC values) did not occur in the presence of any of the extracts at the concentrations tested.

**Table 3.** Antibacterial activity of codfish and sardine extracts against *S. aureus*.

| Sample | MIC* Median (mg/mL) (n = 1/n = 2/n = 3) | MIC Median (mg/mL) (n = 1/n = 2/n = 3) | MBC Median (mg/mL) (n = 1/n = 2/n = 3) |
|---|---|---|---|
| Cf1 | 0.39 (0.39/0.39/0.20) | >50.00 (>50.00/>50.00/>50.00) | >50.00 (>50.00/>50.00/>50.00) |
| Cf2 | 0.78 (0.78/0.78/0.39) | >50.00 (>50.00/>50.00/>50.00) | >50.00 (>50.00/>50.00/>50.00) |
| Cf3 | 0.39 (0.39/0.39/0.20) | >50.00 (>50.00/>50.00/>50.00) | >50.00 (>50.00/>50.00/>50.00) |
| Cf4 | 0.10 (0.10/0.10/0.10) | >50.00 (>50.00/>50.00/>50.00) | >50.00 (>50.00/>50.00/>50.00) |
| S1 | 1.56 (1.56/1.56/0.78) | 25.00 (25.00/25.00/25.00) | 50.00 (50.00/50.00/50.00) |
| S2 | 0.07 (0.07/0.07/0.07) | 1.17 (1.17/2.34/1.17) | 9.38 (9.38/9.38/9.38) |
| S3 | 0.27 (0.27/0.54/0.27) | 68.75 (68.75/68.75/34.38) | 68.75 (68.75/68.75/68.75) |

MIC*—Lowest concentration of a sample at which bacterial growth was visually and differentially affected; MIC—Minimum inhibitory concentration; MBC—Minimum bactericidal concentration.

**Table 4.** Antibacterial activity of codfish and sardine extracts against *K. pneumoniae*.

| Sample | MIC* Median (mg/mL) (n = 1/n = 2/n = 3) | MIC Median (mg/mL) (n = 1/n = 2/n = 3) | MBC Median (mg/mL) (n = 1/n = 2/n = 3) |
|---|---|---|---|
| Cf1 | 12.50 (12.50/12.50/12.50) | >50.00 (>50.00/>50.00/>50.00) | > 50.00 (>50.00/>50.00/>50.00) |
| Cf2 | 25.00 (25.00/25.00/50.00) | >50.00 (>50.00/>50.00/>50.00) | >50.00 (>50.00/>50.00/>50.00) |
| Cf3 | 0.78 (0.39/0.78/0.78) | >50.00 (>50.00/>50.00/>50.00) | >50.00 (>50.00/>50.00/>50.00) |
| Cf4 | 0.39 (0.39/0.39/0.20) | >50.00 (>50.00/>50.00/>50.00) | >50.00 (>50.00/>50.00/>50.00) |
| S1 | 3.13 (3.13/3.13/6.25) | 50.00 (50.00/50.00/50.00) | 50.00 (50.00/50.00/50.00) |
| S2 | 0.07 (0.07/0.07/0.07) | 4.69 (4.69/4.69/9.38) | 18.75 (18.75/18.75/18.75) |
| S3 | 0.54 (0.54/1.07/0.54) | 68.75 (68.75/68.75/68.75) | 68.75 (68.75/>68.75/68.75) |

MIC*—Lowest concentration of a sample at which bacterial growth was visually and differentially affected; MIC—Minimum inhibitory concentration; MBC—Minimum bactericidal concentration.

As S1 and S3 were solubilized in DMSO, assays were performed with this solvent to evaluate its influence on the results obtained for the sardine extracts. The results reveal that the maximum DMSO concentration used in the antibacterial activity testing of the different extracts (up to 12.5%) did not affect the growth of both bacterial strains.

In general, all sardine extracts inhibited the bacterial growth of both gram-positive and gram-negative selected strains, but *S. aureus* was shown to be more susceptible than *K. pneumoniae*, which is expected since the outer membrane of gram-negative bacteria poses an additional barrier to prevent the interference of different molecules with the cell [44]. S2 was shown to be the extract with the most promising antibacterial potential, indicating that higher extraction temperatures favoured the extraction of anti-bacterial compounds. Taking these results into account, S2 was selected to be tested against another gram-positive bacteria, namely *C. acnes*, an aerotolerant anaerobe linked to the acne skin condition [45]. To evaluate the impact of the defatting process on the inhibition capacity of *C. acnes* growth, sample S3 was also selected to be tested in this assay. The results are summarized in Table 5 showing that *C. acnes* behaves similarly to *S. aureus* with its growth being affected by both sardine extracts. Between both samples, S2 presented the highest inhibitory effect (lower MIC and MBC values) suggesting that the defatting process could modify or eliminate some anti-microbial compounds from the sardine samples before the extraction process, especially free fatty acids and lipid oxidation products [22].

**Table 5.** Antibacterial activity of sardine extracts against *C. acnes*.

| Sample | MIC* Median (mg/mL) (n = 1/n = 2/n = 3) | MIC Median (mg/mL) (n = 1/n = 2/n = 3) | MBC Median (mg/mL) (n = 1/n = 2/n = 3) |
|---|---|---|---|
| S2 | 0.29 (0.29/0.29/0.29) | 0.59 (0.59/0.29/0.59) | 2.34 (4.69/1.17/2.34) |
| S3 | 2.15 (2.15/2.15/2.15) | 17.19 (34.38/17.19/17.19) | > 68.75 (>68.75/68.75/>68.75) |

MIC*—Lowest concentration of a sample at which bacterial growth was visually and differentially affected; MIC—Minimum inhibitory concentration; MBC—Minimum bactericidal concentration.

Previous studies showed that some amino acid residues present in peptides can lead to different antibacterial activities [46]. Rodrigues et al. showed that protein derivative-rich extracts obtained by extraction with deep eutectic solvents (DES) from sardine processing waste streams have antibacterial activity toward *S. aureus* and *Escherichia coli* [47]. However, the MIC*, MIC and MBC values of these extracts were lower than those obtained in this work for the same raw material, which could be explained by the synergic and/or additive effect between the extracts' components and DES, as described previously [29]. Nevertheless, the antimicrobial effect of S2 extract is similar to other extracts obtained by subcritical water extraction, namely using kānuka leaves (*S. aureus*—MIC: 0.9 mg/mL, MBC: 3.8–5 mg/mL; *E. coli*—MIC: 3.8–7.5 mg/mL, MBC: 4.4–7.5 mg/mL) [48], that are rich in phenolic compounds already recognized as presenting relevant antimicrobial effects [49].

In this study, the highest correlation coefficient obtained was for MIC values of *S. aureus* and the total glutamic acid content ($R^2$ = 0.6, Table S1).

### 3.3. Cellular Antioxidant and Anti-Inflammatory Effect

In this work, a human keratinocyte cell line (HaCaT) was used to better understand the bioactivity, namely antioxidant and anti-inflammatory effects, of extracts as some of the processes related to the uptake, distribution, and metabolism of bioactive compounds are better addressed [50]. HaCaT cells are one of the predominant cell types encountered in the skin, being responsible for skin integrity and, when affected by senescence or oxidative stress, can accelerate the skin ageing process [51]. Additionally, keratinocytes play an important role in the regulation of skin inflammation, responding to external stimuli, such as bacterial LPS, actively contributing to inflammation pathways especially by releasing proinflammatory cytokines or chemokines [31,52].

In a first approach, cytotoxicity assays were performed to evaluate the safety of samples and to select non-toxic concentrations for further bioactivity studies. Table 6 presents the $IC_{50}$ values of each sample showing that S1 presented the highest cytotoxic effect. In previous studies, these samples showed higher $IC_{50}$ values in Caco-2, a model for crypt enterocytes, than in HaCaT, which indicates that the samples are more toxic for keratinocytes than intestinal cells [11,22].

**Table 6.** $IC_{50}$ values obtained for all the extracts in HaCaT cells, with 24 h incubation. $IC_{50}$ values are expressed as mean with a 95% confidence interval.

| Sample | $IC_{50}$ (mg Extract/mL) |
|---|---|
| Cf1 | 9.7 (9.5, 9.9) |
| Cf2 | 3.6 (3.5, 3.7) |
| Cf3 | 17.3 (15.2, 19.6) |
| Cf4 | 2.0 (1.9, 2.2) |
| S1 | 0.6 (0.5, 0.7) |
| S2 | 3.4 (3.3, 3.6) |
| S3 | >5.5 |

$IC_{50}$—concentration of the sample that leads to a decrease of 50 % of the cell population after 24 h incubation.

The cellular antioxidant activity was then assessed by evaluating the capacity of samples in scavenging intracellular ROS generated by the chemical stressor AAPH. In parallel, anti-inflammatory assays were also performed to investigate the effect of samples in reducing the secretion of IL-8, which has been consistently reported as an important skin inflammation biomarker [53–55], upon pro-inflammatory stimulus with LPS. In these assays, non-cytotoxic concentrations of the extracts were used (0.1875, 0.375 and 0.75 mg/mL for all samples except S1 where 0.75 mg/mL was not tested since this concentration presented a cytotoxic effect, Table 6). In general, all samples inhibited ROS formation in HaCaT, and a dose-dependent effect was observed (Figure 1). Among codfish samples, Cf2 and Cf4 showed the highest cellular antioxidant activities, and amongst sardine samples, S1 showed the highest ROS percentage reduction relative to control.

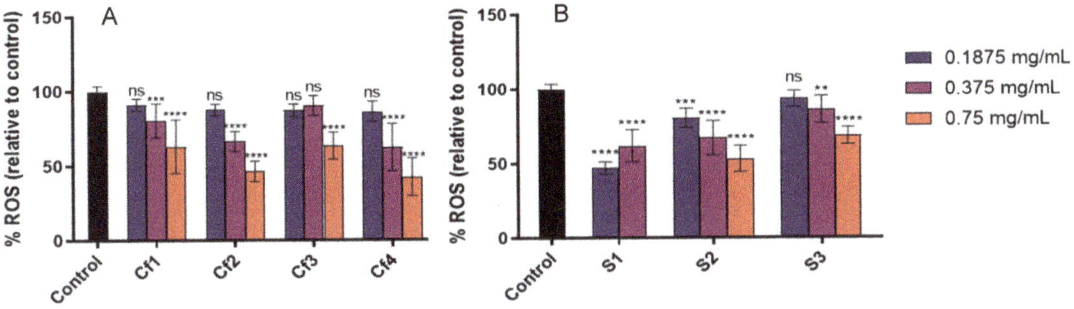

**Figure 1.** Cellular antioxidant capacity, expressed as % of ROS inhibition relative to the control, of each extract. (**A**) Codfish frame extracts; (**B**) Sardine wastes extracts. The results are expressed as mean ROS percentage relative to the control ± SD. The symbol * indicates significance relative to the control ** $p$-value ≤ 0.01, *** $p$-value ≤ 0.001, **** $p$-value ≤ 0.0001); ns—not significant.

Concerning the anti-inflammatory effect, only Cf1 (0.375 and 0.75 mg/mL), S2 (0.75 mg/mL) and S3 (0.375 and 0.75 mg/mL) revealed capacity to inhibit IL-8 secretion by HaCaT cells after LPS-induced inflammation (Figure 2). In contrast, the other samples did not reduce IL-8 and, for some concentrations, Cf4 and S1 showed a pro-inflammatory effect. These two extracts were the ones that demonstrated the highest cytotoxic effect in HaCaT cells and thus the concentrations tested, although not leading to cell death, could induce the activation of inflammatory pathways since injured cells can release danger signals that alert other cells to cell death [56].

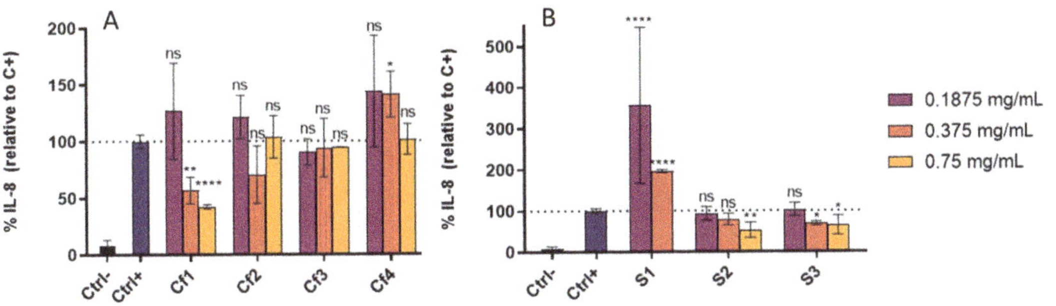

**Figure 2.** IL-8 secreted by HaCaT cells treated for 24 h with different extracts concentrations and 15 µg/mL LPS. (**A**) Codfish residues extracts; (**B**) Sardine residues extracts. Ctrl—cells incubated with only culture media; Ctrl+—Cells incubated with culture media + inflammation inductor (LPS); Cf1, Cf2, Cf3, Cf4—cells incubated with the different extracts from codfish frames + inflammation inductor (LPS); S1, S2, S3—cells incubated with the different extracts from sardine wastes + inflammation inductor (LPS). The results are expressed as mean IL-8 percentage relative to the positive control ± SD. The symbol * indicates significance relative to the positive control (* $p$-value ≤ 0.05, ** $p$-value ≤ 0.01, **** $p$-value ≤ 0.0001); ns—not significant.

The extracts with the highest anti-inflammatory effect, namely Cf1, S2 and S3, were further selected to evaluate their capacity in inhibiting IL-6 secretion, a cytokine also recognized as an important biomarker in skin disorders [53]. Results show that Cf1 was the only sample able to significantly reduce IL-6 secretion by 69.4 ± 3.5 and 53.0 ± 14.7% IL-6 (relative to the positive control, $p$-value < 0.001) at 0.375 and 0.75 mg/mL, respectively, whereas sardine extracts increased in IL-6 levels of the supernatants (Figure 3). Overall, only Cf1 revealed the capacity to reduce both IL-8 and IL-6 secretion by HaCaT cells, suggesting that this codfish extract could be further explored for anti-inflammatory applications in skin conditions.

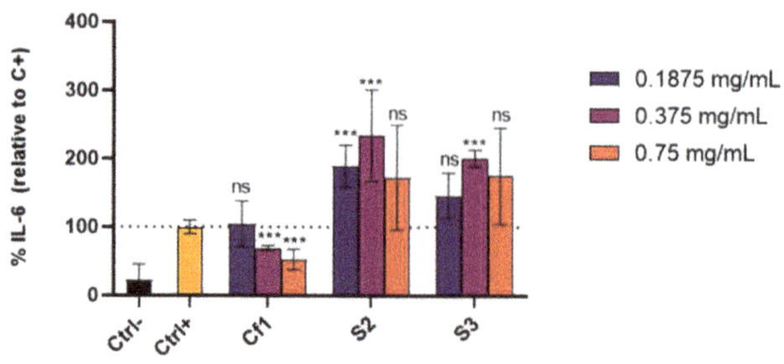


**Figure 3.** IL-6 secreted by HaCaT cells treated for 24 h with different extracts concentrations and 15 μg/mL LPS. Ctrl—cells incubated with only culture media; Ctrl+—Cells incubated with culture media + inflammation inductor (LPS); Cf1– cells incubated with Cf1 extract from codfish frames + inflammation inductor (LPS); S2, S3—cells incubated S2 and S3 extracts from sardine wastes + inflammation inductor (LPS). The results are expressed as mean IL-6 percentage relatively to the positive control ± SD. The symbol * indicates significance relative to the positive control (*** $p$-value $\leq 0.001$); ns—not significant.

The anti-inflammatory activity of codfish extracts was previously evaluated in an intestinal cell line (Caco-2 cells) [11] and the results were in line with the data presented in this study for HaCaT cells. In both cell lines, Cf1 presented the highest anti-inflammatory effect, reinforcing the use of lower temperatures to extract bioactive compounds from codfish frames.

Our results are in accordance with previous studies supporting the idea that fish-derived extracts/hydrolysates display a broad spectrum of bioactivities, including antioxidant, antimicrobial, anti-ageing, anti-hypertensive, anti-human immunodeficiency virus, anti-proliferative, or anticoagulant activities [23,57]. Song et al. showed that enzymatic hydrolysates of the marine fish half-fin anchovy contained antibacterial peptide fractions, with activity against *E. coli* [58]. Fish skin and collagen hydrolysed by subcritical water hydrolysis also showed high antioxidant and antimicrobial activity against *Bacillus cereus*, *S. aureus* and *Pseudomonas putida* [59]. Additionally, Wang and co-workers produced extracts derived from fish side streams of two fish species (rainbow trout and sole) that could inhibit the growth of pathogenic bacteria (*S. aureus* or *Salmonella*) and with anti-inflammatory properties [60].

## 4. Conclusions

In this work, we applied a platform of in vitro bioassays to evaluate for the first time the cosmeceutical potential of protein extracts derived from fishery wastes, namely sardine residues and codfish frames, obtained by high-pressure technology. We demonstrated that both types of extracts showed antioxidant, anti-ageing, and anti-hyperpigmentation potential. Among all, sardine extracts presented the highest anti-bacterial activity, and this effect was more pronounced for samples obtained using higher extraction temperatures (250 °C) and without the defatting step. Codfish samples were the most effective anti-inflammatory agents, and in this case, lower temperatures (90 °C) favoured the extraction of these bioactive compounds. Although further studies are needed to identify which compounds could be responsible for the bioactive effects, total threonine, free valine as well as free and total leucine were identified to highly correlate with the antioxidant activities of samples. Total protein content and free arginine correlated with elastase and tyrosinase inhibition activities, respectively.

This work is a step forward in the development of potential cosmeceutical ingredients with antioxidant, skin whitening, antimicrobial, and anti-inflammatory effects from sardine residues and codfish frames, adding potential high value to these fishery industry wastes.

**Supplementary Materials:** The following supporting information can be downloaded at: https://www.mdpi.com/article/10.3390/antiox11101925/s1, Table S1: Correlation coefficients between protein and amino acids content and different bioactivities of sardine waste and codfish frames extracts.

**Author Contributions:** Conceptualization, I.C.L., F.B.G., P.S., A.P., R.M., N.F. and A.T.S.; methodology, M.C., A.B., I.C.L., F.B.G., N.F. and A.T.S.; investigation, M.C., I.C.L., F.B.G., M.M., R.M., N.F. and A.T.S.; resources, F.B.G., N.F. and A.T.S.; writing—original draft preparation, M.C.; writing—review and editing, I.C.L., F.B.G., R.M., P.S., N.F. and A.T.S.; supervision, F.B.G., P.S., N.F. and A.T.S.; funding acquisition, F.B.G., P.S., N.F. and A.T.S. All authors have read and agreed to the published version of the manuscript.

**Funding:** This work was funded by Fundação para a Ciência e a Tecnologia/Ministério da Ciência, Tecnologia e Ensino Superior (FCT/MCTES, Portugal) through project PTDC/ASPPES/28399/2017 and national funds iNOVA4Health (UIDB/04462/2020 and UIDP/04462/2020) and the Associate Laboratories LS4FUTURE (LA/P/0087/2020) and LAQV (UIDB/QUI/50006/2020). Funding from INTERFACE Programme, through the Innovation, Technology and Circular Economy Fund (FITEC), is gratefully acknowledged. ATS also acknowledges FCT/MCTES for the Individual Grant CEECIND/04801/2017.

**Institutional Review Board Statement:** Not applicable.

**Informed Consent Statement:** Not applicable.

**Data Availability Statement:** The data supporting the findings of this study are available within the article and its supplementary material.

**Acknowledgments:** We acknowledge Pascoal and Filhos S.A. (Gafanha da Nazaré, Portugal) and Conservas A Poveira S.A. (Portugal) for providing us, respectively, the Atlantic codfish (*Gadus morhua*) frames and Sardine (*Sardina pilchardus*) waste used in this work.

**Conflicts of Interest:** The authors declare no conflict of interest.

## References

1. Guillerme, J.B.; Couteau, C.; Coiffard, L. Applications for marine resources in cosmetics. *Cosmetics* **2017**, *4*, 35. [CrossRef]
2. e Silva, S.A.M.; Leonardi, G.R.; Michniak-Kohn, B. An overview about oxidation in clinical practice of skin aging. *An. Bras. Dermatol.* **2017**, *92*, 367–374. [CrossRef] [PubMed]
3. Hoang, H.T.; Moon, J.Y.; Lee, Y.C. Natural antioxidants from plant extracts in skincare cosmetics: Recent applications, challenges and perspectives. *Cosmetics* **2021**, *8*, 106. [CrossRef]
4. Tobin, D.J. Introduction to skin aging. *J. Tissue Viability* **2017**, *26*, 37–46. [CrossRef]
5. Hashizume, H. Skin aging and dry skin. *J. Dermatol.* **2004**, *31*, 603–609. [CrossRef]
6. Matos, M.S.; Romero-Díez, R.; Álvarez, A.; Bronze, M.R.; Rodríguez-Rojo, S.; Mato, R.B.; Cocero, M.J.; Matias, A.A. Polyphenol-rich extracts obtained from winemaking waste streams as natural ingredients with cosmeceutical potential. *Antioxidants* **2019**, *8*, 355. [CrossRef]
7. Nordberg, J.; Arnér, E.S.J. Reactive oxygen species, antioxidants, and the mammalian thioredoxin system. *Free Radic. Biol. Med.* **2001**, *31*, 1287–1312. [CrossRef]
8. Rinnerthaler, M.; Bischof, J.; Streubel, M.K.; Trost, A.; Richter, K. Oxidative stress in aging human skin. *Biomolecules* **2015**, *5*, 545. [CrossRef]
9. Xu, H.; Zheng, Y.-W.; Liu, Q.; Liu, L.-P.; Luo, F.-L.; Zhou, H.C.; Isoda, H.; Ohkohchi, N.; Li, Y.-M. Reactive oxygen species in skin repair, regeneration, aging, and inflammation. *React. Oxyg. Species ROS Living Cells* **2017**, *8*, 69–88. [CrossRef]
10. The State of World Fisheries and Aquaculture 2020. Available online: https://www.fao.org/state-of-fisheries-aquaculture (accessed on 11 January 2022).
11. Melgosa, R.; Marques, M.; Paiva, A.; Bernardo, A.; Fernández, N.; Sá-Nogueira, I.; Simões, P. Subcritical water extraction and hydrolysis of cod (*Gadus morhua*) frames to produce bioactive protein extracts. *Foods* **2021**, *10*, 1222. [CrossRef]
12. Ferraro, V.; Carvalho, A.P.; Piccirillo, C.; Santos, M.M.; Paula, P.M.; Pintado, M.E. Extraction of high added value biological compounds from sardine, sardine-type fish and mackerel canning residues—A review. *Mater. Sci. Eng. C* **2013**, *33*, 3111–3120. [CrossRef] [PubMed]
13. Thuanthong, M.; De Gobba, C.; Sirinupong, N.; Youravong, W.; Otte, J. Purification and characterization of angiotensin-converting enzyme-inhibitory peptides from nile tilapia (*Oreochromis niloticus*) skin gelatine produced by an enzymatic membrane reactor. *J. Funct. Foods* **2017**, *36*, 243–254. [CrossRef]
14. Chi, C.F.; Wang, B.; Hu, F.Y.; Wang, Y.M.; Zhang, B.; Deng, S.G.; Wu, C.W. Purification and identification of three novel antioxidant peptides from protein hydrolysate of bluefin leatherjacket (*Navodon septentrionalis*) skin. *Food Res. Int.* **2015**, *73*, 124–129. [CrossRef]
15. Seo, J.K.; Lee, M.J.; Go, H.J.; Kim, Y.J.; Park, N.G. Antimicrobial function of the GAPDH-related antimicrobial peptide in the skin of skipjack tuna, katsuwonus pelamis. *Fish Shellfish. Immunol.* **2014**, *36*, 571–581. [CrossRef] [PubMed]
16. Wang, T.Y.; Hsieh, C.H.; Hung, C.C.; Jao, C.L.; Chen, M.C.; Hsu, K.C. Fish skin gelatin hydrolysates as dipeptidyl peptidase IV inhibitors and glucagon-like peptide-1 stimulators improve glycaemic control in diabetic rats: A comparison between warm- and cold-water fish. *J. Funct. Foods* **2015**, *19*, 330–340. [CrossRef]

17. Cai, L.; Wu, X.; Zhang, Y.; Li, X.; Ma, S.; Li, J. Purification and characterization of three antioxidant peptides from protein hydrolysate of grass carp (*Ctenopharyngodon idella*) skin. *J. Funct. Foods* **2015**, *16*, 234–242. [CrossRef]
18. Lu, J.; Hou, H.; Fan, Y.; Yang, T.; Li, B. Identification of MMP-1 inhibitory peptides from cod skin gelatin hydrolysates and the inhibition mechanism by MAPK signaling pathway. *J. Funct. Foods* **2017**, *33*, 251–260. [CrossRef]
19. Abdallah, M.M.; Leonardo, I.C.; Krstić, L.; Enríquez-De-salamanca, A.; Diebold, Y.; González-García, M.J.; Gaspar, F.B.; Matias, A.A.; Bronze, M.R.; Fernández, N. Potential ophthalmological application of extracts obtained from tuna vitreous humor using lactic acid-based deep eutectic systems. *Foods* **2022**, *11*, 342. [CrossRef]
20. Rodrigues, L.A.; Pereira, C.V.; Carvalho Partidario, A.M.; Gouveia, L.F.; Simoes, P.; Paiva, A.; Matias, A.A. Supercritical $CO_2$ extraction of bioactive lipids from canned sardine waste streams. *J. $CO_2$ Util.* **2021**, *43*, 101359. [CrossRef]
21. Šližyte, R.; Mozuraityte, R.; Martínez-Alvarez, O.; Falch, E.; Fouchereau-Peron, M.; Rustad, T. Functional, bioactive and antioxidative properties of hydrolysates obtained from cod (*Gadus morhua*) backbones. *Process Biochem.* **2009**, *44*, 668–677. [CrossRef]
22. Melgosa, R.; Trigueros, E.; Sanz, M.T.; Cardeira, M.; Rodrigues, L.; Fernández, N.; Matias, A.A.; Bronze, M.R.; Marques, M.; Paiva, A.; et al. Supercritical $CO_2$ and subcritical water technologies for the production of bioactive extracts from sardine (*Sardina pilchardus*) waste. *J. Supercrit. Fluids* **2020**, *164*, 104943. [CrossRef]
23. Venkatesan, J.; Anil, S.; Kim, S.K.; Shim, M.S. Marine fish proteins and peptides for cosmeceuticals: A review. *Mar. Drugs* **2017**, *15*, 143. [CrossRef] [PubMed]
24. Pedras, B.; Salema-Oom, M.; Sá-Nogueira, I.; Simões, P.; Paiva, A.; Barreiros, S. Valorization of white wine grape pomace through application of subcritical water: Analysis of extraction, hydrolysis, and biological activity of the extracts obtained. *J. Supercrit. Fluids* **2017**, *128*, 138–144. [CrossRef]
25. Huang, D.; Ou, B.; Hampsch-Woodill, M.; Flanagan, J.A.; Prior, R.L. High-throughput assay of oxygen radical absorbance capacity (ORAC) using a multichannel liquid handling system coupled with a microplate fluorescence reader in 96-well format. *J. Agric. Food Chem.* **2002**, *50*, 4437–4444. [CrossRef]
26. Oliveira-Alves, S.C.; Andrade, F.; Prazeres, I.; Silva, A.B.; Capelo, J.; Duarte, B.; Caçador, I.; Coelho, J.; Serra, A.T.; Bronze, M.R. Impact of Drying Processes on the Nutritional Composition, Volatile Profile, Phytochemical Content and Bioactivity of *Salicornia ramosissima* J. Woods. *Antioxidants* **2021**, *10*, 1312. [CrossRef]
27. Wittenauer, J.; MäcKle, S.; Sußmann, D.; Schweiggert-Weisz, U.; Carle, R. Inhibitory effects of polyphenols from grape pomace extract on collagenase and elastase activity. *Fitoterapia* **2015**, *101*, 179–187. [CrossRef]
28. Chan, E.W.C.; Lim, Y.Y.; Wong, L.F.; Lianto, F.S.; Wong, S.K.; Lim, K.K.; Joe, C.E.; Lim, T.Y. Antioxidant and tyrosinase inhibition properties of leaves and rhizomes of ginger species. *Food Chem.* **2008**, *109*, 477–483. [CrossRef]
29. Rodrigues, L.A.; Pereira, C.V.; Leonardo, I.C.; Fernández, N.; Gaspar, F.B.; Silva, J.M.; Reis, R.L.; Duarte, A.R.C.; Paiva, A.; Matias, A.A. Terpene-based natural deep eutectic systems as efficient solvents to recover astaxanthin from brown crab shell residues. *ACS Sustain. Chem. Eng.* **2020**, *8*, 2246–2259. [CrossRef]
30. Serra, A.T.; Matias, A.A.; Frade, R.F.M.; Duarte, R.O.; Feliciano, R.P.; Bronze, M.R.; Figueira, M.E.; de Carvalho, A.; Duarte, C.M.M. Characterization of traditional and exotic apple varieties from portugal. Part 2—Antioxidant and antiproliferative activities. *J. Funct. Foods* **2010**, *2*, 46–53. [CrossRef]
31. Di Caprio, R.; Lembo, S.; Di Costanzo, L.; Balato, A.; Monfrecola, G. Anti-inflammatory properties of low and high doxycycline doses: An in vitro study. *Mediat. Inflamm.* **2015**, *2015*, 329418. [CrossRef]
32. Mukherjee, P.K.; Maity, N.; Nema, N.K.; Sarkar, B.K. Bioactive compounds from natural resources against skin aging. *Phytomedicine* **2011**, *19*, 64–73. [CrossRef]
33. Chalamaiah, M.; Dinesh Kumar, B.; Hemalatha, R.; Jyothirmayi, T. Fish protein hydrolysates: Proximate composition, amino acid composition, antioxidant activities and applications: A review. *Food Chem.* **2012**, *135*, 3020–3038. [CrossRef] [PubMed]
34. Asaduzzaman, A.K.M.; Chun, B.S. Hydrolyzates produced from mackerel *Scomber japonicus* skin by the pressurized hydrothermal process contain amino acids with antioxidant activities and functionalities. *Fish. Sci.* **2014**, *80*, 369–380. [CrossRef]
35. Vázquez, J.A.; Rodríguez-Amado, I.; Sotelo, C.G.; Sanz, N.; Pérez-Martín, R.I.; Valcárcel, J. Production, characterization, and bioactivity of fish protein hydrolysates from aquaculture turbot (*Scophthalmus maximus*) wastes. *Biomolecules* **2020**, *10*, 310. [CrossRef] [PubMed]
36. Upata, M.; Siriwoharn, T.; Makkhun, S.; Yarnpakdee, S.; Regenstein, J.M.; Wangtueai, S. Tyrosinase inhibitory and antioxidant activity of enzymatic protein hydrolysate from jellyfish (*Lobonema smithii*). *Foods* **2022**, *11*, 615. [CrossRef]
37. Fan, Y.F.; Zhu, S.X.; Hou, F.B.; Zhao, D.F.; Pan, Q.S.; Xiang, Y.W.; Qian, X.K.; Ge, G.B.; Wang, P. Spectrophotometric assays for sensing tyrosinase activity and their applications. *Biosensors* **2021**, *11*, 290. [CrossRef] [PubMed]
38. Promden, W.; Viriyabancha, W.; Monthakantirat, O.; Umehara, K.; Noguchi, H.; De-Eknamkul, W. Correlation between the potency of flavonoids on mushroom tyrosinase inhibitory activity and melanin synthesis in melanocytes. *Molecules* **2018**, *23*, 1403. [CrossRef]
39. Zolghadri, S.; Bahrami, A.; Hassan Khan, M.T.; Munoz-Munoz, J.; Garcia-Molina, F.; Garcia-Canovas, F.; Saboury, A.A. A comprehensive review on tyrosinase inhibitors. *J. Enzym. Inhib. Med. Chem.* **2019**, *34*, 279–309. [CrossRef]
40. Strzępek-Gomółka, M.; Gaweł-Bęben, K.; Angelis, A.; Antosiewicz, B.; Sakipova, Z.; Kozhanova, K.; Głowniak, K.; Kukula-Koch, W. Identification of mushroom and murine tyrosinase inhibitors from *Achillea biebersteinii* afan. Extract. *Molecules* **2021**, *26*, 964. [CrossRef]

41. Ji, S.; Qi, X.; Ma, S.; Liu, X.; Min, Y. Effects of dietary threonine levels on intestinal immunity and antioxidant capacity based on cecal metabolites and transcription sequencing of broiler. *Animals* **2019**, *9*, 739. [CrossRef]
42. Cojocaru, E.; Filip, N.; Ungureanu, C.; Filip, C.; Danciu, M. Effects of valine and leucine on some antioxidant enzymes in hypercholesterolemic rats. *Health* **2014**, *6*, 2313–2321. [CrossRef]
43. Nayak, B.N.; Buttar, H.S. Evaluation of the antioxidant properties of tryptophan and its metabolites in in vitro assay. *J. Complement. Integr. Med.* **2016**, *13*, 129–136. [CrossRef] [PubMed]
44. Jones, S. Permeability rules for antibiotic design. *Nat. Biotechnol.* **2017**, *35*, 639. [CrossRef] [PubMed]
45. Mayslich, C.; Grange, P.A.; Dupin, N. *Cutibacterium* acnes as an opportunistic pathogen: An update of its virulence-associated factors. *Microorganisms* **2021**, *9*, 303. [CrossRef]
46. Huan, Y.; Kong, Q.; Mou, H.; Yi, H. Antimicrobial peptides: Classification, design, application and research progress in multiple fields. *Front. Microbiol.* **2020**, *11*, 2559. [CrossRef]
47. Rodrigues, L.A.; Leonardo, I.C.; Gaspar, F.B.; Roseiro, L.C.; Duarte, A.R.C.; Matias, A.A.; Paiva, A. Unveiling the potential of betaine/polyol-based deep eutectic systems for the recovery of bioactive protein derivative-rich extracts from sardine processing residues. *Sep. Purif. Technol.* **2021**, *276*, 119267. [CrossRef]
48. Essien, S.O.; Young, B.; Baroutian, S. The antibacterial and antiproliferative ability of kānuka, *Kunzea ericoides*, leaf extracts obtained by subcritical water extraction. *J. Chem. Technol. Biotechnol.* **2021**, *96*, 1308–1315. [CrossRef]
49. Lu, C.; Li, C.; Chen, B.; Shen, Y. Composition and antioxidant, antibacterial, and anti-HepG2 cell activities of polyphenols from seed coat of amygdalus pedunculata pall. *Food Chem.* **2018**, *265*, 111–119. [CrossRef]
50. Wolfe, K.L.; Rui, H.L. Cellular antioxidant activity (CAA) assay for assessing antioxidants, foods, and dietary supplements. *J. Agric. Food Chem.* **2007**, *55*, 8896–8907. [CrossRef]
51. Csekes, E.; Račková, L. Skin Aging, Cellular senescence and natural polyphenols. *Int. J. Mol. Sci.* **2021**, *22*, 12641. [CrossRef]
52. Li, S.; Xie, R.; Jiang, C.; Liu, M. Schizandrin A Alleviates LPS-Induced Injury in human keratinocyte cell hacat through a microRNA-127-dependent regulation. *Cell. Physiol. Biochem.* **2018**, *49*, 2229–2239. [CrossRef] [PubMed]
53. Zampetti, A.; Mastrofrancesco, A.; Flori, E.; Maresca, V.; Picardo, M.; Amerio, P.; Feliciani, C. Proinflammatory cytokine production in HaCaT cells treated by eosin: Implications for the topical treatment of psoriasis. *Int. J. Immunopathol. Pharmacol.* **2009**, *22*, 1067–1075. [CrossRef] [PubMed]
54. Colombo, I.; Sangiovanni, E.; Maggio, R.; Mattozzi, C.; Zava, S.; Corbett, Y.; Fumagalli, M.; Carlino, C.; Corsetto, P.A.; Scaccabarozzi, D.; et al. HaCaT cells as a reliable in vitro differentiation model to dissect the inflammatory/repair response of human keratinocytes. *Mediat. Inflamm.* **2017**, *2017*, 7435621. [CrossRef]
55. Jeong, S.J.; Lim, H.S.; Seo, C.S.; Jin, S.E.; Yoo, S.R.; Lee, N.; Shin, H.K. Anti-inflammatory actions of herbal formula Gyejibokryeong-hwan regulated by inhibiting chemokine production and STAT1 activation in HaCaT cells. *Biol. Pharm. Bull.* **2015**, *38*, 425–434. [CrossRef] [PubMed]
56. Rock, K.L.; Kono, H. The inflammatory response to cell death. *Annu. Rev. Pathol.* **2008**, *3*, 99–126. [CrossRef]
57. Ngo, D.H.; Vo, T.S.; Ngo, D.N.; Wijesekara, I.; Kim, S.K. Biological activities and potential health benefits of bioactive peptides derived from marine organisms. *Int. J. Biol. Macromol.* **2012**, *51*, 378–383. [CrossRef]
58. Song, R.; Wei, R.B.; Luo, H.Y.; Wang, D.F. Isolation and characterization of an antibacterial peptide fraction from the pepsin hydrolysate of half-fin anchovy (*Setipinna taty*). *Molecules* **2012**, *17*, 2980. [CrossRef]
59. Ahmed, R.; Chun, B.S. Subcritical water hydrolysis for the production of bioactive peptides from tuna skin collagen. *J. Supercrit. Fluids* **2018**, *141*, 88–96. [CrossRef]
60. Wang, M.; Zhou, J.; Pallarés, N.; Bäuerl, C.; Collado, M.C.; Dar, B.N.; Barba, F.J. Role of extracts obtained from rainbow trout and sole side streams by accelerated solvent extraction and pulsed electric fields on modulating bacterial and anti-inflammatory activities. *Separations* **2021**, *8*, 187. [CrossRef]

*antioxidants*

MDPI

*Article*

# Nomilin from Yuzu Seed Has In Vitro Antioxidant Activity and Downregulates Melanogenesis in B16F10 Melanoma Cells through the PKA/CREB Signaling Pathway

**Moon-Hee Choi [1], Seung-Hwa Yang [2], Nam Doo Kim [3] and Hyun-Jae Shin [1,2,*]**

1 Department of Beauty and Cosmetology, Graduate School of Industrial Technology and Entrepreneurship, Chosun University, Gwangju 61452, Korea
2 Department of Chemical Engineering, Graduate School of Chosun University, Gwangju 61452, Korea
3 VORONOI BIO Inc., Incheon 21984, Korea
* Correspondence: shinhj@chosun.ac.kr; Tel.: +82-62-230-7518

**Citation:** Choi, M.-H.; Yang, S.-H.; Kim, N.D.; Shin, H.-J. Nomilin from Yuzu Seed Has In Vitro Antioxidant Activity and Downregulates Melanogenesis in B16F10 Melanoma Cells through the PKA/CREB Signaling Pathway. *Antioxidants* **2022**, *11*, 1636. https://doi.org/10.3390/antiox11091636

Academic Editor: Alessandra Napolitano

Received: 11 July 2022
Accepted: 19 August 2022
Published: 23 August 2022



**Abstract:** Yuzu (*Citrus junos*) is a citrus plant native to Asian countries, including Korea, Japan, and China. Yuzu peel and seed contain abundant vitamin C, citric acid, and polyphenols. Although the antioxidative and antimelanogenic activities of other citrus fruits and yuzu extract have been reported, the tyrosinase inhibitory activity of the limonoid aglycone contained in yuzu seed extract is unknown. We separated yuzu seeds into the husk, shell, and meal and evaluated antioxidant activity of each. The limonoid glucoside fraction of the husk identified nomilin, a novel tyrosinase inhibitor. We performed tyrosinase inhibitory activity and noncompetitive inhibition assays and docking studies to determine nomilin binding sites. Furthermore, we evaluated the antioxidative mechanism and antimelanogenic activity of nomilin in B16F10 melanoma cells. The concentration of nomilin that did not show toxicity was <100 μg/mL. Nomilin suppressed protein expression of TYR, TRP-1, TRP-2, and microphthalmia-associated transcription factor (MITF) in a concentration-dependent manner. Nomilin significantly reduced the levels of p-CREB and p-PKA at the protein level and decreased the levels of skin-whitening-related factors MITF, tyrosinase, TRP-1, and TRP-2 at the mRNA level in a concentration-dependent manner. Thus, nomilin from yuzu seed husk can be used as a skin-whitening agent in cosmetics.

**Keywords:** yuzu tree; seed husk; nomilin; antioxidant; antimelanogenic; tyrosinase inhibitor; skin-whitening agent; reactive oxygen species

## 1. Introduction

Skin aging can be mainly divided into intrinsic and extrinsic aging. Intrinsic aging occurs when the bonds between the skin's epidermis and dermis weaken, the ability of keratinocytes to divide declines, and the ability to form lipids decreases [1]. Extrinsic aging, also known as photoaging, is caused by long-term UV exposure and occurs when UV rays penetrate the epidermis, reaching deep into the dermis to damage collagen and elastin (elastic fibers), which maintain the dermis' elasticity [2]. In response to these challenges, the skin overproduces reactive oxygen species (ROS), including superoxide anions and peroxides [3]. The increase in ROS activity damages DNA and increases transformation signals, ultimately increasing the level of transcription factor activator protein 1 [4]. UV radiation is directly involved in DNA mutagenesis, increases nuclear factor-κB levels, and decreases TGF-β levels [5]. These mechanisms affect the synthesis and degradation of collagen as well as the production of inflammatory cytokines. When the synthesis of collagen and elastin, both components of extracellular spaces, decreases due to UV exposure, the expression of various proteolytic enzymes of the extracellular matrix is promoted. The resulting lack of extracellular matrix proteins has been suggested to be the most important factor in photoaging [6].

Melanin, which determines the color of the skin, hair, and pupils, is produced in melanocytes (pigment cells) [7]. The process of melanin formation in melanocytes is called melanogenesis, and melanocytes are found in the lowermost layer of the skin epidermis. Melanin is a generic term for black or brown pigments in tissues, such as the skin and eyes, and mainly exists as melanin molecule that forms a strong bond with globulin. Melanin is an aggregate of small molecules and is categorized into eumelanin and pheomelanin based on race. Melanin moves through keratinocytes to the epidermis, thereby causing skin pigmentation [8]. Abnormal melanin synthesis mainly occurs through the oxidation of the precursor tyrosine due to excessive UV irradiation, disease, or genetic factors [9]. Tyrosine is oxidized to 3,4-dihydroxy-L-phenylalanine (DOPA) and DOPA quinone by tyrosinase in melanocytes, catalyzed to the intermediate DOPA chrome, and eventually polymerized to melanin via indole-5,6-quinone [10]. The inhibition of tyrosinase activity inhibits the biosynthesis of melanin polymer in the skin, and thus, is extremely important for the development of antioxidants and skin-whitening agents. Additionally, blocking melanosome transfer between melanocytes and keratinocytes can have a skin-whitening effect. Representative tyrosinase inhibitors derived from natural products include kojic acid, (a secondary metabolite from green mold), arbutin (isolated from bearberry leaves), and oxy-resveratrol (isolated from mulberry trees), in addition to other stilbene- and flavonoid-based compounds [11].

Kojic acid and arbutin—representative skin-whitening agents—have been used as additives in cosmetics. However, because they have various side-effects, including skin toxicity and allergies, there is an increasing demand for natural skin-whitening materials in the cosmetics industry. *Citrus junos* (yuzu) is mainly processed to manufacture yuzu syrup using the flesh and skin of yuzu, and due to its excellent flavor and unique acidic taste, is consumed as yuzu tea. However, 10–15% of yuzu fruit is left over as byproduct. Most byproducts remaining after the production of yuzu syrup and juice are discarded, with nearly 1800 tons discarded annually. Yuzu, which is mainly cultivated in northeast Asia, contains large amounts of vitamin C and flavonoids, such as hesperidin and limonin [12]. These flavonoids have been reported to have high anti-inflammatory and antioxidant activity. Flavonoids extracted from citrus species—in particular, hesperidin and naringin were reported to reduce the expression of inflammatory cytokines, such as tumor necrosis factor-α, interleukin-1, and interleukin-6 [13]. However, studies on the skin-whitening effect of yuzu seed byproducts are lacking. To our knowledge, no studies have been conducted on the mechanism of action of nomilin, a component found in yuzu seed byproducts, as a tyrosinase inhibitor and skin-whitening agent. Therefore, in this study, we isolated nomilin from discarded yuzu byproducts and evaluated its suitability as a tyrosinase inhibitor by performing in vitro kinetic and in vivo cell line experiments. In addition, its potential as a functional cosmetic ingredient with skin-whitening and anti-aging effects was discussed.

## 2. Materials and Methods

### 2.1. Chemicals

2,2′-Azino-bis (3-ethylbenzthiazoline-6-sulfonic acid) (ABTS), 2,2-diphenyl-1-picrylhydrazyl (DPPH), Folin–Ciocalteu reagent (for total phenolics), dimethyl sulfoxide-$d_6$ (DMSO), mushroom tyrosinase (EC 1.14.18.1), and ascorbic acid were obtained from Sigma Aldrich (St. Louis, MO, USA). All reagents used were of analytical grade. Phospho-CREB (p-CREB) and phospho-PKA (p-PKA) were purchased from Cell Signaling (Danvers, MA, USA). Antibodies against tyrosinase (TYR), TRP-1, TRP-2, microphthalmia-associated transcription factor (MITF), CREB, PKA, α-melanocyte-stimulating hormone (α-MSH), and β-actin were purchased from Santa Cruz Biotechnology (Dallas, TX, USA). Horseradish peroxidase-conjugated anti-mouse, anti-goat, and anti-rabbit antibodies were purchased from Invitrogen (Carlsbad, CA, USA).

*2.2. Chemical Extracts of Limonoid Aglycones and Limonoid Glucosides*

We separated waste husk, shell, and meal from yuzu seeds. Two hundred grams of separated husk, shell, and meal were placed in 2 L of 100% ethanol at room temperature for 24 h for extraction [4,12]. The filtrate was concentrated using vacuum filtration. Then, 2 L of 100% ethanol was added to the residue, and extraction was repeated three times under the same conditions. Concentrated limonoid aglycones extracted from the husk, shell, and meal were named LA1, LA2, and LA3, respectively (Figure 1). After the limonoid aglycones were extracted, the residues were dried and placed in 2 L of water at 100 °C and 400 rpm for 2 h to extract limonoid glucosides. After extraction, the filtrate was concentrated by vacuum filtration; 2 L of water was added to the filtration residue, and extraction was repeated three times under the same conditions. Concentrated limonoid glucosides extracted from the husk, shell, and meal were named LG1, LG2, and LG3, respectively.

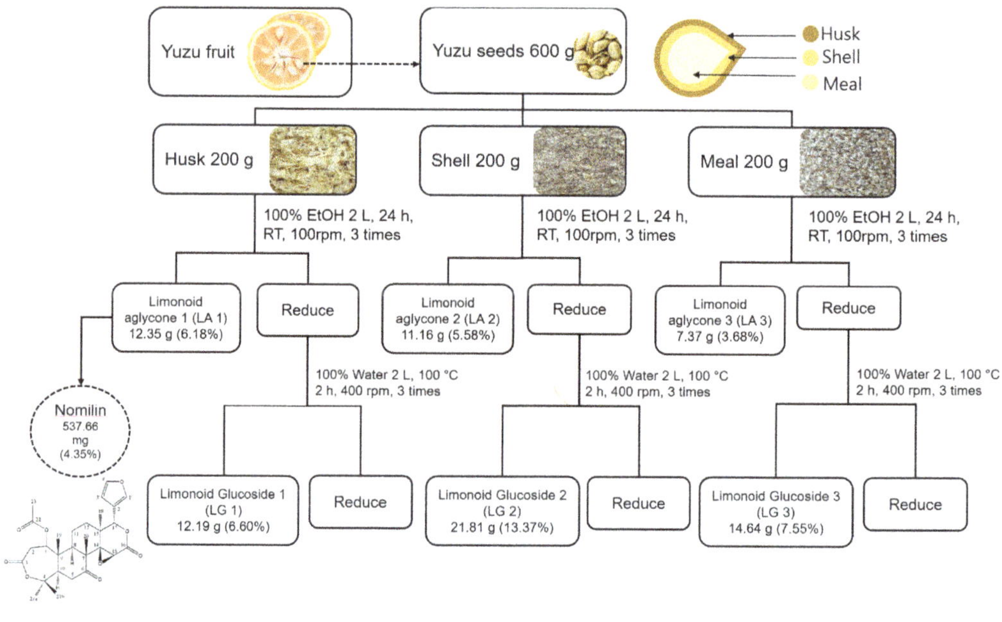


**Figure 1.** Schematic diagram showing the preparation of yuzu seed extracts. Percentages in parenthesis mean extraction yields based on the initial quantity of yuzu seeds. LA: limonoid aglycone, LG: limonoid glucoside.

*2.3. High-Performance Liquid Chromatography with Diode-Array Detection (HPLC–DAD) Analysis*

The most effective extracts of limonoid aglycone (LA1, LA2, and LA3) and limonoid glucoside (LG1, LG2, and LG3) were analyzed quantitatively by HPLC. The HPLC system (DGU-20A; Shimadzu, Kyoto, Japan) consisted of an LC-20AD pump and a diode-array detector (SPD-20A). The compounds were identified by comparison with the retention times of standard materials [14].

*2.4. Extraction of Nomilin*

Medium-pressure liquid chromatography was performed using PREP UV-10V (Yamazen) and PUMP 582 (Yamazen, Osaka, Japan). The experiment was conducted under the following conditions: flow rate, 20 mL/min; UV detection, 254 nm; and column, ULTRA PACK (20 mm × 200 mm, 30 μm; Yamazen). The solvents used were *n*-hexane (A) and EtOAc (B). The gradient program for LA1 was (B) 100%, 0–10 min; (B) 90%, 10–50 min; (B) 80%, 50–90 min; (B) 70%, 90–105 min; (B) 50%, 105–120 min; (B) 0%, 120–125 min; and (B) 0%, 125–135 min. Nomilin was isolated by preparative HPLC with an X-Bridge Prep

OBD $C_{18}$ column (5.0 μm, 19 mm × 150 mm). Elution was performed with a linear gradient of methanol (0 min, 50/50; 30 min, 100/0; 100 min, 100/0) to obtain nomilin as a yellowish powder with 4.35% yield. $^1$H and $^{13}$C NMR spectra were recorded using DMSO-$d_6$ and AVANCE III HD 400 MHz NMR (Bruker, Billerica, MA, USA) (Figures 2, S1 and S2). Coupling constants were expressed in Hz, and chemical shifts were expressed on a d (ppm) scale. $^1$H NMR (DMSO-$d_6$): δ = 1.03 (s, 3H, H-20), 1.15 (s, 3H, H-19), 1.31 (d, 1H, J = 9.32 Hz, H-8), 1.55 (s, 3H, H-18), 1.93 (s, 9H, H-21a, H-21b, H-23), 2.36 (m, 2H, H-12), 2.51 (m, 2H, H-11), 2.81 (dd, J = 16.20 Hz, J = 7.32 Hz, H-10), 3.12 (t, 2H, J = 14.4 Hz, H-5), 3.61 (d, 1H, J = 16.04 Hz, H-1), 3.93 (s, 1H, H-15), 4.84 (d, 1H, J = 7.12 Hz, H-3′), 5.46 (s, 1H, H-17), 6.51 (d, 1H, J = 1.20 Hz, H-3′), 7.67 (t, 1H, J = 1.68 Hz, H-4′), 7.73 (s, H-1′). $^{13}$C-NMR (DMSO-$d_6$): δ = 15.72 (C-20), 16.77 (C-18), 17.04 (C-19), 20.88 (C-21a, C-21b), 23.05 (C-11), 31.95 (C-23), 33.72 (C-12), 35.77 (C-5), 37.47 (C-2), 44.06 (C-8), 44.25 (C-13), 50.90 (C-9), 52.51 (C-10), 53.26 (C-7), 66.19 (C-15), 71.40 (C-14), 77.88 (C-4), 79.63 (C-1), 85.00 (C-17), 110.65 (C-3′), 120.48 (C-2′), 142.13 (C-1′), 143.95 (C-4′), 167.58(C-22), 169.59 (C-3), 169.86 (C-16), 208.41 (C-6).

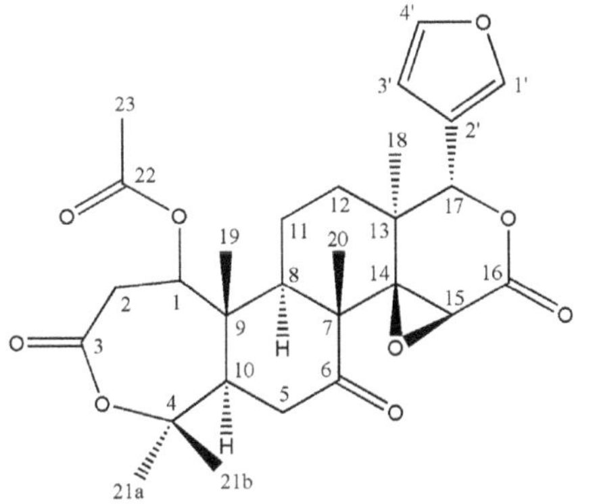

**Figure 2.** Structure of nomilin {7-(furan-3-yl)-1,8,12,17,17-pentamethyl-5,15,20-trioxo-3,6,16-trioxapentacyclo [9.9.0.0$^{2}$,$^{4}$.0$^{2}$,$^{8}$.0$^{12}$,$^{18}$]icosan-13-yl acetate} derived from yuzu seed husk.

*2.5. Antioxidant Activity Assay*

2.5.1. DPPH Radical Scavenging Activity

Radical scavenging activity was determined using the DPPH radical scavenging assay with some modifications [15]. We mixed 200 μL extract with 800 μL of 1 mmol/L methanolic DPPH. Mixtures were left for 15 min in the dark. Then, absorbance was measured at 517 nm with the SCINCO UV-Vis spectrophotometer (S-3100; Seoul, Korea). The scavenging activity of DPPH radicals was calculated using the following equation: Scavenging activity (%) = 100 × ($A_0$ − $A_1$)/$A_0$, where $A_0$ is the absorbance of the MeOH control and $A_1$ is the absorbance in the presence of nomilin extracts. The inhibitory concentration ($IC_{50}$) was defined as the amount of extract required for a 50% reduction of free radical scavenging activity. The $IC_{50}$ values were obtained from the resulting inhibition curves. Results were compared with the activity of ascorbic acid (Sigma Aldrich, St. Louis, MO, USA) as a control.

2.5.2. ABTS Radical Scavenging Activity

A 7 mM solution of ABTS was prepared in water. The ABTS stock solution was reacted with 7 mM potassium persulfate (final concentration), and the mixture was left at room temperature for 12–16 h before use to generate ABTS radicals. Radical scavenging was measured by mixing 200 μL of each sample and 1000 μL ABTS solution [16]. Mixtures were left for 15 min in the dark, and absorbance was measured at 730 nm with the SCINCO UV-Vis spectrophotometer S-3100 (Seoul, Korea). The scavenging activity of ABTS radicals was calculated using the following equation: Scavenging activity = 100 × ($A_0$ − $A_1$)/$A_0$, where $A_0$ is the absorbance of the water control and $A_1$ is the absorbance in the presence of nomilin extract. $IC_{50}$ values were obtained from the resulting inhibition curves. Results were compared with the activity of quercetin (Sigma Aldrich, St. Louis, MO, USA) as a control.

### 2.6. *Total Polyphenol Content*

Total polyphenol content in the fractionated samples was measured using a modified version of the Folin–Ciocalteu method [17]. A total of 500 µL of extract was mixed with 500 µL of Folin–Ciocalteu reagent and 500 µL of 2% sodium carbonate (*w/v*). The mixtures were left for 30 min at 25 °C. Absorbance was measured at 750 nm with a UV-Vis spectrophotometer (S-3100; SCINCO, Seoul, Korea). The extract samples were evaluated at a final concentration of 1 mg/mL. Total phenolic content was expressed as mg/mL of gallic acid equivalents (GAE) using the following equation, which was based on the calibration curve: $y = 19.42x + 0.0541$, $R^2 = 0.996$, where $x$ is the gallic acid equivalent (mg/g) and $y$ is the absorbance.

### 2.7. *Antimelanogenic Activity Assay*

#### 2.7.1. Tyrosinase Inhibition Assay

The tyrosinase inhibition assay was performed according to Macrini et al. [18], with a few modifications. We used 1250 U/mL of tyrosinase (Sigma Aldrich, St. Louis, MO, USA) for the experiment. We added 10 µL of tyrosinase to the wells of 96-well microplates. Then, 70 µL of pH 6.8 phosphate buffer solution and 60 µL of nomilin (10–200 µg/mL), LA1 (10–500 µg/mL), LG1 (10–500 µg/mL), and ascorbic acid (10–100 µg/mL) as a standard were added to the mixture in order. Next, 70 µL of L-tyrosine (Sigma Aldrich) was added at a concentration of 0.3 mg/mL in distilled water (the final volume in the wells was 210 µL). The absorbance of the microplate wells was read using a spectrophotometer (Synergy HT; BIO-TEX, Winooski, VT, USA) at 510 nm ($T_0$). The microplates were incubated at 30 °C ± 1 °C for 60 min, and absorbance was measured ($T_1$). The microplates were further incubated for 60 min at 30 °C ± 1 °C, and absorbance was measured (T2). The inhibitory percentages at the two timepoints ($T_1$ and $T_2$) were obtained according to the following formula: Inhibition activity (IA)% = (C − S)/C × 100, where IA% is the inhibitory activity, C is the absorbance of the negative control, and S is the absorbance of the sample or positive control (absorbance at time $T_1$ or $T_2$ minus absorbance at time $T_0$) [19].

#### 2.7.2. Enzyme Kinetic Assay

Tyrosinase (EC 1.10.3.1) is an enzyme that converts L-tyrosine to DOPA and finally to DOPA quinone. To evaluate inhibition, L-DOPA was used as a substrate at concentrations of 0.5, 1.0, 1.5, and 2.0 mM. Tyrosinase inhibition was detected using a spectrophotometer (Synergy HT; BIO-TEX, Winooski, VT, USA). The $IC_{50}$ assay was performed for tyrosinase according to Fan et al. [20].

For the test, 20 µL aliquots of a solution composed of 500 U/mL mushroom tyrosinase (Sigma Aldrich, St. Louis, MO, USA) were added to 96-well microplates. Then, 100 µL of pH 6.8 phosphate buffer solution and 60 µL of nomilin (0.2–1.0 mM) were added. Absorbance was measured at 510 nm (T0) using a microplate reader (Synergy HT; BIO-TEX, Winooski, VT, USA). The microplates were incubated at 30 °C ± 1 °C for 30 min, and the absorbance was measured again (T1). The microplates were further incubated for 30 min at 30 °C ± 1 °C, and absorbance was measured (T2). The inhibitory percentages at the two timepoints (T1 and T2) were obtained based on the following formula: IA% = (C − S)/C × 100, where IA% is the inhibitory activity, C is the negative control absorbance, and S is the absorbance of the sample or positive control (the absorbance at time T0 subtracted from the absorbance at time T1 or T2) [21].

### 2.8. Molecular Docking Procedure

Molecular docking was performed to predict the binding site of mushroom tyrosinase and TRP-1 to nomilin using the Glide module in the Schrodinger package [22,23]. The X-ray crystal structures of tyrosinase (PDB ID: 2Y9X) and TRP-1 (PDB ID: 5M8O) were retrieved from the Protein Data Bank (http://www.rcsb.org (accessed on 10 October 2020)). The retrieved protein structures were processed using Protein Preparation Wizard in the Schrodinger package to remove crystallographic water molecules, add hydrogen atoms, and assign protonated states and partial charges. The missing side chains and loops were built and refined using the Prime tool of the Schrodinger suite [24]. All protein residues were parameterized using the OPLS3e force field [25,26]. Finally, restrained minimization was performed until the converged average root mean square deviation of heavy atoms was 0.3 Å. Binding mode predictions of nomilin with mushroom tyrosinase and TRP1 were performed using the Glide docking tool in the Schrodinger package. Docking grid boxes were generated considering the catalytic sites of mushroom tyrosinase and TRP-1. Nomilin was docked into the catalytic site of each protein using standard precision scoring modes. The 3D structure of nomilin was minimized using the Macromodel module of the Schrodinger suite.

### 2.9. Cell Culture

B16F10 melanoma cells, a murine melanoma cell line, were purchased from the Korea Cell Line Bank (KCLB80008, Seoul, Korea). Cells were maintained in Dulbecco's modified Eagle's medium (DMEM; HyClone, MA, USA) supplemented with 10% fetal bovine serum (HyClone), 50 units/mL penicillin, and 50 μg/mL streptomycin at 37 °C in a humidified atmosphere with 5% $CO_2$ at 37 °C.

### 2.10. MTT Cell Viability Assay

Cell viability analysis was performed using the 3-(4,5-dimethylthiazol-2-yl)-2,5-diphenyltetrazolium bromide (MTT) assay. B16F10 cells were cultured at $1 \times 10^4$ cells/cm$^3$ in six-well plates. After 24 h, the cells were treated with 25, 50, or 100 μg/mL nomilin for 48 h. At the end of incubation, 100 μL of MTT solution (1 mg/mL in DMEM) was added to each well. After incubation at 37 °C for 1 h, the medium was gently removed, and 400 μL of DMSO was added. The absorbance of each well was measured at 570 nm using a spectrophotometer.

### 2.11. Measurement of Melanin Content

Melanin content was determined according to Hosoi et al. [27]. B16F10 cells were cultured at $1 \times 10^4$ cells/cm in six-well plates. After 24 h, the cells were stimulated with 1 μg/mL α-MSH. Simultaneously, various concentrations of nomilin (62 and 125 μM) were added for 48 h. Then, the cells were washed with phosphate-buffered saline (PBS) and harvested after trypsin treatment. The collected cells were suspended in 100 μL of 1 N NaOH, and absorbance was measured at 405 nm using a spectrophotometer.

### 2.12. Measurement of Intracellular ROS Generation

ROS generated by *t*-BHP as evaluated using 2,7-dichlorodihydrofluorescein (H2DCF-DA) [28]. H2DCF-DA is oxidized to a green, highly fluorescent compound called 2,7-dichlorofluorescein (DCF) upon ROS generation. B16F10 cells were treated with various concentrations of nomilin (10–100 μg/mL) for 24 h. Then, the cells were rinsed with PBS and incubated with 100 μM H2DCF-DA for an additional 30 min at 37 °C. A fluorescence plate reader (Synergy HT; BIO-TEX, Winooski, VT, USA) was used to measure DCF fluorescence intensity at Ex./Em. = 488/525 nm. The experiments were performed three times. The values were expressed as percentage fluorescence relative to control.

### 2.13. Immunoblotting

B16F10 cells were treated with different concentrations of nomilin, disrupted using lysis buffer containing protein inhibitors, and frozen for 24 h in a deep-freezer. The frozen cells were thawed on ice for ~90 min and vortexed 3 to 6 times to disrupt the cells for protein extraction. SDS immunoblotting and polyacrylamide gel electrophoresis were performed as described [29], with a few modifications. Briefly, the samples were separated on 7.5% SDS-PAGE and transferred to a nitrocellulose membrane electrophoretically. The membrane was first incubated with the primary antibodies, and then with horseradish peroxidase-conjugated secondary antibodies. The signal was detected using the Enhanced Chemiluminescence Detection Reagent (Amersham Biosciences, Little Chalfont, UK). β-actin was used as the loading control.

### 2.14. RNA Isolation and Reverse-Transcription Polymerase Chain Reaction (RT-PCR)

Total RNA was extracted using TRIzol reagent (Invitrogen), in accordance with the manufacturer's instructions. To obtain cDNA, total RNA (2 μg) was reverse-transcribed using an oligo(dT) primer. The cDNA was amplified using the High-Capacity cDNA synthesis kit (Bioneer, Daejeon, Korea) in a PCR machine (Bio-Rad, Hercules, CA, USA). PCR was performed using a PCR premix (Bioneer), and real-time RT-PCR was performed using the StepOne model (Applied Biosystems, Foster City, CA, USA) and SYBR Green premix, according to the manufacturer's instructions (Applied Biosystems). Primers were synthesized by Bioneer. The following primer sequences were used: mouse tyrosinase: 5′-ATAACAGCTCCCACCAGTGC-3′ (sense) and 5′-CCCAGAAGCCAATGCACCTA-3′ (antisense); mouse MITF: 5′-CTGTACTCTGAGCAGCAGGTG-3′ (sense) and 5′-CCCGTCTCTGGAAACTTGATCG-3′ (antisense); and mouse TRP-1: 5′-AGACGCTG-CACTGCTGGTCAAGCCTGTAGCCCACGTCGTA-3′ (sense) and 5′-GCTGCAGGAGCCTT-CTTTCT-3′ (antisense). The expression of glyceraldehyde 3-phosphate dehydrogenase was used as an endogenous control for qRT-PCR [30].

### 2.15. Statistical Analysis

For statistical analysis, we used IBM SPSS online version 26.0 (SPSS, Inc., Chicago, IL, USA). For differences between groups, one-way analysis of variance was used, and statistical significance was evaluated. In addition, for each statistically significant effect of treatment, Duncan's multiple range test was used for comparing multiple means. Data were expressed as mean ± standard deviation (SD).

## 3. Results and Discussion

### 3.1. Yuzu Seed Isolation and Yield

Limonoid compounds are secondary metabolites and triterpene derivatives that are mainly present in mature fruits and seeds. A total of 38 limonoid aglycones: 23 neutral limonoids and 15 acidic limonoids, have been isolated from various fruits. Recently, 36 aglycones and 20 glucosides were isolated from limonoids [31]. In citrus fruits, limonoid compounds exist in the form of aglycones or glucosides, with limonin and nomilin being the most abundant [32]. Notably, citrus seeds contain abundant limonoid compounds, with more aglycones than glucosides [4]. In this study, we separated yuzu seed byproducts into husk, shell, and meal and compared results of their composition and material balance. Comparison of material balance (in 200 g each of husk, shell, and meal) revealed that the yields of yuzu seed limonoid aglycones were 12.35, 11.16, and 7.36 g, respectively; yuzu seed husk (LA1) had the highest yield (Figure 1). In another study, yuzu seeds were classified into three types, and their harvest rates were measured. Water content was lowest in yuzu seed. Moreover, high-cost limonoids and yuzu seed oil with high antioxidant activity were extracted from waste yuzu seeds, which had fat-soluble limonoid aglycone (330.6 mg $g^{-1}$ of dry seeds), water-soluble limonoid glucosides (452.0 mg $g^{-1}$ of dry seeds), and oil (40 mg $g^{-1}$ of green seeds) [12]. In another study, yuzu seeds were separated into three parts to measure harvest rate. The findings were consistent with ours, with the shells having the highest yield [33]. Thus, our findings are consistent with those of others—yuzu seed aglycones had the highest yield; the high content of limonoid aglycones was due to the relatively lower water content than that in other parts.

## 3.2. Antioxidant Activity of Yuzu Seed Extracts

### 3.2.1. DPPH Radical Scavenging Activity

Radicals with an odd number of electrons are highly energetic and unstable. They are highly reactive and cause oxidative reactions in vivo, resulting in cell and tissue damage. Lipid peroxidation due to a radical chain reaction is known to be the main cause of accelerated skin aging. Free radicals take electrons from reducing substances and become reduced. The strength of the reducing power is important in protecting skin cells against oxidative damage. The reducing power of antioxidants can be evaluated using the radical scavenging activity assay [34]. Evaluation of the DPPH free radical scavenging ability of limonoid aglycone and limonoid glucoside revealed that the antioxidant activities of limonoid aglycone and limonoid glucoside are similar (Figure S3a). $IC_{50}$ values of limonoid aglycones (LA1, LA2, and LA3) and limonoid glucosides (LG1, LG2, and LG3) were 942.02, 1250.08, and 1240.15 and 1121.84, 1302.20, and 1102.31 μg/mL, respectively (Table 1). The activity of LA1 was highest. In previous studies, the radical scavenging activities of 70% ethanol extracts of yuzu shells and the ethyl acetate fraction were 512.1 and 514.0 μg/mL, respectively [34]. In another study, the sample volume of the free magnetic seed chamber ethanol extract ranged from 20.65% to 57.94% when the sample volume ranged from 100 to 1000 μg/mL, respectively [35]. The electron-donating ability of the citron juice showed 80% or more when a sample solution having a concentration of 0.1% was added at a concentration of $1 \times 10^4$ M DPPH, and an interaction of phenol, hesperidin, and naringin, among others, as active ingredients was reported [36].

**Table 1.** $IC_{50}$ values of the radical scavenging activity and total polyphenol content of limonoid aglycone and limonoid glucoside.

| Sample | DPPH $IC_{50}$ (μg/mL) | ABTS $IC_{50}$ (μg/mL) | TPC (GAE mg/g) |
|---|---|---|---|
| LA1 | $942.02 \pm 45.91$ $^{a,b}$ | $626.34 \pm 14.68$ $^{a,b}$ | $6.72 \pm 0.14$ |
| LA2 | $1250.08 \pm 96.53$ | $808.86 \pm 13.25$ $^{a,b}$ | $5.38 \pm 0.32$ |
| LA3 | $1240.15 \pm 116.90$ | $668.41 \pm 52.53$ | $2.05 \pm 0.04$ |
| LG1 | $1121.84 \pm 31.57$ $^{a,b}$ | $844.78 \pm 13.96$ $^{a,b}$ | $6.17 \pm 0.23$ |
| LG2 | $1302.20 \pm 22.91$ $^{a,b}$ | $972.49 \pm 14.38$ $^{a,b}$ | $4.99 \pm 0.63$ |
| LG3 | $1102.31 \pm 18.80$ $^{a,b}$ | $1077.79 \pm 18.18$ $^{a,b}$ | $1.87 \pm 0.06$ |
| Nomilin | $53.37 \pm 0.83$ | $57.9 \pm 0.75$ | - |
| Ascorbic acid | $56.97 \pm 1.27$ | $39.95 \pm 0.73$ | - |

Each value is the mean ± standard deviation (SD) of three experiments. Differences within and between groups were evaluated by a one-way analysis of variance (ANOVA) followed by a multi-comparison Dunnett's test ($\alpha = 0.05$), $^{a,b}$ $p < 0.01$, compared with the positive controls (a: nomilin, b: ascorbic acid) in the biological assay, whereas multi-comparison Tukey's test was applied to evaluate the extracted yuzu seed fractions ($p < 0.05$). ABTS: 2,2′-azino-bis (3-ethylbenzthiazoline-6-sulfonic acid), DPPH: 2,2-Diphenyl-1-picrylhydrazyl, GAE: gallic acid equivalent, $IC_{50}$: amount of extract required for a 50% reduction of free radical scavenging activity, LA: limonoid aglycone, LG: limonoid glucoside, TPC: total polyphenol content.

### 3.2.2. ABTS Radical Scavenging Activity

The radical scavenging ability is an index of the antioxidant activity of phenolic substances, such as phenolic acid and flavonoids. Greater reducing power indicates higher electron-donating ability [36]. The activity of LA1 and LG1 extracted from the seed coat was the highest (Table 1 and Figure S3b). In another study, the ABTS radical scavenging activity of 70% ethanol extracts of yuzu seeds was found to be below 150 μg/mL [37]. The ABTS radical scavenging activities of limonin and nomilin extracted from sun-dried pomelo seeds were found to be 201.33 and 346.47 μg/mL, respectively [38]. Furthermore, the antioxidant activity of an extract obtained from yuzu seed husk was found to be high. This finding is consistent with the results of our study. Polyphenols, which have high antioxidant activity, were extracted in addition to limonoid aglycones and limonoid glucosides. In addition, the antioxidant activity of yuzu seeds was determined to be higher than that of yuzu peel, probably due to the relatively high content of flavonoids, such as hesperidin, in yuzu seeds.

#### 3.2.3. Total Polyphenol Content

Polyphenolic substances confer special color to plants and act as substrates in redox reactions. Polyphenolic substances, including flavonoids and tannins, are aromatic compounds, with two or more phenol hydroxyl groups in one molecule [39]. Polyphenols play various physiological roles, such as preventing tooth decay; suppressing hypertension; and exerting anti-AIDS, antioxidant, and anti-cancer effects. Our findings revealed that total polyphenol content was consistent with antioxidant activity. The content of LA1 was the highest (Table 1). Furthermore, polyphenol content in limonoid aglycones and limonoid glucosides extracted from yuzu seed husk was high. In a study, the total polyphenol content of *n*-hexane and 70% ethanol extracts of citron seeds were reportedly 201.84 and 246.31 GAE mg/100 g, respectively [38,40]. Consistently, Woo et al. reported [39] a total polyphenol content of 5.67 GAE mg/g. The extraction yield of citron seeds with 75% ethanol was 9.82%, and the total phenol content of the crude extract was 24.44 GAE mg/100 g.

Nomilin has a seven-membered oxepin ring, and its main functional groups are dilactone and acetic ester. The yuzu seed husk used in this study contains an abundance of components, such as naringin, hesperidin, and chlorogenic acid, and thus, is believed to have high antioxidant activity [41]. Antioxidant activity varies depending on the presence or absence of hydroxyl groups. Although nomilin has a small number of hydroxyl groups, the presence of functional groups, such as dilactone and acetic ester ensures that the electron-attractive force is strong and radical scavenging ability is reduced through interactions, such as covalent bonds, hydrogen bonds, and van der Waals' forces.

### 3.3. *HPLC Analysis of Yuzu Seed Parts*

According to previous research, citrus fruits, such as yuzu, contain an abundance of flavonoid compounds, such as hesperidin and naringin, in the peel [42]. Levels of vitamin C, vitamin D, and minerals in yuzu are more than three times higher than those in lemon [43]. In this study, we first extracted and separated the hydrophilic (limonoid glucoside) and hydrophobic (limonoid aglycone) components from yuzu seed byproducts. Then, the limonin and nomilin contents in limonoid aglycone of each part of yuzu seeds (husks, shells, and meal) were detected and quantitatively analyzed by HPLC. The results showed that the limonin contents in LA1 (yuzu seed husks), LA2 (yuzu seed shells), and LA3 (yuzu seed meal) were 641.4, 315.5, and 595.1 mg/g, respectively, and the corresponding nomilin contents were 538.7, 690.7, and 1725.8 mg/g, respectively (Figure S4a, Table S2). HPLC analysis of polyphenol compounds in LG1, LG2, and LG3 revealed the presence of eight standard substances (Figure S4b, Table S3). The levels of chlorogenic acid, naringin, and hesperidin in LG1 were 409.35, 430.17, and 436.76 mg/g, respectively. The levels of these compounds were also high in LG2 and LG3, but lower than that in LG1. These findings explain the difference in antioxidant activity between these fractions. Notably, the higher content of polyphenols in the yuzu seed husk extract explains its high antioxidant activity. In a study, HPLC analysis of yuzu seed extract revealed that levels of naringin, hesperidin, limonin, and nomilin were 100.43, 21.78, 170.98, and 45.36 mg/100 g, respectively [14]. In this study, we detected 3–4 times more polyphenols than other studies, probably because we divided yuzu seeds into three parts.

### 3.4. *Nomilin Tyrosinase Inhibitory Activity*

For testing skin-whitening efficacy in vitro, the in vitro tyrosinase inhibition assay and the in vitro DOPA oxidation inhibition assay are widely used. Tyrosinase is a copper-containing polyphenol oxidase that catalyzes the hydroxylation of monophenols. It is found in microorganisms and animal and plant tissues and contributes to the synthesis of melanin and the production of pigments [44]. Tyrosinase is involved in the initial rate-determining step of the melanin biosynthesis pathway in humans. Many skin-whitening products function by inhibiting tyrosinase. The in vitro tyrosinase inhibition assay evaluates the degree of tyrosinase inhibition in vitro [45]. The DOPA oxidation assay evaluates the effect of skin-whitening compounds by measuring the inhibition of tyrosinase activity, which catalyzes the rate-determining step of the melanin biosynthesis. We isolated and evaluated the potential of nomilin as a skin-whitening agent by reviewing the tyrosinase inhibitory activity of nomilin, which is abundant in the aglycone fraction of yuzu seeds. To our knowledge, this has not previously been attempted. First, based on the antioxidant activity results, the tyrosinase inhibitory activities of LA1 and LG1, which were the highest, were compared. The tyrosinase inhibitory activity of nomilin, the most abundant component in LA1, was confirmed (Figure 3). For the positive control, we used ascorbic acid, a known skin-whitening agent. The $IC_{50}$ values of LA1, LG1, nomilin, and ascorbic acid were 192.63, 688.53, 87.17, and 38.71 μg/mL, respectively. Nomilin showed tyrosinase inhibitory activity higher than that of the extract and lower than that of ascorbic acid. In a study, the tyrosinase inhibitory activities of blue yuzu peel and yellow yuzu peel were compared, and tyrosinase inhibitory activity was found to be related to the content of phenol compounds in yuzu peel [46]. Two types of limonoid aglycones are present in the seed shell, namely, monolactones, such as limonic acid A-ring lactone, and dilactones, such as limonin [47]. Nomilin, isolated from yuzu seed shells, is a limonoid compound with a phenolic structure, and due to its oxidation inhibitory function, is estimated to have high tyrosinase inhibitory activity. In addition, studies have reported that compounds with a furan structure in coffee byproducts exhibited high antioxidant activity [48]. Nomilin extracted from yuzu seed byproducts has a furan structure and an acetate group; therefore, it is proposed to exhibit high antioxidant activity and tyrosinase inhibitory activity.

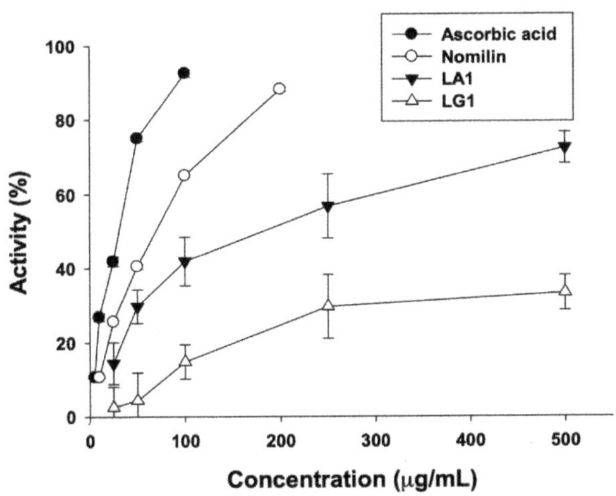


**Figure 3.** The inhibitory effect of ascorbic acid, nomilin, LA1, and LG1 on mushroom tyrosinase in a cell-free system. LA: limonoid aglycone, LG: limonoid glucoside.

### *3.5. Enzyme Kinetic Analysis*

The number of active sites is constant; thus, at high substrate concentrations, enzyme saturation occurs. At high substrate concentrations, the rate of enzyme reactivity is thus independent of substrate concentration; however, at low substrate concentrations, it is proportional to substrate concentration. In a multistep enzymatic reaction, the reaction with the largest $K_m$ value determines the reaction rate of the entire reaction system. In this study, we calculated $K_m$ and $V_{max}$ and identified the type of inhibition using the Lineweaver–Burk equation (Figure 4). The pattern of inhibition of an enzyme depends on the binding site and the type of binding mode. During competitive inhibition, the inhibitor competitively binds to the substrate noncovalently, thereby inhibiting enzyme activity. In noncompetitive inhibition, the inhibitor reversibly binds to both the free enzyme and the enzyme–substrate complex to exhibit inhibitory effects. In this study, values of $K_m$, $V_{max}$, and the inhibition constant ($K_i$) of nomilin against tyrosinase were 0.5049 mM, 0.3931 mmol/min, and 0.6408 mM, respectively. The Lineweaver–Burk plot was linear, confirming that the kinetic behavior was noncompetitive.

(a) (b)

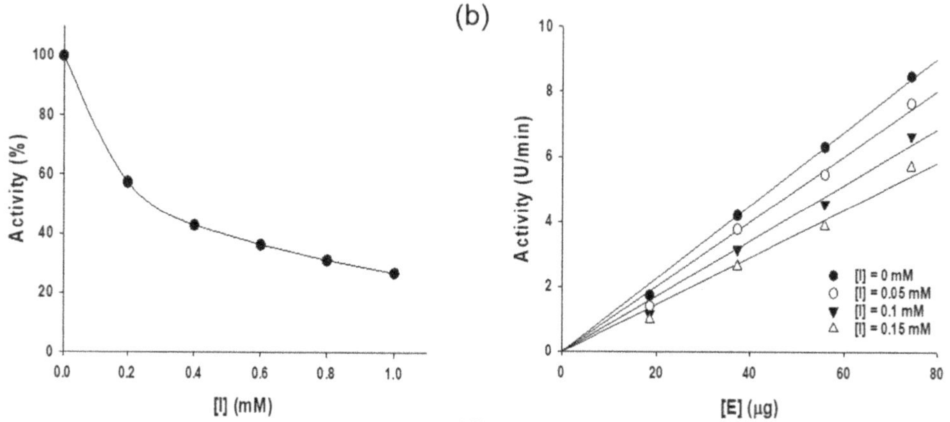


(c) (d)

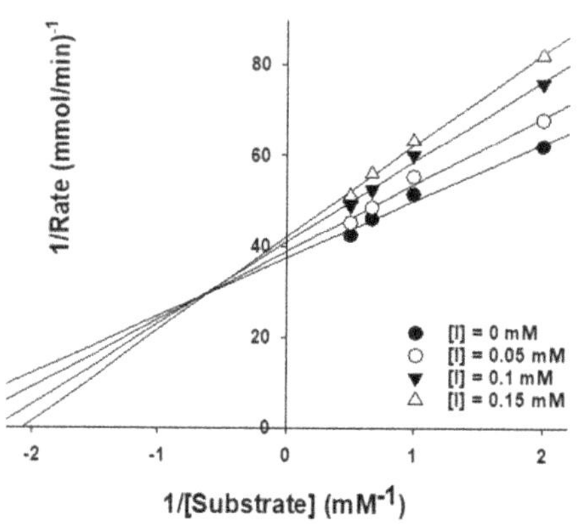


| | Nomilin |
|---|---|
| $IC_{50}$ | 0.2958 mM |
| $K_m$ | 0.5049 mM |
| $V_m$ | 0.3931 mmol/min<br>($2.905\times10^{-2}$ mmol/min) |
| Inhibition | reversible |
| Inhibition type | Noncompetitive |
| $K_i$ | 0.6408 mM |

**Figure 4.** (**a**) Effect of nomilin on the activity of mushroom tyrosinase during the catalysis of L-DOPA (enzyme concentration 4.0 μg/mL). (**b**) Relationship between catalytic activity of mushroom tyrosinase and enzyme concentration at different concentrations of nomilin. Concentrations of nomilin for curves 1-4 were 0, 0.05, 0.1, and 0.15 mM, respectively. (**c**) Lineweaver-Burk plots for the nomilin-mediated inhibition of mushroom tyrosinase during the catalysis of DOPA at 30 °C and pH 6.8. Concentrations of quercetin for curves 1-4 were 0, 0.05, 0.1, and 0.15 mM, respectively; enzyme concentration was 4.0 μg/mL. The inset represents the plot of Kmapp versus quercetin concentration for determining the inhibition constant KI. (**d**) Kinetic parameters and microscopic inhibition rate constants of mushroom tyrosinase in the presence of nomilin.

### *3.6. Molecular Docking Study*

Molecular docking is used to predict binding by modeling the binding and interaction of proteins and ligands at the atomic level [23]. We used molecular docking to predict the binding between nomilin and mushroom tyrosinase and TRP-1 (Figure 5).

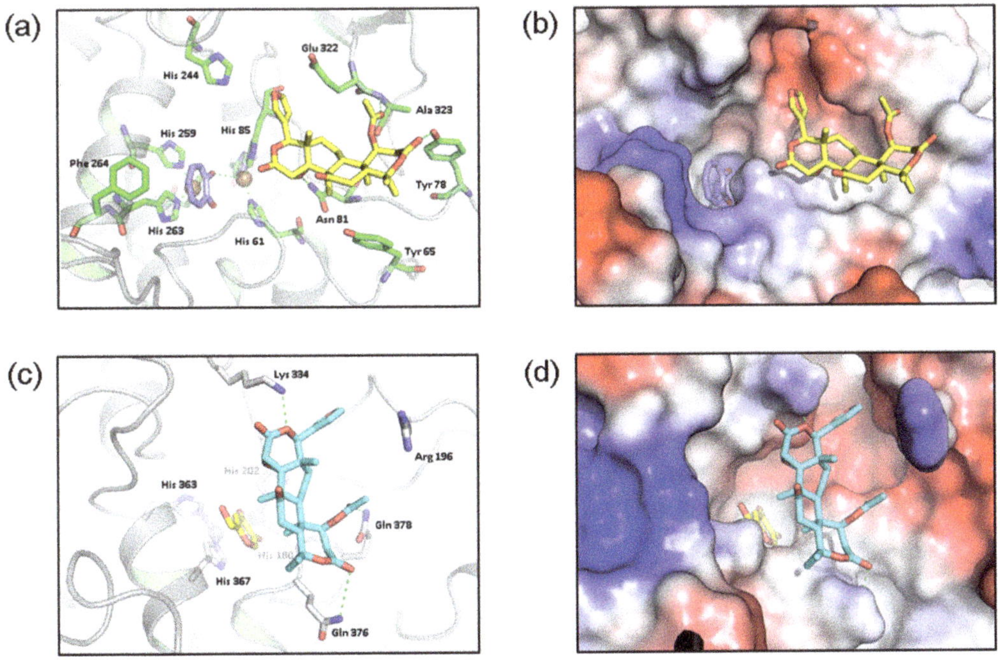


**Figure 5.** Molecular docking study. (**a**) Binding model of mushroom tyrosinase and nomilin; (**b**) surface model of mushroom tyrosinase and nomilin; (**c**) binding model of human tyrosinase and nomilin; (**d**) surface model of human tyrosinase and nomilin.

Molecular docking predicted that nomilin interacts with Tyr 65, Tyr 78, Ala 323, Glu 322, and Asn 81 adjacent to the catalytic site of mushroom tyrosinase. Similarly, nomilin was predicted to interact with Lys 334, Arg 196, Gln 378, and Gln 376 around the catalytic site of TRP-1. Thud, nomilin is predicted to bind adjacent to the catalytic sites of mushroom tyrosinase and TRP-1 (Figure 5c,d).

### *3.7. Cytotoxicity Evaluation and Quantitative Analysis of ROS*

Cell viability and cytotoxicity were measured colorimetrically using the MTT reagent, which turns purple when mitochondrial dehydrogenase and MTT tetrazolium react during cellular metabolism to form MTT formazan rehydrated with DMSO [49]. For each concentration (25–100 μg/mL), the cytotoxicity of nomilin was determined to be <100 μg/mL (Figure 6a). The concentration that did not show the toxicity of nomilin was confirmed to be ≤25 μg/mL, which is lower than the result of this study [50]. Another study evaluated the toxicity of citron peel extract and reported that it was safe at <800 μg/mL. In this study, nomilin was isolated from the yuzu seed husk, and the non-toxic concentration was found to be <100 μg/mL [51]. These results suggest that the range of toxicity may appear differently depending on the yuzu seed extraction method and cell types. Flow cytometry was performed using the fluorescent probe DCF-DA to confirm the effect of nomilin on the reduction of the total amount of intracellular ROS generated during metabolism. B16F10 cells were pretreated with nomilin for 24 h and then with 500 mM *t*-BHP, and the following results were obtained (Figure 6b). When nomilin was applied at concentrations of 10, 25, 50, and 100 μg/mL, the total amount of ROS was reduced by 17%, 23%, 45%, and 84%, respectively.

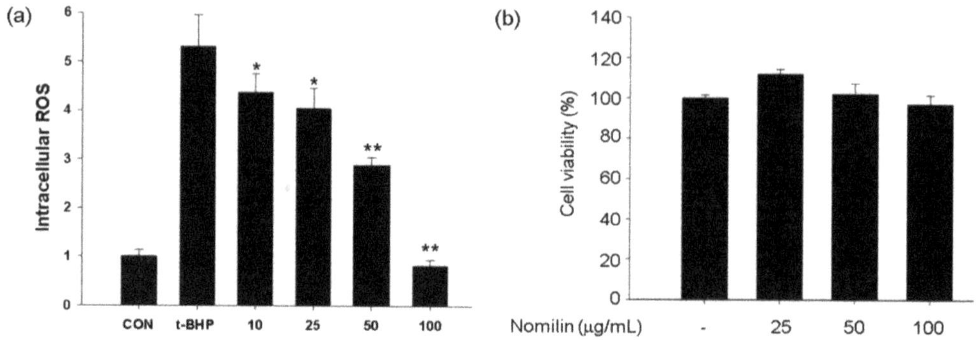


**Figure 6.** Measurement of (**a**) intracellular ROS production in and (**b**) viability of B16F10 melanoma cells treated with nomilin. Nomilin: 25–100 μg/mL (48.59–194.34 μM). * $p < 0.05$, ** $p < 0.01$, compared with t-BHP treatment.

### 3.8. Melanin Content and Cell Morphology in Nomilin-Treated B16F10 Cells

B16F10 melanoma cells were treated with α-MSH (1 μg/mL). Nomilin and limonin isolated from yuzu seeds were added to the medium. Cells were cultured for 48 h, and secreted melanin was measured. The findings revealed that melanin content decreased in a concentration-dependent manner. The skin-whitening activity of nomilin was similar to that of arbutin, a positive control, at 100 μg/mL (Figure 7). Morphologically, B16F10 melanocytes showed a pattern consistent with the results of melanin content. In a study measuring the melanin content of yuzu seed extracts in B16F10 melanoma cells, strong skin-whitening activity was recorded at 0.02% [21]. In another study, a significant decrease in melanin content was seen (1.85-fold) in melanocytes treated with nomilin compared with those treated with arbutin (positive control) [52]. Additionally, melanin content was decreased by yuzu peel extract, which showed better activity than kojic acid, as a positive control, at a concentration of 0.02%. Most studies have not measured melanin content or tyrosinase inhibitory activity. In this study, we went one step further and investigated the skin-whitening activity in B16F10 melanoma cells.

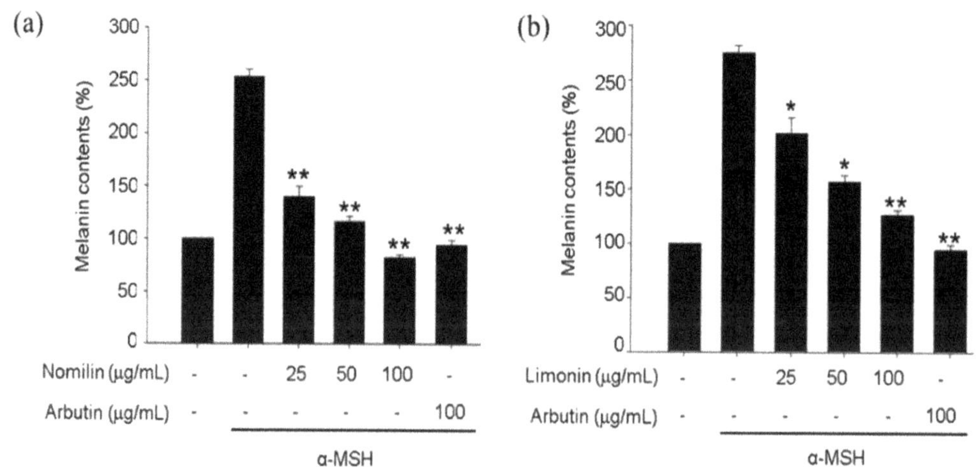


**Figure 7.** *Cont.*

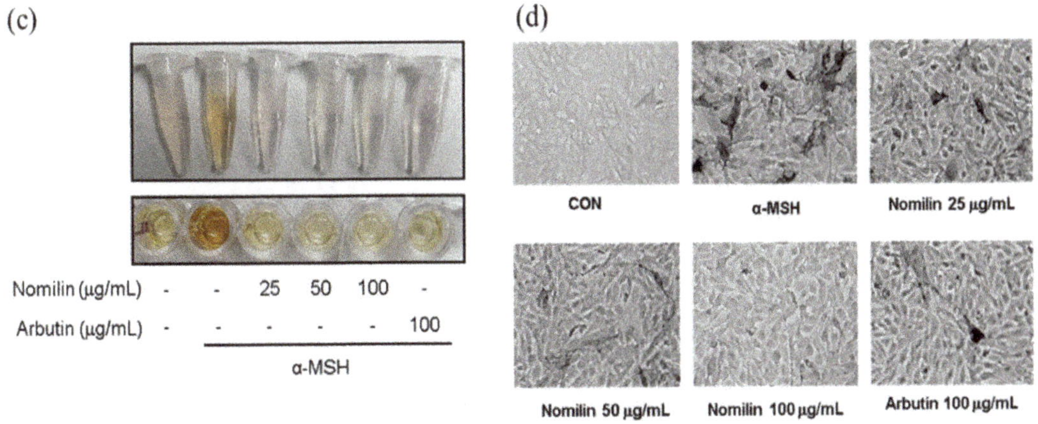


**Figure 7.** Measurement of melanin content in B16F10 melanoma cells treated with nomilin and limonin. (**a**) Measurement of melanin content with 25–100 μg/mL nomilin (48.59–194.34 μM). (**b**) Measurement of melanin content with 25–100 μg/mL limonin (53.13–212.53 μM). Relative melanin content was determined at 72 h after treatment. $n = 3$, error bars, mean ± standard deviation. Effect of 100 μg/mL arbutin (367 μM) on melanin synthesis and tyrosinase activity in B16F10 cells. (**c**) B16F10 melanoma cells. (**d**) Morphological changes due to nomilin. B16F10 cells were cultured for 48 h in the presence of 25–100 μg/mL (48.59–194.34 μM) nomilin and 100 μg/mL (367 μM) arbutin as a positive control or 1 μg/mL α-MSH. Significantly different compared with α-MSH, * $p < 0.05$, ** $p < 0.01$.α-MSH: α-melanocyte-stimulating hormone.

### *3.9. Effect of Nomilin on Anti-Melanogenesis-Related Proteins in B16F10 Cells*

UV rays, cytokines, growth factors, and hormones regulate melanogenesis. In this context, α-MSH is an important hormone [53]. α-MSH is secreted from the middle of the pituitary gland and binds to melanocortin 1 receptor, a membrane receptor expressed only in melanocytes, which activates adenylyl cyclase. This in turn amplifies the intracellular cAMP signal, induces protein kinase A (PKA) activation, and increases the expression of MITF (a transcription factor specific to melanocytes) by activating intracellular cAMP response element binding protein (CREB) [54]. Melanin is synthesized through intracellular signaling mechanisms, among which cAMP/PKA is the main pathway. MITF promotes the transcription of tyrosinase, TRP-1, and TRP-2 during melanin synthesis through CREB [55]. Against this background, the key step in skin-whitening is to inhibit tyrosinase, an enzyme involved in melanin biosynthesis, and to inhibit the synthesis of proteins involved in upstream and downstream signaling processes of the melanin synthesis pathway [56].

α-MSH treatment increased the levels of tyrosinase, TRP-1, and TRP-2, whereas nomilin treatment significantly decreased the levels of tyrosinase, TRP-1, and TRP-2 in B16F10 cells (Figure 8). Moreover, induction of MITF expression by 1 μg/mL α-MSH was inhibited by treatment with 25–100 μg/mL nomilin (Figure 8c,d). Thus, nomilin inhibited melanogenesis by downregulating MITF signaling. As shown in Figure 8e, nomilin preincubation inhibited α-MSH-induced phosphorylation of protein kinase and cAMP response element binding protein (PKA/CREB). Thus, the suppressive mechanism of nomilin was related to the inhibition of PKA/CREB signaling.

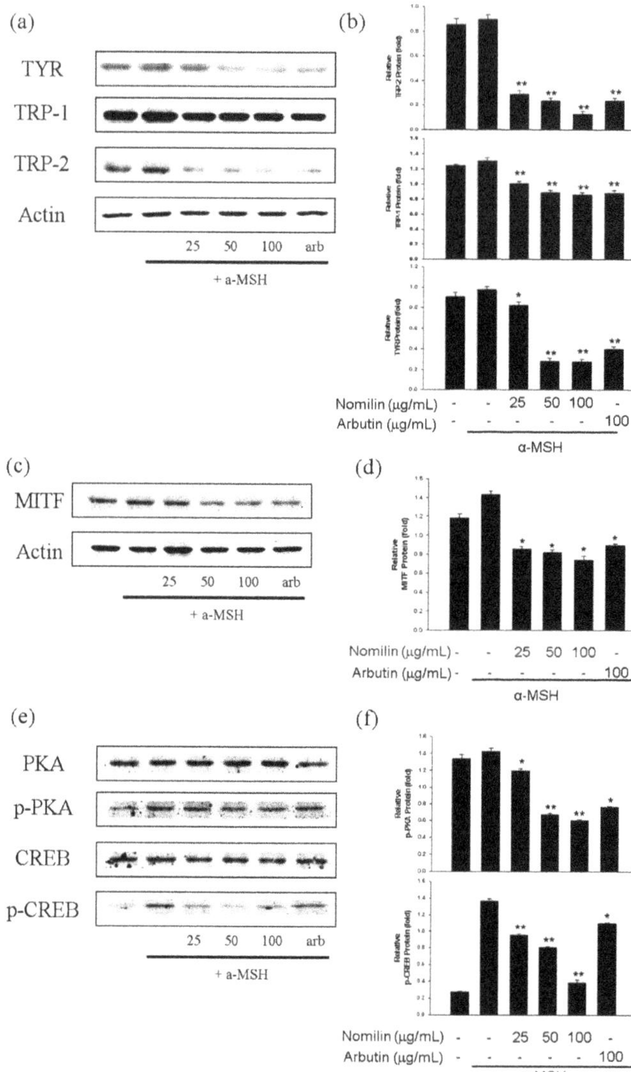


**Figure 8.** (**a**) Effects of nomilin and arbutin on tyrosinase, TRP-1, and TRP-2 expression in B16F10 melanoma cells. B16F10 cells were treated with different concentrations of nomilin and arbutin prior to α-MSH treatment for 24 h. β-actin was used as a loading control antibody. (**b**) Quantitative analysis of tyrosinase, TRP-1, and TRP-2 by Western blotting. Cell lysates were subjected to Western blotting using antibodies against tyrosinase, TRP-1, and TRP-2. (**c**) Effects of nomilin and arbutin on MITF expression in B16F10 cells. B16F10 cells were treated with the indicated concentrations of nomilin and arbutin prior to α-MSH treatment for 4 h. (**d**) Quantitative analysis of MITF by Western blotting. (**e**) Effects of nomilin and arbutin on p-CREB, CREB, p-PKA, and PKA levels in B16F10 cells. B16F10 cells were treated with the indicated concentrations of nomilin and arbutin prior to α-MSH treatment for 3 h. (**f**) Quantitative analysis of p-CREB, CREB, p-PKA, and PKA by Western blotting. Values are significantly different with Duncan's multiple range test (significant compared with the vehicle-treated control, * $p < 0.05$, ** $p < 0.01$, bars indicate SD). CREB: cAMP response element binding protein, p-CREB: phospho-CREB, α-MSH: α-melanocyte-stimulating hormone, MITF: microphthalmia-associated transcription factor, and PKA: protein kinase A, p-PKA: phospho-PKA.

### 3.10. *Effect of Nomilin on Anti-Melanogenesis-Related Genes in B16F10 Cells*

To verify the skin-whitening activity of nomilin, we evaluated the expression of related genes, namely, TYR, TRP-1, TRP-2, and MITF, at the mRNA level using RT-PCR (Figure 9). The results revealed that the expression of genes (tyrosinase, TRP-1, and MITF) related to melanin production via nomilin was induced by α-MSH, whereas nomilin treatment resulted in significantly decreased mRNA expression (25–100 μg/mL).

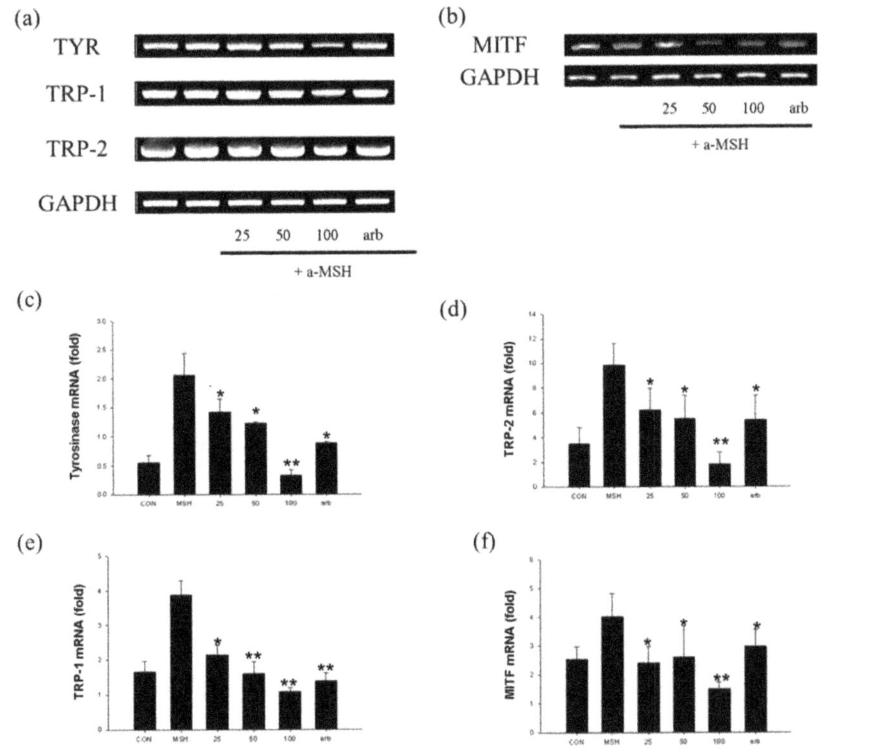


**Figure 9.** RT-PCR analysis of nomilin. (**a**) TYR, TRP-1, and TRP-2 transcripts were analyzed using RT-PCR. The level of GAPDH mRNA was used as control. (**b**) MITF transcripts were analyzed using RT-PCR. The level of GAPDH mRNA was used as control. (**c**) Effects of nomilin on tyrosinase, (**d**) TRP-1, (**e**) TRP-2, and (**f**) MITF mRNA in B16F10 cells. Data from separate experiments are presented (statistically significant compared with the vehicle-treated control, * $p < 0.05$, ** $p < 0.01$, bars indicate SD). α-MSH: α-melanocyte-stimulating hormone, MITF: microphthalmia-associated transcription factor, TYR: tyrosinase.

In particular, nomilin was more effective in inhibiting mRNA expression at 100 μg/mL than at other concentrations. Moreover, nomilin showed more potent inhibitory activity than the positive control arbutin.

## 4. Conclusions

In this study, we separated and extracted limonoid and aglycone fractions from yuzu seed byproducts and confirmed that nomilin is a novel tyrosinase inhibitor. Nomilin was separated from the aglycone fraction as a single substance with strong antioxidant and skin-whitening activities. These effects were mediated by the activation of the PKA/CREB signaling pathway involved in melanogenesis in B16F10 cells. The results of this study suggest that the use of natural ingredients in the development of new skin-whitening materials can resolve the problems associated with using chemicals. Nomilin from yuzu seeds directly inhibited tyrosinase and was involved in protein synthesis and transcription factor regulation during the skin-whitening signal transduction mechanism. Therefore, nomilin has a high potential as a novel natural skin-whitening agent; however, more comprehensive molecular studies are needed to confirm its ability to reduce melanin pigmentation.

**Supplementary Materials:** The following supporting information can be downloaded at: https://www.mdpi.com/article/10.3390/antiox11091636/s1, Figure S1: $^{1}$H-NMR spectrum of nomilin; Figure S2: $^{13}$C-NMR spectrum of nomilin; Figure S3: Antioxidant activity profiles of LA1, LA2, LA3, LG1, LG2, and LG3. (a) DPPH radical scavenging activity and (b) ABTS radical scavenging activity; Figure S4: HPLC chromatogram of limonoid aglycones (LA1, LA2, and LA3) and limonoid glucosides (LG1, LG2, and LG3) using diode-array detection at 220 nm. (a) LA1; (b) LA2; (c) LA3; (d) standard mixture (1: limonin, 2: nomilin); (e) LG1; (f) LG2; (g) LG3; (h) standard mixture (1: chlorogenic acid, 2: caffeic acid, 3: rutin, 4: *p*-coumaric acid, 5: naringin, 6: hesperidin, 7: luteolin, 8: linoleic acid); Table S1: HPLC conditions used for the identification of limonoid aglycones and limonoid glucosides; Table S2: Phenolic compounds identified in limonoid aglycone extracts quantified by HPLC; Table S3: Phenolic compounds identified in limonoid glucoside extracts quantified by HPLC.

**Author Contributions:** Conceptualization, M.-H.C. and H.-J.S.; methodology, M.-H.C. and S.-H.Y.; software, S.-H.Y. and N.D.K.; validation, M.-H.C. and S.-H.Y.; formal analysis, M.-H.C. and H.-J.S.; investigation, M.-H.C. and H.-J.S.; resources, S.-H.Y.; data curation, M.-H.C. and S.-H.Y.; writing—original draft preparation, M.-H.C. and S.-H.Y.; writing—review and editing, H.-J.S.; visualization, N.D.K.; supervision, H.-J.S.; project administration, M.-H.C. and H.-J.S.; funding acquisition, H.-J.S. All authors have read and agreed to the published version of the manuscript.

**Funding:** This research was supported by the Basic Science Research Program through the National Research Foundation of Korea (NRF) funded by the Ministry of Education (NRF-2020R1I1A1A01073631).

**Institutional Review Board Statement:** Not applicable.

**Informed Consent Statement:** Not applicable.

**Data Availability Statement:** The data presented in this study are available in the article and supplementary material.

**Conflicts of Interest:** The authors declare no conflict of interest.

## References

1. Chung, J.H.; Yano, K.; Lee, M.K.; Youn, C.S.; Seo, J.Y.; Kim, K.H.; Cho, K.H.; Eun, H.C.; Detmar, M. Differential effects of photoaging vs intrinsic aging on the vascularization of human skin. *Arch. Dermatol.* **2002**, *138*, 1437–1442. [CrossRef] [PubMed]
2. Fisher, G.J.; Talwar, H.S.; Lin, J.; Voorhees, J.J. Molecular mechanisms of photoaging in human skin in vivo and their prevention by all-trans retinoic acid. *Photochem. Photobiol.* **1999**, *69*, 154–157. [CrossRef] [PubMed]
3. Ishida, T.; Sakaguchi, I. Protection of human keratinocytes from UVB-induced inflammation using root extract of *Lithospermum erythrorhizon*. *Biol. Pharm. Bull.* **2007**, *30*, 928–934. [CrossRef] [PubMed]
4. Minamisawa, M.; Yoshida, S.; Uzawa, A. The functional evaluation of waste yuzu (*Citrus junos*) seeds. *Food Funct.* **2014**, *5*, 330–336. [CrossRef]
5. Christian, F.; Smith, E.L.; Carmody, R.J. The regulation of NF-κB subunits by phosphorylation. *Cells* **2016**, *5*, 12. [CrossRef]
6. McCabe, M.C.; Hill, R.C.; Calderone, K.; Cui, Y.; Yan, Y.; Quan, T.; Fisher, G.J.; Hansen, K.C. Alterations in extracellular matrix composition during aging and photoaging of the skin. *Matrix Biol. Plus* **2020**, *8*, 100041. [CrossRef]
7. Solano, F. Melanins: Skin pigments and much more—Types, structural models, biological functions, and formation routes. *New J. Sci.* **2014**, *2014*, 498276. [CrossRef]
8. D'Mello, S.A.; Finlay, G.J.; Baguley, B.C.; Askarian-Amiri, M.E. Signaling pathways in melanogenesis. *Int. J. Mol. Sci.* **2016**, *17*, 1144. [CrossRef]
9. Chang, T.S. An updated review of tyrosinase inhibitors. *Int. J. Mol. Sci.* **2009**, *10*, 2440–2475. [CrossRef]
10. Wu, T.; Zheng, C.; He, Y.; Guo, M.; Qi, X. Inhibitory effects of polyphenol-rich plant extracts on melanogenesis in B16F10 murine melanoma cell line. *Curr. Dev. Nutr.* **2020**, *4*, 792. [CrossRef]
11. Yu, H.Y.; Park, S.W.; Chung, I.M.; Jung, Y.S. Anti-platelet effects of yuzu extract and its component. *Food Chem. Toxicol.* **2011**, *49*, 3018–3024. [CrossRef] [PubMed]
12. Moon, S.H.; Assefa, A.D.; Ko, E.Y.; Park, S.W. Comparison of flavonoid contents and antioxidant activity of yuzu (*Citrus junos* Sieb. ex Tanaka) based on harvest time. *Hortic. Sci. Technol.* **2015**, *33*, 283–291.
13. Denaro, M.; Smeriglio, A.; Trombetta, D. Antioxidant and anti-inflammatory activity of citrus flavanones mix and its stability after in vitro simulated digestion. *Antioxidants* **2021**, *10*, 140. [CrossRef] [PubMed]

14. Woo, K.L.; Kim, J.I.; Kim, M.C.; Chang, D.K. Determination of flavonoid and limonoid compounds in citron (*Citrus junos* Sieb. et Tanaka) seeds by HPLC and HPLC/MS. *J. South Korean Soc. Food Sci. Nutr.* **2006**, *35*, 353–358.
15. Sak, M.; Al-Faiyz, Y.S.; Elsawy, H.; Shaaban, S. Novel organoselenium redox modulators with potential anticancer, antimicrobial, and antioxidant activities. *Antioxidants* **2022**, *11*, 1231. [CrossRef]
16. Liang, J.H.; Lin, H.R.; Yang, C.S.; Liaw, C.C.; Wang, I.C.; Chen, J.J. Bioactive components from *Ampelopsis japonica* with antioxidant, anti-α-glucosidase, and antiacetylcholinesterase activities. *Antioxidants* **2022**, *11*, 1228. [CrossRef]
17. Dunford, N.T.; Gumus, Z.P.; Gur, C.S. Chemical composition and antioxidant properties of pecan shell water extracts. *Antioxidants* **2022**, *11*, 1127. [CrossRef]
18. Macrini, D.J.; Suffredini, I.B.; Varella, A.D.; Younes, R.N.; Ohara, M.T. Extracts from Amazonian plants have inhibitory activity against tyrosinase: An in vitro evaluation. *Braz. J. Pharm. Sci.* **2009**, *45*, 715–721. [CrossRef]
19. Fan, M.; Zhang, G.; Hu, X.; Xu, X.; Gong, D. Quercetin as a tyrosinase inhibitor: Inhibitory activity, conformational change and mechanism. *Food Res. Int.* **2017**, *100*, 226–233. [CrossRef]
20. Choi, M.H.; Yang, S.H.; Kim, D.S.; Kim, N.D.; Shin, H.J.; Liu, K. Novel quercetin derivative of 3,7-dioleylquercetin shows less toxicity and highly potent tyrosinase inhibition activity. *Int. J. Mol. Sci.* **2021**, *22*, 4264. [CrossRef]
21. Friesner, R.A.; Banks, J.L.; Murphy, R.B.; Halgren, T.A.; Klicic, J.J.; Mainz, D.T.; Matthew, P.; Repasky, M.P.; Knoll, E.H.; Shelley, M.; et al. Glide: A new approach for rapid, accurate docking and scoring. 1. Method and assessment of docking accuracy. *J. Med. Chem.* **2004**, *47*, 1739–1749. [CrossRef] [PubMed]
22. Halgren, T.A.; Murphy, R.B.; Friesner, R.A.; Beard, H.S.; Frye, L.L.; Pollard, W.T.; Banks, J.L. Glide: A new approach for rapid, accurate docking and scoring. 2. Enrichment factors in database screening. *J. Med. Chem.* **2004**, *47*, 1750–1759. [CrossRef] [PubMed]
23. Jacobson, M.P.; Pincus, D.L.; Rapp, C.S.; Day, T.J.; Honig, B.; Shaw, D.E.; Friesner, R.A. A hierarchical approach to all-atom protein loop prediction. *Proteins* **2004**, *55*, 351–367. [CrossRef] [PubMed]
24. Roos, K.; Wu, C.; Damm, W.; Reboul, M.; Stevenson, J.M.; Lu, C.; Dahlgren, M.K.; Mondal, S.; Chen, W.; Wang, L.; et al. OPLS3e: Extending force field coverage for drug-like small molecules. *J. Chem. Theory Comput.* **2019**, *15*, 1863–1874. [CrossRef]
25. Lai, X.; Wichers, H.J.; Soler-Lopez, M.; Dijkstra, B.W. Structure of human tyrosinase related protein 1 reveals a binuclear zinc active site important for melanogenesis. *Angew. Chem. Int. Ed.* **2017**, *56*, 9812–9815. [CrossRef]
26. Hosoi, J.; Abe, E.; Suda, T.; Kuroki, T. Regulation of melanin synthesis of B16 mouse melanoma cells by 1α,25-dihydroxyvitamin D3 and retinoic acid. *Cancer Res.* **1985**, *45*, 1474–1478.
27. Hatiboglu, M.A.; Kocyigit, A.; Guler, E.M.; Akdur, K.; Nalli, A.; Karatas, E.; Tuzgen, S. Thymoquinone induces apoptosis in B16-F10 melanoma cell through inhibition of p-STAT3 and inhibits tumor growth in a murine intracerebral melanoma model. *World Neurosurg.* **2018**, *114*, 182–190. [CrossRef]
28. Yang, J.H.; Choi, M.H.; Yang, S.H.; Cho, S.S.; Park, S.J.; Shin, H.J.; Ki, S.H. Potent anti-inflammatory and antiadipogenic properties of bamboo (*Sasa coreana* Nakai) leaves extract and its major constituent flavonoids. *J. Agric. Food Chem.* **2017**, *65*, 6665–6673. [CrossRef]
29. Hasegawa, S.; Miyake, M.; Ozaki, Y. Biochemistry of citrus limonoids and their anticarcinogenic activity. *ACS Symp. Ser. Am. Chem. Soc.* **1994**, *546*, 198–208.
30. Choi, M.H.; Jo, H.G.; Yang, J.H.; Ki, S.H.; Shin, H.J. Antioxidative and anti-melanogenic activities of bamboo stems (*Phyllostachys nigra* variety henosis) via PKA/CREB-mediated MITF downregulation in B16F10 melanoma cells. *Int. J. Mol. Sci.* **2018**, *19*, 409. [CrossRef]
31. Hasegawa, S.; Bennett, R.D.; Verdon, C.P. Limonoids in citrus seeds: Origin and relative concentration. *J. Agric. Food Chem.* **1980**, *28*, 922–925. [CrossRef]
32. Ohta, H.; Hasegawa, S. Limonoids in pummelos [*Citrus grandis* (L.) Osbeck]. *J. Food Sci.* **1995**, *60*, 1284–1285. [CrossRef]
33. Kim, A.Y.; Jeong, H.J.; Park, S.N. Antioxidant activities of *Citrus junos* seed shell extract and fractions cultivated in South Korea. *JKAST* **2017**, *34*, 236–243.
34. Kwon, O.; Shin, J.H.; Kang, M.J.; Lee, S.J.; Choi, S.Y.; Sung, N.J. Antioxidant activity of ethanol extracts from citron (*Citrus junos* Sieb. ex Tanaka) seed. *J. S. Korean Soc. Food Sci. Nutr.* **2006**, *35*, 294–300.
35. Shin, J.H.; Lee, J.Y.; Ju, J.C.; Lee, S.J.; Cho, H.S.; Sung, N.J. Chemical properties and nitrite scavenging ability of citron (*Citrus junos*). *J. S. Korean Soc. Food Sci. Nutr.* **2005**, *34*, 496–502.
36. Choi, I.W.; Choi, S.Y.; Nam, B.R.; Kim, Y.S.; Choi, H.D. Contents of polyphenols and limonoids in citron (*Citrus junos* Sieb. ex Tanaka) seed extracts and their antioxidant properties. *Food Sci. Biotechnol.* **2008**, *17*, 373–378.
37. Han, C.H.; Kim, M.A.; Kim, M.J. Antioxidant properties and oxidative stability of oregano seed ethanol extract. *Korean J. Food Preserv.* **2019**, *26*, 165–173. [CrossRef]
38. Yoshizawa, S.; Horiuchi, T.; Fujiki, H.; Yoshida, T.; Okuda, T.; Sugimura, T. Antitumor promoting activity of (−)-epigallocatechin gallate, the main constituent of "Tannin" in green tea. *Phytother. Res.* **1987**, *1*, 44–47. [CrossRef]
39. Woo, K.S.; Jeong, J.Y.; Hwang, I.G.; Lee, Y.J.; Lee, Y.R.; Park, H.J.; Jeong, H.S. Antioxidant activity of ethanol extraction on citron seed by response surface methodology. *J. S. Korean Soc. Food Sci. Nutr.* **2009**, *38*, 384–390. [CrossRef]
40. Ozaki, Y.; Fong, C.H.; Herman, Z.; Maeda, H.; Miyake, M.; Ifuku, Y.; Hasegawa, S. Limonoid glucosides in citrus seeds. *Agric. Biol. Chem.* **1991**, *55*, 137–141.

41. Nogata, Y.; Sakamoto, K.; Shiratsuchi, H.; Ishii, T.; Yano, M.; Ohta, H. Flavonoid composition of fruit tissues of citrus species. *Biosci. Biotechnol. Biochem.* **2006**, *70*, 178–192. [CrossRef] [PubMed]
42. Sir Elkhatim, K.A.; Elagib, R.A.; Hassan, A.B. Content of phenolic compounds and vitamin C and antioxidant activity in wasted parts of Sudanese citrus fruits. *Food Sci. Nutr.* **2018**, *6*, 1214–1219. [CrossRef] [PubMed]
43. Zaidi, K.U.; Ali, A.S.; Ali, S.A.; Naaz, I. Microbial tyrosinases: Promising enzymes for pharmaceutical, food bioprocessing, and environmental industry. *Biochem. Res. Int.* **2014**, *2014*, 16. [CrossRef]
44. Momtaz, S.; Mapunya, B.M.; Houghton, P.J.; Edgerly, C.; Hussein, A.; Naidoo, S.; Lall, N. Tyrosinase inhibition by extracts and constituents of *Sideroxylon inerme L.* stem bark, used in South Africa for skin lightening. *J. Ethnopharmacol.* **2008**, *119*, 507–512. [CrossRef]
45. Matsuura, R.; Ukeda, H.; Sawamura, M. Tyrosinase inhibitory activity of citrus essential oils. *J. Agric. Food Chem.* **2006**, *54*, 2309–2313. [CrossRef]
46. Montoya, C.; González, L.; Pulido, S.; Atehortúa, L.; Robledo, S.M. Identification and quantification of limonoid aglycones content of Citrus seeds. *Rev. Bras. Farmacogn.* **2020**, *29*, 710–714. [CrossRef]
47. Bedoya-Ramírez, D.; Cilla, A.; Contreras-Calderón, J.; Alegría-Torán, A. Evaluation of the antioxidant capacity, furan compounds and cytoprotective/cytotoxic effects upon Caco–2 cells of commercial Colombian coffee. *Food Chem.* **2017**, *219*, 364–372. [CrossRef]
48. Kim, D.S.; Kim, D.H.; Oh, M.J.; Lee, K.G.; Kook, M.C.; Park, C.S. Antiaging and skin-whitening activities of ethanol extract of Yuza (*Citrus junos* Sieb. ex Tanaka) by-product. *J. S. Korean Soc. Cosmetol.* **2010**, *36*, 137–143.
49. Denaro, M.; Smeriglio, A.; Xiao, J.; Cornara, L.; Burlando, B.; Trombetta, D. New insights into *Citrus* genus: From ancient fruits to new hybrids. *Food Front.* **2020**, *1*, 305–328. [CrossRef]
50. Pratheeshkumar, P.; Raphael, T.J.; Kuttan, G. Nomilin inhibits metastasis via induction of apoptosis and regulates the activation of transcription factors and the cytokine profile in B16F-10 cells. *Integr. Cancer Ther.* **2012**, *11*, 48–60. [CrossRef]
51. Lee, D.H.; Woo, J.K.; Heo, W.; Huang, W.Y.; Kim, Y.; Chung, S.; Lee, G.H.; Park, J.W.; Han, B.K.; Shin, E.C.; et al. *Citrus junos* Tanaka peel extract and its bioactive naringin reduce fine dust-Induced respiratory injury markers in BALB/c male mice. *Nutrients* **2022**, *14*, 1101. [CrossRef] [PubMed]
52. Bertolotto, C.; Abbe, P.; Hemesath, T.J.; Bille, K.; Fisher, D.E.; Ortonne, J.P.; Ballotti, R. Microphthalmia gene product as a signal transducer in cAMP-induced differentiation of melanocytes. *J. Cell Biol.* **1998**, *142*, 827–835. [CrossRef] [PubMed]
53. Cheli, Y.; Luciani, F.; Khaled, M.; Beuret, L.; Bille, K.; Gounon, P.; Ortonne, J.P.; Bertolotto, C.; Ballotti, R. αMSH and cyclic AMP elevating agents control melanosome pH through a protein kinase A-independent mechanism. *J. Biol. Chem.* **2009**, *284*, 18699–18706. [CrossRef] [PubMed]
54. Gao, D.; Kim, J.H.; Kim, C.T.; Jeong, W.S.; Kim, H.M.; Sim, J.; Kang, J.S. Evaluation of anti-melanogenesis activity of enriched *Pueraria lobata* stem extracts and characterization of its phytochemical components using HPLC–PDA–ESI–MS/MS. *Int. J. Mol. Sci.* **2021**, *22*, 8105. [CrossRef]
55. Gaggiol, C.; Buscà, R.; Abbe, P.; Ortonne, J.P.; Ballotti, R. Microphthalmia-associated transcription factor (MITF) is required but is not sufficient to induce the expression of melanogenic genes. *Pigment. Cell Res.* **2003**, *16*, 374–382. [CrossRef]
56. Qian, W.; Liu, W.; Zhu, D.; Cao, Y.; Tang, A.; Gong, G.; Su, H. Natural skin-whitening compounds for the treatment of melanogenesis. *Exp. Ther. Med.* **2020**, *20*, 173–185. [CrossRef]

*antioxidants*

MDPI

*Review*

# The Prospects of *Swietenia macrophylla* King in Skin Care

**Camille Keisha Mahendra [1], Khang Wen Goh [2], Long Chiau Ming [3,*], Gokhan Zengin [4], Liang Ee Low [5,6], Hooi-Leng Ser [7,8] and Bey Hing Goh [1,9,10,*]**

1 Biofunctional Molecule Exploratory Research Group, School of Pharmacy, Monash University Malaysia, Bandar Sunway 47500, Malaysia; camille.mahendra@monash.edu
2 Faculty of Data Science and Information Technology, INTI International University, Nilai 71800, Malaysia; khangwen.goh@newinti.edu.my
3 PAP Rashidah Sa'adatul Bolkiah Institute of Health Sciences, Universiti Brunei Darussalam, Gadong BE1410, Brunei
4 Biochemistry and Physiology Research Laboratory, Department of Biology, Science Faculty, Selcuk University, Konya 42130, Turkey; gokhanzengin@selcuk.edu.tr
5 Chemical Engineering Discipline, School of Engineering, Monash University Malaysia, Bandar Sunway 47500, Malaysia; low.liangee@monash.edu
6 Advanced Engineering Platform, Monash University Malaysia, Bandar Sunway 47500, Malaysia
7 Department of Biological Sciences, School of Medical and Life Sciences, Sunway University, Bandar Sunway 47500, Malaysia; hooilengs@sunway.edu.my
8 Novel Bacteria and Drug Discovery Research Group, Microbiome and Bioresource Research Strength Jeffrey Cheah School of Medicine and Health Sciences, Monash University Malaysia, Bandar Sunway 47500, Malaysia
9 College of Pharmaceutical Sciences, Zhejiang University, 866 Yuhangtang Road, Hangzhou 310058, China
10 Health and Well-Being Cluster, Global Asia in the 21st Century (GA21) Platform, Monash University Malaysia, Bandar Sunway 47500, Malaysia
* Correspondence: longchiauming@gmail.com (L.C.M.); goh.bey.hing@monash.edu (B.H.G.)

**Abstract:** The importance of cosmetics in our lives is immeasurable. Covering items from daily personal hygienic products to skincare, it has become essential to consumers that the items that they use are safe and effective. Since natural products are from natural sources, and therefore considered "natural" and "green" in the public's eyes, the rise in demand for such products is not surprising. Even so, factoring in the need to remain on trend and innovative, cosmetic companies are on a constant search for new ingredients and inventive new formulations. Based on numerous literature, the seed of *Swietenia macrophylla* has been shown to possess several potential "cosmetic-worthy" bioproperties, such as skin whitening, photoprotective, antioxidant, antimicrobial, etc. These properties are vital in the cosmetic business, as they ultimately contribute to the "ageless" beauty that many consumers yearn for. Therefore, with further refinement and research, these active phytocompounds may be a great contribution to the cosmetic field in the near future.

**Keywords:** cosmeceutical; photoaging; anti-pigmentation; *Swietenia macrophylla*; natural product; skin whitening; skin repair and regeneration; wound healing; acne vulgaris; anti-microbial

**Citation:** Mahendra, C.K.; Goh, K.W.; Ming, L.C.; Zengin, G.; Low, L.E.; Ser, H.-L.; Goh, B.H. The Prospects of *Swietenia macrophylla* King in Skin Care. *Antioxidants* **2022**, *11*, 913. https://doi.org/10.3390/antiox11050913

Academic Editors: Irene Dini and Sonia Laneri

Received: 9 April 2022
Accepted: 3 May 2022
Published: 6 May 2022



## 1. Introduction

Cosmetics, since time immemorable, have always been a large part of our lives. With global sales revenue projected to achieve $429.8 billion by 2022, the cosmetic industry is a lucrative, fast-paced, and innovative business that constantly seeks to fulfill consumer's expectations and preferences [1,2]. Although many do not know it, the umbrella of cosmetics actually covers a large range of products categorized as personal hygiene, makeup, fragrances, skincare, and hair care [3]. Nevertheless, in the last few decades, the trend for "natural", "green", "sustainable", and "safe" cosmetics is on the rise. With rising awareness and concerns regarding the ingredients within a product, there is an inclination for cosmetics that are developed from a natural source. This forces the cosmetic industry to be quick in its search for new, inventive, and potent natural ingredients that not only entice the consumers but also produce the desired effect as marketed [4]. One category that is highly affected by this recent trend is skincare. Natural product-based skincare products containing bioproperties, such as antioxidant, antibacterial, anti-inflammatory, whitening, and much more, are high in demand to combat skin aging and acne blemishes [5,6]. Based on these criteria, the properties of *Swietenia macrophylla* that were previously reported may prove to be an interesting, novel natural ingredient that can contribute to the development of skincare cosmetics. For an introduction, *S. macrophylla*, along with its congeners *Swietenia mahagoni* (L.) Jacq and *Swietenia humilis* Zucc., belong to the *Meliaceae* family. Easily found within the tropical region from North to South America, it is well known for its high-quality wood, used in the creation of furniture, artisanal crafts, decorative materials, and musical instruments [7,8]. Other than being a prized timber, its seed is also employed as a traditional medicine against various ailments and discomforts, such as hypertension, persistent pain, diabetes, malaria, etc. [9].

## 2. An Antimicrobial Remedy against Skin Disorders

Our skin is home to diverse microorganisms, such as bacteria, viruses, yeast, mites, fungi, etc. It is also through the mutual benefit and balance of the skin microbiota that our skin remains healthy. A healthy skin microbiota predominately consists of the following bacterial phyla: *Actinobacteria*, *Firmicutes*, *Proteobacteria*, and *Bacteroidetes* [10,11]. However, the imbalance in cutaneous microbiota, also known as dysbiosis, is frequently associated with the development of skin diseases, such as acne vulgaris [12]. By definition, acne vulgaris is a form of dermatoses that forms excess production of sebum, inflamed pilosebaceous units, and follicular hyperkeratinization, in the presence of selected phylotypes of *Cutibacterium acnes* (which was formerly known as *Propionibacterium acnes*) [12,13]. These bacteria thrive on sebum and are responsible for the release of proinflammatory cytokines that induce the occurrence of acne vulgaris [14]. In the meantime, another commensal bacterium, *Staphylococcus epidermidis*, fails to perform its role as a regulator of the skin's homeostasis, hypothetically due to its decreased abundance that intensifies the disequilibrium in the skin microbiota [15]. Finally, accompanying the disruption of the skin barrier, opportunistic pathogens, such as *Staphylococcus aureus*, then aggravate the symptoms by promoting skin inflammation [15]. Thus, it is essential to retain the delicate balance of microbiota on our skin. This is where antibacterial cosmetic products come into play in the treatment of skin pathology. Besides treatment against the aforementioned bacteria during dysbiosis, it should also restore the equilibrium of the skin's microbiota.

In a study conducted by Suliman and the team, it had been shown that the crude alkaloid extract of the seed can inhibit the growth of *S. aureus*, *Escherichia coli*, and *Pseudomonas aeruginosa*, with minimum bactericidal concentration (MBC) values of 12.5 mg/mL against *S. aureus* and 50 mg/mL against *E. coli* and *P. aeruginosa*. An almost similar value was obtained for the minimum inhibition concentration (MIC) assay, indicating that the extract is both inhibitory and bactericidal [16]. Interestingly, another study conducted on *P. aeruginosa* using *Caenorhabditis elegans* as a model showed that, although the seed extract improved its survival, it was unable to inhibit its growth and colonization in the worm's gut. It was suggested that this observation could be attributed to additional effects of the extract, such as modifying the virulence factors of the associated bacteria or improving the host's tolerance (as lysozyme-like protein 7 gene expression was increased) [17]. Although different outcomes were observed in both studies, it should be noted that different extraction solvents were used in the extraction process. Therefore, different types or amounts of phytocompounds could be extracted or excluded based on their polarity and solubility. Similarly, a study on the oil of the *S. macrophylla* seed, obtained through n-hexane and diethyl ether, also exhibited different levels of antagonistic effects towards the same bacterial species, with the addition of *Salmonella* Typhimurium [18]. In this case, there were smaller inhibition zones for both *S. aureus* and *P. aeruginosa* and no signs of inhibition against *E. coli* after treatment with the oil. Next, methanol extract of *S. macrophylla* seeds inhibited the growth of five different bacteria species and one yeast.

These microorganisms were *Acinobacter anitratus*, *S. aureus*, *Bacillus cereus*, *Bacillus subtilis*, *Shigella boydii*, and *Candida utilis* [19]. Again, *E. coli* was not reported to be inhibited by this particular extract, which is the opposite to another study that also used methanol as the extracting solvent [19,20]. In an article written by Durai and team [20], both *S. aureus*, *E. coli*, and three other fungi (*Fusarium* sp., *Alternaria* sp., and *Helminthosporium* sp.) were inhibited by the methanol extract of *S. macrophylla*. The probable answer to this varying inhibitory activity on *E. coli* could be either the difference in strain or location of seed harvest. Regardless, even though methicillin-resistant *S. aureus* (MRSA) growth was not affected by *S. macrophylla* methanol extract, its virulence was suspected to decrease as the survival of *C. elegans* was increased after treating with 200 μg/mL of either methanol, ethyl acetate, or butanol extract [19,21].

After the purification of the extract, the antibacterial activity of *S. macrophylla* can be accredited to the presence of specific phytocompounds, as can be observed in Figure 1. One such phytocompound is swietenolide, a limonoid phytocompound of the triterpenoid class that can be found in the seed of *S. macrophylla*. In an early study conducted in 2009, the compound demonstrated antibacterial activity against eight different multiple-drug-resistant bacterial strains. These bacterial strains were group A β-haemolytic *S. aureus* (code 312), *S. aureus* (code 477), *Streptococcus pneumonia* (code 32215), *Haemophilus influenza* (code 32139), *E. coli* (code 169), *Klebsiella pneumonia* (code 32140), *Salmonella typhi* (code 1467), and *Salmonella paratyphi* (code 1272) [22]. Besides swietenolide, swietenine and 3-O-tigloylswietenolide were also reported to exhibit antibacterial activity against several Gram-positive and -negative bacteria. These bacteria included *B. subtilis*, *Bacillus megaterium*, *Sarcina lutea*, *S. aureus*, *E. coli*, *P. aeruginosa*, *S. typhi*, *S. boydii*, *Shigella dysenteriae*, *Shigella sonnei*, *Shigella shiga*, and *Klebsiella* sp. [23]. Looking at the similarities in the bacterial species tested across the different studies, it can be concluded that these three phytocompounds are indeed responsible for the antibacterial action of *S. macrophylla*. Interestingly, proceranolide, another limonoid that can be found in the seed, was discovered to have moderate inhibitory effects against two *Mycobacterium tuberculosis* strains (low virulent H37Rv ATCC27294 strain and high virulent M299 Beijing strain), with a $MIC_{50}$ value of $37.6 \pm 0.4$ μg/mL and $44.9 \pm 1.0$ μg/mL for the H37Rv and M299 strains, respectively [24,25]. Although both bacteria strains are not of cosmeceutical importance, it could be beneficial for proceranolide to be included in future tests, in order to study its protective effects against pathogenic skin bacteria.

Overall, it is clear that the *S. macrophylla* seed does possess antimicrobial properties, as can be observed in Table 1. However, the strength and coverage of its antimicrobial activity are very much dependent on the presence of certain phytochemicals and their amount in the extract. That in itself pertains to many uncontrollable factors, such as extraction method, solvents used, location of the plant, and harvest seasons that affect the abundance of antimicrobial phytochemicals [26]. On the contrary, the inconsistency in the data reported could also be due to the dissimilar bacteria strains used. Among the articles reviewed, some studies did not reveal the strain number used, which challenges any form of direct comparison across articles. Therefore, for future improvements, it is encouraged that similar strains of bacteria, especially clinical strains, should be used in experiments to ease comparison across studies. It would also be useful to set a standard set of antimicrobial tests in the antimicrobial testing for cosmetic products. For example, the natural product incorporated in cosmetic creams and gels must be able to exhibit a certain degree of activity before these claims can be placed on the product labels. This is to ensure that the reported antimicrobial activity is up to par and that consumers would benefit from the desired effect (as claimed in the product description/labelling). Lastly, as of yet, *S. macrophylla* extracts have not been tested against some bacteria, such as *C. acnes* and *S. epidermidis*. To truly explore its use in cosmetics, it would be indispensable that the extracts and phytocompounds extracted are not only analyzed for their antimicrobial activity against them, but also if the extract and compounds can restore and promote the balance of the skin microbiota. Incidentally, it would be worthwhile to study the effect of *S. macrophylla* on bacteria-induced inflammation in skin diseases. Thus, this calls for antimicrobial studies beyond just the bacteria itself, such as in vivo analysis of the effect of *S. macrophylla* on the skin microbiome among the human population and clinical trials of *S. macrophylla* based skin care products.

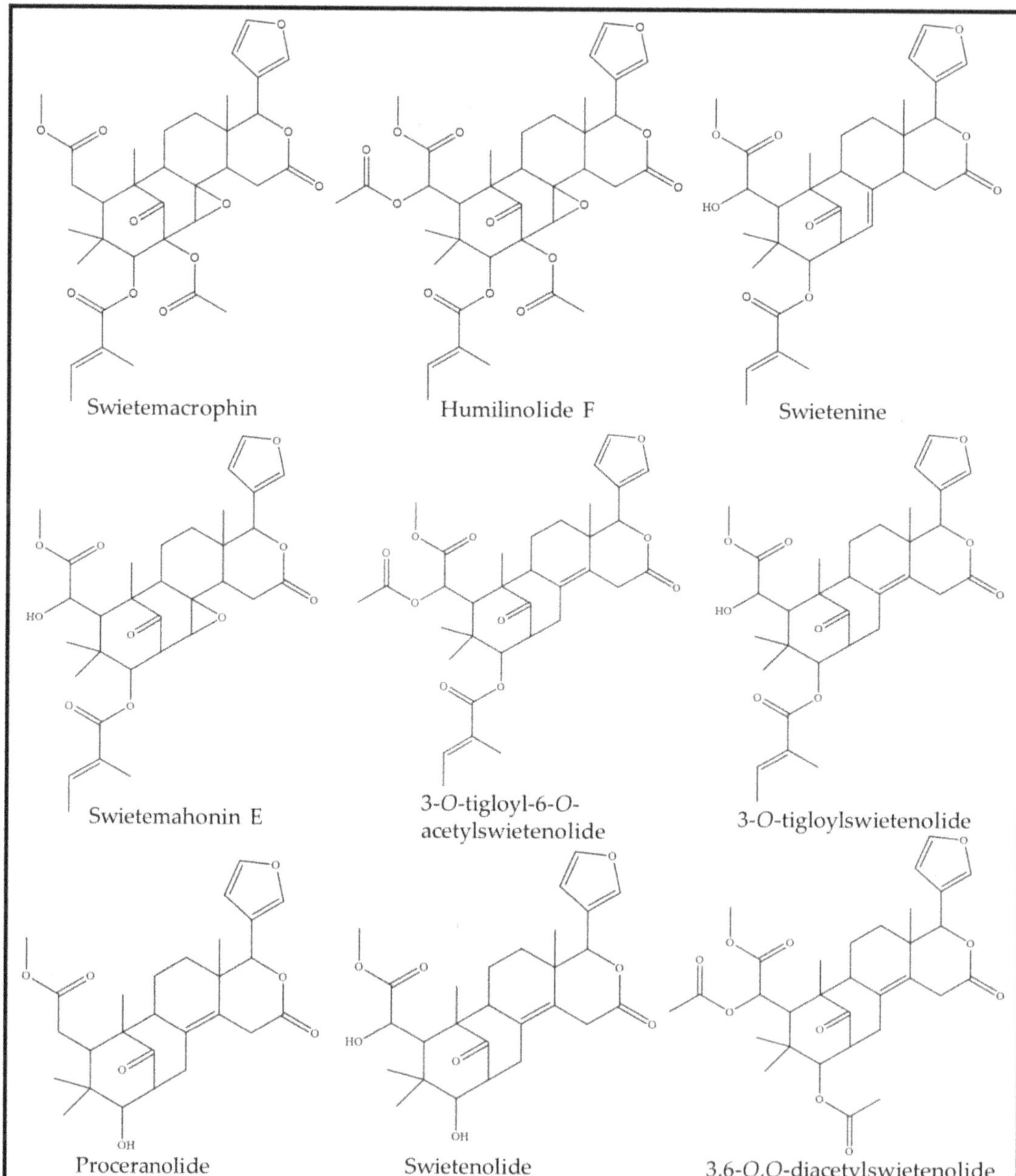


**Figure 1.** Limonoid phytocompounds extracted from *S. macrophylla* seed with cosmeceutical bioproperties.

**Table 1.** The antimicrobial activity of *S. macrophylla* seed against various antimicrobial strains.

| Seed Extract | Extraction Method | Antimicrobial Strains | Antimicrobial Activity | References |
|---|---|---|---|---|
| Alkaloid crude extract Rate of alkaloid extracted from the seeds: 2.85% | (A) The seeds were cleaned, sun-dried and grinded to fine powder.<br>(B) Ethanol (300 mL; 70%) was mixed with 20 g powder of the seeds in an orbital shaker at 150 rpm for 24 h at 25 °C.<br>(C) The extract was then evaporated with a rotary evaporator until one fifth of the initial volume was reached, before adding 20 mL of 0.1 N hydrochloric acid.<br>(D) The extract was filtered and extracted twice with 20 mL of chloroform.<br>(E) The extract was treated twice with 10 mL of 0.1 N hydrochloric acid and 0.1 N ammonia was added to obtain pH 9, before adding 30 mL of chloroform.<br>(F) The operation was repeated three times, before evaporating and dissolving the residue in 20 mL of 0.02 N hydrochloric acid.<br>(G) The extract was titrated with 0.02 N potassium hydroxide, with methyl red as an indicator. | *S. aureus* ATCC1026, *E. coli* ATCC10536, *P. aeruginosa* ATCC15442. | (A) Inhibition activity (disk diffusion method):<br>(i) *S. aureus*: 17 mm (50 mg/mL); 21 mm (100 mg/mL);<br>(ii) *E. coli*: 10 mm (50 mg/mL); 12 mm (100 mg/mL);<br>(iii) *P. aeruginosa*: 12 mm (50 mg/mL); 15 mm (100 mg/mL).<br>(B) MIC and MBC value:<br>(i) *S. aureus*: not available (MIC); 12.5 mg/mL (MBC);<br>(ii) *E. coli*: 25 mg/mL (MIC); 50 mg/mL (MBC);<br>(iii) *P. aeruginosa*: 25 mg/mL (MIC); 50 mg/mL (MBC). | [16] |
| Methanol, ethyl acetate and butanol crude extract | The sample was air-dried, powdered and extracted with methanol, ethyl acetate and butanol. | *C. elegans* infected with MRSA ATCC33591 or methicillin-sensitive *S. aureus* (MSSA) NCTC83254. | (A) Survival of 72 h *S. aureus* infected *C. elegans* after treatment (200 µg/mL):<br>(i) Methanol: 70% survival (3-fold change vs. untreated);<br>(ii) Ethyl acetate: 65% (2.8-fold change vs. untreated);<br>(iii) Butanol: 96.7% (4.2-fold change vs. untreated).<br>(B) None of the extracts had bacteriostatic or bactericidal activity on *S. aureus* growth in vitro.<br>(C) All three extracts significantly increased the survival of infected *C. elegans* but did not affect replication of *S. aureus*. There was no difference in CFU of the intestinal bacterial loads.<br>(D) All three extracts promoted survival in 72 h MRSA infected *C. elegans* (>70% survival). | [21] |

**Table 1.** *Cont.*

| Seed Extract | Extraction Method | Antimicrobial Strains | Antimicrobial Activity | References |
|---|---|---|---|---|
| Methanol extract and ethyl acetate extract | (A) The sample was air-dried, powdered and extracted with methanol.<br>(B) The crude extract was "defattened" using hexane and then further extracted with ethyl acetate. | *C. elegans* infected with *P. aeruginosa* (Bacteria strain was not specified). | (A) Ethyl acetate extract (200 and 400 µg/mL) and methanol extract (200 µg/mL) significantly improved the survival of *C. elegans* 48 h after *P. aeruginosa* infection.<br>(B) Treatment with 200 µg/mL of ethyl acetate extract had the highest survival of 59.5 ± 1.65%.<br>(C) MIC assay demonstrated that the ethyl acetate extract (25–1000 µg/mL) had no inhibition against *P. aeruginosa*.<br>(D) There was no significant difference in CFU of *P. aeruginosa* in *C. elegans* intestines after both extract treatment.<br>© Ethyl acetate extract (200 µg/mL) increased expression of lys-7 gene that was suppressed by *P. aeruginosa* infection. | [17] |
| Seed oil extract by diethyl ether (39%) and n-hexane (42.7%) | (A) The seeds were dried and ground to small pieces.<br>(B) The seeds (10 g) were extracted using n-hexane and diethyl ether to recover at least 10 mL of oil using the Soxhlet apparatus.<br>(C) The solvent was evaporated using the rotary evaporator, before further drying under the open air. | *S. aureus*, *E. coli*, *P. aeruginosa*, *Salmonella* Typhimurium. (All bacteria are in-house bacteria strains). | (A) Antibacterial activity of seed oil (disk diffusion method):<br>(i) 10 µg/mL: 5 mm (*S. aureus*); 4 mm (*Salmonella* Typhimurium); 5 mm (*P. aeruginosa*);<br>(ii) 20 µg/mL: 8 mm (*S. aureus*); 6 mm (*Salmonella* Typhimurium); 5 mm (*P. aeruginosa*);<br>(iii) 50 µg/mL: 8 mm (*S. aureus*); 9 mm (*Salmonella* Typhimurium); 6 mm (*P. aeruginosa*);<br>(iv) 100 µg/mL: 9 mm (*S. aureus*); 10 mm (*Salmonella* Typhimurium); 6 mm (*P. aeruginosa*);<br>(v) 1000 µg/mL: 11 mm (*S. aureus*); 20 mm (*Salmonella* Typhimurium); 11 mm (*P. aeruginosa*);<br>(vi) There was no effect on *E. coli* after the seed oil treatment. | [18] |
| Methanol extract | (A) The seeds were washed and dried at 50 °C for 3 days, before grinding to a coarse powder.<br>(B) The powder was then soaked in methanol at a ratio of 1:5 (*w*/*v*) for 24 h.<br>(C) The extract was filtered and concentrated using the rotary evaporator at 60 °C. | Gram-positive: *S. aureus*, *B. cereus*, *B. subtilis*, MRSA<br>Gram-negative: *Proteus mirabilis*, *Yersinia* sp., *E. coli*, *Klebsiella pneumoniae*, *S. boydii*, *A. anitratus*<br>Fungal: *Aspergillus niger*, *Microsporum fulvum*, *Rhizopus* sp.<br>Yeast: *Candida utilis*.<br>(Microbial strains are clinical isolates). | (A) Antimicrobial activity of methanol extract (disk diffusion method):<br>(i) *S. aureus*: 13.0 ± 0.6 mm;<br>(ii) *B. cereus*: 15.0 ± 0.6 mm;<br>(iii) *B. subtilis*: 15.0 ± 1.5 mm;<br>(iv) *S. boydii*: 15.0 ± 1.0 mm;<br>(v) *A. anitratus*: 16.0 ± 1.2 mm;<br>(vi) *C. utilis*: 10.0 ± 1.2 mm;<br>(vii) No inhibitory activity was observed for MRSA, *P. mirabilis*, *Yersinia* sp., *E. coli*, *K. pneumoniae*, *A. niger*, *M. fulvum*, or *Rhizopus* sp.<br>(B) MIC and MBC of methanol extract:<br>(i) *B. cereus*: 3.13 mg/mL (MIC); 6.25 mg/mL (MBC);<br>(ii) *S. aureus*: 6.25 mg/mL (MIC); 25 mg/mL(MBC);<br>(iii) *B. subtilis*: 1.56 mg/mL (MIC); 3.13 mg/mL (MBC);<br>(iv) *S. boydii*: 12.5 mg/mL (MIC); 25 mg/mL (MBC);<br>(v) *A. anitratus*: 12.5 mg/mL (MIC); 25 mg/mL (MBC);<br>(vi) *C. utilis*: 12.5 mg/mL (MIC); 25 mg/mL (MBC). | [19] |

**Table 1.** *Cont.*

| Seed Extract | Extraction Method | Antimicrobial Strains | Antimicrobial Activity | References |
|---|---|---|---|---|
| Methanol extract | (A) The seeds were dried for 2 days and pulverized to powder form.<br>(B) The powdered seeds (30 g) were then extracted by reflux using Soxhlet apparatus for 10 h with successive organic solvent (80% methanol) and concentrated through oven drying.<br>(C) Each fraction was collected, distilled and dried in the incubator. | Gram-positive: *S. aureus*<br>Gram-negative: *E. coli*<br>Fungi: *Fusarium* sp, *Helminthosporium* sp, *Alternaria* sp.<br>(Bacteria and fungi are lab strains). | (A) Agar diffusion method (inhibition zone):<br>(i) *S. aureus:*<br>• 20 μg/mL: 14.4 ± 1.0 mm;<br>• 30 μg/mL: 19.3 ± 0.5 mm;<br>• 50 μg/mL: 27.0 ± 1.0 mm.<br>(ii) *E. coli:*<br>• 20 μg/mL: 21.8 ± 1.6 mm;<br>• 30 μg/mL: 31.0 ± 1.0 mm;<br>• 50 μg/mL: 40.6 ± 1.5 mm.<br>(B) Anti-fungal activity (inhibition zone):<br>(i) *Fusarium* sp.:<br>• 20 μg/mL: 43.3 ± 2.0%;<br>• 30 μg/mL: 54.0 ± 4.0%;<br>• 50 μg/mL: 43.3 ± 2.0%.<br>(ii) *Alternaria* sp.:<br>• 20 μg/mL: 42.6 ± 2.5%;<br>• 30 μg/mL: 52.6 ± 1.5%;<br>• 50 μg/mL: 76.0 ± 2.0%.<br>(iii) *Helminthosporium* sp.:<br>• 20 μg/mL: 29.3 ± 2.5%;<br>• 30 μg/mL: 37.0 ± 2.0%;<br>• 50 μg/mL: 63.3 ± 2.8%. | [20] |

## 3. Photoprotective, Skin Whitening, and Skin Repair Cosmeceutical Prospects of *S. macrophylla*

Aside from its antimicrobial properties, *S. macrophylla* seed extract has the potential for cosmetic use against skin aging. In spite of the many factors that can expedite the rate of skin aging, one main factor is the constant exposure of one's skin to sunlight. Sunlight that can penetrate through the ozone layer contains ultraviolet A (UVA) and B (UVB). The differences between both types of ultraviolet rays (UVR) are their wavelength and penetration level through the skin. UVA, which has a longer wavelength of 320–400 nm, can penetrate deeply into the dermal layer of the skin. On the other hand, UVB, which has a shorter wavelength of 280–320 nm, can only penetrate up to the epidermal layer of our skin [27]. Regardless of their characteristics, overexposure to both UVR causes dysregulation in the skin's circadian rhythm and activates the production of reactive oxygen species (ROS) and reactive nitrogen species (RNS) [28–30]. The overproduction of both ROS and RNS then quickly overwhelms the skin's antioxidant defense system, causing nitrosative and oxidative stress in the skin [29]. In response, the skin begins producing pro-inflammatory cytokines, signaling the start of skin inflammation [31,32]. Following this, neutrophils quickly infiltrate the skin and secrete neutrophil elastase, which cleaves extracellular matrices and activates matrix metalloproteases [33–35]. Skin melanogenesis is also activated in response to UVR exposure by the activation of p53 expression, which goes on to initiate the transcription of the pro-opiomelanocortin (POMC) gene, eventually beginning the conversion of tyrosine into melanin [36]. Other events, such as the migration of Langerhans cells from the epidermis to the lymph nodes, were noted after UVR exposure. A side effect of this is that the skin's hypersensitivity to allergens will be lowered during the event [37]. Direct DNA damage by UVB, through the formation of cyclobutene pyrimidine dimers (CPD) and pyrimidine-pyrimidone (6–4) photoproducts, also causes cell death and increases the risk of skin cancer [38–40]. A summary of the mentioned UVR-induced damage can be observed in Figure 2. According to this evidence, it becomes even more necessary to protect our skin against overexposure to UVR.

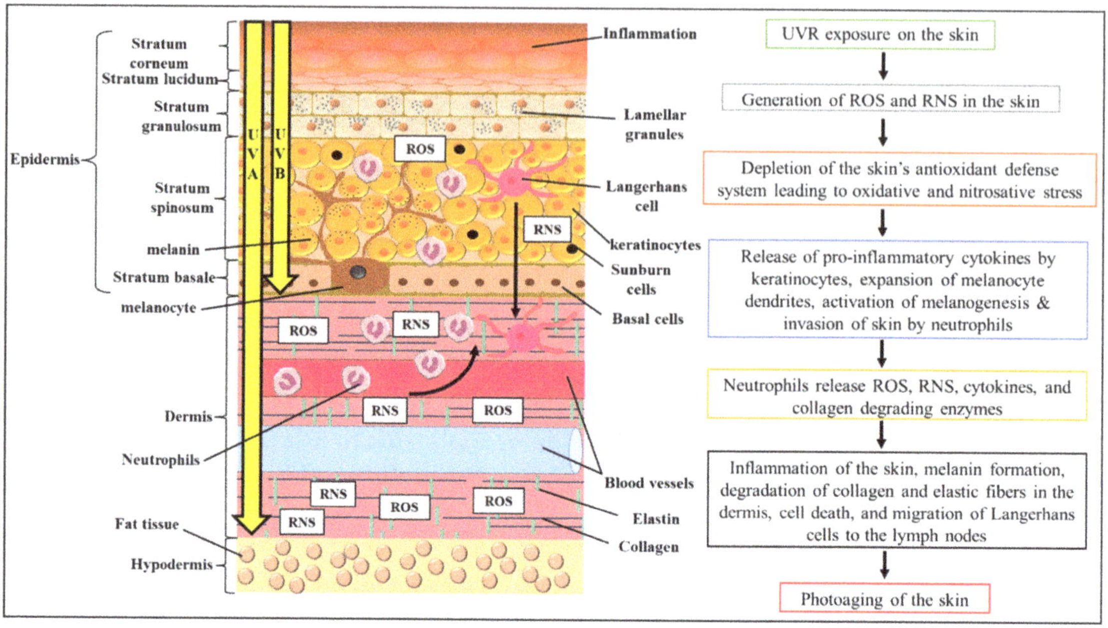


**Figure 2.** The impact of UVR exposure on the skin and how it mediates photoaging.

Photoprotection against UVR can be achieved through either immediate "blocking" of UVR penetration through the skin, or by reversing the activated pathways instigated through UVR exposure. The "blocking" of UVR penetration can be achieved through the application of either physical or chemical sunscreen. Physical sunscreens are sunscreens that contain titanium dioxide or zinc oxide nanoparticles that reflect and scatter UVR, whereas chemical sunscreens are made of organic molecules that absorb UVR energy and emit them as less dangerous wavelengths [41–44]. Nevertheless, regardless of their classification, the validation of the "blocking" ability of these sunscreens was set by ISO and FDA to have at least a critical wavelength of 370 nm when claiming the title of being a broad-spectrum sunscreen [45–47]. This ensures that the sunscreens on the market can cover from 280–370 nm of the sun's wavelength. Furthermore, despite sun protective factor (SPF) being more commonly known to the public, it is, in fact, the measurement of protection against UVB. Instead, UVA is recommended to be greater than one third of the SPF value and is measured on a scale from PA+ (low) to PA++++ (high) [48,49]. As an alternative to sunscreens, the reversal or disruption of UVR-induced photoaging can also be achieved by interrupting the UVR-induced pathways in the skin. At the early stages of the UVB-induced pathway, suppressing the generation of ROS and RNS, by boosting the skin's antioxidant defense and negating the emergence of pro-inflammatory cytokines, can aid in reducing UVR damage [50,51]. Meanwhile, hampering the production of unwanted pigmentation and encouraging the production of collagen repair at the later stages will help deter premature skin aging [52,53]. Although melanin by nature in a sense is our protector against UVR damage, through its chemical and optical filtrating properties, uneven hyperpigmentation of the skin, such as freckles, solar lentigines or melasma, are still undesirable [54,55].

Returning to the context, it has been noted that the extract from the seed has a marked aptitude as an anti-aging and photoprotective cosmetic agent (Table 2). As a photoprotective agent, ethanolic extract of the seed was reported to have a critical wavelength of 347.6 nm, when tested across the UVA-UVB spectrum. Subsequently, its hexane, ethyl acetate, and water fraction had critical wavelengths of 345, 341.6 and 362.4 nm, respectively [56]. Whilst it is not definite, the discovery proposes that the extract and fractions of the seed may contain photoprotective properties similar to sunscreen. Following this, in vitro treatment of epidermal skin (HaCaT) cells with *S. macrophylla* ethanolic extract had successfully reversed the expression of tumor necrosis factor alpha (TNF-α) by UVB, leading to the subsequent decrease in matrix metalloprotease (MMP)-1 (collagen degrading metalloprotease). Next, solvent fractionation of the ethanolic extract into both hexane and ethyl acetate fractions further enhanced its "anti-UVB" effect. The ethyl acetate fraction additionally repressed the Bcl-2 associated X protein (Bax) and nuclear factor kappa B (NF-κB), alongside TNF-α and MMP-1. This suggests that the phytocompounds that were actively attenuating the UVR-induced inflammation and cell apoptosis in the ethanolic extract can be found in the ethyl acetate fraction and, according to the solvent, is semi-polar by nature. On another note, LCMS data of the cells treated with the hexane fraction revealed a completely different effect compared to the ethanolic extract. Irradiated cells treated with the hexane fraction activated various pathways, involving the redox system, transcription to translation processes, cell growth, proliferation and migration, glycolysis, and DNA maintenance and repair, effectively attenuating the impact UVB has on human skin [57]. This ability of the *S. macrophylla* seed in reversing UVB-induced damage could be due to its antioxidant and anti-inflammatory properties. In support of these findings, several works have demonstrated that the extract of *S. macrophylla*, be it via methanol, aqueous or ethyl acetate extraction, had significant free radical scavenging and antioxidant activity [56,58–60]. With regard to anti-inflammation, pure phytocompounds that can be found in the seed, such as swietenine, were able to inhibit the production of nitric oxide (NO), downregulate several pro-inflammatory cytokines, and upregulate antioxidant proteins, like nuclear factor erythroid 2-related factor 2 (NRF2), and haem-oxygenase (HO-1), in lipopolysaccharide (LPS) stimulated RAW264.7 macrophage cells. The treatment of murine hepatoma (Hepa-1c1c7) cells with swietenine also induces NAD(P)H quinone oxidoreductase 1 (NQO1) dose-dependently, supporting the activation of NRF2 observed in RAW264.7 cells [61]. Other limonoids, such as swietemacrophin, 3-*O*-tigloylswietenolide, and swietemahonin E, are also potent inhibitors of inflammation. Tested against NO-generating macrophages, these phytocompounds produced an $IC_{50}$ of less than 36.32 μM, which is suggestive of their potency in suppressing the onset of inflammation [62]. These compounds, together with humilinolide F and 3,6-*O*,*O*-diacetylswietenolide, also effectively inhibit the generation of superoxide anions in neutrophils with significant $IC_{50}$ values of less than 45.44 μM, demonstrating their anti-inflammatory ability [62]. The same was also reported for 3-*O*-tigloyl-6-*O*-acetylswietenolide, another limonoid that is present in the seed, whereby the phytocompound effectively inhibited the generation of superoxide anion in neutrophils at an $IC_{50}$ of 27.9 ± 2.4 μM [63,64]. Furthermore, proceranolide also showed considerable inhibition against the generation of NO in RAW264.7 macrophage cells, with an $IC_{50}$ of 26.9 ± 0.6μg/mL, comparable to the positive control NG-methyl-L-arginine acetate salt (IC50: 13.2 ± 0.6 μg/mL) [25]. Based on these findings, this confirms the notion that *S. macrophylla* seeds contain both antioxidant and anti-inflammatory phytocompounds. Together, both these properties then function to remedy the damaging effects of UVR. Subsequently, a study further exploited the anti-inflammatory abilities of the seed [65,66]. By developing nanoemulsions and a nanoemulgel of *S. macrophylla* oil, Eid and his team achieved an even greater anti-inflammatory effect on the carrageenan-induced paw edema of rats, as compared to just the oil itself in raw form [65,66]. This sort of enhancement and modification that improves extract or phytocompound delivery to the targeted area and magnification of its bioproperties is beneficial in the cosmetic industry. It would certainly bode well, in future studies, if the research on these seed extracts could move beyond the discoveries of its bioproperties towards formulating efficient drug transport and active use of its potential in cosmetic formulation for the public.

Apart from this, other anti-premature aging properties of the seed also lie in its ability to impede the formation of irregular skin pigmentation and induce skin regeneration. In the case of skin pigmentation, one way to hinder the buildup of unwanted melanin is to inhibit the enzymatic activity of tyrosinase. Tyrosinase is a rate-limiting enzyme that converts L-tyrosine to L-3,4-dihydroxyphenylalanine (L-DOPA), which then continues through a series of conversions involving tyrosinase-related protein 1 and 2 to form melanin in the melanogenesis process [67]. Thus, the inhibition of both ethyl acetate and methanolic extract of the seed against mushroom tyrosinase suggests it might have the ability to suppress the formation of melanin [59]. Even though currently this is the only published research on the whitening effect of the seed, it is a good indicator and prompts further research regarding its anti-pigmentation properties on human tyrosinase. On another note, the wound healing properties of the seed have been explored by the authors of both [56,68]. The treatment with its ethanolic extract significantly promoted wound closure over 24 h, as tested in scratched immortalized human epithelial (HaCaT) cells. Moreover, fractionation of the ethanolic extract with water and hexane further increased its capabilities in wound closure, while the ethyl acetate fraction showed a slightly lower healing capacity, as compared to the ethanolic extract. This suggests that the extract has skin regenerative properties, and it is likely that the phytocompounds responsible are either polar or non-polar compounds [56]. Subsequently, in a study conducted by Nilugal et al. [68], an in vivo approach was used instead. Albino rats were excised on the dorsal thoracic region and ointment with ethanolic extract of the seed was applied to the wound. Wound area reduction was then measured for 20 days. The results demonstrate that the extract induced complete wound closure by day 15, instead of day 20 as observed in the negative control. In the histological images, it can be observed that the extract induced higher proliferation in the fibroblast, formed new blood capillaries, and increased collagen fibers. In short, as photoaging brings about the degradation of collagen fibers, the discovery of wound healing and skin re-epithelization properties in the seeds are promising for future cosmetic development. Tying in with the ability of the seed in inhibiting UVB-mediated production of MMP-1 (as mentioned earlier), these positive prospects should incite an increased fervor in the search for the compounds responsible for these properties. Finally, studies involving human skin and 3D culture will also increase our knowledge of the deeper mechanisms involved in the process of skin repair.

**Table 2.** The bioproperties and bioactivities of *S. macrophylla* seed that has cosmeceutical potential.

| Seed Extract | Bioproperties | Extraction Method | In Vitro/In Vivo Studies | Bioactivity | References |
|---|---|---|---|---|---|
| Aqueous extract | Antioxidant | (A) The seeds were washed, dried at room temperature, powdered, and sieved through 40 meshes.<br>(B) For 200 mg of powder, 1 mL of distilled water was added, before centrifuging for 15 min at 3000 rpm.<br>(C) The supernatant was filtered and collected as the extract. | (A) Winstar rats (200–250 g) injected intraperitoneally with 65 mg/kg of streptozotocin in a volume of 1 mL/kg body weight to induce diabetes. The rats were fed with glucose 30 min prior to treatment with the extract. Blood was drawn from the tail.<br>(B) Antioxidant testing on the extract itself. | (A) Antioxidant level in the blood demonstrated dose-dependent increase in antioxidant activity, using modified free oxygen radical defense (FORD) assay.<br>(B) Antioxidant activity of pure aqueous extract also showed increasing antioxidant activity dose-dependently via modified FORD assay. | [58] |
| Aqueous extract | Antioxidant | (A) The seeds were washed, dried at room temperature, powdered, and sieved.<br>(B) The powdered seeds were added to 200 mL of boiled distilled water, before centrifuging at 3500 rpm for 8 min and filtered. | (A) FORD assay.<br>(B) Free radical 2,2-diphenyl-1-picrylhydrazyl (DPPH) + H2 assay.<br>(C) Ferric-reducing antioxidant power (FRAP) assay.<br>(D) Oxidative stress test on *Saccharomyces cerevisiae* (*S. cerevisiae*). | (A) FORD assay: increase in antioxidant activity over time.<br>(B) DPPH assay: $56.2 \pm 0.97\%$.<br>(C) FRAP assay: $34.8 \pm 0.13$ µmol Fe + 2/g PM.<br>(D) Oxidative stress test: I extract promotes the growth of *S. cerevisiae* over time, even in the presence of hydrogen peroxide that causes oxidative stress. | [60] |

**Table 2.** *Cont.*

| Seed Extract | Bioproperties | Extraction Method | In Vitro/In Vivo Studies | Bioactivity | References |
|---|---|---|---|---|---|
| Ethyl acetate extract; methanol extract | Antioxidant and anti-pigmentation | (A) The seeds were dried and then ground to powder, before being subjected to extraction. (B) The powdered seeds were extracted with ethyl acetate and methanol in increasing order of their polarity. (C) Each extract was concentrated using a rotary evaporator at 40–50 °C. | (A) DPPH assay. (B) Tyrosinase inhibitory activity. | (A) DPPH radical scavenging capacity (500 μg/mL): (i) Ethyl acetate extract: 30.30 ± 1.63%; $IC_{50}$ detected; (ii) Methanol extract: 56.82 ± 2.67%; $IC_{50}$: 200 μg/mL. (B) Percentage of inhibition on tyrosinase activity: (i) Ethyl acetate extract: 14.44 ± 2.45%; (ii) Methanol extract: 15.95 ± 1.27%. | [59] |
| Ethanol extract (SMCE); hexane fraction (SMHF); ethyl acetate fraction (SMEAF); aqueous fraction (SMWF) | Photoprotection, antioxidant, wound healing | (A) The seeds (3 kg) were finely grounded and soaked in ethanol for 72 h at room temperature. (B) The extract was filtered and concentrated with a rotary evaporator at 40 °C to obtain SMCE. (C) SMCE was then dissolved in hexane to obtain the hexane fraction. The supernatant was dried with anhydrous sodium sulphate, before concentrating with a rotary vacuum evaporator to obtain SMHF. (D) The insoluble residues of hexane were subjected to ethyl acetate and water portioning in a 1:1 ratio. (E) The ethyl acetate fraction was dried via rotary evaporation to obtain SMEAF, while the water fraction was freeze-dried to obtain SMWF. | (A) DPPH radical scavenging assay. (B) 2′azino-bis (3, -ethylbenzothaizoline-6-sulfonic acid). (ABTS) radical scavenging assay. (C) Ferrous ion chelating assay. (D) Critical wavelength measurement. (E) Scratch wound assay on HaCaT cells. | (A) Antioxidant assays: (i) SMCE: not significant for DPPH, ABTS and ion chelating activity; (ii) SMHF: significant only for iron chelating activity at 2000 μg/mL with 14.073 ± 0.18% activity. Not significant for DPPH and ABTS assays; (iii) SMEAF: not significant for DPPH, ABTS and ion chelating activity; (iv) SMWF: (a) Significant for DPPH activity: 6.332 ± 0.80% at 2000 μg/mL; (b) Significant and dose-dependent increase for ABTS activity: 12.796 ± 2.01% at 125 μg/mL; (c) Significant and dose-dependent increase for iron chelating activity: 8.014 ± 2.51% at 125 μg/mL. (B) Critical wavelength of SMCE, SMHF, SMEAF, and SMWF are 347.6, 345, 341.6, and 362.4 nm, respectively. (C) Percentage of wound closure after 24 h of treatment: (i) SMCE (6.25 μg/mL): 54.10 ± 2.59%; (ii) SMHF (100 μg/mL): 59.45 ± 5.72%; (iii) SMEAF (12.5 μg/mL): 41.48 ± 3.91%; (iv) SMWF (50 μg/mL): 74.68 ± 5.16%. | [56] |

**Table 2.** *Cont.*

| Seed Extract | Bioproperties | Extraction Method | In Vitro/In Vivo Studies | Bioactivity | References |
|---|---|---|---|---|---|
| SMCE, SMHF, SMEAF, SMWF | Photoprotection against UVB irradiation | (A) The seeds (3 kg) were finely ground and soaked in ethanol for 72 h at room temperature.<br>(B) The extract was filtered and concentrated with a rotary evaporator at 40 °C to obtain SMCE.<br>(C) SMCE was then dissolved in hexane to obtain the hexane fraction. The supernatant was dried with anhydrous sodium sulphate, before concentrating with a rotary vacuum evaporator to obtain SMHF.<br>(D) The insoluble residues of hexane were subjected to ethyl acetate and water portioning in a 1:1 ratio.<br>(E) The ethyl acetate fraction was dried via rotary evaporation to obtain SMEAF, while the water fraction was freeze-dried to obtain SMWF. | HaCaT cells treated with SMCE (6.25 μg/mL), SMHF (100 μg/mL), SMEAF (12.5 μg/mL), and SMWF (50 μg/mL) in PBS, while being exposed to 50 mJ/cm$^2$ UVB. Cells were then rinsed and incubated for 24 h at 37 °C, 5% $CO_2$. Protein and gene expression changes were taken 24 h post exposure.<br>Negative control: non-irradiated cells.<br>Inducer control: irradiated but non-treated cells. | Comparison of treatment with negative and inducer controls:<br>(A) SMCE:<br>(i) Gene expression changes: downregulation of TNF-α and MMP-1 (vs inducer control);<br>(ii) Protein expression changes: downregulation of ribosomes and Filamin Bβ (vs negative control).<br>(B) SMHF:<br>(i) Gene expression changes: downregulation of NF-κB and cyclin D1 (vs inducer control);<br>(ii) Protein expression changes: multiple changes across the redox system, RNA to protein processing, DNA maintenance and repair, glycolysis process, and cell growth, proliferation and migration. All changes demonstrated reversal against UVB induced damage.<br>(C) SMEAF:<br>(i) Gene expression changes: downregulation of TNF-α, NF-κB, MMP-1 and Bax;<br>(ii) Protein expression changes: downregulation of PRDX-3, PDI-A3, and fascin (vs negative control).<br>(D) SMWF:<br>(i) Gene expression changes: no significant changes in TNF-α, NF-κB, COX-2, MMP-1, cyclin D1 and Bax;<br>(ii) Protein expression changes: multiple changes across the redox system, RNA to protein processing, DNA maintenance and repair, glycolysis process, and cell growth, proliferation and migration, in which majority are opposite to SMHF. | [57] |
| Pure compounds<br>(i) swietemacrophin<br>(ii) humilinolide F<br>(iii) 3,6-*O*,*O*-diacetylswietenolide<br>(iv) 3-*O*-tigloylswietenolide<br>(v) swietemahonin E;<br>(vi) swietenine | Anti-inflammation | (A) Dried seeds (380 g) were pulverized and extracted with methanol for 3 days at room temperature.<br>(B) The extract was concentrated at 35 °C with reduced pressure, before being partitioned between ethyl acetate and water in a 1:1 ratio.<br>(C) The water fraction was further extracted with n-butanol to produce a butanol soluble fraction and water fraction.<br>(D) The ethyl acetate fraction was further fractionated and purified to produce six pure compounds. | (A) Human neutrophils obtained from the venous blood of healthy, adult volunteers aged 20–30 years old.<br>(B) RAW264.7 (murine macrophage) cells. | (A) Suppression of superoxide anion generation by human neutrophils ($IC_{50}$):<br>(i) Swietemacrophin: 45.44 ± 3.76 μM ($p < 0.05$);<br>(ii) Humilinolide F: 27.13 ± 1.82 μM ($p < 0.01$);<br>(iii) 3,6-O,O-diacetylswietenolide: 29.36 ± 1.75 μM ($p < 0.05$);<br>(iv) 3-O-tigloylswietenolide: 35.58 ± 2.12 μM;<br>(v) Swietemahonin E: 33.64 ± 2.05 μM ($p < 0.05$);<br>(vi) Swietenine: >100.<br><br>(B) Inhibition of NO generation by RAW264.7 cells:<br>(i) Swietemacrophin: 33.45 ± 1.88 μM ($p < 0.01$);<br>(ii) Humilinolide F: 49.36 ± 4.01 μM;<br>(iii) 3,6-O,O-diacetylswietenolide: 64.21 ± 5.67 μM;<br>(iv) 3-O-tigloylswietenolide: 32.62 ± 3.27μM ($p < 0.01$);<br>(v) Swietemahonin E: 29.70 ± 2.11μM ($p < 0.05$);<br>(vi) Swietenine: 36.32 ± 2.84. | [62] |

**Table 2.** *Cont.*

| Seed Extract | Bioproperties | Extraction Method | In Vitro/In Vivo Studies | Bioactivity | References |
|---|---|---|---|---|---|
| Swietenine | Antioxidant and anti-inflammation | (A) The seeds were dried for 24 h in a drying oven at 30 °C.<br>(B) Oil from the seeds were removed using an oil press machine.<br>(C) The pressed seed was then sequentially extracted with hexane, ethyl acetate and methanol via a Soxhlet extractor.<br>(D) The ethyl acetate extract was concentrated with a rotary evaporator and dried in a vacuum dryer, before further purification to isolate swietenine. | (A) RAW264.7 cells induced by lipopolysaccharide.<br>(B) Hepa1c1c7 (murine hepatoma) cells. | (A) Swietenine dose-dependently inhibited NO production in induced RAW264.7 cells with 65.97 ± 0.7% at 0.78 μM and 21.03 ± 1.4% at 25 μM.<br>(B) Swietenine dose-dependently significantly inhibited production of pro-inflammatory cytokine IL-1β, IFN-γ, TNF-α, and IL-6 in induced RAW264.7 cells. At 25 μM, RAW264.7 experienced a reduction in fold change in IL-1β by 1.3 ± 0.13, IFN-γ by 3.40 ± 0.07, TNF-α by 1.45 ± 0.06, and IL-6 by 1.60 ± 0.20.<br>(C) Swietenine dose-dependently inhibited the expression of COX-2 and NF-κB of induced RAW264.7 cells. At 25μM, RAW264.7 experienced a reduction in fold change in COX-2 by 1.73 ± 0.06 and NF-κB by 2.90 ± 0.09.<br>(D) Swietenine dose-dependently upregulated NRF2 and HO-1 in induced RAW264.7 cells. At 25μM, RAW264.7 experienced an increase in fold change in NRF2 by 2.57 ± 0.02 μM and HO-1 by 2.46 ± 0.03.<br>(E) Swietenine induced NQO1 activity in Hepa1c1c7 cells. The CD value of swietenine was 15.8 ± 0.23 μM. | [61] |
| (A) *Swietenia* oil extracted from *Swietenia macrophylla*<br>(B) Nanoemulsion *Swietenia* oil<br>(C) Nanoemulgel of *Swietenia* oil | Anti-inflammation | Not available | Male Sprague–Dawley rats (180–200 g) with induced edema in the right hind paw. The rats were treated with the *Swietenia* oil, before being induced to have an edema. | (A) Significant dose-dependent inhibition of inflammation in the paw of the rats across 4 h of *Swietenia* oil treatment.<br>(B) Nanoemulsion of *Swietenia* oil improved the percentage of inflammation inhibition from 54% to 76.4%, at 4 h of 4 mg/kg treatment.<br>(C) Nanoemulgel of *Swietenia* oil improved the percentage of inflammation inhibition from 27% to 69.6%, at 4 h of 20% concentration treatment. | [65] |
| Ethanol extract in the form of ointment (10% *w/w*) | Wound healing | (A) The seeds were dried and homogenized before extracting with 95% ethanol at room temperature for 6 days.<br>(B) The extract was filtered and concentrated. | Adult male Sprague–Dawley albino rats (200–250 g) were excised on the shaved dorsal thoracic region. The wound size was 200 mm and 2 mm deep. The wound was then blotted with a cotton swab soaked in normal saline to achieve hemostasis, before leaving it open. | (A) The wound area was closed by the ethanolic ointment by day 15 as compared to the control, which took 21 days.<br>(B) The ethanolic ointment demonstrated higher fibroblast proliferation and increased formation of blood capillaries. There was also presence of collagen fibers and collagen deposition. As compared to the control, the control sample had disorganized fibroblasts, fewer blood capillaries and reduced collagen deposition. | [68] |

## 4. Conclusions

In summary, the potential of *S. macrophylla* seeds as a cosmeceutical product is positive. Its natural antioxidant, anti-inflammatory, antibacterial, skin whitening, and wound healing phytocompounds make it an ideal active ingredient to be present in a cosmetic product. Nonetheless, there is still a large research gap that entails the transformation of the seed extract into usable cosmetic products. Besides this, studies on skin whitening and wound healing properties of S. macrophylla are still in their rudimentary form. Furthermore, analyses on purer fractions and the phytocompounds have also not been carried out yet.

On that account, additional in vivo or clinical study research to validate its effects would certainly boost its potential and development as a cosmeceutical agent.

**Author Contributions:** Conceptualization, C.K.M. and B.H.G.; literature search, C.K.M. and H.-L.S.; writing, C.K.M. and H.-L.S.; editing of manuscript, H.-L.S. and B.H.G.; review of manuscript, B.H.G. and L.E.L.; insights to manuscript, G.Z., L.E.L., K.W.G. and, L.C.M.; supervision, B.H.G. All authors have read and agreed to the published version of the manuscript.

**Funding:** This research was supported by Monash University Malaysia, School of Pharmacy Pilot Research Grant (SOP/SRG-Pilot/02/2022); Monash Global Asia in the 21st Century (GA21) research grant (GA-HW-19-L01 and GA-HW-19-S02) and Fundamental Research Grant Scheme (FRGS/1/2019/WAB09/MUSM/02/1).

**Institutional Review Board Statement:** Not applicable.

**Informed Consent Statement:** Not applicable.

**Data Availability Statement:** Not applicable.

**Conflicts of Interest:** The authors declare no conflict of interest.

## References

1. Haddara, M.; Hsieh, J.; Fagerstrøm, A.; Eriksson, N.; Sigurðsson, V. Exploring customer online reviews for new product development: The case of identifying reinforcers in the cosmetic industry. *MDE Manag. Decis. Econ.* **2019**, *41*, 250–273. [CrossRef]
2. Feetham, H.J.; Jeong, H.S.; McKesey, J.; Wickless, H.; Jacobe, H. Skin care and cosmeceuticals: Attitudes and trends among trainees and educators. *J. Cosmet. Dermatol.* **2018**, *17*, 220–226. [CrossRef] [PubMed]
3. Kumar, S. Exploratory analysis of global cosmetic industry: Major players, technology and market trends. *Technovation* **2005**, *25*, 1263–1272. [CrossRef]
4. Dini, I.; Laneri, S. The new challenge of green cosmetics: Natural food ingredients for cosmetic formulations. *Molecules* **2021**, *26*, 3921. [CrossRef] [PubMed]
5. Ribeiro, A.S.; Estanqueiro, M.; Oliveira, M.B.; Sousa Lobo, J.M. Main benefits and applicability of plant extracts in skin care products. *Cosmetics* **2015**, *2*, 48–65. [CrossRef]
6. Ashawat, M.S.; Banchhor, M.; Saraf, S.; Saraf, S. Herbal cosmetics: Trends in skin care formulation. *Pharmacogn. Rev.* **2009**, *3*, 82–89.
7. He, T.; Marco, J.; Soares, R.; Yin, Y.; Wiedenhoeft, A.C. Machine learning models with quantitative wood anatomy data can discriminate between *Swietenia macrophylla* and *Swietenia mahagoni*. *Forests* **2020**, *11*, 36. [CrossRef]
8. Bergo, M.C.J.; Pastore, T.C.M.; Coradin, V.T.R.; Wiedenhoeft, A.C.; Braga, J.W.B. NIRS identification of *Swietenia macrophylla* is robust across specimens from 27 countries. *IAWA J.* **2016**, *37*, 420–430. [CrossRef]
9. Moghadamtousi, S.Z.; Goh, B.H.; Chan, C.K.; Shabab, T.; Kadir, H.A. Biological activities and phytochemicals of *Swietenia macrophylla* King. *Molecules* **2013**, *18*, 10465–10483. [CrossRef]
10. Grice, E.A.; Kong, H.H.; Conlan, S.; Deming, C.B.; Davis, J.; Young, A.C.; Bouffard, G.G.; Blakesley, R.W.; Murray, P.R.; Green, E.D.; et al. Topographical and temporal diversity of the human skin microbiome. *Science* **2009**, *324*, 1190–1192. [CrossRef]
11. Johnson, T.R.; Gómez, B.I.; McIntyre, M.K.; Dubick, M.A.; Christy, R.J.; Nicholson, S.E.; Burmeister, D.M. The cutaneous microbiome and wounds: New molecular targets to promote wound healing. *Int. J. Mol. Sci.* **2018**, *19*, 2699. [CrossRef] [PubMed]
12. Fournière, M.; Latire, T.; Souak, D.; Feuilloley, M.G.J.; Bedoux, G. *Staphylococcus epidermidis* and *Cutibacterium acnes*: Two major sentinels of skin microbiota and the influence of cosmetics. *Microorganisms* **2020**, *8*, 1752. [CrossRef] [PubMed]
13. Dréno, B. What is new in the pathophysiology of acne, an overview. *J. Eur. Acad. Dermatol. Venereol.* **2017**, *31*, 8–12. [CrossRef] [PubMed]
14. Kim, M.-J.; Kim, K.-P.; Choi, E.; Yim, J.-H.; Choi, C.; Yun, H.-S.; Ahn, H.-Y.; Oh, J.-Y.; Cho, Y. Effects of *Lactobacillus plantarum* CJLP55 on clinical improvement, skin condition and urine bacterial extracellular vesicles in patients with acne vulgaris: A Randomized, double-blind, placebo-controlled study. *Nutrients* **2021**, *13*, 1368. [CrossRef]
15. Claudel, J.P.; Auffret, N.; Leccia, M.T.; Poli, F.; Corvec, S.; Dréno, B. *Staphylococcus epidermidis*: A potential new player in the physiopathology of acne? *Dermatology* **2019**, *235*, 287–294. [CrossRef]
16. Suliman, M.B.; Nour, A.H.; Yusoff, M.; Nour, A.; Mazza, A.S. Growth inhibitory effect on bacteria of *Swietenia macrophylla* king seeds and leaves crude alkaloid extracts. *Int. J. Pharm. Sci. Res.* **2014**, *5*, 1000–1004. [CrossRef]
17. Dharmalingam, K.; Tan, B.-K.; Mahmud, M.Z.; Sedek, S.A.M.; Majid, M.I.A.; Kuah, M.-K.; Sulaiman, S.F.; Ooi, K.L.; Khan, N.A.K.; Muhammad, T.S.T.; et al. *Swietenia macrophylla* extract promotes the ability of *Caenorhabditis elegans* to survive *Pseudomonas aeruginosa* infection. *J. Ethnopharmacol.* **2012**, *139*, 657–663. [CrossRef]
18. Suliman, M.B.; Nour, A.; Yusoff, M.; Nour, A.; Kuppusamy, P.; Ar, Y.; Adam, M. Fatty acid composition and antibacterial activity of *Swietenia macrophylla* king seed oil. *Afr. J. Plant Sci.* **2013**, *7*, 300–303. [CrossRef]

19. Gopalan, H.K.; Md Hanafiah, N.F.; Leong, C.R.; Tan, W.-N.; Wahidin, S.; Hway, T.S.; Yenn, T.W. Chemical composition and antimicrobial efficiency of *Swietenia macrophylla* seed extract on clinical wound pathogens. *Nat. Prod. Sci.* **2019**, *25*, 38–43. [CrossRef]
20. Durai, M.V.; Balamuniappan, G.; Geetha, S. Phytochemical screening and antimicrobial activity of leaf, seed and central-fruit-axis crude extract of *Swietenia macrophylla* King. *J. Pharmacogn. Phytochem.* **2016**, *5*, 181–186.
21. Kong, C.; Yehye, W.A.; Abd Rahman, N.; Tan, M.W.; Nathan, S. Discovery of potential anti-infectives against *Staphylococcus aureus* using a *Caenorhabditis elegans* infection model. *BMC Complement. Altern. Med.* **2014**, *14*, 1–17. [CrossRef] [PubMed]
22. Rahman, A.K.M.S.; Chowdhury, A.K.A.; Ali, H.-A.; Raihan, S.Z.; Ali, M.S.; Nahar, L.; Sarker, S.D. Antibacterial activity of two limonoids from *Swietenia mahagoni* against multiple-drug-resistant (MDR) bacterial strains. *J. Nat. Med.* **2009**, *63*, 41–45. [CrossRef]
23. Kader, M.A.; Haque, M.E.; Khondkar, P.; Islam, M.M.; Rahman, M.M. Antibacterial and Cytotoxic Limonoids from the Seeds of *Swietenia mahagony*. *Dhaka Univ. J. Pharm. Sci.* **2010**, *8*, 141–145. [CrossRef]
24. Mootoo, B.S.; Ali, A.; Motilal, R.; Pingal, R.; Ramlal, A.; Khan, A.; Reynolds, W.F.; McLean, S. Limonoids from *Swietenia macrophylla* and S. aubrevilleana. *J. Nat. Prod.* **1999**, *62*, 1514–1517. [CrossRef] [PubMed]
25. Nogueira, T.S.R.; Passos, M.d.S.; Calixto, S.D.; Ventura, T.L.B.; Lassounskaia, E.; Muzitano, M.F.; Braz-Filho, R.; Vieira, I.J.C. Anti-Mycobacterium tuberculosis Activity of Compounds from *Cedrela fissilis* Vell Seeds (*Meliaceae*). *Rev. Virtual Quim.* **2021**, *13*, 1116–1121. [CrossRef]
26. Suliman, M.B. Preliminary phytochemical screening and thin layer chromatography analysis of *Swietenia macrophylla* King methanol extracts. *Chem. Adv. Mater.* **2018**, *3*, 1–7.
27. García-Bores, A.M.; Espinosa-González, A.M.; Reyna-Campos, A.; Cruz-Toscano, S.; Benítez-Flores, J.C.; Hernández-Delgado, C.T.; Flores-Maya, S.; Urzúa-Meza, M.; Peñalosa-Castro, I.; Céspedes-Acuña, C.L.; et al. *Lippia graveolens* photochemopreventive effect against UVB radiation-induced skin carcinogenesis. *J. Photochem. Photobiol. B* **2017**, *167*, 72–81. [CrossRef]
28. Herrling, T.; Jung, K.; Fuchs, J. Measurements of UV-generated free radicals/reactive oxygen species (ROS) in skin. *Spectrochim. Acta A Mol. Biomol. Spectrosc.* **2006**, *63*, 840–845. [CrossRef]
29. Terra, V.A.; Souza-Neto, F.P.; Pereira, R.C.; Silva, T.N.X.; Costa, A.C.C.; Luiz, R.C.; Cecchini, R.; Cecchini, A.L. Time-dependent reactive species formation and oxidative stress damage in the skin after UVB irradiation. *J. Photochem. Photobiol. B* **2012**, *109*, 34–41. [CrossRef]
30. Mahendra, C.K.; Ser, H.-L.; Pusparajah, P.; Htar, T.T.; Chuah, L.-H.; Yap, W.H.; Tang, Y.-Q.; Zengin, G.; Tang, S.Y.; Lee, W.L.; et al. Cosmeceutical therapy: Engaging the repercussions of UVR photoaging on the skin's circadian rhythm. *Int. J. Mol. Sci.* **2022**, *23*, 2884. [CrossRef]
31. Pillai, S.; Oresajo, C.; Hayward, J. Ultraviolet radiation and skin aging: Roles of reactive oxygen species, inflammation and protease activation, and strategies for prevention of inflammation-induced matrix degradation—A review. *Int. J. Cosmet. Sci.* **2005**, *27*, 17–34. [CrossRef] [PubMed]
32. Muthusamy, V.; Piva, T.J. The UV response of the skin: A review of the MAPK, NFκB and TNFα signal transduction pathways. *Arch. Dermatol. Res.* **2009**, *302*, 5–17. [CrossRef] [PubMed]
33. Lee, P.L.; van Weelden, H.; Bruijnzeel, P.L.B. Neutrophil Infiltration in normal human skin after exposure to different ultraviolet radiation sources. *Photochem. Photobiol.* **2008**, *84*, 1528–1534. [CrossRef]
34. Takeuchi, H.; Gomi, T.; Shishido, M.; Watanabe, H.; Suenobu, N. Neutrophil elastase contributes to extracellular matrix damage induced by chronic low-dose UV irradiation in a hairless mouse photoaging model. *J. Dermatol. Sci.* **2010**, *60*, 151–158. [CrossRef]
35. Rijken, F.; Bruijnzeel-Koomen, C.A.F.M. Photoaged skin: The role of neutrophils, preventive measures, and potential pharmacological targets. *Clin. Pharmacol. Ther.* **2011**, *89*, 120–124. [CrossRef]
36. Tran, T.-N.T.; Schulman, J.; Fisher, D.E. UV and pigmentation: Molecular mechanisms and social controversies. *Pigment Cell Melanoma Res.* **2008**, *21*, 509–516. [CrossRef]
37. Hamakawa, M.; Sugihara, A.; Okamoto, H.; Horio, T. Ultraviolet B radiation suppresses Langerhans cell migration in the dermis by down-regulation of α4 integrin. *Photodermatol. Photoimmunol. Photomed.* **2006**, *22*, 116–123. [CrossRef]
38. Wong, W.C.; Wu, J.Y.; Benzie, I.F.F. Photoprotective potential of Cordyceps polysaccharides against ultraviolet B radiation-induced DNA damage to human skin cells. *Br. J. Dermatol.* **2011**, *164*, 980–986. [CrossRef]
39. Toriyama, E.; Masuda, H.; Torii, K.; Ikumi, K.; Morita, A. Time kinetics of cyclobutane pyrimidine dimer formation by narrowband and broadband UVB irradiation. *J. Dermatol. Sci.* **2021**, *103*, 151–155. [CrossRef]
40. Mahendra, C.K.; Tan, L.T.-H.; Yap, W.H.; Chan, C.K.; Pusparajah, P.; Goh, B.H. An optimized cosmetic screening assay for ultraviolet B (UVB) protective property of natural products. *Prog. Drug Discov. Biomed. Sci.* **2019**, *2*, 1–6. [CrossRef]
41. Smijs, T.G.; Pavel, S. Titanium dioxide and zinc oxide nanoparticles in sunscreens: Focus on their safety and effectiveness. *Nanotechnol. Sci. Appl.* **2011**, *4*, 95–112. [CrossRef]
42. Lu, P.-J.; Huang, S.-C.; Chen, Y.-P.; Chiueh, L.-C.; Shih, D.Y.-C. Analysis of titanium dioxide and zinc oxide nanoparticles in cosmetics. *J. Food Drug Anal.* **2015**, *23*, 587–594. [CrossRef]
43. Scheuer, E.; Warshaw, E. Sunscreen allergy: A review of epidemiology, clinical characteristics, and responsible allergens. *Dermatitis* **2006**, *17*, 3–11. [CrossRef]
44. Lee Granger, K.; Brown, P.R. The chemistry and HPLC analysis of chemical sunscreen filters in sunscreens and cosmetics. *J. Liq. Chromatogr. Relat. Technol.* **2001**, *24*, 2895–2924. [CrossRef]

45. ISO. ISO 24444:2010 Cosmetics—Sun Protection Test Methods—In vivo Determination of the Sun Protection Factor (SPF). Available online: https://www.iso.org/obp/ui/#iso:std:iso:24444:ed-1:v1:en (accessed on 3 April 2022).
46. FDA. Guidance for Industry Labelling and Effectiveness Testing: Sunscreen Drug Products for Over-the Counter Human Use—Small Entity Compliance Guide. Available online: https://www.fda.gov/drugs/guidancecomplianceregulatoryinformation/guidances/ucm330694.htm (accessed on 3 April 2022).
47. Mahendra, C.K.; Mahendra, C.K.; Pusparajah, P.; Htar, T.-T.; Chuah, L.-H.; Duangjai, A.; Khan, T.M.; Yow, Y.Y.; Kumari, Y.; Goh, B.H. Simplified, cost effective, and accurate calculation of critical wavelength via the MATLAB software. *Prog. Drug Discov. Biomed. Sci.* **2021**, *4*, 1–16. [CrossRef]
48. Osterwalder, U.; Herzog, B. Sun protection factors: World wide confusion. *Br. J. Dermatol.* **2009**, *161*, 13–24. [CrossRef]
49. Bernstein, E.F.; Sarkas, H.W.; Boland, P.; Bouche, D. Beyond sun protection factor: An approach to environmental protection with novel mineral coatings in a vehicle containing a blend of skincare ingredients. *J. Cosmet. Dermatol.* **2020**, *19*, 407–415. [CrossRef]
50. Pandel, R.; Poljšak, B.; Godic, A.; Dahmane, R. Skin photoaging and the role of antioxidants in its prevention. *ISRN Dermatol.* **2013**, *2013*, 930164. [CrossRef]
51. Cui, B.; Wang, Y.; Jin, J.; Yang, Z.; Guo, R.; Li, X.; Yang, L.; Li, Z. Resveratrol treats UVB-induced photoaging by anti-MMP expression, through anti-inflammatory, antioxidant, and antiapoptotic properties, and treats photoaging by upregulating VEGF-B expression. *Oxid. Med. Cell Longev.* **2022**, *2022*, 6037303. [CrossRef]
52. Draelos, Z.D. Novel approach to the treatment of hyperpigmented photodamaged skin: 4% hydroquinone/0.3% retinol versus tretinoin 0.05% emollient cream. *Dermatol. Surg.* **2005**, *31*, 799–805. [CrossRef]
53. Liu, Z.; Li, Y.; Song, H.; He, J.; Li, G.; Zheng, Y.; Li, B. Collagen peptides promote photoaging skin cell repair by activating the TGF-β/Smad pathway and depressing collagen degradation. *Food Funct.* **2019**, *10*, 6121–6134. [CrossRef]
54. Slominski, A.; Tobin, D.J.; Shibahara, S.; Wortsman, J. Melanin pigmentation in mammalian skin and its hormonal regulation. *Physiol. Rev.* **2004**, *84*, 1155–1228. [CrossRef]
55. Levy, L.L.; Emer, J.J. Emotional benefit of cosmetic camouflage in the treatment of facial skin conditions: Personal experience and review. *Clin. Cosmet. Investig. Dermatol.* **2012**, *5*, 173–182. [CrossRef]
56. Mahendra, C.K.; Tan, L.T.H.; Mahendra, C.K.; Ser, H.-L.; Pusparajah, P.; Htar, T.T.; Chuah, L.-H.; Yap, W.H.; Tang, S.Y.; Ming, L.C.; et al. The potential of sky fruit as an anti-aging and wound healing cosmeceutical agent. *Cosmetics* **2021**, *8*, 79. [CrossRef]
57. Mahendra, C.K.; Abidin, S.A.Z.; Htar, T.T.; Chuah, L.H.; Khan, S.U.; Ming, L.C.; Tang, S.Y.; Pusparajah, P.; Goh, B.H. Counteracting the ramifications of UVB irradiation and photoaging with *Swietenia macrophylla* king seed. *Molecules* **2021**, *26*, 2000. [CrossRef]
58. Dutta, M.; Biswas, U.K.; Chakraborty, R.; Banerjee, P.; Maji, D.; Mondal, M.C.; Raychaudhuri, U. Antidiabetic and antioxidant effect of *Swietenia macrophylla* seeds in experimental type 2 diabetic rats. *Int. J. Diabetes Dev. Ctries.* **2013**, *33*, 60–65. [CrossRef]
59. Wan Hassan, W.N.A.; Zulkifli, R.; Farediah, A.; Yunus, M. Antioxidant and tyrosinase inhibition activities of *Eurycoma longifolia* and *Swietenia macrophylla*. *J. Appl. Pharm. Sci.* **2015**, *5*, 006–010. [CrossRef]
60. Coello, F.P.; Azuaje, D.R.; Catari, I.P.; Marrero, M.P.; Vargas, C.O. Evaluation of the antioxidant activity of aqueous extracts of leaves and seeds of *Swietenia macrophylla* King by chemical and biological methods. *J. Drug Res. Dev.* **2020**, *6*, 1–4. [CrossRef]
61. Mak, K.-K.; Shiming, Z.; Balijepalli, M.K.; Dinkova-Kostova, A.T.; Epemolu, O.; Mohd, Z.; Pichika, M.R. Studies on the mechanism of anti-inflammatory action of swietenine, a tetranortriterpenoid isolated from *Swietenia macrophylla* seeds. *Phytomed. Plus* **2021**, *1*, 100018. [CrossRef]
62. Chen, L.-C.; Liao, H.-R.; Chen, P.-Y.; Kuo, W.-L.; Chang, T.-H.; Sung, P.-J.; Wen, Z.-H.; Chen, J.-J. Limonoids from the seeds of *Swietenia macrophylla* and their anti-inflammatory activities. *Molecules* **2015**, *20*, 18551–18564. [CrossRef]
63. Chen, J.-J.; Huang, S.-S.; Liao, C.-H.; Wei, D.-C.; Sung, P.-J.; Wang, T.-C.; Cheng, M.-J. A new phragmalin-type limonoid and anti-inflammatory constituents from the fruits of *Swietenia macrophylla*. *Food Chem.* **2010**, *120*, 379–384. [CrossRef]
64. Ma, Y.-Q.; Jiang, K.; Deng, Y.; Guo, L.; Wan, Y.-Q.; Tan, C.-H. Mexicanolide-type limonoids from the seeds of *Swietenia macrophylla*. *J. Asian Nat. Prod. Res.* **2018**, *20*, 299–305. [CrossRef]
65. Eid, A.M.; El Enshasy, H.A.; Aziz, R.; Elmarzugi, N.A. Preparation, characterization and anti-inflammatory activity of *Swietenia macrophylla* nanoemulgel. *J. Nanomed. Nanotechnol.* **2014**, *5*, 1–10. [CrossRef]
66. Eid, A.M.; El-Enshasy, H.A.; Aziz, R.; Elmarzugi, N.A. The preparation and evaluation of self-nanoemulsifying systems containing Swietenia oil and an examination of its anti-inflammatory effects. *Int. J. Nanomed.* **2014**, *9*, 4685–4695. [CrossRef]
67. Dolinska, M.B.; Wingfield, P.T.; Young, K.L.; Sergeev, Y.V. The TYRP1-mediated protection of human tyrosinase activity does not involve stable interactions of tyrosinase domains. *Pigment Cell Melanoma Res.* **2019**, *32*, 753–765. [CrossRef] [PubMed]
68. Nilugal, K.C.; Fattepur, S.; Asmani, F.M.; Abdullah, I.; Vijendren, S.; Ugandar, R.E. Evaluation of wound healing activity of *Swietenia macrophylla* (Meliaceae) seed extract in albino rats. *Am. J. PharmTech Res.* **2017**, *7*, 113–124.

MDPI
St. Alban-Anlage 66
4052 Basel
Switzerland
www.mdpi.com

*Antioxidants* Editorial Office
E-mail: antioxidants@mdpi.com
www.mdpi.com/journal/antioxidants

www.ingramcontent.com/pod-product-compliance
Lightning Source LLC
Chambersburg PA
CBHW040806250926
51850CB00007BA/1094
* 9 7 8 3 7 2 5 8 0 0 5 5 1 *